A MANUAL OF INORGANIC CHEMISTRY,

ARRANGED TO FACILITATE THE

EXPERIMENTAL DEMONSTRATION

OF THE

FACTS AND PRINCIPLES OF THE SCIENCE.

BY

CHARLES W. ELIOT,
PROFESSOR OF ANALYTICAL CHEMISTRY AND METALLURGY,

AND

FRANK H. STORER,
PROFESSOR OF GENERAL AND INDUSTRIAL CHEMISTRY,

IN THE MASSACHUSETTS INSTITUTE OF TECHNOLOGY.

FIFTH THOUSAND.

NEW YORK:
IVISON, BLAKEMAN, TAYLOR, & COMPANY,
138 & 140 GRAND STREET.
CHICAGO: 133 & 135 STATE STREET.
1876.

THE AUTHORS

INSCRIBE THIS BOOK TO THEIR TEACHER IN CHEMISTRY,

PROF. JOSIAH P. COOKE,

OF HARVARD COLLEGE,

IN TOKEN OF

GRATITUDE AND FRIENDSHIP.

PREFACE TO THE FIRST EDITION.

In preparing this manual, it has been the authors' object to facilitate the teaching of chemistry by the experimental and inductive method. The book will enable the careful student to acquaint himself with the main facts and principles of chemistry, through the attentive use of his own perceptive faculties, by a process not unlike that by which these facts and principles were first established. The authors believe that the study of a science of observation ought to develop and discipline the observing faculties, and that such a study fails of its true end if it become a mere exercise of the memory.

The minute instructions, given in the descriptions of experiments and printed in the smaller type, are intended to enable the student to see, smell, and touch for himself; these detailed descriptions are meant for laboratory use. In order to mark as clearly as possible the distinction between chemistry and chemical manipulation, the necessary instructions on the latter subject have been put in an Appendix. In cases in which it is impossible for every student to experiment for himself, the authors hope that this manual will make it easy for the teacher, even if he be not a professional chemist, to exhibit to his class, in a familiar and inexpensive manner, experiments enough to supply ocular demonstration of the leading facts and generalizations of the science. Judging from their own experience, the authors venture to hope that even professional chemists may find it convenient to have always at hand the details of several hundred experiments, covering the ground of an extensive course of chemical lectures.

The student of this manual is supposed to be already acquainted with the rudiments of physics. The chemist must often depend upon physical properties for his means of characterizing the numerous substances with which he deals, and he is nearly concerned with the physical properties of gases and vapors; but chemistry has now so wide a scope and so great an importance as to deserve to be studied by itself, and not merely as an appendix to the subject of molecular physics.

Like all elementary text-books, this manual is a mere compilation; it embodies in a somewhat new form previously existing statements of well-recognized facts and principles which have become the common stock of the science. There is little original in the book, except its arrangement and method, in part, and its general tone. The authors have, of course, drawn largely from the invaluable compilations made by Gmelin, Otto, and Watts, and they have also availed themselves freely of the text-books of Stoeckhardt and Miller and the writings of Hofmann.

The book is not written in the interest of any particular theory or system of notation, but is intended to exhibit, so far as is possible within the limits proper to an elementary manual, the present state of the science.

The authors will feel very grateful to any one who will communicate to them errors, detected in using the book, or suggestions for its improvement.

BOSTON, June 1867.

PREFACE TO THE SECOND EDITION.

The authors have thoroughly revised this second edition of their manual. Practical experience in using the book with two classes in the laboratory, the questions of students, and the suggestions of friends have enabled them to improve some of the detailed directions for experiments, and to make a few other changes and additions calculated to smooth difficulties or supply defects. Special pains have been taken to make the printing of this edition as accurate as possible.

Boston, December 1867.

TABLE OF CONTENTS.

PAGE

Introduction.—Subject matter of Chemistry. Chemical change. Analysis and synthesis. Fact and theory 1-4

Chap. I.—Air. Atmospheric pressure. Properties. Analysis. Air a mixture. Composition of air 4-10

Chap. II.—Oxygen. Preparation and properties of oxygen. Oxygen supports combustion. Oxides. Oxidation. Wide diffusion of oxygen . . . 11-15

Chap. III.—Nitrogen. Preparation and properties of nitrogen . . . 16-18

Chap. IV.—Water. Properties of water. The gramme. Specific gravity. Specific heat. Ice. Steam. Analysis, electrolysis, and synthesis of water. Hydrogen. Atoms and molecules. Molecular hypothesis. Atomic weights. Chemical combination. The chemical force. Water in nature. Distillation. Preparation of pure water. Solution. Solution and chemical combination compared 18-39

Chap. V.—Hydrogen. Preparation and properties of hydrogen. Symbols. Diffusion of gases. Diffusive power and inflammability of hydrogen. Heat from burning hydrogen. Unit of heat. Oxyhydrogen blowpipe. Form of gas-flames. Explosive mixtures of hydrogen and oxygen. Oxygen burns in hydrogen as well as hydrogen in oxygen 39-50

Chap. VI.—Compounds of oxygen, hydrogen, and nitrogen. Peroxide of hydrogen. Nitric acid. One of the meanings of the terms acid and alkaline. Neutralization. Nitrous oxide. Composition of nitrous oxide. Atomic weight of nitrogen. Properties of nitrous oxide. Nitric oxide. A test for free oxygen. The terminations *ous* and *ic*. Hyponitric acid. Nitrous acid. Analysis of nitric acid. Synopsis of the oxides of nitrogen. Law of multiple proportions. Definite and obscure chemical action. Air a mixture. Anhydrous and hydrated nitric acid. Oxidizing and reducing agents. Atomic weights and combining weights. Molecular formulæ. Combining weight of nitric acid. Nitric acid reactions. Nitrogen and hydrogen. Ammonia. Analysis and synthesis of ammonia. Nascent state. Composition of ammonia. Ammonium. Salts of ammonium. Sources of ammonia. Ammonia-water. Empirical and rational formulæ. Dualistic formulæ. Typical formulæ. Uses of symbolic formulæ 50-91

Chap. VII.—Chlorhydric acid. Properties, analysis, and composition of chlorhydric acid. Atomic weight of chlorine. Synthesis of chlorhydric acid. Manufacture of chlorhydric acid. Practical application of chemical equations. Chemical affinity. Preparation and uses of chlorhydric acid. Aqua regia . . 91-105

Chap. VIII.—Chlorine. Preparation and properties of chlorine. Chlorine-water. Metals burn in chlorine. Chlorine burns in hydrogen, and hydrogen in chlorine. Combustion in chlorine. Uses of chlorine. How chlorine acts in bleaching and disinfecting. Action of chlorine on ammonia. Chloride of nitrogen. Oxides of chlorine. Hypochlorous acid. Chlorous acid. Hypochloric acid. Chloric acid. Perchloric acid 105-119

Chap. IX.—Bromine. Bromhydric acid. Bromic acid. Hypobromous acid. Chloride of bromine. Bromide of nitrogen 119-123

Chap. X.—Iodine. Extraction and properties of iodine. Tests for iodine. Iodohydric acid. Iodic acid. Periodic acid. Iodide of nitrogen. Chlorides of iodine. Bromides of iodine. The chlorine group of elements 123-135

Chap. XI.—Fluorine. Fluorhydric acid. Etching glass 135-139

Chap. XII.—Ozone and Antozone. Allotropism. Preparation and properties of ozone. Tests for ozone. Ozone a disinfectant. Ozone in the atmosphere. Preparation and properties of antozone. The antozone cloud. Antozone formed in all processes of combustion. Antozone oxidizes water. Differences between ozone and antozone 139-154

Chap. XIII.—Sulphur. Crystallization of sulphur. Crystalline structure. The six systems of crystallization. Sulphur a dimorphous element. Change of prismatic into octahedral sulphur. Methods of obtaining crystals. Soft sulphur. Milk of sulphur. Metals burn in sulphur as in oxygen. Uses of sulphur. Sulphydric acid. Preparation, analysis, and properties of sulphydric acid. Sulphuretted-hydrogen water. Sulphydric acid as a reagent. Ready decomposition of sulphydric acid. Persulphide of hydrogen. Greek and Latin numeral prefixes. Oxides of sulphur. Sulphurous acid. Preparation, properties, and composition of sulphurous acid. Liquid sulphurous acid. Bleaching-power of sulphurous acid. Oxidation of sulphurous acid. Sulphites. Sulphuric acid. Manufacture and properties of sulphuric acid. Its behavior towards water and ice. Hydrates of sulphuric acid. Sulphates. The terms monobasic, bibasic, &c. Fuming sulphuric acid. Anhydrous sulphuric acid. Preparation and properties of anhydrous sulphuric acid. Hyposulphurous acid. Hyposulphites. Chlorides of sulphur 154-198

Chap. XIV.—Selenium and Tellurium. Properties of selenium. Isomorphism. Atomic volume. Tellurium. Compounds of tellurium. The sulphur group of elements 198-203

Chap. XV.—Combination by volume. Synopsis of the gaseous compounds previously studied. Condensation ratios. Combining-weight and volume-weight. Double or product volume. Molecular condition of elementary gases. Molecular formulæ 203-209

Chap. XVI.—Phosphorus. Allotropic modifications of phosphorus. Friction matches. Phosphorescence. Solutions of phosphorus. Poisonous properties of phosphorus. Manufacture of phosphorus. Red phosphorus. Amorphous and crystallized red phosphorus. Safety-matches. Phosphorus with oxidizing agents. Phosphuretted hydrogen. Preparation and analysis of phosphuretted hydrogen gas. Composition of phosphuretted hydrogen as compared with that of ammonia. Unit-volume weight of phosphorus twice its atomic weight. Liquid and solid phosphuretted hydrogen. Compounds of phosphorus and oxygen. Red oxide of phosphorus. Hypophosphorous acid. Phosphorous acid. Spontaneous combustion. Phosphoric acid. Hydrates of phosphoric acid. Meta- and pyro-phosphoric acids. Chlorides of phosphorus. Dissociation. Bromides, iodides, and sulphides of phosphorus 209–241

Chap. XVII.—Arsenic. Sources and properties of arsenic. Arseniuretted hydrogen. Oxides of arsenic. Manufacture of arsenious acid. Isomerism. Properties and uses of arsenious acid. Reduction of arsenious acid. Solubility of arsenious acid. Arsenic acid. Salts of arsenic acid; their analogy to the phosphates. Detection of arsenic in cases of poisoning. Diffusion of liquids. Dialysis. Crystalloids and colloids. Destroying the organic matters. Precipitation and reduction of sulphide of arsenic. Marsh's test. Chloride of arsenic. Sulphides of arsenic. Sulphur-salts. Sulpharsenites and Sulpharseniates 241–265

Chap. XVIII.—Antimony. Sources, properties, and alloys of antimony. Antimoniuretted hydrogen. Testing for antimony. Distinguishing between arsenic and antimony. Teroxide of antimony. Antimoniate of antimony. Antimonic acid. Metantimonic acid. Chlorides of antimony. Chloridizing agents. Sulphides of antimony. Sulphur-salts 266–281

Chap. XIX.—Bismuth. Sources and properties of bismuth. Fusible metal. Teroxide of bismuth. Bismuthic acid. Chloride of bismuth. Sulphide of bismuth. The nitrogen group of elements 281–287

Chap. XX.—Carbon. Allotropic modifications of carbon. Diamond. Graphite or plumbago. Graphitic acid. Gas-carbon. Coke. Anthracite. Charcoal. Preparation of charcoal. Distillation of wood and of coal. Illuminating gas. Lampblack. Properties of charcoal. Reducing-power of charcoal. Deflagration. Stability of charcoal. Charcoal absorbs gases. Induces combinations. Disinfects. Decolorizes. Compounds of carbon and hydrogen. Organic chemistry. Homologous series. Marsh-gas or light carburetted hydrogen. Atomic weight of carbon. Typical hydrogen compounds. Composition of illuminating gas. Carbonic acid. Preparation and properties of carbonic acid. Ventilation of wells. Diffusion of gases. Solubility of carbonic acid. Carbonic acid produced in the processes of fermentation, respiration, decay, and combustion. Liquid and solid carbonic acid. Decomposition, analysis, and synthesis of carbonic acid. Carbonates. Carbonic oxide. Preparation and properties of carbonic oxide. Carbonic oxide a deoxidizing agent. Heat evolved in the combustion of carbonic oxide. Dissociation of carbonic oxide. Composition of carbonic oxide. Combustion. Luminosity of flames. The Bunsen burner. Quantity and intensity flames. All flames gas-flames. Form of luminous flames. Blast-lamps and blowpipes. Oxidizing and reducing flames. Chimneys. Indestructibility of matter. Kindling-temperature. Flames and fires extinguished by wire gauze and other good conductors. Safety-lamps. Flaming fires. Loss of heat from incomplete combustion. Chlorides of carbon. Compounds of carbon and nitrogen. Bisulphide of carbon 287–364

Chap. XXI.—Boron. Sources of boron compounds. Allotropism of boron. Boracic acid. Chloride of boron. Fluoride of boron. Fluoboric acid. Sulphide and nitride of boron 365–372

Chap. XXII.—Silicon. Abundance of silicon. Preparation of silicon. Modifications of silicon. Silicon and hydrogen. Oxide of silicon. Silicic acid. Soluble silicic acid. Varieties of silicic acid. Silica in natural waters. Silicates. Formulæ of silicates. Decomposition of silicates. Composition and formula of silicic acid. Chloride of silicon. Composition of chloride of silicon. Compound of silicon, hydrogen, and chlorine. Fluoride of silicon. Fluosilicic acid. Fluosilicates. Sulphide of silicon. The carbon group 372–391

Chap. XXIII.—Sodium. Sources of sodium salts. Chloride of sodium. Salt-works. Properties and uses of salt. Bromide and iodide of sodium. Sulphate of sodium. Glauber's salt. Supersaturated solutions. Bisulphate of sodium. Carbonate of sodium. Leblanc's process. Soda-ash. Soda-crystals. Bicarbonate of sodium. Yeast powders. Sulphides of sodium. Metallic sodium. Hydrate of sodium. Caustic soda. Bases and acids. Replacement. Direct combination. Phosphates of sodium. Borax. Borax as a blowpipe test. Silicates of sodium. Waterglass. Glass. Devitrified and colored glass. Hyposulphite of sodium 391–414

Chap. XXIV.—Potassium. Sources of potassium. Carbonate of potassium. Bicarbonate of potassium. Hydrate of potassium. Caustic potash. Strength of bases. Alkalimetry. Volumetric analysis. Valuation of potash and soda-ash. Metallic potassium. Cream of tartar. Chloride of potassium. Bromide of potassium. Iodide of potassium. Cyanide of potassium. Sulphides of potassium. Liver of sulphur. Sulphates of potassium. Nitrate of potassium. Refining of saltpetre. Saltpetre not explosive. Gunpowder. Chlorate of potassium 414–436

Chap. XXV.—Ammonium-salts. The hypothetical metal ammonium. Hydrate of ammonium. Chloride of ammonium. Sulphate of ammonium. Nitrate of ammonium. Carbonates of ammonium. Sulphides of ammonium . . 437–443

Chap. XXVI.—Lithium. Spectrum analysis. Rubidium and Cæsium. Thallium 443–448

Chap. XXVII.—Silver. Occurrence of silver. Extraction of silver. Metallic silver. The term metal. Acids and bases. Silver coin. Nitrate of silver. Indelible ink. Oxides of silver. Fulminating silver. Photography. The daguerreotype. Photography on glass. Photography on paper. Chloride of silver. Atomic weight of silver. Bromide of silver. Iodide of silver. Cyanide of silver. Sulphide of silver. Sulphate of silver. The alkali group. Quantivalence. Atomicity 448-466

Chap. XXVIII.—Calcium—Strontium—Barium—Lead. Calcium. Carbonate of calcium. Calcareous petrifactions. Calc-spar and arragonite. Oxide of calcium. Hydrate of calcium. Milk of lime. Slaked lime. Mortar. Caustic lime. Uses of lime. Sulphate of calcium. Plaster of Paris, or calcined gypsum. Plaster casts. Incrustation of steam-boilers. Hard water. Testing water. Phosphates of calcium. Chloride of calcium. Hypochlorite of calcium. Bleaching-powder. Preparation of chlorine and oxygen from bleaching-powder. Chlorimetry. Nitrate of calcium. Sulphydrate of calcium. Sulphite of calcium. Metallic calcium.—Strontium and barium. Peroxide of barium. Making oxygen with peroxide of barium. Strontium salts. Barium salts. Uses of strontium and barium salts. The calcium group. The alkaline earths.—Lead. Extraction of lead. Separation of silver from lead. Oxides of lead. Corrosion of lead by water. Leaden water-pipes. Suboxide of lead. Protoxide of lead or litharge. Cupellation. Peroxide of lead. Red lead or minium. Sesquioxide of lead. Sulphides of lead. Chloride of lead. Sugar of lead. White lead. Silicate of lead . 467–500

Chap. XXIX. — Magnesium—Zinc—Cadmium. Metallic magnesium. Oxide of magnesium. Calcined magnesia. Hydraulic magnesia. Chloride of magnesium. Sulphate of magnesium or Epsom salts. Carbonate of magnesium. Citrate of magnesium. Phosphate of magnesium and ammonium.—Zinc. Granulated zinc. Galvanized iron. The galvanic current. Correlation of forces. Replacement of one metal by another. Oxide of zinc. Chloride of zinc. Sulphate of zinc. Alloys of zinc.—Cadmium. Sources and properties of cadmium. Unit-volume weight of cadmium half its atomic weight. The magnesium group. . 501–511

Chap. XXX. —Aluminum—Glucinum—Chromium—Manganese—Iron—Cobalt—Nickel—Uranium. Metallic aluminum. Alloys of aluminum. Aluminum bronze. Atomic weight of aluminum. Oxide of aluminum. Hydrate of aluminum. Aluminates. Mordants in dyeing. Lakes. Chloride of aluminum. Sulphate of aluminum. Alum. Acetate of aluminum. Silicates of aluminum. Clay. Earthenware, bricks, and pottery. Glazes. Hydraulic cement. Concrete.—Glucinum.—Chromium. Oxides of chromium. Chlorides of chromium. Sulphate of chromium. Chromic acid. Chromates. Photolithography.—Manganese. Protoxide of manganese. Sesquioxide of manganese. Alums. Binoxide of manganese. Manganic acid. Manganates. Chameleon mineral. Permanganic acid. Permanganate of potassium.—Iron. Ores of iron. Extraction of iron. The bloomary process. The blast furnace. Fluxes. Cast iron. Impurities of iron. Wrought iron. The puddling process. Steel. The Bessemer process. Protoxide of iron or ferrous oxide. Sesquioxide of iron or ferric oxide. Oxidation by iron-rust. Ferric hydrate. Sulphide of iron. Iron pyrites. Chlorides of iron. Ferrous sulphate or copperas. Ink. Dyeing. Ferric sulphate. Nitrates of iron. Silicates of iron. Cyanides of iron. Prussian blue. Oxidizing ferrous and reducing ferric salts.—Cobalt and nickel. The terminations *ous* and *ic*.—Uranium. Salts of the sesquioxides. The sesquioxide group. Atomic volume of the alums. Rare elements allied to aluminum and iron 512–561

Chap. XXXI.—Copper—Mercury. Extraction of copper from its ores. Properties of copper. Alloys of copper. Dioxide of copper. Protoxide of copper. Hydrate of copper. Sulphides of copper. Chlorides of copper. Sulphate of copper. Nitrate of copper. Verdigris.—Mercury. Its extraction and properties. Unit-volume weight of mercury half its atomic weight. Oxides of mercury. Red oxide of mercury as a source of oxygen. Sulphides of mercury. Vermilion. Calomel. Corrosive sublimate. Mercuric iodide. Sulphates and nitrates of mercury. Amalgams. Detecting mercury 562–577

Chap. XXXII.—Titanium. Tin. Its ore. Its extraction and properties. Tinning. Protoxide of tin. Binoxide of tin. Stannic acid. Metastannic acid. Stannates. Sulphides of tin. Protochloride of tin. Its reducing-power. Bichloride of tin. Alloys of tin. The tin group 577–585

Chap. XXXIII.—Molybdenum. Bisulphide of molybdenum. Oxides of molybdenum.—Vanadium. Its occurrence. Its oxides.—Tungsten. Its occurrence and properties. Oxides of tungsten. Wolfram. Tungstate of sodium 585–587

Chap. XXXIV.—Gold. Its wide diffusion. Its physical properties. Indestructibility of gold. Refining of gold. Chloride of gold. Gilding. Alloys of gold.—Platinum. Chlorides of platinum. Platinum sponge. Platinum black. The platinum group 587-596

Chap. XXXV.—Atomic weights and symbols of the elements. Classification. Natural groups. Atomic heats of the elements. Atomic heats of compounds. Value of specific heats. Two sets of atomic weights in use. Barred symbols. The student's view of discussions concerning theories 597-605

Appendix.—Glass tubing. Cutting and cracking glass. Bending, drawing, and closing glass tubes. Blowing bulbs. Lamps. Blast-lamps and blowers. Caoutchouc stoppers, tubing, and sheets. Corks. Cork-cutters. Putting tubes through corks. Supports for vessels—iron stand, sand-bath, wire gauze. Pneumatic trough. Collecting gases. Safety-tubes. Gas-holders. Deflagrating-spoons. Platinum foil and wire. Filtering. Filters. Drying of gases. Drying-tubes. Chloride-of-calcium tubes Spring clip. Screw compressor. Water-bath. Iron retort. Self-regulating gas-generator. Glass retorts. Flasks. Beakers. Test-tubes. Test-glasses. Measuring-glasses. Burettes. Reading-off Burettes. Pipettes. Wash-bottle. Porcelain dishes and crucibles. Rings to support round-bottomed vessels. Crucibles. Tongs. Pincers. Mortars. Spatulæ. Thermometers. Furnaces. The metrical system of weights and measures. Table for the conversion of degrees of the centigrade thermometer into degrees of Fahrenheit's thermometer. Table for the conversion of grammes into grains, and centimetres into inches . . . . i-xlii

EMENDATIONS TO BE MADE IN THE TEXT.

Page 2, line 25. Strike out the words "both copper and sulphur have disappeared," and insert the words, "it suddenly glows vividly, at the instant when the copper and sulphur combine."

Page 16, line 11. Insert after the word "air," the words "See Appendix, Fig. XVII."

Page 61, line 3 from foot. Add the words "See Appendix, § 17."

Page 173, Exp. 94. Instead of the first sentence (three and a half lines) substitute the following: "Mix 75 or 100 grms. of flowers of sulphur and an equal weight of slaked lime. Heat half a litre of water in a capacious flask until it boils vigorously. Pour the mixture of lime and sulphur into the water, little by little, slowly enough not check the boiling. The movement of ebullition prevents the solid matter from becoming impacted upon the flask."

Page 255, line 23. Strike out the words "consists of" and insert "may be made by folding a circular piece of parchment paper into the form of a cone, like the filter represented in Fig. XX. of the Appendix, and cementing the upper edges of the paper with a drop of a thick solution of shellac in alcohol. When nearly filled with the liquid to be dialyzed and placed in a tumbler of water, the cone floats upright and presents a large surface for the diffusion of the liquids. A more elaborate form of dialyzer may be made with"

Page 267, line 6 from foot. Between the words "the" and "globule," insert the words "space about."

Page 267, line 12 from foot. Strike out the words "Chapter XX," and insert "§ 431 and Exp. 220."

Page 299, line 1. After the word "lamp," insert "see Exp. 193."

Page 315, line 17. For "potassium," read "sodium."

Page 397, line 4. Insert after "sodium," the words "in bleaching."

Page 498, lines 12 and 13 from foot. Instead of "SbO_5" and SbS_5" read "Sb_2O_5" and "Sb_2S_5."

Page 518, line 20. Strike out the words "is still," and insert "was until lately."

Page 518, line 21. Strike out the words "in Europe."

Page 518, line 22. Strike out the word "there."

Page 518, line 25. Insert after "clay," the words "or by passing ammonia gas into a mixture of dilute sulphuric acid and roasted shale."

Page 554, line 17 from foot. After "iron," insert the words "containing a small percentage of sulphur."

Page 554, line 16 from foot. Insert after "formulated," the word "approximately."

Page 590, line 16. Instead of "AuO" and "AuO_3," read "Au_2O" and "Au_2O_3."

Appendix, page xiii., line 10. After the word "motion," insert "(6) The tube should always be thrust completely through the cork, so that it may project a centimetre beyond the lower end of the cork."

ADDITIONAL EMENDATIONS.

Page 221, line 3 of Exp. 116. After "No. 2" insert "about 25 C. M. long."

Page 291, line 2 of Exp. 147. After "powder" read "Hold a pane of window-glass before the eyes and stir the mixed powders into 14 grms.," *etc.*

Page 361, line 9 from foot. For "needed" read "theoretically necessary."

Page 369, line 1. Before "as" insert "something."

Appendix, page x., lines 16 and 17. Strike out the words "the tube *g* is closed by inserting a glass rod."

MANUAL

OF

INORGANIC CHEMISTRY.

INTRODUCTION.

1. THE various objects which constitute external Nature present to the observing eye an infinite variety of quality and circumstance. Some bodies are hard, others soft; some are brittle, others tough or elastic; some natural objects are endowed with life,—they grow; others are lifeless,—they may be moved, but never move themselves; some bodies are in a state of incessant change, while others are so immovable and unchangeable that they seem everlasting. In the midst of this infinite diversity of external objects, where lies the domain of Chemistry? What is the subject-matter of this science?

When air moves in wind, when water moves in tides, or in the fall of rain or snow, the air and water remain air and water still; their constitution is not changed by the motion, however frequent or however great. A bit of granite, thrown off from the ledge by frost, is still a bit of granite, and no new or altered thing. If a solid piece of iron be reduced to filings, each finest morsel is metallic iron still, of the same substance as the original piece, as will appear to the eye if a morsel be sufficiently magnified under the microscope. The melted, fluid lead in the hot crucible, and the solid lead of the cold bullet cast from it, are the same in substance, only differing in respect to temperature. In all these cases the changes are external and non-essential, not intimate and constitutional; they are called *physical* changes.

2. When iron is exposed to the weather, it becomes covered with a brownish, earthy coating which bears no outward resemblance to the original iron; and, if exposed long enough, the metal completely disappears, being wholly changed into this very different substance, *rust*. A piece of coal burns in the grate, and soon it vanishes, leaving nothing behind but a little ashes. Dead vegetable or animal matters, buried in the ground, soon putrefy, decay, and disappear. So, too, the fragment of granite which frost has broken from the ledge, exposed for centuries to the action of air and rain, becomes changed; it "weathers," and after a time could no longer be recognized as granite. All these changes involve alterations in the intimate constitution of the bodies which undergo them; they are called *chemical* changes.

Experiment 1.—Mix thoroughly 3 grammes (for Tables of the Metrical System of Weights and Measures, see Appendix) of coarsely powdered sulphur with 8 grammes of copper filings, or fine turnings. Put the mixture into a tube of hard glass, No. 3, about 12 centimetres long, and closed at one end. (For the manipulation of tubing, see Appendix, §§ 1–4.) Hold the tube by the open end, with the wooden nippers, as in Fig. 1, and heat the mixture over the gas-lamp (Appendix, § 5), until both copper and sulphur have disappeared.

Fig. 1.

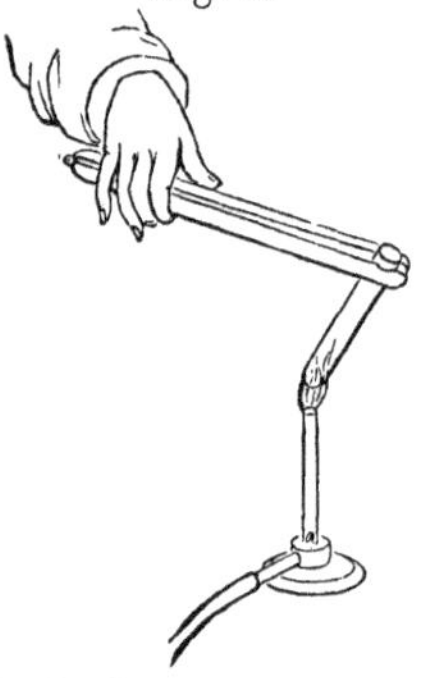

Before heat was applied, the mixture of the two substances was simply mechanical, and the copper might have been completely separated from the sulphur, by due care and patience; but during the ignition the copper and sulphur have united chemically, and there has been formed a substance which, while containing both, has no external resemblance to either. In the new body, the eye can detect neither copper nor sulphur.

If 10 grammes of metallic iron be allowed to rust away completely in moist air, the pile of rust which remains when the metal has disappeared, will weigh much more than 10 grammes. The iron has combined with two of the constituents of the atmosphere, the gas called oxygen and the vapor of water. The weight of the rust is the sum of the weights of the metallic iron, and of the water and oxygen with which the iron has combined.

Processes by which the whole character and appearance of the bodies concerned are changed, as in these experiments, so that essentially new bodies are formed from the old, are chemical processes. It is the function of the chemist, on the one hand, to investigate the action of each substance upon every other, to take note of the phenomena which attend this action, and to study the properties of the combinations which result from it; and on the other, to contrive the resolution of compound bodies into their simpler constituents. He further seeks the general laws by which the intimate combinations of matter are controlled. With these ends in view, the chemist endeavors to pull to pieces, or in technical language, to *analyze*, every natural substance on which he lays hands. Having thus found out the composition of the substance, he seeks to put it together again, or to recompose it out of its constituent parts. By one or both of these two processes, analysis (unloosing) and synthesis (putting together), the chemist studies all substances.

3. There are two questions which the chemist asks himself concerning every natural substance. First, *Of what is it composed?* With the means at his disposal the chemist may, or may not, be able to resolve the substance into simpler constituents. If he succeeds in decomposing it, he obtains the answer to this first question; if the body cannot be decomposed by any known method of analysis, the substance must be regarded as being already at its simplest. Such simple bodies are called *elements.* Secondly, the chemist asks, *How does this new substance comport itself when brought into contact with other substances already familiar?* There are sixty-five substances which are, at present, admitted to be simple, primary substances, or elements. Of compound bodies, formed by the union of these elements with each other, we find a series, numbering many thousands, in the inorganic kingdom of Nature, comprising all the diversified mineral constituents of the earth's crust; while another series, far more complex in composition, and almost innumerable in multitude, exists in the vegetable and animal world. The task of the chemist in thoroughly answering his second question would

clearly be endless, were it not for the existence of general properties common to extensive groups of both elementary and compound bodies, and of general laws which chemical processes invariably obey.

While, therefore, the chemist seeks the answers to the two fundamental questions above stated, he is at the same time inquiring what relations exist between the properties of a body and its composition, and he is also studying that natural and invariable sequence of chemical phenomena, which, when fully known, will constitute the perfect science of chemistry.

4. Generalizations from observed facts, so long as they are uncertain and incomplete, are called hypotheses and theories; when tolerably complete and reasonably certain, they are called laws. The attention of the student should be constantly directed to the keen discrimination between *facts* and the *speculations* founded upon those facts—between the actual evidence of our trained senses, brought intelligently to bear upon chemical phenomena, and the reasonings and abstract conclusions based upon this evidence—between, in short, that which we may know, and that which we may believe.

CHAPTER I.

AIR.

5. We are everywhere surrounded by an atmosphere of invisible gas, called air. All objects upon the earth's surface are bathed with it. In motion, it is wind, and we recognize its existence by its powerful effects; but in the stillest places it exists as well.

The presence of air in any ordinary bottle, flask, or other hollow vessel which is empty, in the sense in which this word is ordinarily applied, can be shown very simply by attempting to put some other substance into the vessel, under such conditions that the air cannot pass out from it. Or the air can actually be pumped out of the bottle; and it can be removed by other means, both mechanical and chemical.

Exp. 2.—Wrap around the throat of a funnel, with narrow outlet, a strip of moistened cloth or paper, so that the funnel shall fit tightly into the neck of a bottle. After the funnel has been fitted to the bottle, as is shown in Fig. 2, fill it with water, and observe that this water does not run into the bottle. The bottle which we have called empty is in reality filled with air, and it is this air which prevents the water from entering the bottle. If, now, the funnel be lifted slightly, so that the mouth of the bottle shall no longer be completely closed by it, the air within the bottle will pass out, and the water in the funnel will instantly flow down.

Fig. 2.

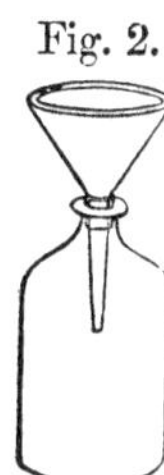

6. In order to pump the air out of any hollow vessel, an apparatus known as the air-pump is commonly employed. Descriptions of this machine may be found in the text-books on physics. For purposes of illustration, a portion of the air can always be removed by suction, or by mechanically displacing the air by some other substance, as when the finger thrusts the air out of a thimble.

Exp. 3.—Fit to any small flask or phial a perforated cork (for the manipulation of corks, see Appendix, § 8), to which has been adapted a short piece of glass tubing, No. 7. Tie upon this glass tube a short piece of caoutchouc tubing. Suck part of the air out of the flask, and then nip the caoutchouc tube with thumb and finger, so that no air shall reënter. Immerse the neck of the flask in a basin of water, and release the caoutchouc tube. Water will instantly rise into the flask to take the place of the air which has been sucked out.

7. The water, in this experiment, is forced into the flask by the pressure of the superincumbent atmosphere. Air has weight. It is subject to the law of gravitation, and is attracted towards the earth's centre in the same way as other ponderable matter. It has been found by experiment that a litre of dry air, at the temperature of 0°, and under a pressure of 760 millimetres, weighs 1·2932 gramme. It has also been determined by the physicists that the force with which the air is attracted to the earth is on an average equal to a weight of 1·033 kilogramme to the square centimetre of surface. That is to say, the ocean of air above us presses down upon every square centimetre of the earth's surface with a force equal to that which would be

exerted by a bar of metal, or other substance, a centimetre square in section, and long enough to weigh 1·033 kilogramme.

If such a bar were constructed of iron, it would be 1·3 metre long; if of water (and a bar of this substance can readily be made by enclosing the water in a tube), it would be 10·33 metres long; if of mercury, 76 centimetres long. A clear conception of the atmospheric pressure is important to the chemist, because of its bearing upon the mechanical collection, manipulation, and measurement of gases.

8. In addition to the qualities already mentioned, we find air to be tasteless, and odorless, colorless when in small depths, but exhibiting a blue tint when seen in large masses, as when in the absence of clouds we look at the sky, or at a distant mountain. Several of its properties are not peculiar, but are common to all gases. Dry air obeys precisely the law of Mariotte up to a pressure of several atmospheres; that is to say, its volume diminishes or increases in proportion to the pressure to which it is subjected; but it has never yet been condensed to a liquid. Upon being heated one degree centigrade, it expands $\frac{1}{273}$, or in other terms 0·003665, its volume at 0°. At the temperature of 0°, air is 773 times lighter than water at 4°; that is, than water at its maximum density.

These physical properties of air have been enumerated in order to a distinct acquaintance with this gas, and that we may more clearly comprehend its chemical properties as we now proceed to study them.

9. *Of what is air composed?* When a bar of iron is heated in the air, as at a blacksmith's forge, it becomes covered with a coating, which flies off in scales when the iron is beaten upon the anvil. If a bit of tin be melted in a shallow crucible, its surface soon becomes covered with a white earth, or ashes. At a high temperature both iron and tin slowly *burn* in the air, and are converted into earth or ashes. With the exception of gold, silver, platinum, and a few other exceedingly rare metals, all the metals burn, or rust, when heated in the air. If no air be present, this rust or ashes will not be formed, however long or intensely the metal may be heated. But in what manner is the rust formed? Is something driven out of the metal into the air, or does some-

thing come out of the air and unite with the metal? Experiment shall answer.

Exp. 4.—Place 20 grammes of tin-foil in a porcelain dish, 4 or 5 c.m. in diameter. Counterbalance the dish, with its contents, upon scales which will turn promptly with 0·5 grm. when thus loaded. Heat the dish, placed upon the wire gauze of the iron-stand (see Appendix, § 9), over the gas-lamp moderately and continuously for two or three hours, or until a large part of the metal has been converted into ashes. To facilitate the change, move the dish frequently, so that a new surface of metal shall be often exposed to the air. Replace the dish, when perfectly cool, upon the pan of the balance, and observe that the weight of the dish and its contents has very decidedly increased.

A more rapid demonstration may be obtained by substituting for the tin-foil a piece of magnesium wire or ribbon. This metal takes fire when touched with a lighted match.

10. It is possible that during the heating the metal may have lost something, but it is certain that it has gained more. This additional matter has not come from any alteration in the dish, for it is made of materials expressly adapted to resist such treatment, and a little cleaning will restore it to exactly its original condition. We have, then, taken something out of the air, which, gaseous in the air, has become solid in the white ashes of the tin.

If this something with which the tin has united can be separated again from the metal, and restored to the gaseous condition, it will be easy to compare it with common air, and so learn whether the matter which combined with the heated tin is air itself, or only a part of the air. It is quite possible to recover the gas which enters into the composition of the rust of iron, or copper, or tin; but the processes required are too circuitous for the present purpose. From the rust of other common metals, such as lead, manganese, or mercury, the absorbed gas can be very easily expelled. The rust of mercury is the most easily decomposed. Mercury-rust may be prepared by the long heating of the metal in the air, in a manner strictly analogous to the method already applied to the preparation of tin-rust.

Exp. 5.—Put into a tube of hard glass, No. 3, about 12 c.m. long, and closed at one end as in Exp. 1, 10 grammes of the rust of mercury, a substance which is sold under the name of red oxide of mercury. Such tubes of hard glass will be hereafter designated as "ignition-tubes." Attach to this ignition-tube, by means of a perforated cork, or caoutchouc stopper, a delivery-tube of glass, No. 8, of such shape and length that it shall reach beneath the inverted saucer in the pan of water, as represented in Fig. 3. The point of departure, for the construction of the apparatus, is the top of the gas-lamp placed upon the foot of the iron-stand; the end of the ignition-tube should be about 4 c.m. above the top of the lamp. The other details of the apparatus may be learned from the figure.

Fig. 3.

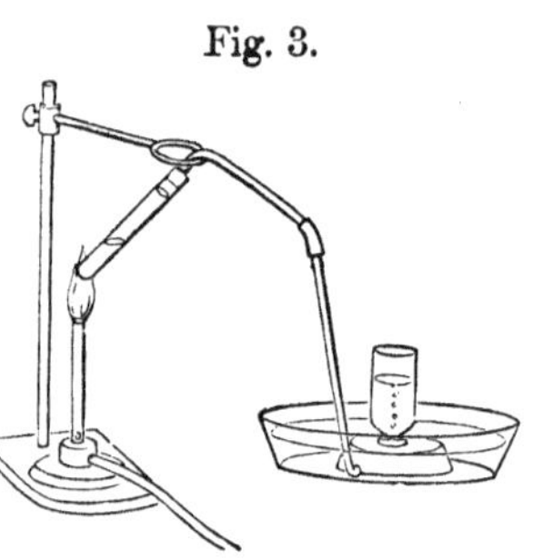

Upon lighting the lamp, the air within the ignition-tube will expand, and a portion of it will pass out through the delivery-tube. This air should be collected in a small bottle by itself, and thrown away. The volume of air thus thrown away should of course not be much greater than that of the tubes. (For a description of the pneumatic trough, see Appendix, § 10.)

As the ignition-tube becomes hot, gas will be freely given off from the red oxide of mercury contained in it. It is necessary to avoid heating intensely a single small spot of the ignition-tube, lest the glass soften, and, yielding to the pressure from within, blow outward, and so spoil the tube and arrest the experiment. The gas-flame should be so placed and regulated as to heat 3 or 4 c.m. of the tube at once. Collect the escaping gas in bottles of 100 to 150 c. c. capacity.

As soon as the disengagement of gas slackens, lift the iron-stand up, and take the delivery-tube out of the water, taking care that no water shall remain in the end of the tube. Then, *and not till then*, extinguish the lamp. (See Appendix, § 10.) In the upper part of the ignition-tube, and sometimes in the delivery-tube also, metallic mercury will be found condensed in minute globules. The liquid metal is volatile at the temperature to which it has been subjected, and has distilled away from the hot part of the tube, and condensed upon the cooler part.

Thus is recovered the metallic mercury from which the red mercury-rust was originally prepared by long heating in contact with air. Is the gas, which the mercury originally took from the atmosphere, air itself, or something different?

Exp. 6.—Introduce a lighted splinter of soft wood into a bottle of the gas collected in the last experiment. It will burn with much greater brilliancy than in the air. Attach a bit of wax taper to a piece of wire; light the taper, blow it out, and while the wick still glows, introduce it into a second bottle of the gas of Exp. 5. The glowing wick will burst into flame, and the taper will burn with extraordinary brilliancy.

11. It is very obvious, from these experiments, that the gas which enters into the composition of mercury-rust is not air itself. But since it came originally from the air, if it is not the whole of air, it must be a part or constituent of air. This gas, which causes combustible substances to burn with such intensity, is indeed a constant constituent of the air, and a thorough study of all its properties will hereafter convince us that it is a chemical element of very various powers and great importance. It is called *oxygen*, and under this name it will form the subject of the next chapter.

12. But if oxygen be not air itself, but only a constituent of air, it follows that air must have other constituents, or at least one other constituent. If mercury be long heated in contact with a certain confined portion of air, it will abstract from this air, as we have seen, one of its ingredients, namely, oxygen, and there will be left behind whatever of air is not oxygen. This experiment, the original one by which the illustrious chemist Lavoisier demonstrated that air is a constant mixture of two different gases, deserves careful study, both for its philosophical value and its historical importance. The actual experiment lasts several days, and is therefore unsuitable for repetition by the student. A description of it will suffice.

Into a flask, provided with a long neck, some metallic mercury was introduced; the neck of the flask was then bent, as shown in Fig 4, the flask placed upon a furnace, and the end of the neck plunged into a basin of mercury; a jar was then placed over the end of the tube, and a portion of the air within the jar was sucked out by means of a bent tube; the mercury thereupon rose in the jar, and the point at which it stood was accurately noted. The thermometer and barometer were also simultaneously observed. Fire was then lighted in the furnace, and the heat maintained for twelve days at a point just below that required to make the mercury boil. The mercury became gradually covered with red scales, and the air in the jar, which at first expanded from the action of the heat, slowly decreased in bulk until fresh scales were no longer formed. From these red scales Lavoisier obtained, by the method already exhibited (Exp. 5), the element oxygen. The residual air in the jar proved, on examination, to be unfit for the support of combustion and of animal life; a candle was instantly extinguished by it, as if plunged in water, and small animals, thrust into the gas, died in a few seconds. The gas is, in reality, a second elementary substance, distinguished by marked chemical and physical peculiarities. It is called *nitrogen,* and under this name will be more completely studied in another chapter.

Fig. 4.

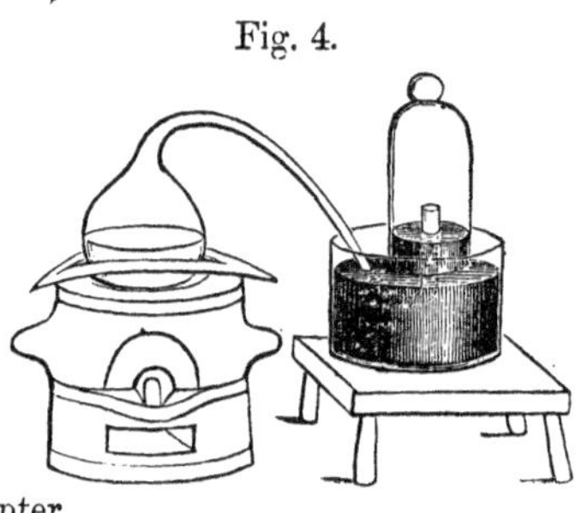

13. The experiment of Lavoisier not only affords the means of separating the two different gases of which air is composed, but also determines the proportions in which they are mixed in air. If the diminution in bulk which the air in the jar undergoes during the whole progress of the experiment be accurately measured, it will be found that the bulk of the residual gas, the nitrogen, is only four-fifths of the original volume of air. The lost fifth is the oxygen which has combined with the mercury. The air, then, is not an element, but is compound, and its two principal ingredients are the elementary bodies oxygen and nitrogen, mixed in the proportion of four measures of nitrogen to one of oxygen. It is quite possible to prove by synthesis what analysis has thus taught. On putting together four measures of nitrogen and one measure of oxygen, a mixture is obtained, which, except by very refined experiments, is not to be distinguished from pure air. Aqueous vapor is another normal constituent of the actual atmosphere, and small traces of other gases than nitrogen and oxygen are always present in it, as will be set forth hereafter.

CHAPTER II.

OXYGEN.

14. Oxygen gas may be prepared by heating red oxide of mercury, as described in Exp. 5 ; or, far more conveniently, by heating a mixture of chlorate of potassium and black oxide of manganese. For the present we have to regard these substances merely as materials suitable for the preparation of oxygen. Their constitution will be studied hereafter.

Exp. 7.—Mix intimately 5 grms. of chlorate of potassium with 5 grms. of black oxide of manganese, which has been previously well dried. Place the mixture in a tube of hard glass, No. 1, 12 or 15 c.m. in length. Attach to this ignition-tube, by means of a perforated cork or caoutchouc stopper, a delivery-tube of glass, No. 7, as represented in Fig 3, and described upon page 8. Heat the mixture in the ignition-tube and collect the gas which will be given off, in bottles or jars of the capacity of about 250 c. c. The first 100 c. c. or so of gas should be rejected, since it will be contaminated with the air originally contained in the apparatus.

It is easy to determine the moment at which the evolution of oxygen commences, by noting the increased size of the bubbles of this gas as contrasted with those of the expanded air, and the greater rapidity with which the bubbles of oxygen come over. For every grm. of chlorate of potassium taken, about 230 c. c. of oxygen gas should be obtained.

Besides the general precautions described under Exp. 5, the following should here be observed. 1. Both the chlorate of potassium and the oxide of manganese should be perfectly dry and pure—that is, free from moisture, dust, or particles of organic matter. 2. So soon as the oxygen begins to be delivered, the heat beneath the ignition-tube should be diminished, if need be, and so regulated that the evolution of gas shall be tranquil and uniform. 3. The uppermost portions of the mixture should be heated before the lower. 4. An ignition-tube should never be filled to more than one-third its total capacity, lest portions of its contents be projected into the delivery-tube. The chief danger to be guarded against in using ignition-tubes is the stoppage of their outlets with solid matter. 5. The ignition-tube should always be inclined as represented in Fig. 3, and never placed upright in the flame.

The oxygen thus obtained is to be employed in the experiments shortly to be described. In case large quantities of oxygen are needed, a similar mixture of equal weights of chlorate of potassium and black oxide of manganese is heated in a retort of iron or copper, and the gas is collected in large metallic vessels, called gas-holders, such as are described in § 11 of the Appendix.

Besides the methods above described, there are many other ways of preparing oxygen. Several of these methods will be described hereafter, when the materials employed can be intelligently studied.

15. Oxygen is a transparent and colorless gas, not to be distinguished by its aspect from atmospheric air. Like air, it has neither taste nor smell. It is, however, somewhat heavier than air. If the weight of a measure of air be taken as 1, then the weight of the same measure of oxygen is found to be 1·1056. At 0°, and a pressure of 760 m.m. of mercury 1 litre of oxygen gas weighs 1·4298 grm.

Since oxygen is thus heavier than air, it is not absolutely necessary, in collecting it, or transferring it from one vessel to another, that we should operate over water, as has been directed. When a gas is much heavier, or much lighter, than atmospheric air, it may often be conveniently collected by *displacement*. A bottle can readily be filled with oxygen from the gas-holder by carrying the delivery-tube to the bottom of the upright bottle, and allowing the gas to flow in slowly, as if it were water. In a short time the air will be wholly displaced, and the bottle filled with oxygen. The progress of the operation can be followed by testing the contents of the upper part of the bottle from time to time, with a glowing match; when this bursts sharply into flame the gas may be assumed to be pure enough for all ordinary purposes.

16. Oxygen has never yet been reduced to the liquid condition. Of all known substances it exerts the smallest refracting power upon the rays of light. Compared with that of atmospheric air, its refractive power is as 0·830 to 1·000. The specific heat of oxygen, compared with that of an equal weight of water taken as unity, is 0·2182; it has a lower capacity than other gases for absorbing and radiating heat. Water dissolves it in small proportion; 100 volumes of water dissolve, at the ordinary tempera-

ture, about 3 volumes of oxygen. It exhibits decided magnetic properties. A glass tube filled with oxygen and suspended by a fibre of silk between the ends of a strong horse-shoe magnet soon comes to rest in such a position that its long axis is parallel to a line drawn between the two poles. As with iron, its susceptibility to magnetization is diminished, or even temporarily suspended, by a sufficient elevation of temperature. Its magnetic force, compared with that of the air, is as 17·5 to 3·4, that of a vacuum being taken as 0.

17. A striking characteristic of oxygen is its power of supporting combustion. This has already been illustrated in Exp. 6, and may be further exhibited by a great variety of experiments.

Fig. 5.

Exp. 8.—Place in a deflagrating spoon (see Appendix, § 12), a bit of sulphur as large as a pea. Light the sulphur, and thrust it into a bottle of oxygen. It will burn with a beautiful blue flame, and much more brilliantly than in air. An acid, suffocating gas is produced.

Exp. 9.—Place a piece of charcoal, that of bark is best, in a deflagrating spoon. Kindle the charcoal by holding it in the flame of a lamp, and then introduce it into a bottle of oxygen. It will burn vividly, throwing off brilliant sparks if bark charcoal has been employed.

In this experiment, as in the preceding, the products of the combustion are obviously gaseous, no solid substance being formed.

Exp. 10.—A piece of phosphorus, the size of a small pea, having been well dried between pieces of blotting-paper, is placed in a deflagrating spoon, touched with a hot wire or a lighted match, and then thrust into a jar of oxygen. It will burn with intense brilliancy and the formation of a dense white smoke.

It should be observed that phosphorus is a substance which inflames very readily in the air, when subjected to friction or any slight elevation of temperature. It is hence so dangerous that it must always be kept under water. It should also be cut under water.

18. Many substances commonly called incombustible, because they do not burn readily in ordinary air, burn vigorously in oxygen. Of these, metallic iron may be taken as an example.

Exp. 11.—Convert two-thirds of a piece of fine piano-wire 20 or 30 c.m. long into a spiral, by winding it tightly around a glass rod, No. 6, or lead pencil, and then withdrawing the rod or pencil. Thrust the straight end of this wire into a cork or piece of thin board large enough to cover one of the bottles of oxygen obtained in Exp. 7. Upon the lower end of the wire coil tie a small piece of waxed thread, or touch with a bit of melted sulphur the end of the spiral, so that a small bead of sulphur shall adhere to the wire. Kindle the thread or sulphur, and quickly place the wire in a bottle of oxygen, at the bottom of which has been spread a layer of sand.

The burning sulphur heats the iron to redness, which then burns brilliantly, with scintillation. From time to time, glowing balls of molten matter fall off from the wire, and bury themselves in the sand at the bottom of the bottle, or even melt into the glass.

If an abundance of oxygen be at hand, this experiment had better be performed in a jar of 2 or 3 litres capacity,—a watch-spring which has been rendered flexible by igniting it, and then allowing it to cool slowly, being the best material with which to form the spiral coil. In this case it is well to tie a bit of tinder upon the end of the coil as the kindling material, or to attach a piece of twine to the wire, and soak this in sulphur. But the experiment succeeds well even in very small bottles of oxygen, provided the wire be fine, and that the quantity of sulphur, or other matter, employed for kindling be not too large.

Fig. 6.

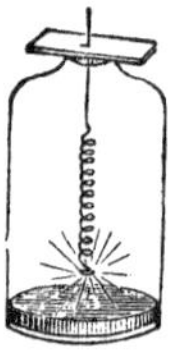

19. This experiment clearly proves what has been already stated, that iron, when red-hot, combines with oxygen. It is the burnt or oxidized iron which falls in globules to the bottom of the bottle. This substance is called oxide of iron. The compounds which are formed by the union of oxygen with other elements are called *oxides*. The substances which have been heretofore mentioned under the more familiar name of rust, like iron-rust, tin-rust, mercury-rust, are called in chemistry oxides —as the oxide of iron, oxide of tin, and oxide of mercury.

20. It will have been observed that the combinations obtained in the foregoing experiments are of very various quality. Some of these compounds are solid, others gaseous; some are acid and caustic, while others are tasteless and innocuous. They agree only in this, that they all contain oxygen. All these bodies will be studied in detail hereafter. It concerns us now more particularly to realize the number and variety of the bodies into which oxygen enters as an essential ingredient. In fact, the most important quality of oxygen is that, with a single exception, it unites with all the other elements to form compounds.

This act of combination is often accompanied by development of light and heat, as in the foregoing experiments; in the affairs of common life we daily witness similar effects. All the ordinary phenomena of fire and light depend upon the union of the body burned with the oxygen of the air. Indeed, the term *combustion* may for all ordinary purposes be regarded as synonymous with *oxidation*.

Combustion is less vivid in air than in pure oxygen, because of the nitrogen with which the oxygen of the air is diluted. If a substance combines slowly with oxygen, it may often happen that no evolution of heat or light can be detected. Thus, when a small piece of iron rusts at the ordinary temperature of the air, there is perceived neither light nor heat, although a combination of iron and oxygen has been formed, as in Exp. 11. Heat is really disengaged in the slow rusting of iron, as in every act of chemical combination, but it is taken up and carried away by the circumambient air at the moment of its formation, so that it cannot usually be perceived. Slow oxidation, such as this, is often spoken of as slow combustion.

21. Many of the compounds of oxygen are very familiar bodies. Indeed, oxygen is the most widely diffused and the most abundant of all known substances. Not only does it occur in the air, of which it constitutes about one-fifth the volume, as has been already remarked, but at least one-third of the solid crust of the globe is composed of it. It is the chief ingredient of water, as will appear in a subsequent chapter. It enters largely into the composition of plants and animals. Silica, in all its varieties of sand, flint, quartz, rock-crystal, &c., contains about half its weight of oxygen, and the same is true of the various kinds of clay, and of chalk, limestone, and marble. Oxygen is absolutely essential to the maintenance of animal and vegetable life. The chemistry of the respiration of animals depends upon the absorption of oxygen from the air respired. In the absence of oxygen suffocation ensues.

CHAPTER III.

NITROGEN.

22. The common modes of obtaining nitrogen depend upon the removal of oxygen from the air.

Exp. 12.—Fill a tube of hard glass, No. 2, about 35 centimetres long, with bright, not too coarse, copper turnings; place this tube upon a semicylindrical trough of sheet iron, and support it upon a ring of the iron stand, as shown in Fig. 7. It is well to interpose a thin layer of asbestos between the tube and the iron trough, in order to prevent the glass from adhering to the metal when it becomes soft by heat. By means of corks, attach to one end of this tube a delivery-tube leading to the water-pan, as shown in the figure, and connect the

Fig. 7.

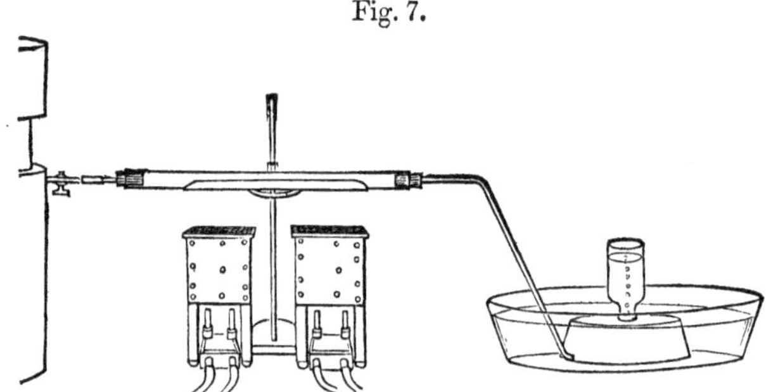

other end with a gas-holder filled with air. Light the lamps (for a description of these, see Appendix, § 5) beneath the tube which contains the copper turnings, and wait until the copper has become red-hot; then allow air to flow slowly out of the gas-holder over the hot metal. The heated copper will combine with the oxygen of this air, and retain the whole of it, so that nothing but nitrogen will be delivered at the water-pan. This nitrogen may be collected in small bottles and tested with lighted splinters of wood, which should be instantly extinguished on being immersed in it.

23. A still simpler method of preparing nitrogen is to burn out the oxygen from a confined portion of air, by phosphorus, or by a jet of hydrogen.

Exp. 13.—Float a small porcelain capsule upon the surface of the water-pan; a large cork must be placed beneath the capsule if the latter will not float of itself. In the capsule put about a cubic centimetre of phosphorus, and set it on fire. Invert over the whole a wide-mouthed bottle, of the capacity of a litre or more, and hold this bottle so that its mouth shall dip beneath the surface of the water. During the first moments of the combustion, the heat developed thereby will cause the air within the bottle to expand to such an extent that a few bubbles of the air will be expelled; but after several seconds water will rise into the bottle to take the place of the oxygen which has united with the phosphorus.

The dense white cloud, which fills the bottle at first, is a compound of phosphorus and oxygen which is soluble in water. It will, therefore, soon be absorbed by the water in the pan, and will disappear, so that at the close of the experiment nearly pure nitrogen will be left in the bottle. But, as the phosphorus ceases to burn before the last traces of oxygen are exhausted (compare § 180), the nitrogen obtained by this method is never absolutely pure.

As soon as the phosphorus goes out, the bottle should be shaken in such a way that the porcelain capsule may be upset, and sunk in the water-pan. The properties of the nitrogen may now be examined.

24. Nitrogen is a transparent, colorless, tasteless, odorless, incondensable gas. It is not quite so heavy as air. If a measure of air weigh 1 gramme, then an equal measure of nitrogen will weigh 0·9714 gramme. At 0°, and a pressure of 760 millimetres of mercury, 1 litre of nitrogen weighs 1·256 gramme. The specific heat of the gas is 0·244, that of an equal weight of water being 1·000. A litre of water at 0° dissolves only 20 cubic centimetres of nitrogen. Its refractive power in regard to light is to that of atmospheric air as 1·034 to 1·000.

25. In its chemical deportment towards other substances, nitrogen is remarkably different from oxygen. Whilst oxygen is active and, as it were, aggressive, nitrogen, at least when in the condition in which it exists in air, is remarkably inert and indifferent as regards entering into combination with other bodies. It is marked rather by the absence of salient characteristics than by any active properties of its own. Many of the metals, sulphur, phosphorus, and numerous other substances, may be kept unchanged for any length of time in vessels filled with nitrogen. A burning candle will instantly be extinguished when

thrust into a jar of nitrogen gas, for with the nitrogen the constituents of the candle have no tendency to combine.

As it extinguishes flame, so it destroys life. Animals cannot live in an atmosphere of pure nitrogen. It may, indeed, be breathed for a short time with impunity, but it does not support respiration. It is not poisonous; if it were, it could not be breathed in such large quantities as it is in air. An animal immersed in it dies, simply from want of oxygen.

26. As a diluent of the oxygen in the air, nitrogen is essential to the existing balance and order of Nature. All animal and vegetable life—most inanimate matter, even—stands in harmonious relations with the chemical composition of the atmosphere. The presence of so large a proportion of nitrogen in the air prevents the too rapid action, as regards combustion and respiration, that would take place in an atmosphere of unmixed oxygen.

Nitrogen is widely diffused in nature. Besides occurring in the air, it is a constituent part of all animals and vegetables, and of many of the products resulting from their decomposition. Notwithstanding the indisposition of nitrogen in the free state to enter into combination, a very large number of interesting and important compounds can be formed by resorting to indirect methods of effecting its union with other elements.

CHAPTER IV.

WATER.

27. Another natural substance, quite as common as air, is water. Three-fourths of the earth's surface is covered with it. It is diffused through the atmosphere in the form of vapor, and is a constituent of all animal and vegetable substances, and of many minerals. We take up next this familiar substance, in order that we may gain, through the study of it, a deeper insight into chemical principles, and enlarge our experience by making acquaintance with a new element. Let us first define with precision the external and physical properties of water, and then

apply the two chemical methods, of analysis and synthesis, to the closer investigation of its essential nature.

28. At the ordinary temperature of the air, pure water is a transparent liquid, devoid of taste or smell. In thin layers it appears to be colorless, but large masses of it are distinctly blue, as seen in mid-ocean, in many deep lakes of pure water, and in masses of ice, such as icebergs and some glaciers where it is possible to look through the ice from below.

This color can be seen upon the small scale by looking down through a column of pure water, 2 metres long, upon pieces of white porcelain. The water may be held in a glass tube, 5 c.m. wide, which has been blackened internally with lamp-black and wax to within 1·25 c.m. of the end, which is closed by a cork. Fill the tube with chemically pure water, throw into it a few pieces of porcelain, and place it in a vertical position, on a white plate. On now looking through the column of water at the bits of porcelain, which can only be illumined by light reflected from the white plate through the rim of clear glass, it will be observed that they exhibit a pure blue tint, the intensity of which will diminish in proportion as the column of water is shortened. The blue coloration may also be recognized when a white object is illuminated through the column of water, by sunlight, and seen at the bottom of the tube through a small lateral opening in the black coating.

29. At 4°, the temperature at which it is densest, water is 773 times heavier than air at 0°; at 15° it is 819 times heavier than air of the same temperature. A cubic centimetre of water at its greatest density, that is, at 4°, weighed in a vacuum, is our unit of weight—a gramme. One litre of water, which measures 1000 cubic centimetres, therefore, weighs a kilogramme. Water is compressible and elastic; by the pressure of one atmosphere it can be reduced to the extent of about 47-millionths of its original volume, and this is true for every added atmosphere of pressure so far as experiment has extended. Water expands upon being heated, though at a less rate than other liquids; the rate of expansion increases with the temperature. Notwithstanding the fact that water expands when cooled below 4°, as well as when warmed above that temperature, its refractive power on light continues to increase regularly below 4°, as though it contracted. The refractive index increases continuously between +5·2°

and $-1{\cdot}3°$, the direction of the variation not changing in the passage through the point of maximum density. At 0° the index is 1·333.

30. Pure water at 0°, a temperature always to be obtained by melting ice, is taken as a standard to which the weights of equal bulks of other substances, liquid or solid, are referred. In other words, the *specific gravity* of water is taken as 1; and in terms of this unit the specific gravities of all other liquid and solid substances are expressed. The specific gravity of gold, for example, is 19·3; that is to say, the weights of equal bulks of water and of gold are to one another as 1 to 19·3.

31. Water is also the standard of specific heat. By specific heats are meant the relative capacities for heat of the same weights of different substances, at the same temperature. For example, to raise 1 kilogramme of mercury from 0° to 1° requires only one-thirtieth of the quantity of heat necessary to raise 1 kilogramme of water from 0° to 1°. Water having been made the standard of specific heat, its capacity for heat is denoted by 1, and that of mercury will accordingly be 0·033. At the same temperature, and for equal weights, water has a greater capacity for heat than any solid or liquid known. Hence it results that the specific heats of all solid and liquid substances are expressed by fractions.

Water conducts heat very slowly; it may be boiled many minutes at the top of the test-tube, which is held all the while by the lower end, in the fingers, without inconvenience.

32. When exposed to a certain degree of cold, water crystallizes, with formation of ice, or snow, according to circumstances; and upon being heated sufficiently it is transformed into an invisible gas, called steam. Both these changes, however, are purely physical, and therefore do not fall within the province of this manual. The chemical composition of the water is the same, whether it be liquid, solid, or gaseous. The temperature at which ice melts is one of the fixed points of the centigrade thermometer, numbered 0°, and the temperature at which water boils, under a pressure of 76 c.m. of mercury, is the other fixed point, numbered 100°. Water evaporates at all temperatures, and is therefore a constant ingredient of the atmosphere. Even

ice slowly evaporates, at temperatures far below 0°, without first passing into the liquid condition.

In crystallizing, that is to say, in assuming the solid form, water increases in volume. The specific gravity of ice is only 0·916, which is equivalent to saying that, in the act of freezing, 916 c. c. (cubic centimetres) of water will be changed into a litre of ice. From this fact result many familiar phenomena, such as the floating of ice, the upheaving and disintegrating action of frost, and the bursting of pipes and other hollow vessels when water is frozen in them. The crystals of ice belong to the so-called hexagonal system; they are six-sided prisms, with regular faces; by agglomeration they produce stellar and fern-like forms of infinite variety and great beauty. Ice is a slow conductor of heat, and a non-conductor of electricity. It becomes electric by friction.

Steam is a colorless, transparent gas, as invisible as atmospheric air. It is lighter than air, the weight of any given volume of steam, at the ordinary temperature, being to that of the same volume of air as 0·622 to 1, a ratio deduced by calculation from the composition of steam. At 100°, the boiling-point of water, the ratio of the weights of equal volumes of steam and air is 0·455 to 1, and one volume of water furnishes about 1700 volumes of steam of 100°. When steam is heated by itself, without the presence of any liquid water, it is called *superheated* steam; but when there is water present, so that no excess of heat can accumulate in the steam, above the quantity needed for its formation under the pressure at which it exists, the steam is called *saturated,* meaning saturated with water. When steam escapes into the air, there is formed a multitude of little bubbles or vesicles, composed of a film of water filled with air, precisely similar to the vesicles seen in clouds and fogs. This steam-cloud is sometimes improperly spoken of as steam or vapor, an error against which the student should be upon his guard. Similar fogs of air-filled vesicles are formed whenever the atmosphere is cooled to a temperature so low that the aqueous vapor contained in it can no longer exist in the gaseous state.

33. Let us pass now to the analysis of water. Of what is water composed? We can determine this point by methods

similar to those which were adopted in the examination of air. There are several metals which, upon being brought into contact with water, will abstract from it one of its ingredients, in the same way that we have seen them abstract oxygen from the air. Some metals can abstract this ingredient even at the ordinary temperature. Thus the metal called sodium, on being brought into contact with water, decomposes it, and, uniting with one of its constituents, sets free another as a gas. This new gas is called Hydrogen.

Exp. 14.—Make a small cylinder of wire gauze by rolling a piece of fine gauze, about 6 c.m. square, around a thick piece of No. 6 glass tubing. Twist fine wire around the cylinder in order to preserve its form, then slip the cylinder off the glass, and close one end of it by pressure with a stout pair of pincers. Within this cylinder of wire gauze place a piece of metallic sodium as large as a small pea, and then close the upper end of the cylinder by pressure with the pincers, as before. Quickly place the wire gauze cylinder under the mouth of a small bottle of 100 or 200 c. c. capacity, which has previously been filled with water and left inverted in the water-pan.

As soon as water comes in contact with the sodium, bubbles of gas will issue from the wire gauze cage, and, rising through the water, will collect at the top of the inverted bottle. If no gas is generated during the first moment after the wire gauze has been placed in the water, move the bottle to and fro, in such manner that the gauze cylinder may be shaken about, and water forced through its interstices. As soon as the evolution of gas has ceased, close the mouth of the bottle with a small plate of glass, turn it mouth uppermost, remove the plate, and touch a lighted match to the gas. The gas will take fire with a sudden flash.

34. At a low red heat water can be decomposed by several of the metals, such as iron, tin, zinc, and magnesium. Iron is well adapted for this experiment.

Exp. 15.—Provide a piece of iron gas-pipe, about 35 c.m. long, and 1 c.m. or more in internal diameter; fill it with small, bright iron-turnings, and support it upon a ring of the iron stand over one or two wire-gauze gas-lamps. By means of perforated corks, connect with the iron tube, on the one hand, a glass delivery-tube leading to the water-pan, as shown in the figure, and, upon the other, a delivery-tube coming from a thin-bottomed glass flask, half full of water, and supported upon a ring of a second iron stand. Light the lamps beneath the iron tube, and wait until its contents have become red-hot; then heat the water in the flask until it boils slowly. As the aqueous

Fig. 8.

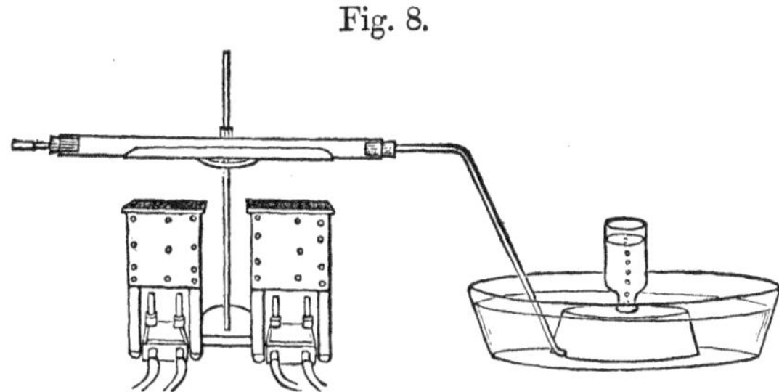

vapor passes over the hot iron-turnings it will be decomposed, one of its constituents will unite with the iron, and hydrogen will pass off through the delivery-tube and may be collected in bottles at the water-pan, so soon as the air originally contained in the tubes and flask has all been expelled.

If, at the close of this experiment, and after the tube has become cold, the iron be removed from the tube, it will be found to be covered with a black coating similar to that which forms on iron heated in the air.

35. By these experiments it has been proved that one of the components of water is a gas called hydrogen. But with the exception of the coating upon the iron of Exp. 15, we have as yet nothing to indicate what other ingredients the water may contain. Such evidence can, however, be readily obtained by resorting to another method of analysis. If a galvanic current is caused to flow through water, the force by which the constituents of the water are held together will be overcome, and the water will be resolved into the elements of which it is composed. On immersing the platinum poles of a galvanic battery in water, to which a little sulphuric acid has been added for the purpose of increasing its conducting-power, minute bubbles of gas will immediately be given off from these poles, and will be seen rising through the liquid. We have here abundant proof of the powerful action exerted by the battery upon the water. But the experiment will be much more satisfactory if it be made in a vessel so arranged that the evolved gases may be collected for examination.

Fig. 9.

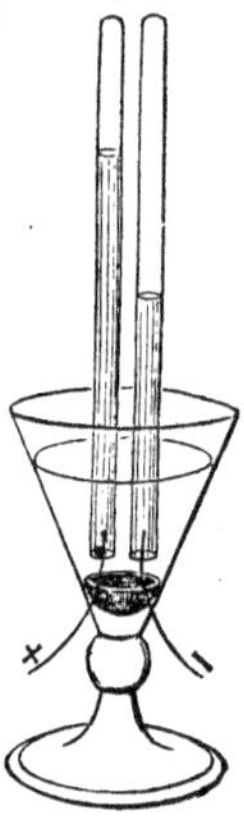

For this purpose the apparatus shown in Fig. 9 can be conveniently employed. The test-glass, nearly full of water which has been mixed with from $\frac{1}{10}$ to $\frac{1}{6}$ of sulphuric acid, carries two platinum wires cemented with shellac into the glass. These wires terminate above in thin plates of platinum; over each of these plates there is inverted a long, narrow test-tube full of water, acidulated in the same way as that in the test-glass. Upon connecting the wires with a galvanic battery,— two Bunsen's cells of medium size will be sufficient,—the water will be decomposed, and the resulting gases, as they are given off at the platinum plates, will rise, transparent and colorless, into the test-tubes. On comparing the bulks of the two gases, it will be found that twice as much gas has collected in the one tube as in the other. If the test-tube containing the larger volume of gas be now closed with the thumb, turned mouth uppermost, and the gas within touched with a lighted match, it will take fire and burn with the characteristic flame of hydrogen. It is, in fact, hydrogen.

If the smaller volume of gas in the other tube be examined in the same way, it will not inflame, although it gives intense brilliancy to the combustion of the match. If a splinter of wood, retaining but a single ignited spark, be immersed in the gas, it instantly exhibits a vivid incandescence, and in a moment bursts into flame. This gas is oxygen. It is thus proved that out of water may be unloosed two volumes of hydrogen and one volume of oxygen.

If now the platinum plates be pressed so near together that a single large test-tube, full of acidulated water, can be placed over both, the gas obtained by passing the galvanic current will exhibit properties differing from those of either hydrogen or oxygen. It is in fact a mechanical mixture of these gases in the proportions in which they would unite chemically to form water. The mixture is precisely similar to that which would have been obtained if the two volumes of hydrogen and one volume of oxygen, previously collected in two separate tubes, had been mixed in one. On touching a lighted match to the mixed gas it instantly explodes with great violence, the hydrogen burning suddenly, so that for a moment a flash of flame fills the whole interior of the tube. Incited by the burning match, the hydrogen and oxygen have combined chemically to form water, a portion of which is deposited as dew upon the inner walls of the tube.

At the temperature of the air, and under ordinary circumstances, oxygen and hydrogen do not combine chemically. Upon being brought together they simply mix with one another mechanically in conformity with the physical laws which govern the diffusion of gases. But under the influence of heat, of electricity, and of certain other agents, the two gases will unite chemically, and will thus again be converted into the water from which they were derived.

36. It remains to be investigated whether hydrogen and oxygen, during their conversion into water, undergo any change of volume, or merely combine to produce a measure of gaseous water exactly equal to the sum of the measures of the constituents. To determine this point it is necessary to compare the joint volumes of the constituents of the water with the volume of the product formed, at a temperature high enough to maintain the latter in the purely gaseous condition known as dry steam.

Through the closed end of a U-tube (Fig. 10) two platinum wires are passed, and welded tightly to the glass walls of the tube. The outer ends of these wires are formed into loops for the attachment of appro-

Fig. 11.

Fig. 10.

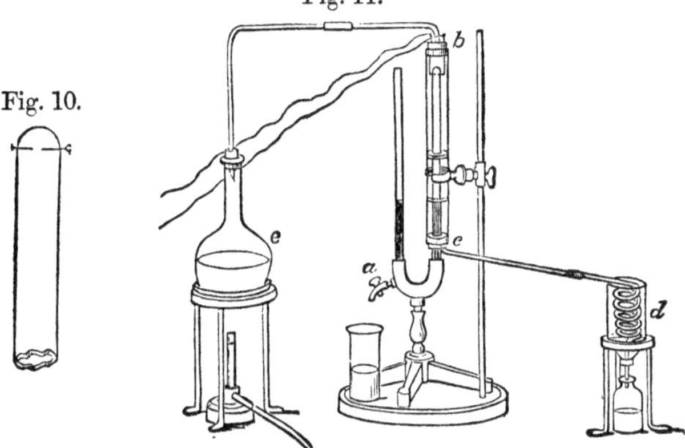

priate battery wires; their inner ends are separated by a distance of two millimetres. The general arrangement of the apparatus to be employed is shown in Fig. 11. The U-tube is first completely filled with mercury, and then the screw-compressor (Appendix, § 16) at *a* is opened so as to afford a gradual exit to the metal in the open limb. By means of a delivery-tube reaching down the open limb to the bend of the tube, we introduce from a gas-holder (see Appendix, § 11) a quantity of a mixture of oxygen and hydrogen, made in the proportions in which they form water,—namely, two-thirds hydrogen, and one-third oxygen,—in such a manner that the gas shall bubble up through the mercury into the sealed limb, from which, of course, the mercury escapes as the gas enters. A column of gas 25 or 30 c.m. high is thus admitted.

It must be borne in mind that this mixture of hydrogen and oxygen is very explosive; fire should be carefully kept away from the vicinity of the gas-holder which contains it, and any remnant of the mixture which is not used should be thrown away at the end of the experiment.

The gas-filled limb of the U-tube is next surrounded by a high glass cylinder, *b c*, the ends of which are fitted with corks; through the lower cork pass the U-tube, and a small glass tube, which is connected with a condensing worm, *d*, kept cool with water; through the upper cork pass the wires which are to carry the electric current to the platinum points at the top of the U-tube, and a bent glass tube coming from the flask, *e*. The top of the cylinder *b c* rises about 5 c.m. above the sealed extremity of the U-tube. In the flask *e*, fusel oil, a liquid which boils at 132°, a point much higher than the temperature at which water becomes a gas, is kept in constant ebullition. The vapor rising from the flask penetrates the annular space between the U-tube and the enclosing cylinder, and quickly raises the tube to its own temperature. These strong-smelling vapors are not allowed to escape into the atmosphere, but are carried out from the bottom of the cylinder *b c*, into the condenser *d*.

When thus heated, the column of mixed oxygen and hydrogen in the tube expands, and its height is marked by a caoutchouc ring, previously slipped over the cylinder *b c*. Care must be taken, before doing this, to bring the mercury to the same level in both limbs of the U-tube, by adding or withdrawing mercury as may be required. A few centimetres of mercury are next poured into the open limb, which is then closed with a good cork. Between this cork and the mercury intervenes a column of air, 8 or 10 c.m. in length, which will act as a spring, and break the shock caused by the explosion of the mixed gases. This mixture is now to be inflamed by causing an electric spark to pass between the platinum points within the tube. This spark may be obtained from a Ruhmkorff coil, or from an electrical machine. The gases instantly rush into combination, with an intense energy which produces the phenomena called explosive, and at the high temperature which exists within the tube (132°) the water formed retains the gaseous condition. On removing the cork, and allowing the mercury to flow through the screw-compressor until it is level in both limbs of the U-tube, it becomes obvious that the original volume of the gaseous mixture is diminished by one-third; the residuary two-thirds are dry steam. If the U-tube is allowed to cool, this steam will condense into liquid water.

Figs. 12 and 13 represent another form of apparatus for performing this important experiment. In Fig. 12 the U-tube of Fig. 11 is replaced

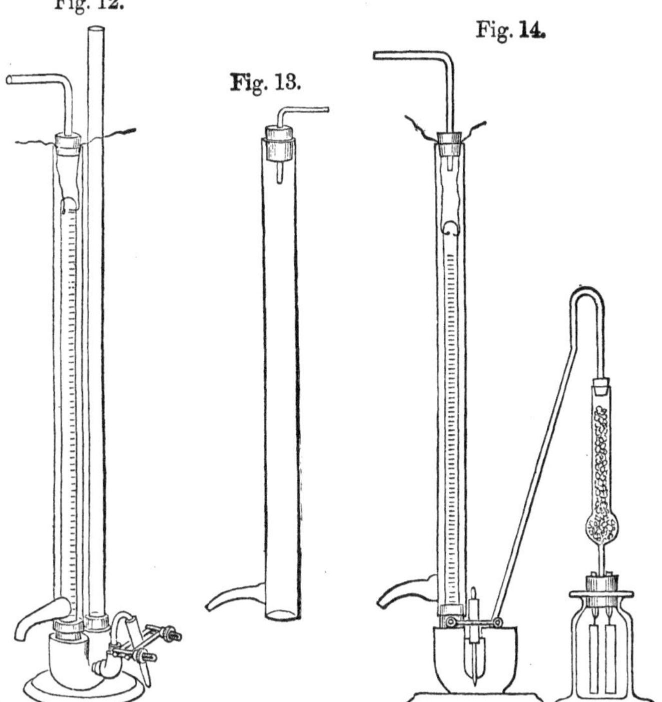
Fig. 12. Fig. 13. Fig. 14.

by an iron U, such as is used in connecting parallel steam-pipes, into which two straight glass tubes are fitted by means of caoutchouc stoppers. One of these tubes is closed at one end, carries two platinum wires, and may be graduated with advantage; the other is a plain tube open at both ends. The iron U is fastened to a solid iron foot, and into its side a small iron tube is inserted, to which is attached a caoutchouc connector with a screw compressor, as in the other apparatus. The jacket-tube (Fig. 13) is a large tube, open at the bottom and closed at the top with a caoutchouc stopper carrying the delivery-tube for the hot vapor; near the bottom an exit-tube for the vapor is welded into the glass. This jacket-tube, when in position, slips tightly *over* the same caoutchouc stopper *through* which passes the graduated tube. The advantages of this apparatus are that its parts are not rigid, that it may be taken apart for cleaning, and that it may be made with greater ease and fewer resources than the other apparatus. It is, moreover, capable of very general application in the analysis of gases. Fig. 14 shows the ease with which the closed limb of the U-tube in this apparatus may be charged. The closed limb and the iron U being full of mercury, and the apparatus placed in an iron pan suited to catch the overflow of mercury, the long open tube is taken away, and the bent gas-delivery tube is inserted beneath the closed limb through the iron U. The iron U acts simply like the water-pan or pneumatic trough. When the required amount of gas has been introduced, the open glass tube is replaced, and the level of mercury in the two tubes is adjusted at will. For the experiment now under consideration, it is well to employ the actual mixture of gases obtained by the electrolysis of water; and Fig. 14 represents a simple bottle provided with two platinum plates and a single delivery-tube for this purpose. The gas evolved is dried by passing over an absorbent called chloride of calcium in a "drying-tube" (Appendix, § 15). The only noticeable feature in this apparatus for electrolysis is the manner in which the wires from the battery are connected with the platinum plates. This is more clearly shown in Fig. 15. There passes through the caoutchouc stopper a short piece of glass tubing, open at the top and drawn to a point at the lower end, so as to enclose and hold tightly a piece of platinum wire, as large as a common knitting-needle, previously placed within it. With the wire thus welded to the glass is connected a thin plate of platinum, which hangs in the liquid to be decomposed; this plate may be folded or rolled up. A little mercury is poured into the glass tubes, and the battery-wires are simply placed in the mercury when the operator desires to start the decomposition.

Fig. 15.

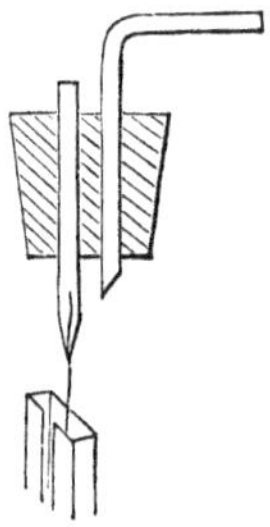

This experiment demonstrates that two volumes of hydrogen and one volume of oxygen are compacted, when chemically united, into two volumes of steam.

37. We have thus established the composition of water by

analysis, having, through the agency of the electric current, resolved water into two gaseous constituents, hydrogen and oxygen, and we have also demonstrated, by the converse or *synthetical* method, that hydrogen and oxygen are its only constituents, since we have reproduced water by effecting the chemical union of these two elementary materials mixed in due proportion.

If equal volumes of hydrogen and oxygen be represented by equal squares, having the initials of the elements inscribed therein, the composition of water by volume, and the condensation which occurs when the chemical union of the elements takes place, may be thus expressed to the eye:

Each smallest possible or greatest conceivable volume of steam will invariably yield, on decomposition, its own volume of hydrogen, and half its volume of oxygen.

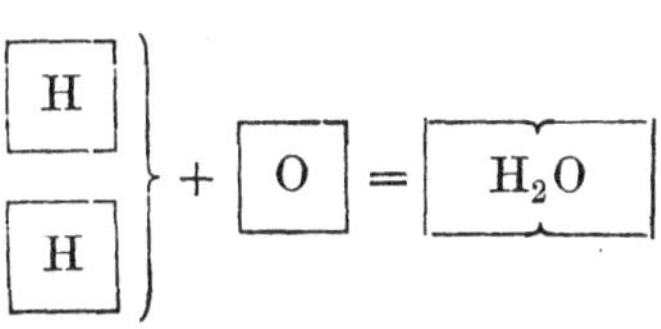

$$\left.\begin{array}{c}\boxed{H}\\ \boxed{H}\end{array}\right\} + \boxed{O} = \boxed{\;H_2O\;}$$

38. It has been agreed to call by the name "*atom*" the smallest quantity of an element which can be conceived to exist in combination; this technical term is applied only to the chemical elements, and to certain chemical knots, or groups, of elements, which, under conditions hereafter to be studied, play the part of an element.

It has further been agreed among chemists to call by the name "*molecule,*" the least quantity of a compound, or of an element, which can exist by itself uncombined, or take part in any chemical process; a *molecule* always contains more than one atom, but these atoms may be either of one, two, or of several kinds.

39. Physical experiments upon the expansion and contraction of numerous gases, simple and compound, have proved that all gases comport themselves in sensibly the same manner under like variations of temperature and pressure; whence it has been inferred that the intimate mechanical structure of all gases, compound as well as simple, is the same. This theoretical conception is expressed in the following propositions, of which the second is the more general and includes the first:—

The *elementary* gases contain, under like conditions of temperature and pressure, equal numbers of *atoms* in equal volumes.

Equal volumes of *all* gases, whether simple or compound, contain under like conditions, the same number of *molecules.*

The idea of an atom is complete and independent in itself; the idea of a molecule is partly a consequent of the idea of an atom, and partly of the physical facts which the definition helps to formulate.

These definitions and hypotheses have found acceptance, partly on the strength of experimental evidence, partly because of their adaptation to the mathematical mode of investigating physical problems which border on the domain of chemistry, but chiefly on account of the clearness and formal consistency which they have imparted to chemical language and modes of thought. Chemical symbolization and nomenclature are mainly based on the above definitions and hypotheses, which therefore justly demand the student's closest attention. Let us apply them to the chemical compound, water.

40. The *molecule* of water, or least quantity of water which is conceived to exist by itself, must yield, like any other quantity when resolved into its elements, twice as large a volume of hydrogen as of oxygen. In accordance with the physical hypothesis above explained, the molecule must consequently contain twice as many atoms of hydrogen as of oxygen. The bulk and weight of the molecule and atom are not *absolute* quantities, on account of their assumed infinitesimal character. None but *relative* facts can be known touching these hypothetical quantities, which are both less than any assignable quantity, although one must be smaller than the other. We shall express in the simplest terms all our actual knowledge of the matter, and shall at the same time conform to our definitions, in saying that a molecule of water contains two atoms of hydrogen and one atom of oxygen. The symbol H_2O which we have already used to indicate the volumetric composition of water (§ 37) will now receive an added meaning; the H will represent for us an *atom* of hydrogen, and the O an atom of oxygen.

When the proportions in which two bodies combine by volume, and their specific gravities, or equal-volume weights, are known, it is a matter of easy calculation to determine the proportions in which they combine by weight. The specific gravity of oxygen, or its density compared with that of air, has already been given, namely, 1·1056. The specific gravity of hydrogen likewise referred to air as the term of comparison, has been found by the most exact experiments yet made to be 0·06926. Oxygen is therefore 16 times heavier than hydrogen. If hydrogen be made the standard of specific gravity for gases, its specific gravity will be denoted by 1, and that of oxygen will be 16. Now any measure of dry steam is, as we have seen, resolvable into its own measure of hydrogen and half that measure of oxygen ; the weights of equal measures of hydrogen and oxygen are as 1 to 16 : but there is twice as much hydrogen as oxygen in bulk, therefore the weight of the hydrogen generated from any quantity of water, small or great, is to the weight of the oxygen simultaneously produced, as 2 to 16. In 18 parts by weight of steam, water, or ice, there are then 2 parts by weight of hydrogen and 16 of oxygen : and it matters not what the absolute weight of these *parts* may be ; the proposition is as true of kilogrammes as of grammes, of the milligramme as of the millionth of the milligramme of water, in either of its physical states.

Applying these facts of observation to our abstract definitions of molecule and atom, it will appear that the molecule of water, the least proportional weight in which it is conceived to exist uncombined, must be composed, like any other mass of water, of 2 parts by weight of hydrogen, and 16 parts by weight of oxygen; but in conformity with our definitions and hypotheses we conceive of the molecule as consisting of two atoms of hydrogen and one of oxygen ; one proportional part by weight of hydrogen is then, in chemical language, synonymous with one atom of hydrogen, and 16 of the same parts by weight is the relative quantity of the atom of oxygen. As for volume, so for weight, absolute quantities are entirely unattainable ; the numbers express proportions only. The numbers 1 and 16 are called the *atomic weights* of hydrogen and oxygen respectively ; they express the *proportions* by weight in which these two elements enter into combination. If these numbers be borne in mind, the symbol of water, H_2O, will always remind us that water consists of 1 part by weight of hydrogen and 8 parts of oxygen.

That any given weight of water, as for example one gramme, is one-ninth hydrogen and eight-ninths oxygen, is a fact capable of experimental demonstration. It is not difficult to decompose a convenient weight of water, and to actually weigh separately the hydrogen and the oxygen which are produced; the weights of the two gases will invariably be to each other as 1 to 8 or as 2 to 16. The great value of the symbols used in chemistry may be well illustrated by the amount of information condensed into the concise expression H_2O: we learn from it the number and names of the elements entering into the composition of water, and the ratios in which the elements are united by volume and by weight.

41. This discussion of the constitution of water rests upon two solid facts of observation, namely the composition of water by volume and its composition by weight; all else is plausible hypothesis and convenient theory. The strong chemical compound, water, admirably illustrates the essential changes which the elements undergo, when they are joined together by that peculiar force whose play it is the object of chemistry to study. Nothing can be more striking than the contrast between the properties of hydrogen and oxygen gases, or of a mechanical mixture of these elements, and those of the liquid water which is produced by their chemical union; even in dry steam, a prominent property of hydrogen (inflammability) and a marked characteristic of oxygen (power of supporting combustion) entirely disappear. In mechanical mixtures the constituents may be mingled in any proportions; in chemical compounds the elements are forcibly united in definite volumetric and ponderal proportions, and the individuality of the elements is lost in the formation of a new substance with new properties. The CHEMICAL FORCE is a peculiar power, distinct from, though akin to, the forces of Light, Heat, and Electricity; it is the province of chemistry to investigate the conditions, modes, and effects of its action.

42. Having thus succeeded in determining the constituents of air and water, we are naturally led to inquire whether it be not possible to resolve oxygen, nitrogen, and hydrogen themselves into simpler forms of matter. To this question but one answer can be made—the result of the accumulated experience of many philosophers of this and former generations—namely, that oxygen, nitrogen, and hydrogen are incapable of decomposition by any means as yet at our disposal. They resist the most powerful influences of electricity and heat, and they issue unchanged from every variety and form of chemical reaction hitherto devised in the hope of resolving them into simpler forms of matter. We are, therefore, justified in regarding these gases as *simple* bodies, or *elements*, in contradistinction to decomposable bodies, such as air and water.

43. The water which occurs in nature is never absolutely pure. In the form of ice, and as it falls from the clouds as rain or snow, it is, indeed, tolerably free from foreign substances; but after having once soaked into the ground, it becomes charged with a variety of mineral and other substances, which, being soluble in water, are dissolved by it as it trickles through the earth.

Where the proportion of soluble matter contained in the water is unusually large, and particularly if it possesses marked medicinal properties. the water is called mineral water, and the springs from which it issues are known as mineral springs. Sea-water may be regarded as a variety of mineral water.

44. For the conduct of chemical investigations it is often necessary to purify natural water. This is done by a process called distillation. As a general rule distilled water is employed in all delicate chemical operations.

Exp. 16.—Into a retort of 500 c. c. capacity, put 200 or 300 c. c. of well-water. Thrust the neck of the retort into a half-litre receiver placed in a pan of cold water. Cover the receiver with a cloth, or with coarse paper, and upon this pour cold water from time to time, or pile upon it fragments of ice. Place the retort upon wire gauze, on a ring of the iron lamp-stand, and adjust the distance of the retort from the lamp as described in Exp. 5, Fig. 3. Light the lamp beneath the retort, and bring the water to boiling. As fast as the water in the retort is converted into steam, this vapor will pass over into the cold receiver, and will there be condensed again to the liquid condition. Continue to boil until about three-quarters of the water in the retort has evaporated.

The earthy and saline ingredients of well-water are for the most part not volatile; very few of them are capable of accompanying the water as it goes off in vapor; hence the greater part of the original impurity of the water will remain behind in the retort.

Fig. 16.

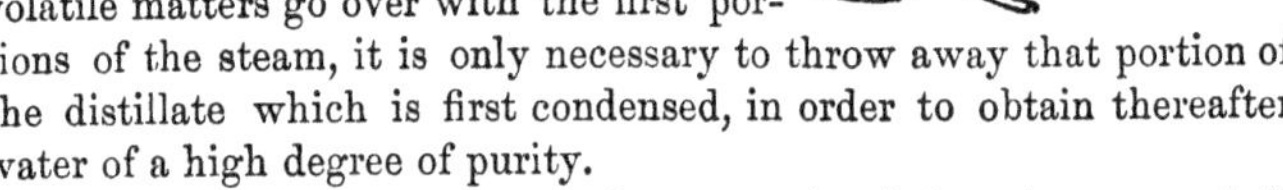

Besides the non-volatile impurities, there are often contained in well-water certain volatile substances, such as ammoniacal salts and organic matter, which pass over into the receiver with the aqueous vapor; but since it has been found that most of these volatile matters go over with the first portions of the steam, it is only necessary to throw away that portion of the distillate which is first condensed, in order to obtain thereafter water of a high degree of purity.

This experiment must not only be so regulated that the retort shall not boil over, but care must be taken that vapor alone shall pass off. The ebullition should be so moderate that none of the particles of water which are thrown up mechanically from the surface of the liquid can be projected into the neck of the retort, or carried thither by the current of steam.

45. In the operation of distillation the substance to be distilled must in the first place be converted into the condition of vapor, this vapor must next be transferred to another vessel, and there, by refrigeration, be again condensed to the liquid state. As will appear from the foregoing experiment, the vaporization is effected in the *retort* or *still*, and the refrigeration in the *condenser*. In the experiment above given the receiver acts at once as receiver and condenser; but a far more efficient apparatus can be constructed by interposing a long tube between the retort and the receiver. This tube may be wrapped with cloths upon which bits of ice are laid, or water is poured; or better, the tube may be enclosed in a larger tube, or a metallic pipe, through which a current of cold water is made to circulate. The water, which may be iced if need be, is poured in through the funnel at the lower end of the tube, and passes out at the top (Fig. 17). This

exceedingly convenient and valuable form of condenser was invented by Weigel; it is, however, commonly called Liebig's.

Exp. 17.—Place a few drops of the distilled water obtained in the preceding experiment upon a piece of platinum foil (Appendix, § 13). Hold the foil with iron pincers above the gas-flame in such a manner that the liquid may slowly evaporate without boiling or spirting. After the water has disappeared, no residue will be found upon the foil. Take out of the retort the same number of drops of water, and evaporate them upon the foil as before. A very decided residue of earthy matter will be left upon the foil.

Fig. 17.

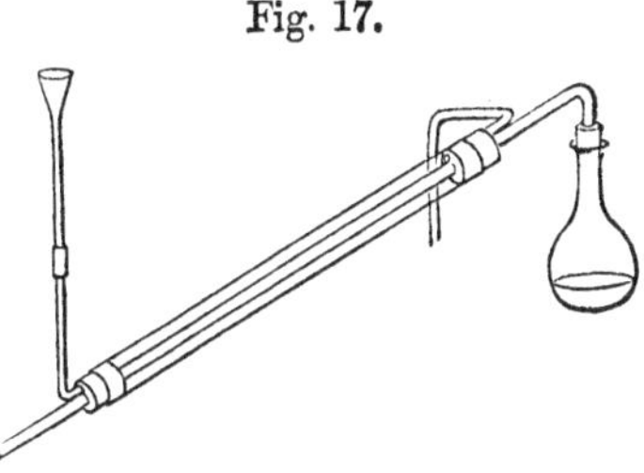

46. In this country, where ice can be had in abundance at all times, it may often be employed as a convenient substitute for distilled water. In freezing, that is, in crystallizing, water rejects a great part of the foreign substances which were dissolved in it. Hence, by collecting ice and remelting it, there can be obtained water which is nearly pure.

Rain-water, also, especially that which has been collected in the open country, is often pure enough to be used for chemical purposes.

47. But even after all the mineral and all the organic matters have been removed, the water is not yet absolutely pure. It still contains oxygen and nitrogen in solution. Both of these gases are soluble in water to a certain extent; and since the water upon the surface of the earth is all the while in contact with air, it must necessarily become charged with the constituents of the air. A method of collecting these gases for examination will be described in a subsequent chapter; we are here more particularly concerned with their removal. This may be effected by long-continued boiling.

Exp. 18.—In a common long-necked medicine-phial of thin "blown" glass, and of the capacity of about half a litre, place 300 or 400 c. c. of recently distilled water. Draw out and bend the neck of the phial, as shown in Fig. 18, and tie upon its point a short piece of caoutchouc tubing. Boil the water steadily during half an hour. Finally, nip the open end of the caoutchouc tube, at the same moment remove the lamp from beneath the flask, and instantly seal the neck of the phial by directing the flame of a blowpipe against the narrow spot. This water can now be preserved for an indefinite length of time, without undergoing change. Upon inverting the phial, the water, which has been thus thoroughly freed from air, will fall about as if it were a solid body, and will strike against the glass with a sudden shock. The apparatus is in fact nothing else than the so-called water-hammer of the physicists.

Fig. 18.

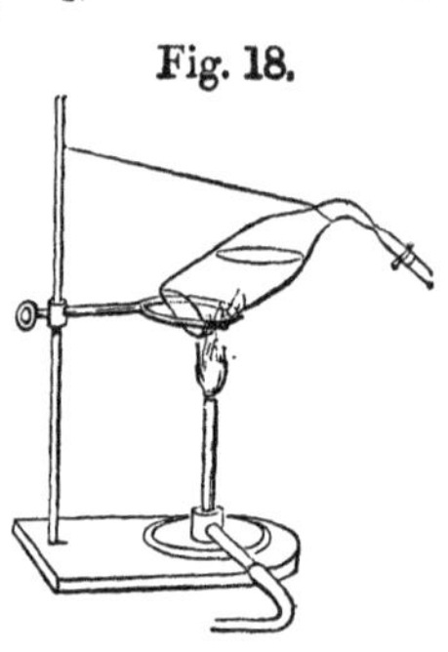

It follows, then, that whenever absolutely pure water is needed for chemical investigations, natural water must first be distilled, with the precautions above indicated, and this distilled water must subsequently be thoroughly boiled, in order to expel the gases which it holds in solution. Water so purified, though necessary for chemical purposes, is unfit to support the life of fishes or other animals which breathe in water, and is not suitable for drinking. It is not only insipid and unpalatable, but is not refreshing like ordinary water. Even if only a part of the dissolved gases have been removed, as is the case with water which has been recently distilled, the taste of the water is still flat, and repugnant. Hence, on board vessels where fresh water is prepared by distilling sea-water, the distillate should be left for some time in contact with air, in order that by absorbing the constituents of the air it may become fit for drinking.

48. As might be inferred from the foregoing, water has the property of dissolving many substances, solid, liquid, and gaseous. Sugar, for example, dissolves readily in water; but sand is insoluble therein.

A substance is said to be soluble when it is capable of being divided in, and dispersed through water so intimately and completely that its particles become invisible, and can no longer be separated by filtration; the result of this coalescence, or the *solution* as it is termed, is a transparent liquid, as a general rule scarcely less mobile than the water itself.

Of the various substances soluble in water, some dissolve in far larger proportion than others. With some liquids (as alcohol, for example) water can be mixed in any proportion; but of ether it dissolves but little, and of oil none. The proportion of any substance that can be dissolved in a given quantity of water is usually limited, and under fixed conditions is definite and peculiar for each substance. When a given quantity of water has dissolved as much of a substance as it is capable of dissolving at the temperature and pressure to which it happens to be exposed, the solution is said to be *saturated*. Generally speaking, solid substances dissolve in far larger quantity in hot than in cold water, though with gases and some exceptional solids the contrary obtains. From the saturated hot solution of any saline substance, crystals are usually deposited during the process of cooling. But so long as a solution is neither exposed to variations of temperature, nor changed by the addition of another substance or by the abstraction of either of its parts, it will usually deposit nothing, and will remain unaltered during an indefinite period of time.

During the act of solution, the first portions of the solid dissolve with comparative rapidity, the subsequent portions dissolving more and more slowly, until complete saturation is attained. In preparing a solution of any solid, at the ordinary temperature of the air, it is therefore inadvisable to add a large quantity of water all at once; a much more satisfactory result will usually be obtained if the substance be rubbed in a mortar with repeated small portions of water, the several portions of the solution being poured off into a common receptacle as fast as the water becomes nearly saturated.

There are many other liquids besides water which are commonly used as solvents; but as water is the commonest solvent of all, and the most universally applicable, some of the general principles of solution may here be appropriately set forth.

49. Solution, though in many cases closely allied to chemical action, is usually treated of as a distinct process. From the best-marked chemical action it differs in several particulars. In true

chemical combination the union of the several ingredients is so close and intimate, that their properties are merged and lost in those of the compound; while in the solvents proper, such as water, alcohol, and benzine, the particles of the dissolved matter appear to be merely mechanically divided and diffused through the liquid. The chemical properties of the dissolved matter undergo no essential change during the act of solution, but remain unimpaired. When common salt is dissolved in water, the brine retains the peculiar taste of the salt, and behaves like salt itself towards most chemical agents; moreover the salt can readily be recovered unchanged by evaporating the water. But if a piece of chalk be placed in muriatic acid, chemical decomposition and combination will immediately occur, the first signalized by a violent effervescence, the second resulting in the formation of a liquid which contains neither chalk nor muriatic acid, if the materials have been mixed in due proportion, but which yields, on evaporation, a solid chemical compound, containing one of the constituents of each.

Chemical combination, as usually defined, occurs in fixed proportions only, whereas solution takes place in indefinite proportions; not only may many substances, as alcohol and glycerine, be mixed with water in every proportion, but where the solubility of a substance is limited in one direction, as that of common salt, of which only about 0·355 part is dissolved by one part of water at 15°, the substance can nevertheless be dissolved in every possible proportion below this maximum. Chemical action is most marked between substances of unlike character; but with solution the rule is different. In general, solution occurs most readily when the solvent is not far removed in composition and properties from the body dissolved.

Extreme cases of chemical action upon the one hand and of solution on the other, are readily distinguishable. But there is a wide range between these extremes, and it is well nigh impossible to find a point at which the line of demarcation shall be drawn. Many cases which at first sight seem to be examples of simple solution can readily be shown to depend in part upon chemical force.

The majority of chemists are now inclined to regard most

instances of solution as feeble exhibitions of the chemical force, or at all events as intermediate between purely chemical and merely mechanical action. Solution facilitates chemical action between heterogeneous materials, both by overcoming the force of cohesion by which the particles of homogeneous solids are held together, and also by bringing the particles of the unlike bodies into intimate contact with one another through the vehicle of the common solvent. Cohesion resists chemical action as it does gravity; but solution overcomes cohesion, frees the particles from the bonds which held them, and, as we may imagine, leaves them free to enter into other combinations.

CHAPTER V.

HYDROGEN.

50. The commonest method of preparing hydrogen is by treating zinc or iron with dilute sulphuric or muriatic acid. Unless very large quantities of the gas are needed, this method is cheaper and more convenient than either of those heretofore mentioned.

Exp. 19.—To a bottle 18 or 20 c.m. high, and of 500 or 600 c. c. capacity, the mouth of which has an internal diameter of 2·5 to 3 c.m., fit a caoutchouc stopper or a sound cork, furnished with a thistle-tube (Fig. 19) and a gas delivery-tube, of No. 6 glass. Within the bottle put 15 or 20 grms. of granulated zinc, or small scraps of the sheet metal, and as much water as will cover the zinc and close the end of the thistle-tube. Replace the cork in the bottle, taking care to press it in tightly, and gradually pour in common muriatic acid through the thistle-tube. The thistle-

Fig. 19.

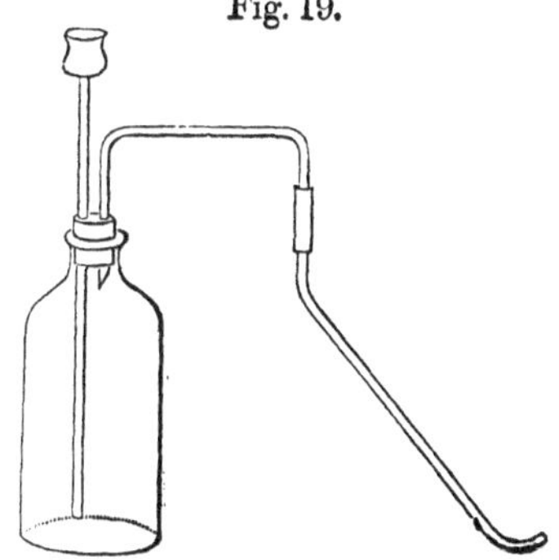

tube must reach nearly to the bottom of the bottle, so that its point may dip beneath the water; and the muriatic acid must be added by small successive portions, not more than a large thimbleful at a time.

On the addition of the first portions of the acid, chemical action will ensue, the contents of the bottle will become warm, and gas will be seen to escape from the liquid. This gas is hydrogen.

After all the air has been expelled from the bottle the hydrogen may be collected over the water-pan, in inverted bottles filled with water, or it may be passed into a gas-holder (Appendix, § 11). The rapidity with which the gas shall be evolved is easily controlled by regulating the supply of acid; and the moment at which the hydrogen ceases to be contaminated with air can be determined by collecting small portions of the escaping gas, in wide-mouthed bottles of about 50 c. c. capacity, and testing its quality by means of a lighted match. In doing this the small bottle filled with gas must not be turned over, but should be carefully lifted from the water without changing its vertical position, and the lighted match should then be applied to the mouth of the bottle. If the hydrogen be pure, it will burn tranquilly at the mouth of and within the bottle; but in case the gas is still mixed with much air, a sharp explosion will occur at the moment when the match is applied to it. In order to avoid these explosions, which would be exceedingly dangerous if the volume of mixed gases were large, it is indispensably necessary, in preparing hydrogen, to take care that none of the gas shall be admitted into the gas-holder until all the atmospheric air has been expelled from the bottle in which the gas is generated. So, too, in experimenting with hydrogen, no light should ever be brought into contact with the contents of the bottle into which it is generated, or with any large quantity of the gas, until the purity of the sample, or rather its non-explosive character, has been demonstrated by applying to a very small volume of the gas the test above described.

This experiment, which has here been executed with zinc, can be equally well performed with iron-filings, and with several other of the less common metals.

Muriatic acid, or, in chemical nomenclature, chlorhydric acid, is a compound of hydrogen and another element, called chlorine, which will shortly be described. The chemical composition of this substance can be represented by the symbol H Cl, in which H represents, as before, the least proportional weight of hydrogen which exists in combination, and Cl the least proportional weight of chlorine. We may likewise abbreviate the word zinc to the symbol Zn; and the chemical process, or ***reaction***, by which the hydrogen is liberated may then be symbolized by the *equation*

$$2\,H\,Cl + Zn = Zn\,Cl_2 + 2\,H.$$

Since hydrogen is a gas, it escapes as such, and there remains dissolved in the water within the bottle a compound of the elements chlorine and zinc, called chloride of zinc ($Zn\,Cl_2$). The zinc which was free, enters into combination, and the hydrogen which was in combination, is set free; in other words, the zinc has been *substituted* for, or has *replaced*, the hydrogen. It may here be stated that chemists of all nations have agreed to represent each of the elements by a symbol, which consists either of the initial letter of the Latin name of the element, or, when the names of two or more elements begin with the same letter, of the initial letter together with the first of the succeeding letters in the Latin name which is distinctive. Thus Fe (*Ferrum*) is the symbol of iron, W (*Wolframium*) of tungsten, C of carbon, Ca of calcium, Cd of cadmium, Cl of chlorine, and Cr of chromium.

51. Hydrogen is a transparent, colorless, and tasteless gas, odorless when pure. It is not poisonous, though animals die from suffocation when immersed in it, as they do in an atmosphere of nitrogen. It has never been condensed to a liquid. It is the lightest substance known, being about 14½ times as light as air. If 1 volume of air weighs 1 gramme, an equal volume of hydrogen will weigh only 0·0693 grm. It is 11,160 times as light as water, 151,700 times as light as quicksilver, and 236,000 times as light as platinum. 1 litre of hydrogen at 0°, and a pressure of 76 c.m. mercury, weighs 0·089578 grm. Hydrogen is the most suitable standard of specific gravity for gases, as water is for liquids and solids; when thus used as the standard, its specific gravity is, of course, unity. The student should, however, be informed that air is used by many writers as the standard of specific gravity for gases.

52. The exceeding lightness of hydrogen can be illustrated in various ways. From an inverted bottle, even though it be open below, hydrogen will escape but slowly; but if a bottle of hydrogen be opened in the air, with the mouth upward, the gas will

quickly escape. Hence it can readily be poured or decanted upwards from one vessel to another.

Exp. 20.—Lift from the water-pan a thick, strong, wide-mouthed bottle, of 200 to 300 c. c. capacity, full of hydrogen, taking care to hold it in a perpendicular position, with the mouth downward. With the other hand place another bottle of equal size and strength, but containing only air, beside the hydrogen-bottle, so that the mouths of the bottles shall touch at one end. Gradually turn down the hydrogen bottle, and at the same time push its mouth beneath that of the air-bottle in such manner that the bottle which originally contained the hydrogen shall at last stand upright beneath the inverted bottle. During this operation, the lighter hydrogen flows up into the upper bottle, while the heavier air sinks into the lower. If a burning match be now thrust into the upper bottle, the hydrogen within it will take fire; but upon applying the match to the lower bottle, originally full of hydrogen, there will be found in it nothing but air.

In like manner, hydrogen may be collected by displacement—an upright delivery-tube being carried from the bottle in which the gas is generated, to the top of an inverted recipient. The student will do well to remember as a general rule, that in manipulating with hydrogen we must operate in a manner precisely opposite to that which would be adopted if we were at work with water. Where water would flow down, hydrogen will flow up.

Owing to its lightness, hydrogen is well adapted for filling balloons; and it is still sometimes employed for this purpose in military operations, being prepared by means of hot iron, as in Exp. 15. For purposes of illustration, soap-bubbles filled with hydrogen will serve as well as balloons of more costly construction.

Exp. 21.—By means of a caoutchouc tube, attach an ordinary tobacco pipe to a gas-holder containing hydrogen. (See Appendix § 11, Fig. xvii.) Dip the pipe in a solution of soap for a moment, then turn it mouth uppermost, and slowly open the stopcock of the gas-holder so that hydrogen may flow out and inflate the film of water upon the mouth of the pipe. The bubble will soon break away from the pipe and rise rapidly through the air. If a burning match be applied to the bubble the hydrogen within it will, of course, burst into flame.

53. There is another noticeable peculiarity of hydrogen which is directly connected with its extreme lightness. It possesses in a

high degree the power of diffusion. This diffusive power is a physical property common to all gases and vapors; in the case of hydrogen it is only the intensity of the diffusive power which is remarkable.

Fig. 20.

When different gases, which have no chemical action upon each other, are brought into contact, they will not remain separate, but will commingle. This tendency of the gases to intermingle is so strong that it will not only overcome the greatest differences of specific gravity, but even cause the spread of gases directly against powerful currents of air or vapor. If a bottle of oxygen, standing upright, be connected with an inverted bottle full of hydrogen, by means of a tube a metre in length, and no more than 8 or 10 m.m. in diameter, both the bottles will be found to be filled with a uniform mixture of the two gases after the lapse of a very few hours. Upon now touching a lighted match to the open mouth of either bottle, the gaseous mixture will explode. As a precautionary measure it is best in this experiment to employ the thick, strong bottles in which soda-water is kept—or, in lack of these, strong wide-mouthed bottles enveloped in thick towels.

The velocities with which gases diffuse are in the inverse ratio of the square roots of their specific gravities. Hence it happens that hydrogen being the most attenuated of all gases, diffuses with the greatest rapidity. Compared with that of oxygen, its rate of diffusion is as 4 to 1; that is to say, the relative rates of diffusion of the two gases are inversely as the square roots of the numbers 1 and 16, which represent the specific gravities of hydrogen and oxygen respectively.

Fig. 21.

Exp. 22.—A glass tube, 3 or 4 c.m. in diameter, and 30 or 40 c.m. long, is closed at one end with a plug of plaster of Paris 1 or 2 c.m. thick. The tube is then set aside for a day or two, in order that the plaster may become dry. When the plug is dry, fill the tube with hydrogen by displacement, and set it upright in a glass of water. Water will rise rapidly in the tube, since hydrogen escapes through the plaster more rapidly than air can enter the tube through this porous plug. That some air does enter, however, can be shown by exploding the contents of the tube by applying a lighted match, after the lapse of some time. Of course, if the tube be left to itself, air will slowly enter through the plaster, so that the water within the tube will in due time sink to the level of the outside liquid.

On account of its high diffusive power, hydrogen can be kept only in perfectly tight vessels. It has been found that it will leak rapidly under a pressure of 27 or 28 atmospheres through stopcocks that are perfectly tight for nitrogen at a pressure of even 50 or 60 atmospheres; from the same cause it cannot be kept for any length of time in bladders or rubber bags. If a sheet of paper be held a short distance in front of the opening of a gas-holder from which hydrogen is escaping, the current of gas will pass directly through the paper, and can be inflamed upon the other side of the sheet. The high diffusive power of hydrogen, which is to some extent shared by its compounds also, is an obstacle to be overcome before ballooning can be made practicable.

Sound is propagated in hydrogen but little better than in a vacuum. The specific heat of hydrogen is 3·4046, that of an equal weight of water being 1·000; it is 0·2356, that of an equal volume of air being 0·2377. It refracts light very powerfully.

54. Hydrogen is exceedingly inflammable, as has been already seen; that is to say, the temperature at which it takes fire is comparatively low. But, as a matter of course, it extinguishes any burning body which is immersed in it, since oxygen is necessary for the support of combustion.

Exp. 23.—Carefully lift from the water-pan a bottle of 200 or 300 c. c. capacity, completely full of hydrogen, slowly carry the bottle, the mouth of which is of course held downward, to a burning candle or splinter of wood, and depress the bottle over this flame. The hydrogen will take fire and burn, below, at the mouth of the bottle where it is in contact with the oxygen of the atmosphere; but the flame of the candle will be extinguished the moment it becomes completely enveloped by the hydrogen. The candle can easily be relit by slowly lifting the bottle until the wick is brought into contact with the air and the burning hydrogen.

Exp. 24.—Fill a bottle of the capacity of 400 or 500 c. c. with hydrogen, close the mouth with a cork or a plate of glass, stand the bottle upon the table with the mouth upward, remove the stopper, inflame the hydrogen, and immediately pour out of a pitcher into the bottle a large quantity of water. The flame will instantly be increased, since the water will force the gas out of the bottle into the air. Within the bottle the hydrogen can only burn gradually, since it takes time for the outside air to enter; but if the gas be pushed out of the bottle into the air, it will burn at once.

In the familiar instances where water extinguishes fire, it does so by reducing the temperature of the combustible—that is to say, by cooling it to below the temperature at which it will take fire. It would act thus in this case, were it not for the lightness and mobility of the hydrogen, by virtue of which this gas immediately escapes from contact with the water.

55. It has been seen that the hydrogen-flame affords only an exceedingly feeble light; but it would be a grave error to infer that but little heat is developed by the combustion.

The temperature of the hydrogen-flame is in reality very high. Indeed it has been found that when a given weight of hydrogen enters into chemical union with oxygen, more heat is developed than in the burning of the same weight of any other substance.

In order to determine the amount of heat which is developed in any act of combination, this heat can be transferred to water, and there estimated either by the quantity of water heated, or the amount of steam produced. A *unit of heat* is that amount of heat which will raise 1 gramme of water from 0° centigrade to 1°.

The amount of heat evolved during the combustion of a body is as constant and unvarying as any other direct consequent of its properties, and the quantity of heat evolved is absolutely the same, no matter whether the combustion occurs in air or pure oxygen, or whether it be slow or rapid. The actual amount of heat developed during the most vivid combustion is no greater than when the same combustible combines with oxygen, by gradual oxidation, without visibly burning. From 1 gramme of hydrogen, as it unites with oxygen, there are evolved 34462 units of heat.

Although the same amount of heat is developed, in the aggregate, when a litre of hydrogen burns in the air, as when it burns in pure oxygen, it is none the less true that a far hotter flame is obtained by burning the hydrogen in oxygen than in air. If the combustion be complicated by the presence of nitrogen from the air, a great deal of heat will expand its energy in expanding this useless nitrogen. Moreover, since the nitrogen occupies much room, it will, as it were, keep asunder the particles of hydrogen and oxygen; the flame will thus be made longer and more dispersed, or, in other words, the heat evolved by the union of the hydrogen and oxygen will be spread over more space than if the nitrogen were absent. Where nothing but hydrogen and oxygen are present, and in the precise proportions in which these gases unite to form water, the flame produced by their union will be to all intents and purposes solid, and the heat will be concentrated in the smallest possible space.

Exp. 25.—Provide two gas-holders (see Appendix, § 11), one full of hydrogen, the other full of oxygen, also a metallic jet so constructed that the tube which carries the oxygen shall pass through the centre of the hydrogen tube as shown in Fig. 22. Screw the jet on to the oxygen gas-holder, and connect the other opening with the hydrogen gas-holder by means of a caoutchouc tube. Open the cock of the hydrogen gas-holder, and inflame the gas at the point of the jet, then slowly open the cock of the oxygen gas-holder until the flame of the burning hydrogen has been reduced to a fine pencil. This apparatus is known as the compound, or oxy-hydrogen blowpipe. In order to insure a steady flame, care must be taken that a constant and sufficient pressure be maintained upon the contents of the gas-holders.

Fig. 22.

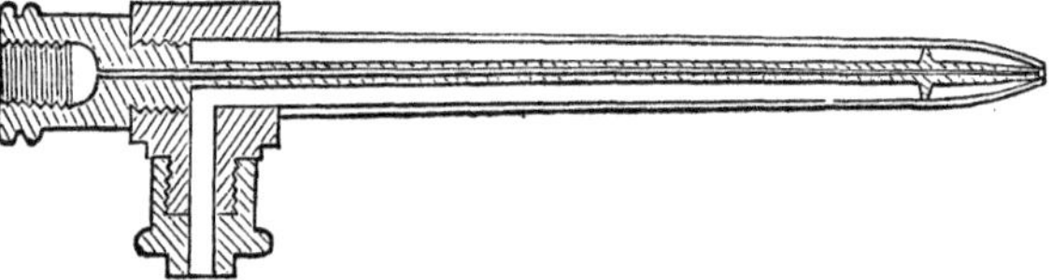

Exp. 26.—In the flame of the oxy-hydrogen blowpipe, described in the foregoing paragraph, hold the end of a piece of platinum wire, about 10 c.m. long and less than one m.m. in diameter. The platinum will melt and fall down in drops.

The intense heat of the oxy-hydrogen flame is thus admirably illustrated; for platinum is an exceedingly infusible metal, which can scarcely be softened in the hottest furnace. The drops of platinum should be caught upon sand, or in water. In case the melted platinum falls into water, a portion of the water will be decomposed into hydrogen and oxygen, bubbles of which will be seen issuing from the water. Some of this gas can easily be collected, by an assistant, in an inverted ignition-tube filled with water, and its explosive character demonstrated by applying a lighted match.

Exp. 27.—If a piece of chalk or lime, scraped to a fine point, be held in the flame of the oxy-hydrogen blowpipe, it will quickly become white-hot and evolve light of great brilliancy, almost comparable with that of the sun. If the bit of lime be long exposed to the intense heat, it will undergo incipient fusion, and afford less light than at first.

Where a constant light is desired, cylinders or plates of chalk are kept continually moving before the flame by mechanical power, so that fresh portions shall continually be brought into the flame. This is the so-called Drummond or Calcium light, often employed for night signals and optical experiments.

56. No matter in what way hydrogen is burned, whether in the pure state or in combination with other materials, whether in pure oxygen or in the air, the product of the combustion is always water. At the high temperature of the flame, this water must of course remain in the condition of a gas, but it can readily be brought to the liquid state by reducing the temperature.

Exp. 28.—Over a jet of burning hydrogen, best obtained from a gas-holder, hold a dry, cold bottle. The glass soon becomes covered with a film of dew, as the water generated by the union of hydrogen and oxygen condenses in droplets upon the cold sides of the bottle.

57. As the burning jet of hydrogen is the simplest instance of combustion with flame, some exact knowledge of the form and quality of flames may here be gained.

As hydrogen, or any other gas, issues from a small orifice into the air by force of pressure from behind, the escaping gas assumes a certain definite shape in accordance with the physical conditions to which it is exposed, just as a fountain of water takes form in accordance with the size and shape of the orifice from which the water is expelled, the pressure by which it is expelled, the gravity of the water, the resistance of the air, and the force and direction of the wind, and so forth.

If a lighted match be brought into contact with the column or fountain of gas, the gas will take fire and burn; that is to say, the hydrogen will enter into combination with oxygen as fast as the latter can be furnished from the air. But, in any event, the column of gas will burn only upon the outside, for there alone can the oxygen of the air come into contact with the hydrogen. Neither the interior of the flame nor the contents of the reservoir from which the gas is flowing can burn; for they consist of pure hydrogen, which, as has been shown in Exp. 23, will by itself immediately extinguish combustion.

All ordinary gas-flames are, of necessity, hollow. They are visible to us through the evolution of light which is an accompaniment of the chemical action. But even if no combustion were going on upon its surface, the escaping column of gas could still be made visible by causing it to pass through a quantity of dust or other fine powder before coming into the air. A portion of the solid matter would be transported by the current of gas, and the form assumed by the latter would be made manifest.

The shape of the unignited gas-column would of course be somewhat different from that of the burning flame; in the latter, not only is the outer edge of the column sharply defined by the zone of combustion, but the actual form of the column itself is modified through the expansion of the gas as it becomes heated by the enveloping fire.

58. If, instead of burning pure hydrogen as it flows into the air, as in the foregoing experiments, the gas be first mixed with air or oxygen and then ignited, a very different result will be obtained. The hydrogen being now in contact with oxygen at all points, the entire mass of gas will burn with a violent explosion the instant a light touches it.

Exp. 29.—Introduce 2 volumes of hydrogen and 5 volumes of air into a strong round-bottomed bottle such as is used for soda-water. Close the mouth of the bottle with a cork, and shake violently in order that the gases shall be mixed. A small quantity of water should be left in the bottle to act as a stirrer. Grasp the bottle firmly in one hand, remove the cork with the other, and apply the open mouth of the bottle to a lighted candle. An explosion will immediately ensue.

Exp. 30.—Into a gas-holder, bladder, or rubber bag, introduce a mixture of 2 vols. hydrogen and 1 vol. oxygen. Connect therewith, by means of caoutchouc tubing, a tobacco-pipe, or bit of glass tube. Press the gas through the pipe into a dish of soap-suds, in such manner that there shall be formed upon the surface of the suds a mass of foam as large as an egg. Close the gas-holder and remove it from the vicinity of the suds. On now touching the foam with a long, lighted stick an exceedingly violent explosion will occur.

It is well to avoid the formation of a large quantity of foam in this experiment, since the concussion is in any event deafening. In this experiment the explosive mixture is purposely confined in an exceedingly flimsy envelope, in order that no harm may be done by the fragments.

Care should be taken to throw away any remnant of the mixture of hydrogen and oxygen which may have been left in the gas-holder at the close of the experiment, and upon no account should fire ever be brought into its vicinity.

59. The cause of these loud explosions is twofold. By the act of combination water is formed, and at the same time intense heat is emitted ; the water, or rather steam, is thereby enormously expanded, so that for a moment there is violent motion outwards in all directions. This outward motion would scatter about in a most dangerous manner the fragments of any vessel in which, through carelessness, a mixture of hydrogen and oxygen might be ignited—unless, indeed, the vessel were very strong, small, and of large aperture. But in the next instant, as the steam condenses, there is an even more violent motion inwards.

The original mixture of hydrogen and oxygen occupies about 2000 times as much space as the liquid water which results from the combination of these gases. Hence a partial vacuum is formed, into which air rushes from all sides ; and it is the heavy and sudden undulations thus communicated to the air which occasion the noise. The outward and inward shocks follow one another so quickly that the ear cannot distinguish between them.

Mixtures of hydrogen and air produce less violent explosions than mixtures of hydrogen and oxygen, because of the inert nitrogen in the air, which acts as an elastic pad, or cushion, to break the force of the shock.

60. Since air is everywhere about us, and since all ordinary combustions occur in it, it has become customary to speak of it and of oxygen as *supporters of combustion*, in contradistinction to

the so-called *combustibles*, such as hydrogen. These terms are often convenient, but, as will appear from the following experiment, they have only a relative, and no absolute significance.

Exp. 31.—Provide a tube of thin glass, the neck of a broken retort, for example, 30 or 40 c.m. long, and 3 or 4 c.m. in diameter; fix it in a vertical position, so that the lower opening shall be 20 or 30 c.m. above the table, and connect the upper opening with a gas-holder filled with hydrogen.

To a second gas-holder, containing oxygen, attach a caoutchouc tube, and to the end of this fit a piece of glass tubing, No. 7, 25 or 30 c.m. long, bent at a right angle at 5 or 10 c.m. from the end which is attached to the caoutchouc tube, and drawn out to a fine open point at the other. The caoutchouc tube must be long enough to reach as far as the lower mouth of the wide vertical glass tube above mentioned.

Open now the stopcock of the hydrogen gas-holder so that the vertical tube shall be filled with the gas, then apply a lighted match to the mouth of this tube, and regulate the flow of gas so that the latter may continue to burn slowly at the lower edge of the tube. Finally, open the stopcock of the oxygen gas-holder, so that a current of this gas shall flow through the pointed delivery-tube, and thrust this tube up into the middle of the wide vertical tube which has been filled with hydrogen.

As the stream of oxygen passes through the burning hydrogen at the bottom of the vertical tube, it takes fire, and afterwards continues to burn in the atmosphere of hydrogen within the tube.

CHAPTER VI.

OTHER COMPOUNDS OF THE ELEMENTS ALREADY STUDIED.

61. Oxygen and hydrogen do not unite directly in any other proportion than that in which they form water; but by indirect means, too complex for profitable study at this stage, a molecule of water can be made to combine with an atom of oxygen, forming a new substance, known by the name of *peroxide of hydrogen.* Its formula is, in accordance with this statement, H_2O_2, and it

would yield, if completely decomposed into its elements, equal volumes of hydrogen and oxygen; its composition by weight must be 2 parts of hydrogen to 32 of oxygen; it contains just twice as large a proportion of oxygen as water. The best method of preparing this substance does not yield the pure thing itself, but only a concentrated solution of it in water; the specific gravity of this solution is 1·45: it is colorless, transparent, and somewhat syrupy, has a metallic taste, corrodes the skin, and bleaches vegetable colors.

How different a substance this is from water, appears at once from this enumeration of some of its properties. It is very unstable, being readily decomposed by heat, and by contact with various substances at the ordinary temperature, into oxygen and water. Its instability and the intense chemical activity of which it is capable, emphatically distinguish this, as yet obscure, body from the neutral, stable, inactive compound of hydrogen and oxygen, common water.

62. But though the peroxide of hydrogen is not water, it is nevertheless a true chemical compound of the same elements which are united in water; it is definite and constant in composition, and its properties are as unlike those of its elementary constituents as are those of water. A new fact of great significance here comes plainly into view. Two of the elements are evidently capable of combining in *two* definite proportions to form two chemical compounds, each differing from the other and from its primary constituents. The study of the compounds of nitrogen and oxygen will bring into clearer view the general principle of which this fact is a single illustration. These compounds form a series of five members, all derived, more or less directly, from common nitric acid.

63. *Nitric Acid.*—Two abundant sources of this material are found in nature, and are familiar as articles of commerce. Saltpetre or nitre, a whitish, saline, crystallized substance, now mainly brought from India, is one of these sources; a similar substance, known as Chili-saltpetre, or soda-nitre, is collected on a desert tract in Chili and Peru, and forms a valuable article of export from those countries. These two substances only differ from each other in this,—that the first contains the metal potas-

sium, the second the very similar metal sodium, in either case combined with definite proportions of the elements nitrogen and oxygen. By the reaction of sulphuric acid (oil of vitriol) on either of these two substances, nitric acid is obtained. On a small scale, for laboratory purposes, saltpetre, or, as it is called in chemistry, nitrate of potassium, is generally employed; on a manufacturing scale, soda-saltpetre or nitrate of sodium is used, because this salt costs less than nitrate of potassium, and also contains a larger proportion of nitric acid, which it yields up with greater facility.

Exp. 32.—Into a tubulated, glass-stoppered retort of 250 c. c. capacity put 40 grammes of powdered nitrate of potassium, or, better, 34 grammes of powdered nitrate of sodium if it can be obtained, and through the tubulature pour 50 grammes of strong sulphuric acid, which has been weighed out in a bottle previously counterpoised upon the balance with shot or coarse sand. Imbed the bottom of the retort in sand, contained in a small iron pan placed over the gas-lamp on a ring of the iron-stand. Thrust the neck of the retort into the receiver with two tubulatures; the retort-neck should fit the tubulature of the receiver with tolerable accuracy. The second tubulature of the receiver should be left open, or loosely covered with a bit of glass, in order to avoid the possibility of any pressure being created within the retort during the operation. Place the receiver in a pan of cold water, and cover it with cloth or bibulous paper, which must be kept constantly wet during the distillation. (See Fig. 16, p. 34.) Apply a moderate heat to the sand-bath; reddish vapors will appear for a moment, then disappear, and a yellowish fuming liquid will begin to condense in the receiver. Towards the end of the operation the red vapors reappear; after this has happened, and the saline matter in the retort has attained a state of tranquil fusion, while very little liquid passes over into the receiver, the lamp may be put out, for the process is finished.

The very acid, corrosive, and poisonous liquid in the receiver is *nitric acid*; its faint color is not its own, but is due to the presence of a compound of nitrogen and oxygen shortly to be described. Transfer the liquid to a glass-stoppered bottle, and keep it for future use. In all manipulations with nitric acid it is necessary to avoid getting it upon the skin, since it produces rather permanent yellow stains.

As the retort cools, the residue solidifies into a white, saline mass, which must be dissolved out of the vessel by heating it with water after the apparatus has become thoroughly cold. It will be observed that the liquid sulphuric acid which was used has disappeared, though the saline residue is still intensely acid.

64. Pure nitric acid is colorless, and is about half as heavy again as water. It may be mixed with water in all proportions.

Exp. 33.—To one-third of the nitric acid obtained in the last experiment add an equal bulk of water. The solution thus obtained will be still intensely acid, as may be proved by its action on vegetable colors. Litmus is a blue coloring-matter, prepared from various lichens, and used in dyeing. Unsized paper, colored with a solution of litmus in water, is a convenient test for many acids, which, as a rule, change the color of the paper from blue to red. If the acidity of this diluted nitric acid be now destroyed or neutralized by the addition of some other substance of opposite quality, the point at which the liquid ceases to be acid may be determined by observing when the blue paper remains blue on immersion in the liquid.

To the diluted nitric acid, placed in an evaporating-dish, add cautiously ammonia-water (the Liquor Ammoniæ of the apothecaries), which has been previously diluted with its own bulk of water, until the liquid no longer turns the litmus-paper red. The ammonia-water must be added, slowly at first, and at last drop by drop, and the mixture must be constantly stirred with a glass rod. The ammonia-water has the property of turning litmus-paper, which has been reddened by an acid, back again to blue, as direct experiment may prove; this property is possessed by a class of bodies called *alkalies*, and this reaction with litmus is termed the *alkaline reaction*, in contradistinction to the change from blue to red, which is the reaction characteristic of acids. When mixed with nitric acid, ammonia-water produces a compound which, when fresh and pure, has no action on vegetable colors, and being soluble in water is not visible at this stage of the experiment. Place the evaporating-dish on the wire gauze over the gas-lamp, and evaporate the liquid, taking care to avoid actual ebullition, until a drop of the solution taken out on a glass rod becomes almost solid on cooling. Extinguish the lamp, let the dish become perfectly cold, separate the semitransparent crystals which have formed during the cooling from the fluid, if any, which remains in the dish, allow them to drain, dry them by gentle pressure between folds of bibulous paper, and reserve them for use in the next experiment. Besides water, these crystals are the sole product of the reaction; they must therefore contain both all of the nitric acid and all of the ammonia which is not water. The chemical name of the substance is nitrate of ammonium.

65. We find in this experiment a striking illustration of what is meant by chemical combination. From two fuming liquids, of very intense but opposite properties, there has come forth a neutral solid, as unlike its constituents in taste, smell, and all physical attributes, as could well be imagined. The idea of neutralization, well exemplified in this experiment, has in the history of chemistry been very fruitful both of names and hypotheses. Bearing in mind the fiery constituents of the cooling, harmless salt which we have thus synthetically prepared, we proceed to learn by experiment whatever its decomposition may teach.

Exp. 34.—Introduce into a Florence oil-flask, or other suitable flask of thin glass, and of about 300 c. c. capacity, the nitrate of ammonium obtained in the last experiment. From the mouth of the flask, placed upon the wire gauze on the iron-stand, carry a gas delivery-tube, No. 6, beneath the saucer in the water-pan; but interrupt the tube, at some convenient point to interpose (by means of a cork or caoutchouc stopper with two holes) a small bottle which can be kept cool with water, as shown in the figure.

Fig. 23.

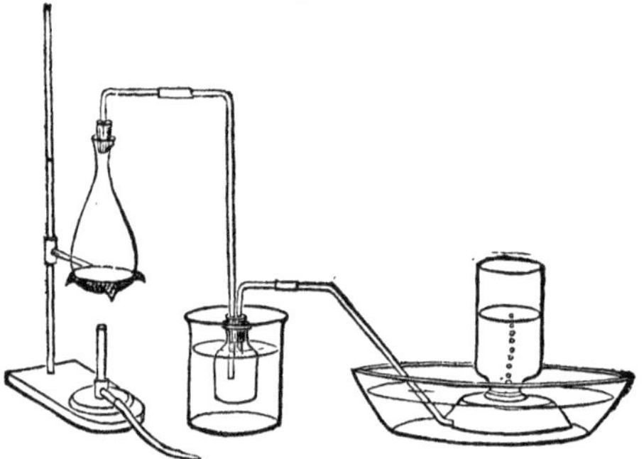

Heat the flask moderately and cautiously, to avoid breaking it. The nitrate of ammonium will melt, and little bubbles will soon begin to escape from the fused mass. The heat must now be so controlled that the evolution of the gas shall not be tumultuous. The gas is to be collected in bottles of 300 to 400 c. c. capacity. In the cooled bottle, through which the gas passes, a clear and colorless liquid will condense, which will be found on examination, if the process has been successfully conducted, to be neither acid nor alkaline, to have neither taste nor smell, and to be wholly volatile when heated on platinum foil,—in short, to possess all the properties of water, and none other. When two bottles of gas have been filled, and enough water for testing condensed, the delivery-tube may be withdrawn from the water and the lamp extinguished, although the nitrate of ammonium be not all decomposed. The nitrate of ammonium may be entirely resolved into water and the gas which now awaits examination, but it is difficult to push the decomposition to actual completion without breaking the flask in which the operation is performed. That the nitrate leaves no residue behind, when sufficiently heated, may be proved by heating a crystal of it on platinum foil over the gas-lamp.

66. *Nitrous Oxide.*—It is obvious that the colorless and transparent gas, which is the most voluminous product of the decomposition of nitrate of ammonium just accomplished, must contain all the elements, besides those of water, which enter into the composition of nitrate of ammonium, and therefore of its constituents, nitric acid and ammonia-water; much interest, therefore, attaches to the determination of the composition of this gas.

Exp. 35.—Insert a glowing splinter of wood into a bottle of the gas. It will reinflame with almost as much energy as in oxygen.

If oxygen be really a constituent of this gas, it may be possible to mix the gas with hydrogen, and effect the chemical combination into water of the oxygen in the gas and the added hydrogen, by heating the mixture, or passing through it an electric spark (see pp. 24, 26). If just enough hydrogen can be added to exactly convert all the oxygen contained in a given volume of the gas into water, the constituents other than oxygen will be left behind for separate examination, and we shall have determined how much oxygen a given volume of the gas contains by observing how much hydrogen has been necessary to convert it into water, the volumetric composition of water being already known. Now it has been found that when any volume of this gas is mixed with an equal volume of hydrogen, in a strong tube, provided with platinum points, like those of the U-tube already used (p. 25), and the mixture is fired by the electric spark, a violent explosion takes place, a dew of water condenses upon the walls of the tube, and there remains a volume of colorless gas, precisely equal to that with which the hydrogen was originally mixed. On studying the properties of this residual gas, it is found to be tasteless, odorless, a little lighter than air, and to be neither inflammable, nor yet a supporter of combustion; it is recognized as the pure element, nitrogen.

It follows from this experiment, and the knowledge of the composition of water previously gained, that any measure of the gas obtained from nitrate of ammonium contains its own measure of nitrogen and half that measure of oxygen. The constitution of this gas is strictly analogous to that of steam; as two volumes of hydrogen and one volume of oxygen are compacted into two volumes of steam, so two volumes of nitrogen and one volume of oxygen are condensed into two volumes of this transparent gas. As the chemical formula or symbol of water is H_2O, so the formula of this new gas is N_2O, and its volumetric composition may be represented by a diagram similar to that by which we conveyed to the eye the composition of water.

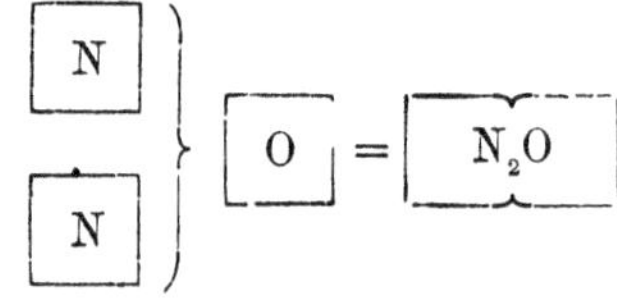

67. As has been already said, a combination of oxygen with another element is called an *oxide*; the name of the second element is given either by an adjective which precedes the word oxide, as in the case of this gas N_2O, whose name is *nitrous oxide,* or by connecting the name of the second element with the word oxide by the preposition *of,* as in case of *oxide of iron.*

From the above composition by volume, and from the known specific gravities of nitrogen and oxygen, the composition of nitrous oxide by weight is readily deduced. The specific gravity of nitrogen, referred to hydrogen, is 14; that of oxygen 16; since there are two volumes of nitrogen for each volume of oxygen, the two elements must, in any given weight of the gas, be combined in the proportion of 28 parts by weight of nitrogen to 16 of oxygen. The molecule of nitrous oxide, N_2O, must be composed, like any other quantity of the gas, of 28 parts by weight of nitrogen and 16 of oxygen; but precisely as in the case of water, we conceive of the molecule as made up of two atoms of nitrogen and one atom of oxygen, and we have already learned that if the atomic weight of hydrogen be represented by 1, that of oxygen must be 16. It follows, from the constitution of nitrous oxide, that, if 16 represent the smallest proportional weight of oxygen which exists in combination, 14 must be the corresponding smallest weight of nitrogen when thus united with oxygen. Nitrous oxide contains 16-44ths, or 36·36 per cent. of oxygen.

68. Nitrous oxide is almost without odor, but has a distinctly

sweet taste; its specific gravity referred to hydrogen is 22; it is quite soluble in water, which at 0° dissolves more than its own volume of the gas, and more than half its volume at 24°. Owing to this solubility there is a trifling loss incurred by collecting it in the usual manner over water; this loss may be partially avoided by using warm water in the pan. Nitrous oxide may be obtained in the liquid state by submitting it to a mechanical pressure of about 30 atmospheres in an apparatus cooled to 0°. The liquid is very mobile, boils at −88°, and crystallizes at about −100° when allowed to evaporate spontaneously under the exhausted receiver of an air-pump. It is noteworthy that the *permanent* gases, such as nitrogen, oxygen, and hydrogen, are but slightly soluble in water, while all the soluble gases are liquifiable, and often the more readily liquifiable in proportion to their solubility. A drop of liquid nitrous oxide blisters the skin like a hot iron. By mixing the solid, snow-like nitrous oxide with the volatile liquid called bisulphide of carbon, and evaporating the mixture in a vacuum, the lowest temperature which has hitherto been attained is produced; it is estimated at −140°.

Small animals immersed in gaseous nitrous oxide die after some time, but it may be respired for a few minutes with entire impunity by the healthy human being. The physiological effects of this gas, when respired, vary somewhat, according to the quality of the gas and the mode of administration; sometimes it produces a lively intoxication, attended with a disposition to muscular exertion and violent laughter, whence its trivial name of *laughing-gas*; sometimes, on the contrary, it produces a complete insensibility, during which surgical operations may be performed without pain. When intended for respiration, great attention should be paid to the purity of the gas; carefully prepared and judiciously administered, it is advantageously used as an anæsthetic agent, especially for operations lasting but a few seconds.

As the gas contains nearly twice as much oxygen as atmospheric air, it does not seem strange that it should make common combustibles burn with great intensity; it forms explosive mixtures with many inflammable gases; it causes glowing charcoal to burst into flame, and sulphur and phosphorus burn in the gas with great brilliancy, if well on fire when immersed in it.

Exp. 36.—Place a bit of sulphur in a deflagrating spoon, and ignite it with the least possible application of heat; then thrust it into a bottle of nitrous oxide. It will be extinguished. Yet it would continue to burn in the air. Heat the sulphur much hotter, and again introduce it into the bottle of nitrous oxide. It will now burn far more brilliantly than in the air.

The air is not a chemical compound, but only a mechanical mixture of nitrogen and oxygen; so that a body burning in the air has only to take oxygen, which is perfectly free to join it. The case is entirely different when the substance at whose expense the oxygen is furnished is a chemical compound; to dismember the compound will require a force superior to that which binds its elements together. Before the sulphur, in this experiment, can unite with oxygen, it must detach the oxygen from the nitrogen with which it is combined. To accomplish this, the sulphur must be hotter than it need be for simple burning in the air. We shall soon learn that there are many chemical compounds, much richer in oxygen than either the air or nitrous oxide, which nevertheless cannot support combustion at all, in the ordinary sense of the term, and this simply because the common combustibles are quite unable to detach the oxygen from the elements with which it is already combined.

69. *Nitric Oxide.*—The nitrous oxide which we have thus studied, is a derivative of nitric acid, or, more exactly, of the compound of nitric acid with ammonia-water, but it is only one of several such derivatives. We proceed to investigate another substance still more directly obtained from nitric acid.

Fig. 24.

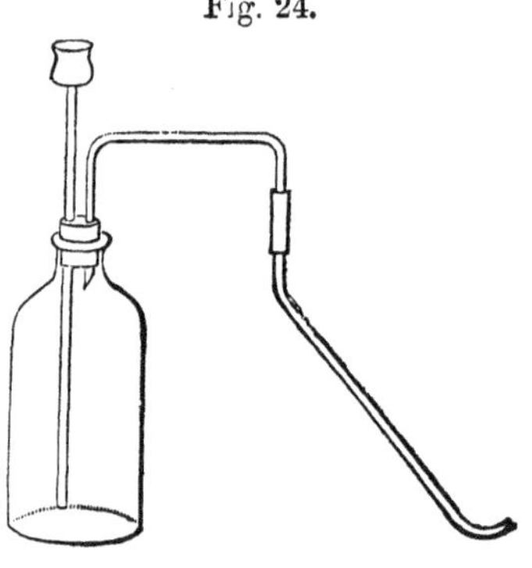

Exp. 37.—Place some copper turnings or filings in a bottle arranged precisely as for generating hydrogen (see Exp. 19), and pour upon them one half of the nitric acid still remaining from Exp. 32, previously diluting this portion of acid with twice its bulk of water. Brisk action will immediately occur. The bottle becomes filled with red fumes, but when the gas disengaged is collected over water, it is found to be colorless. Collect four bottles, of 300 to 400 c. c. capacity, of this gas. Save the blue solution (nitrate of copper) which remains in the generator for future use.

Exp. 38.—Dip a lighted candle into a bottle of the gas. It goes out. Into the same bottle thrust a glowing splinter of wood. It will not inflame.

Exp. 39. Lift a bottle of the gas from the water so that the air may enter the bottle, and the gas may escape into the air. Red fumes, of very disagreeable smell, and very irritating when inhaled, are abundantly produced. Bring in contact with these fumes a piece of moistened litmus-paper. It will be immediately reddened.

Exp. 40.—Thoroughly ignite a bit of sulphur in a deflagrating-spoon, and introduce it into a bottle of the gas. It will not burn.

Exp. 41.—Into the same bottle thrust a piece of phosphorus as big as a pea, burning *actively*. The combustion will be continued with great brilliancy.

70. The new gas is transparent and colorless, and that it is sparingly soluble in water may be inferred from the fact that bottles of it may stand indefinitely over water without appreciable loss. It differs from nitrous oxide, and from all the other gases thus far studied, in its relation to combustibles. The commonest combustibles will not burn in it at all; phosphorus may be melted in the gas without inflaming; but when its combustion is once started, phosphorus burns with a vividness which recalls its burning in oxygen.

When the gas touches the air, a new compound, red, acid, and irritating, is immediately produced; the question arises, Is it the nitrogen or the oxygen of the air which gives rise to this new combination? Experiment would answer this question in favor of oxygen. If into a bottle of this new gas nitrogen were introduced, the result would be simply negative; no visible change and no chemical combination would take place. The introduction of oxygen into a bottle of the gas would, on the other hand, produce the red vapors in question, only more vividly than air, because dilution with the inert nitrogen of the air would have been avoided. So visible and trustworthy is this reaction, that the gas we are studying may be used to exhibit the presence of free oxygen in gaseous mixtures. For example, both oxygen and nitrous oxide reinflame a glowing splinter, and we cannot distinguish between these two gases by this test; but the gas we are now studying supplies us with a means of discrimination, since it produces no red fumes with nitrous oxide.

If a sufficient quantity of the metal potassium be heated in the dry gas till it burns, and the experiment be so executed as to allow the volume of gas to be measured both before and after the combustion, it will be found that one half of the volume of gas used has disappeared, and that the half which remains possesses few or none of the qualities of the original gas; a slight examination would convince us that we had set free the well-known element nitrogen. The other half of the original volume of gas has united with the potassium to form a body which we shall hereafter be familiar with under the name of oxide of potassium. The other half of the gas is then oxygen, and we have found the elements which enter into the composition of this gas, and the volumetric proportions in which they are united. One volume of nitrogen is combined with one volume of oxygen to form two volumes of the compound gas.

We meet here the first case of chemical combination between two gases unattended by any condensation of the ingredients. The molecule of the gas will be represented by the formula NO, and its elements are united by weight in the proportion of 14 parts of nitrogen to 16 of oxygen, because equal volumes of nitrogen and oxygen weigh respectively 14 and 16 times as much as the same volume of hydrogen. The gas is another oxide of nitrogen, and is distinguished by the name *nitric oxide.*

When there are two or more oxides of one element, the termination *ous* implies less oxygen than the termination *ic*, as in this case; nitr*ous* oxide contains half as much oxygen for its nitrogen as nitr*ic* oxide.

Nitric oxide is one of the permanent gases; it has never been liquified. It is a very stable compound, and if perfectly dry is not decomposed by a red heat or by the action of electric sparks. Owing to its rapid union with oxygen and the formation of acid products, its taste, smell, and respirability cannot be ascertained. Bearing in mind the fact that certain red acid fumes, the like of which we remember to have seen in making nitric acid, are formed by adding oxygen to nitric oxide, we proceed to a further study of still other oxides of nitrogen.

71. *Hyponitric Acid.*—When a mixture of two volumes of nitric oxide and one volume of oxygen, thoroughly stirred to-

gether and perfectly dried, is submitted to the action of a freezing mixture of salt and ice, transparent, colorless crystals are condensed from the mixed gases; but if the least trace of moisture has been present, the product will be an almost colorless liquid. The vapor of this new substance has a brownish red color; from a mixture of two measures of nitric oxide and one measure of oxygen, two measures of the new vapor are produced. Since two measures of nitric oxide contain one measure of nitrogen and one of oxygen, the composition of the new substance may be represented by the accompanying diagram; the molecule will be represented by the formula NO_2, and the composition of the substance by weight will be 14 parts of nitrogen to 32 of oxygen. The name of this new body is Hyponitric acid,

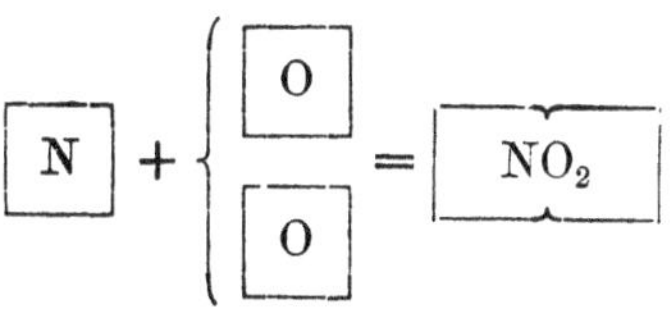

a name derived from nitric acid by prefixing the Greek ὑπο, "below." The term is used to indicate that the substance to which it is applied contains less oxygen than the other substance from which the name is derived. Thus hyponitric acid contains less oxygen than nitric acid, hyposulphurous acid less than sulphurous, and so forth. It remains to justify this assertion respecting the comparative oxygen-contents of hyponitric and nitric acids.

Exp. 42.—Add to the nitric acid which remains from Exp. 32, previously diluted with twice its bulk of water and warmed over the lamp, finely powdered litharge in small portions so long as it readily dissolves. The operation may be best performed in an evaporating-dish, which should be only very moderately heated. The substance sold under the name of litharge is a simple combination of the two elements, lead and oxygen; the formula of its molecule is PbO, in which Pb represents the least proportional weight of metallic lead (plumbum); its composition by weight is accurately known. When the litharge no longer dissolves with promptness, no more should be added, and the liquid in the dish should be evaporated to dryness, at first on the wire gauze over the lamp, but towards the end of the operation on a water-bath.

During this evaporation there escape into the air unchanged, water and the excess of nitric acid which was not neutralized by the oxide of lead. There remains a white, saline substance, which has resulted from the combination of the oxide of lead with a portion of the nitric acid; it is called nitrate of lead. Experience has proved that it is a perfectly dry substance, containing no water whatever; it is the raw material, so to speak, of important experiments shortly to be described.

Exp. 43.—Heat one or two teaspoonfuls of the nitrate of lead, of Exp. 42, in an ignition-tube, and observe that the compound decomposes; deep red fumes are produced, and oxide of lead (litharge) is left behind. If these red vapors be carried through a U-tube immersed in a mixture of ice and salt, a portion of the vapors condense into a liquid, while an uncondensable gas passes through the cold tube, and may be collected by a suitable arrangement beyond the U-tube. The condensed liquid is hyponitric acid; the gas is oxygen. If the complete absence of moisture has been secured, crystals of hyponitric acid may often be obtained by replacing the first U-tube by a second towards the end of the distillation. In fact, this is by far the most convenient method of preparing hyponitric acid. Into the U-tube containing the liquid hyponitric acid drop a bit of ice or a little snow; the color of the solution changes from reddish to a greenish blue, and two layers become visible in the liquid. The explanation of this change will be found in the next section.

The name of hyponitric acid is now justified; for it is apparent that when the litharge was dissolved in nitric acid, it found there, and combined with, an oxide of nitrogen containing more oxygen than hyponitric acid does, since when this oxide of nitrogen is decomposed by heat, as in the experiment just described, hyponitric acid and oxygen are evolved, the litharge remaining behind unaltered. Reserving the further discussion of the composition of nitric acid, we may here speak of the properties and products of hyponitric acid.

72. Hyponitric acid occurs in the solid, liquid, and gaseous states; at $-9°$ it crystallizes; above that temperature and below $22°$ it is a mobile liquid of sp. gr. 1·451, and of various colors at various temperatures; it boils at $22°$. The liquid acid gives off red, acid, irrespirable vapors at the ordinary temperature. If to liquid hyponitric acid, cooled by ice and salt, a proportionally small quantity of ice-water be added, two layers of liquid are formed, as has been above illustrated, the upper and least-colored of which consists principally of nitric acid, the lower and darker of a fluid which yields, on cautious distillation at a low tempe-

rature, a very volatile blue liquid. This blue liquid is so unstable that its composition and qualities are not certainly known; but it is supposed to be, or to contain, an oxide of nitrogen differing from all those heretofore studied,—an oxide capable of direct derivation from nitric oxide by adding, to any volume of this last, one-fourth that volume of pure oxygen, and therefore answering to the formula N_2O_3, and being composed of 28 parts by weight of nitrogen and 48 parts of oxygen. This obscure substance is known by the name of *nitrous acid*; though itself but imperfectly known, some of its compounds are not unfamiliar bodies.

73. We have already learned that the white, saline substance called nitrate of lead can be made from oxide of lead and nitric acid, that it contains no water in its composition, and that when heated it is decomposed into oxygen, an oxide of nitrogen, and the original oxide of lead. On the basis of these facts, a method has been constructed of determining the composition of nitric acid *by weight*.

To 10 grammes of oxide of lead add something more than enough nitric acid to transform it completely into nitrate of lead; evaporate the excess of acid, dry the residual nitrate of lead completely, and weigh it. If *W* represent this weight in grammes, W – 10 is the weight of the unknown oxide of nitrogen which has combined with 10 grammes of oxide of lead to form W grammes of nitrate of lead. From these data the percentage composition of nitrate of lead can be calculated; 10 grammes of pure nitrate of lead invariably contain

Oxide of lead	6·738 grammes.
Oxide of nitrogen . .	3·262 "

If now 10 grammes of pure nitrate of lead be decomposed by heat under such conditions that the vapors which it yields shall pass over some substance capable of abstracting all the oxygen from the vapors without affecting the nitrogen, it will be possible to collect all the nitrogen contained in 3·262 grammes of the unknown oxide of nitrogen, and so to determine its volume and thence its weight.

This determination is actually made by heating the nitrate of lead in a long glass tube, in such a manner that all the vapors evolved from it pass over a large surface of red-hot copper. The copper absorbs the oxygen, the nitrogen passes over it unaltered and is collected in a suitable vessel and accurately measured. As the specific gravity of nitrogen at any given temperature and pressure is known with precision, the exact weight of the nitrogen can be deduced from this volume.

Such experiments often repeated have led to the conclusion that the oxide of nitrogen, which with oxide of lead makes up nitrate of lead, contains

Nitrogen	25·93 per cent.
Oxygen	74·07 „

If these two numbers be divided by the specific gravities, or unit-volume-weights, of the two gases respectively, the quotients will express the composition by measure of the oxide of nitrogen ; the quotients are 1·862 and 4·63, numbers which are to each other as 1 to 2·5 or, avoiding fractions, as 2 to 5.

The molecule of this compound is therefore considered to contain two volumes of nitrogen and five volumes of oxygen, and is represented by the formula N_2O_5 ; the weight of this molecule will be 108, of which 28 parts will be nitrogen and 80 oxygen,—a composition precisely corresponding, of course, to the percentage composition above given. This new oxide of nitrogen may be called for the present *nitric acid*, though we shall soon learn to distinguish between this body and common nitric acid.

74. The combining proportions of nitrogen and oxygen in this oxide have been experimentally determined by weight, and not by volume, as was the case with all the preceding oxides of nitrogen. The reason of this different treatment is to be found in the fact that nitric acid cannot be converted into vapor without suffering decomposition—not indeed into its elements, but into oxygen and a lower oxide of nitrogen. It is therefore, in the present state of science, impossible to obtain a volume of nitric acid gas capable of experimental resolution into nitrogen and oxygen.

Since a large majority of the elementary bodies are non-volatile under any treatment in our power to employ, and since the greater number of chemical compounds either are non-volatile, or are, like nitric acid, decomposed by a temperature high enough to volatilize them, the proportions in which the elements unite by weight are of much more general value than the proportions in which they unite by volume ; and the methods of determining the *atomic weights* of the elements have been studied with a thoroughness, and brought to a perfection commensurate with the fundamental importance of these proportional numbers.

75. Nitric acid completes the series of oxides of nitrogen ; no

higher oxide is known. We are now prepared to exhibit this series in a diagram which shall present to the eye at once the volumetric composition of the oxides, the resultant volumes after the condensation of the ingredients, the atomic weights of the elements, and the combining weights of the compounds. Since the atomic weights of the gaseous elements are at the same time their specific gravities referred to hydrogen, it will be easy to deduce the specific gravities, or equal-volume weights, of the compound gases, N_2O, NO, and NO_2, from their combining weights by dividing these weights by two. As the resultant volumes after condensation are not known for the two members of this series N_2O_3 and N_2O_5, we abstain from figuring hypothetical volumes which analogy may point to as probable, but which experiment has never demonstrated as fact.

THE OXIDES OF NITROGEN.

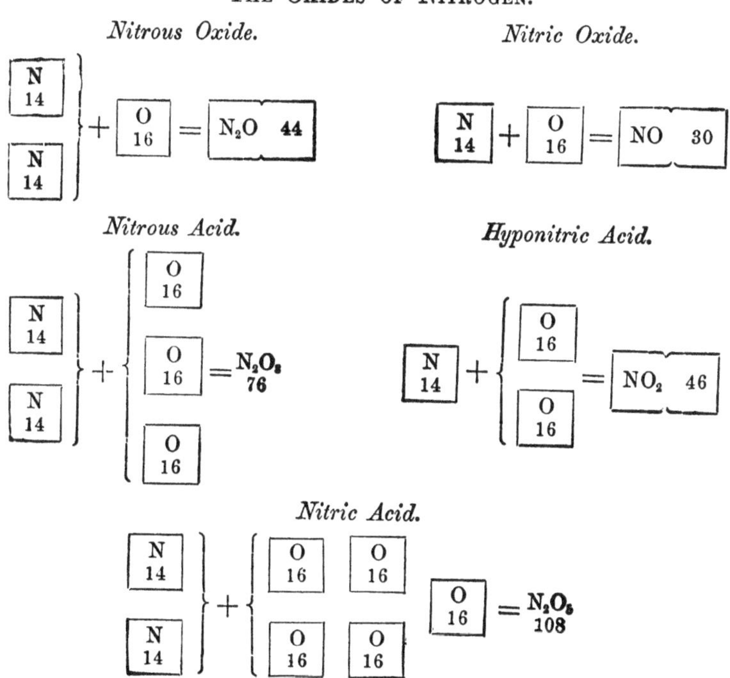

76. These five bodies are all chemical compounds; they are definite and constant in composition; and all differ essentially from their elementary constituents and from each other, as the experiments we have performed with them have abundantly demonstrated. It is, therefore, obvious that two of the elements are capable of combining in several proportions to form definite chemical compounds; and what is here proved of two of the elements we shall hereafter find to be true of all, though not of every couple; so that the series of oxides of nitrogen is but one illustration of a most comprehensive law. The difference between a mechanical mixture and a chemical compound does not on this account become less marked. The possible *mixtures* of nitrogen with oxygen are innumerable; the known *combinations* of these two elements are only five,—two volumes of nitrogen combining chemically either with one, two, three, four, or five volumes of oxygen, and with no other proportions whatsoever. As for volumes, so for weights; the proportional weight of oxygen in these oxides rises by definite leaps from the first member of the series to the last.

This definite, step by step mode of forming chemical compounds is one of the most characteristic, as it is one of the most general facts of chemistry; no other science offers a parallel to it; but long experience and patient labor with the balance and measuring-glass have established it as the habitual mode in which the force called chemical ordinarily acts. The abstract results of observation and experiment may be expressed in the following proposition, often called the Law of Multiple Proportions: *If two bodies combine in more than one proportion, the ratios in which they combine in the second, third, and subsequent compounds are definite multiples of those in which they combine to form the first.*

While the mode of action of the chemical force set forth in this proposition is that which has long been uppermost in the minds of chemists, most prominent in chemical treatises, and perhaps most important to the progress of the science in the direction in which it has thus far been cultivated, it should be remembered that in the phenomena of solution, in the formation of metallic alloys by fusion, and in the crystallization of minerals and other substances with constant forms but variable composition, the chemical force has a part to play which, if more obscure than its ordinary manifestations, is not less real.

77. *Air a mixture.*—All that has been said of the distinction between a mechanical mixture and a chemical combination finds perfect illustration in the differences between air and the definite oxides of nitrogen which have just been studied. Air is not, like these, a chemical combination. The evidence that it is a mechanical mixture merely may here be appropriately presented.

The statement that air contains 1 volume of oxygen to 4 volumes of nitrogen is not absolutely true. These proportions are never actually found; the gases are not combined in any simple ratio either by volume or weight; they are always mixed in the proportion of 20·81 measures of oxygen to 79·19 measures of nitrogen, or 23·10 parts by weight of oxygen to 76·90 parts of nitrogen. The experimental processes by which these numbers have been fixed are so perfect, that it is impossible to entertain the idea that the gases are really mixed in the ratio of 20 measures to 80, or 1 volume to 4 volumes, or the proportion of 20 parts by weight of oxygen to 70 of nitrogen, as the formula N_4O would require. When 20·81 parts of oxygen are mixed with 79·19 of nitrogen, there is no development either of light, heat, or electricity, such as usually attends the formation of a chemical compound; and the specific gravity, magnetism, and refractive power of the mixture are such as calculation would directly deduce from the numbers expressing these properties for the two constituents.

We have seen that when nitric oxide is added to nitrous oxide no red fumes are produced (§ 70), but that when the nitric oxide is brought in contact with air these suffocating fumes are abundantly formed, though the air contains only half as much oxygen as the nitrous oxide. These experiments go to show that while in nitrous oxide the oxygen is held in chemical combination, in air it is free.

Strong positive evidence that air is a mere mixture is afforded by its behavior towards water. All gases are soluble in water to a greater or less extent, each one dissolving in a certain fixed and definite proportion at any given temperature and pressure. Thus at 15° and a pressure of 76 c.m. of mercury, 1 volume of water dissolves 0·0193 volume of hydrogen and 0·7778 volume of nitrous oxide. As a general rule, the pressure to which the liquid is exposed being constant, the quantity of gas dissolved by water is less in proportion as the temperature of the water is high; in many cases boiling water is altogether incapable of retaining gases in solution, particularly those which are only sparingly soluble at any temperature.

Hence, by prolonged boiling, many gases can be completely expelled from the water which held them in solution. For example, if a solution of nitrous oxide be boiled, and the gas be collected as it escapes, this gas will be found to exhibit the characteristic properties of nitrous oxide. The nitrous oxide has not been altered by the act of solution. It still remains a definite chemical compound as before. But a result very different from this is obtained upon boiling water which has become charged with the ingredients of atmospheric air. The gas collected in this case is composed of oxygen and nitrogen, it is true, but not in those proportions in which the elements are united in air.

Water does not dissolve air directly as such, as it should do were air a chemical compound; but it dissolves out from it a quantity of oxygen, just as if no nitrogen were present; at the same time it dissolves nitrogen in accordance with the solubility of this element, and to precisely the same extent that it would absorb it if there were no oxygen in the air. Oxygen is dissolved by water in larger proportions than nitrogen; 1 volume of water at 15° and under a pressure of 76 c.m. of mercury dissolves 0·02989 volume of oxygen, but only 0·0148 volume of nitrogen.

Exp. 44.—By means of a sound perforated cork or caoutchouc stopper, adapt to a flask of the capacity of 1 or 2 litres a gas-delivery tube, No. 6, long enough to reach to the water-pan in the usual way. Upon the outer end of the delivery-tube tie a short piece of caoutchouc tubing, to which a stopper, made of a bit of glass rod or a wooden plug, has been fitted. Fill the flask completely with ordinary well or river water; fill also the delivery-tube with water, and close it by putting the stopper in the caoutchouc tube. Carefully place the cork of the delivery-tube in the neck of the flask in such manner that no air shall be entangled by the cork; at the same moment remove the plug from the delivery-tube, and finally press the cork firmly into the flask. Both flask and tube will now be completely full of water. Place the dried flask upon a ring of the iron stand, and invert a small bottle filled with water over the end of the delivery-tube. Then slowly bring the contents of the flask to boiling.

As the water gradually becomes warm, numerous little bubbles of gas will be seen to separate from the liquid and to collect upon the sides of the flask; these subsequently coalesce to larger bubbles, which collect in the neck of the flask. As soon as the water actually boils, the steam will force this air out of the flask, and it will collect in the inverted bottle at the end of the delivery-tube, the steam being meanwhile condensed as fast as it comes in contact with the cold water in the pan. By continuing to boil moderately during ten or fifteen minutes, nearly all the air can be swept out from the flask by means of the escaping steam. The gas delivery-tube may then be lifted from the water-pan and the lamp extinguished.

From a litre of ordinary water about 50 c. c. of gas can usually be obtained. This contains, of course, besides oxygen and nitrogen, a certain amount of carbonic acid; but careful analyses have shown that it is much richer in oxygen than ordinary air, the proportion of oxygen to nitrogen in the gas from water being as 32 to 68 in 100 volumes, instead of 20·81 to 79·19 as in air.

78. *Nitric Acid.*—The nitric acid above referred to (§ 73), whose composition is represented by the formula N_2O_5, is an unstable solid which melts at 30°; the liquid produced boils at 47° and is decomposed at 80°. The crystals of this substance are transparent and colorless; they undergo spontaneous decomposition into hyponitric acid and oxygen, even when preserved in closed tubes. It is obvious that this is not the common nitric acid with which we are already familiar. It remains to demonstrate that commercial nitric acid contains *water* and the oxide of nitrogen N_2O_5. A single experiment may be made use of to demonstrate the presence of water in common nitric acid, and at the same time to determine the proportion in which it enters into the composition of the sample of acid examined.

By adding to a known weight of common nitric acid (10 grammes, for example) a weighed quantity of the oxide of lead much larger than the acid is able to dissolve (100 grammes, for instance), and then heating the mixture, with suitable precautions against over-heating, in a weighed flask, the vapor of water, and nothing else, is given off; this vapor may of course be condensed, and proved to be common water. Since the oxide of lead contains no water, and the nitrate of lead is also anhydrous, it follows that whatever water escapes from the flask during the heating must be derived from the 10 grammes of nitric acid, and, further, that if the heating be long enough maintained, all the water which the acid contained will be expelled. By weighing the flask and its contents after all water has been thus driven out, the loss of weight will be the quantity of water contained in the 10 grammes of acid.

It has been already proved (Exp. 42) that the nitrate of lead, which, with a quantity of unused litharge, constitutes the residue in the flask, yields the oxide of nitrogen N_2O_5. If this oxide of nitrogen be called nitric acid, then is the common acid *hydrated* nitric acid; but if the shorter name, nitric acid, be applied to the commoner substance the commercial acid, then the oxide N_2O_5 must be distinguished as *anhydrous* nitric acid.

79. Anhydrous nitric acid unites with water in at least two definite proportions; its molecule combines with *one* molecule of water, or with *four*, to form the two hydrates represented by the formulæ H_2O, N_2O_5, and $4H_2O$, N_2O_5 respectively, wherein the symbols of one and four molecules of water are simply placed beside the formula of the molecule of anhydrous nitric acid. Ponderal analysis has given us this knowledge of the composition of these two hydrates, by actually weighing the proportion of water combined with the oxide of nitrogen in the two cases.

When *pure* nitric acid is spoken of, the acid containing one combining proportion of water to one of the oxide of nitrogen is generally referred to. This acid is often called the *monohydrated* acid, an adjective which may be applied to any substance which is coupled with a single molecule of water. The monohydrated acid is a colorless, transparent, mobile liquid of specific gravity 1·52, which boils at 86° and freezes at about −50°. Light slowly decomposes it, and a very moderate heat resolves it, not indeed into its elements, but into less complex compounds. It exerts a highly corrosive action on organic bodies, and stains tissues containing nitrogen of a bright orange colour. It should be handled with great care, as it burns the skin like a hot iron. It absorbs water from the air. When mixed with water, heat is developed from the mixture, and the second definite hydrate is formed, $4H_2O$, N_2O_5,—a colorless, strongly acid liquid, having a specific gravity of 1·42, and containing 60 per cent. of anhydrous nitric acid. To this last hydrate all weaker and stronger acids are alike converted by boiling. The ordinary nitric acid of commerce has a specific gravity of either 1·40 or 1·42, and therefore contains either 56 or 60 per cent. of anhydrous nitric acid. Sometimes its specific gravity is as low as 1·33 or 1·28. What is known in commerce as *aqua fortis* is a still weaker nitric acid containing some nitrous acid and other impurities.

80. Nitric acid, especially when hot, gives up a part of its oxygen with great facility to substances capable of combining with oxygen. When a substance habitually and readily imparts oxygen to other bodies with which it is brought in contact, it is called an *oxidizing agent*; and, on the other hand, a substance which habitually and readily takes oxygen out of other substances with which it is brought in contact, is called a *reducing agent*. When concentrated, nitric acid acts with more energy than when diluted with water; and when mixed with strong sulphuric acid (a substance which tends to take water from other compounds), it becomes an oxidizing agent of intense power. We have seen liquid nitric acid yield a part of its oxygen to copper with evolution of nitric oxide, a lower member of the series (Exp. 37), and we have also learned that the vapor of anhydrous nitric acid will give all its oxygen to red-hot copper, nitrogen being set free (§ 73). Most of the metals are dissolved by nitric acid, with evolution of one or other of the lower oxides of nitrogen; and sulphur, phosphorus, arsenic, carbon, and many other less familiar elements are converted by it into oxides. Organic substances are oxidized by nitric acid to very various degrees and with very various products, according to the strength and temperature of the acid employed.

81. *Combining Weights of Chemical Compounds.*—The *atomic weight* of oxygen, or the weight of the least proportional quantity of oxygen which enters into combination, is the same when it unites with nitrogen as with hydrogen. It is a general fact, that each element has but one *least* combining weight with each and all of the other elements. The atomic hypothesis is based upon this important fact. This hypothesis attributes to the imagined atom of each element a constant proportional weight,

expressed by the same number which experiment proves to be the combining proportion by weight of the element taken in finite, ponderable quantities. In this sense the atom of oxygen is said to be 16 times as heavy as the atom of hydrogen, and the atom of nitrogen 14 times as heavy as the atom of hydrogen. The combining weight of a chemical compound is always equal to the sum of the atomic weights of the elementary atoms contained in its molecule. Thus the combining weight of water, H_2O, is $18=2+16$; of anhydrous nitric acid, N_2O_5, is $108 = 2\times14+5\times16$; of monohydrated nitric acid, H_2O, N_2O_5, is $126=2+2\times14+6\times16$.

It must never be forgotten that these combining weights are not *absolute* weights, but simply express the *proportions* by weight in which the elements and their compounds invariably unite. The combining weight of a compound is directly deduced from the composition of the *molecule*, or least quantity of the compound which can exist by itself uncombined, or take part in any chemical process. To correctly determine this least proportional quantity often requires a large accumulation of facts, and a just collation of these facts, such as is possible only when a wide experience is added to a keen insight into the principles of chemical philosophy.

82. The discussion of the true molecular formulæ of chemical compounds presents difficulties which render it entirely unsuitable for our present stage of progress; nitric acid, however, will enable us to illustrate one of the difficulties appertaining to the subject. When nitric acid dissolves oxide of lead (Exp. 42) the reaction which occurs may be thus symbolized:—

PbO	+	H_2O, N_2O_5	=	PbO, N_2O_5	+	H_2O
Litharge.		*Nitric acid.*		*Nitrate of lead.*		*Water.*

The resulting molecular formulæ of nitrate of lead is not divisible by any number but unity, and therefore cannot be made simpler and still express the same proportional combination of its elements.

If instead of the oxide of lead we employ the oxide of silver, we shall find it possible to express the resulting molecule of nitrate of silver by two formulæ, either of which will represent correctly the proportions by weight in which the elements have combined:—

Ag_2O	+	H_2O, N_2O_5	=	Ag_2O, N_2O_5	+	H_2O; or
Oxide of silver.		*Nitric acid.*		*Nitrate of silver.*		*Water.*
Ag_2O	+	H_2O, N_2O_5	=	$2AgNO_3$	+	H_2O.

Now there is a class of metals which enter into reactions in the manner of silver in the above equations; and there is another class, of which lead may be taken as an example; and the difference between these two classes is put into the form of a definition by the statement that silver and its analogues are *uni-valent* or *mono-atomic*, and that lead and its analogues are *bi-valent* or *bi-atomic*,—terms which are intended to express the fact that one atom of any metal of the lead class is capable of uniting with twice as many atoms of oxygen, chlorine, or any other element as one atom of any metal of the silver class.

The use of the common algebraic signs in these formulæ requires no explanation. The sign of equality denotes the equality of the sum of the atomic weights on either side of it; a numeral on the left of a group of symbols is intended to multiply the whole group, unless a comma divides the group, in which case the numeral multiplies that part of the group on the left of the comma; brackets are sometimes used, as in algebra, to mark the extent of the multiplication. The formula of nitric acid itself admits of simpler expression:

$$H_2O, N_2O_5 = H_2N_2O_6 = 2HNO_3.$$

The first formula reminds us that the nitric acid, of which we speak, may, *by indirect means, be made to yield anhydrous nitric a id and water*; from the second we easily learn the proportions in which the three elements are united by weight; but the third, HNO_3, expresses these same proportions with precision, and is the most concise of the three. Now, although the two formulæ $H_2N_2O_6$ and HNO_3 express precisely the same compound of the same three elements in the same fixed proportions by weight, the *combining weight* of nitric acid is 63 if the last formula be correct, and 126 if the first represents the real molecule of the acid. In the great majority of chemical processes in which nitric acid is involved, that proportional weight of nitric acid is necessary which is implied by the molecular formula $H_2N_2O_6$; but there are not a few cases in which the proportional weight represented by the simpler formula HNO_3 completely accomplishes the actual reaction, and is capable of representation in the algebraic form.

To illustrate the first class of cases we have the formula, just given, of the reaction which occurs when oxide of lead is dissolved in nitric acid; a less proportional weight of nitric acid than $H_2N_2O_6$ or $2HNO_3$ will not answer the conditions of the reaction. Let us draw from the preceding sections some further illustrations of this class of nitric-acid reactions. When copper (whose symbol is *Cu*, from *cuprum*) is used to set free nitric oxide from nitric acid (see Exp. 37), the reaction is symbolized as follows:—

$$3Cu + 4H_2N_2O_6 = 2NO + 3CuN_2O_6 + 4H_2O.$$

Copper. Nitric acid. Nitric oxide. Nitrate of copper. Water.

When nitrate of lead is decomposed by heat into litharge, hyponitric acid, and oxygen, the following equation represents the chemical change:—

$$PbN_2O_6 = PbO + 2NO_2 + O.$$

In these two reactions the nitrates of copper and lead contain two combining weights of nitrogen and six of oxygen for each one of the metal.

To illustrate the second class of nitric-acid reactions, we have only to explain more fully the nature of some of the substances with which we have already experimented. As the nitrate of lead may be made by bringing together oxide of lead and nitric acid, water being eliminated, so the nitrates of potassium and sodium, from which we originally prepared nitric acid (see Exp. 32), may be formed by an analogous, though not identical, reaction. The composition of common caustic potash, or caustic soda, may be expressed in two ways. Some chemists represent these substances as consisting of oxide of potassium, or sodium, and water, and therefore prefer the formula K_2O,H_2O; while other chemists divide this molecular formula by two, and represent caustic potash by the briefer formula KHO, and caustic soda by the corresponding symbol NaHO. In these formulæ K stands for *Kalium*, the Latin name of potassium, and Na for *Natrium*, the Latin name of sodium. If we adopt for the moment these shorter formula, which of course express precisely the same proportional composition by weight as the longer, the reaction by which nitrate of potassium, or sodium, may be prepared will be written as follows:—

KHO	+	HNO_3	=	KNO_3	+	H_2O
Caustic potash.		*Nitric acid.*		*Nitrate of potassium.*		*Water.*
$NaHO$	+	HNO_3	=	$NaNO_3$	+	H_2O
Caustic soda.		*Nitric acid.*		*Nitrate of sodium.*		*Water.*

From this nitrate of potassium, or sodium, nitric acid is prepared (see Exp. 32) by treating it with sulphuric acid, a substance whose composition by weight, as we shall hereafter learn, may be correctly expressed by the formula H_2SO_4. This reaction may be thus symbolized:—

KNO_3	+	H_2SO_4	=	$KHSO_4$	+	HNO_3
Nitrate of potas.		*Sulph. acid.*		*Acid sulph. of potas.*		*Nitric acid.*

By substituting Na for K the reaction with nitrate of sodium would be represented. It thus appears that the molecule of nitric acid, which will represent in the simplest way its reactions with caustic potash, will not represent at all its reactions with oxide of lead, unless two molecules are assumed to enter into every reaction with this latter substance. The formula $H_2N_2O_6$ is the more comprehensive, because nitrate of potassium can be represented by the formula $K_2N_2O_6$ as accurately, if not as simply, as by the formula KNO_3.

It is noteworthy that, in the reactions above formulated, one atom or combining proportional weight of potassium, or sodium, changes place with one atom of hydrogen, while one atom of lead or copper replaces two atoms of hydrogen. These different capacities of the other elements to replace hydrogen are of great importance in chemical philosophy, and will be more fully treated of hereafter.

83. *Nitrogen and Hydrogen.*—Ammonia-water, such as we made use of in Exp. 33, when gently heated, evolves a very pungent, colorless gas, which now claims our attention.

This gas may readily be prepared as follows:—Fill a flask of 250 to 500 c. c. capacity about half full of the strongest ammonia-water to be had at the druggist's. Close the flask by a cork provided with a funnel-tube and an exit-tube; carry the delivery-tube to the bottom of a tall bottle, having a capacity of at least a litre and filled with fragments of quick-lime. When the ammonia-water in the flask is gently boiled, the gas which passes off will be deprived of moisture by the quick-lime and will issue dry from the bottle; it may be collected either over mercury, or by displacement, as shown in the figure (Fig. 25). The gas is so extremely soluble in water that it cannot be collected over the ordinary water-pan; as it has little more than half the density of atmospheric air, it can be readily collected by displacement. When thus collected, the gas should be allowed to pass into the very loosely corked bottle, until a piece of turmeric paper, held at the mouth, is immediately turned brown; the delivery-tube is then withdrawn, and the mouth of the bottle is tightly closed with a caoutchouc or glass stopper.

Fig. 25.

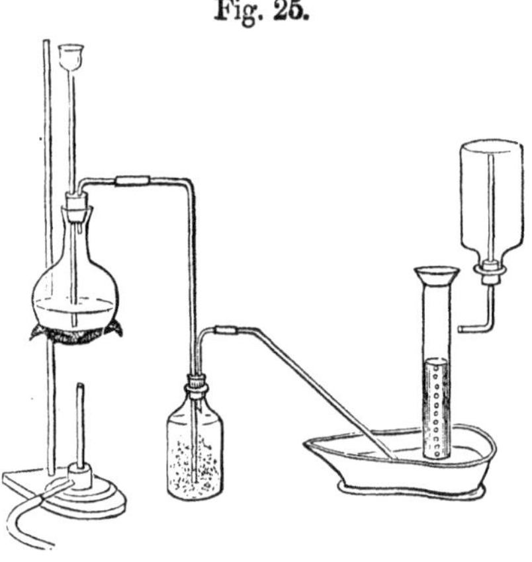

The gas thus obtained is transparent and colorless, possesses an extraordinarily pungent odor which provokes tears, and has an acrid, alkaline taste. It will be found to be uninflammable, and is, of course, irrespirable. It turns red litmus to blue most energetically. Its specific gravity as deduced from actual experiment is 8·62; a litre of the gas weighs 0·7625 grm. One measure of water at 0° dissolves 1049 measures of the gas.

The ready solubility of ammonia gas may be exhibited as follows:—Fill a stout glass tube, an ignition-tube for example, over mercury with the gas; grasp the tube by the top, and, holding it upright, dip its mouth into a vessel of water. The water will rush up the tube, if the gas be pure, with a force which might break the tube, if too thin.

84. The solution of ammonia exposed to the air, or placed in a vacuum, or simply boiled, loses all its gas. As its ready solubility in water suggests (compare § 68), the liquefaction of the gas

is not only possible but easy; the gas becomes a colorless, transparent, mobile liquid at 0°, under a pressure of 4½ atmospheres, or at —40° at the ordinary pressure. This liquid freezes at about −80°. An excellent *freezer*, applicable on both the large and small scale, is now constructed, in which the cold is produced by the rapid evaporation of *liquefied* ammonia-gas. It is the low pressure at which ammonia becomes a liquid which renders this machine possible.

85. But of what is this gas, whose properties are so strikingly unlike those of any gas previously studied, composed? The question may be answered by the following experiment:—

Fig. 26.

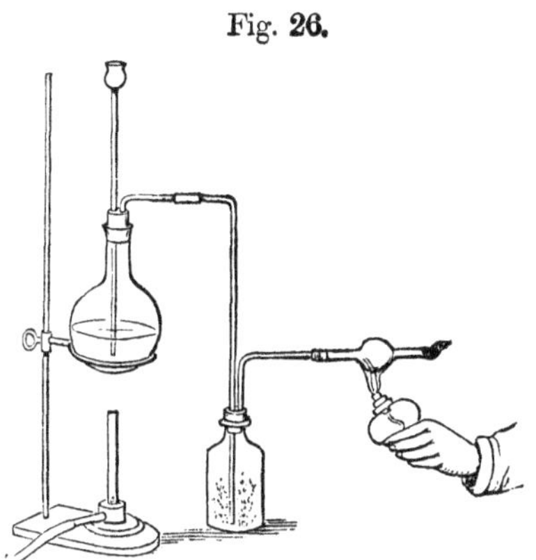

With the exit-tube of the drying-bottle of an apparatus fitted for the generation of ammonia gas as shown in the figure (Fig. 26), connect a tube of hard glass, in which a bulb has been blown (see Appendix, § 4). Thrust into this bulb a piece of the metal potassium; cause ammonia gas to flow through the bulb by heating the contents of the flask, and then warm the glass bulb. As soon as the potassium melts, it becomes covered with a brownish-green film, and a gas begins to escape which we recognize as hydrogen by lighting it at the mouth of the tube. The burning gas is certainly not ammonia, for ammonia is not inflammable, but, as the odor proves, it is mixed with some ammonia which has escaped decomposition by the potassium.

To separate this ammonia, and collect the pure hydrogen, fill a test-tube, about 14 c.m. long, three-quarters full of mercury; pour water upon the mercury till the tube is full, close the tube with the thumb, and invert it into a cup of mercury; with the outer end of the bulb-tube (Fig. 26) connect a suitable delivery-tube which shall dip into the cup of mercury and deliver the gas into the test-tube, whose upper quarter is full of water. The gas must pass through this water, which frees the hydrogen from intermixed ammonia.

It is absolutely necessary to use mercury in this experiment, because if the delivery-tube were allowed to dip into water, the extreme solubility of ammonia in water might cause the water to suck back into the bulb containing the heated metal, whereupon an explosion would inevitably ensue.

86. Having thus learned that potassium will set free hydrogen from ammonia, just as the analogous metal sodium eliminates hydrogen from water, we shall be inclined to try upon ammonia the same powerful agent by which we resolved water into its elements—the galvanic current.

A glass tube (No. 1), 60 to 80 c.m. long, open at one end and closed at the other, is bent into the form of a V; the closed limb is provided with a platinum-wire fused into the glass, through which the wire passes, to terminate near the bend of the V, in a slip of platinum foil. Fill the whole of the closed limb and nearly half of the open limb of this tube with ammonia-water, to which has been added a teaspoonful of a strong solution of sulphate of ammonium in order to increase its conducting-power; support the tube as shown in the figure (Fig. 27), and connect with the platinum-wire in the closed limb of the tube the negative or zinc pole of two medium-sized Bunsen cells, at the same time inserting a platinum-wire and plate, attached to the positive or carbon pole, in the open limb. Gas quickly collects in the closed limb. Disconnect the battery-wires, fill the open limb with water, close it with the thumb, and by inclining the tube transfer the gas to the open limb. On applying a match to the gas, it proves to be inflammable, and we recognize it without difficulty as hydrogen.

Fig. 27.

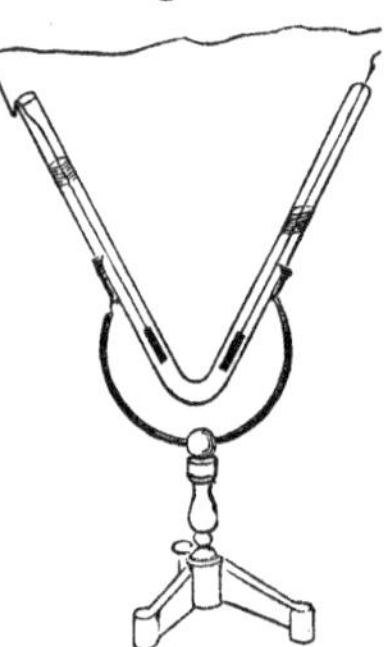

The experiment is now repeated with the electrodes reversed; the positive pole is connected with the sealed, and the negative with the open limb. The hydrogen, which is disengaged at the negative pole, now escapes through the open end of the tube into the air, while a transparent and colorless gas, previously evolved at the positive pole in the open limb and consequently lost, is now collected in the sealed end of the apparatus. The quantity of gas evolved at the positive pole is comparatively small, but in half an hour enough for examination will probably have been collected. By the same manipulation as before, transfer this gas to the open limb, and thrust into it a lighted match. It is neither inflammable nor will it support combustion, and has in itself neither taste nor smell; it is the inert *nitrogen*.

A cheaper and perfectly effective form of the apparatus used in this experiment may be made by closing one end of a common U-tube with a good caoutchouc stopper, through which the conducting wire is thrust.

87. These experiments have conclusively proved hydrogen and nitrogen to be constituents of ammonia; and it now becomes desirable to prove synthetically that these two gases are the *only* constituents of ammonia; which would be done, if ammonia could be experimentally produced by the direct union of the two gases. Unfortunately, no process has been discovered whereby ammonia can be directly reproduced from free hydrogen and free nitrogen. The following experiments, however, will demonstrate that ammonia is actually produced from materials which are known to generate a mixture of hydrogen and nitrogen—or, more strictly, which are known to be capable of generating both hydrogen and nitrogen:—

Exp. 45.—Place in an ignition-tube an intimate mixture of 3 grammes of fine iron filings with 0·2 gramme of caustic potash; adapt a delivery-tube (No. 7) to the ignition-tube, heat the contents of the tube over the gas-lamp, and collect the gas which escapes in a test-tube over the water-pan. Examine this gas, which will prove to be the inflammable hydrogen. Caustic potash, as we have already learned (p. 74), consists of potassium, hydrogen, and oxygen; at a high temperature, metallic iron is able to seize upon a portion of the oxygen in this compound, setting free hydrogen, which finds no place in the new combinations.

Exp. 46.—Heat in a second ignition-tube, similarly disposed, a mixture of 3 grammes of fine iron filings and 0·2 gramme of nitrate of potassium, and collect the gas as before, over water. This gas has neither taste nor smell, and when tested with a lighted splinter it is found to be uninflammable, and in fact to extinguish the taper. It is nitrogen. Nitrate of potassium contains, as has been already stated (p. 75), potassium, nitrogen, and oxygen; at the high temperature employed the salt is partially decomposed, the metallic iron combines with the oxygen of the nitrous vapors formed, and their nitrogen is set free.

Exp. 47.—In a third ignition-tube, heat the same quantities of the same materials which have been used in the last two experiments, at once and together. A delivery-tube is not necessary in this case; the tube may be held in the wooden nippers by the open end. Neither hydrogen nor nitrogen will be evolved as before, but instead of them we have ammonia, whose presence may be manifested by holding a bit of reddened litmus-paper at the mouth of the tube. The intense alkaline reaction of the gas, and its odor, sufficiently distinguish it from both hydrogen and nitrogen.

88. It is to be observed that the hydrogen and nitrogen, which refuse to unite when once actually in the free state, will form a chemical compound, as in the last experiment, at the precise instant when they issue from other compounds of which they formed part. This fact is expressed by saying that these two gases will enter into combination when in the *nascent state*—that is, at the moment of birth.

There are numerous cases in which bodies which do not unite under ordinary conditions are capable of chemical combination at the instant when they are disengaged from other compounds; and the phrase "*in the nascent state*" is one of some convenience, though it must not be supposed to explain, or in any way to account for, the phenomena with reference to which it is used.

89. The exact quantitative analysis of ammonia gas will afford proof that hydrogen and nitrogen are the sole constituents of this gas, and will further show the proportions in which they are combined. Ammonia is completely decomposed into its elements by heat—the heat of a furnace or the heat produced by a continuous discharge of electric sparks. When, therefore, 50 c. c. of the gas are placed in a eudiometer (gas-measure) (Fig. 28), and sparks are passed between the platinum points by means of a Ruhmkorff apparatus, the gas is finally resolved into its elements; its volume increases until it reaches 100 c. c., or double its original bulk, when it remains constant. We know that a part, at least, of these 100 c. c. of gas is hydrogen, and that this hydrogen can be eliminated from the gaseous mixture by introducing oxygen in sufficient quantity to convert the hydrogen into water, and then exploding the mixture. Were the 100 c. c. of gas all hydrogen, 50 c. c. of oxygen would convert it into water. That we may be sure of having enough oxygen, let us introduce into

Fig. 28.

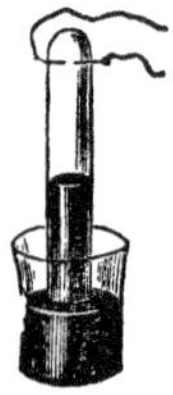

the eudiometer 50 c. c. of oxygen, and then pass a spark to explode the mixture.

After the explosion, only 37·5 c. c. of gas will remain; 112·5 c. c. have disappeared, having been converted into water. But of these lost 112·5 c. c. we know that 75 c. c. must have been hydrogen and 37·5 c. c. oxygen, in accordance with the known volumetric composition of water; consequently, the 100 c. c. of gas, into which the original 50 c. c. of ammonia were dilated, contained 75 c. c. of hydrogen. After the explosion, there remained 37·5 c. c. of gas, and since we used only 37·5 c. c. of oxygen out of 50 c. c. added, we may infer that 12·5 c. c. of oxygen still remain in the residue from the explosion. On introducing into the eudiometer a little pyrogallic acid (an acid used in photography) dissolved in water, and a few drops of a solution of caustic potash, these 12·5 c. c. of oxygen will all be absorbed by these liquids, and there will remain 25 c. c. of a colorless gas, which may readily be recognized as pure nitrogen. The original 50 c. c. of ammonia have therefore yielded 75 c. c. of hydrogen and 25 c. c. of nitrogen, and the composition of ammonia, both by weight and measure, may be fully expressed by the diagram:

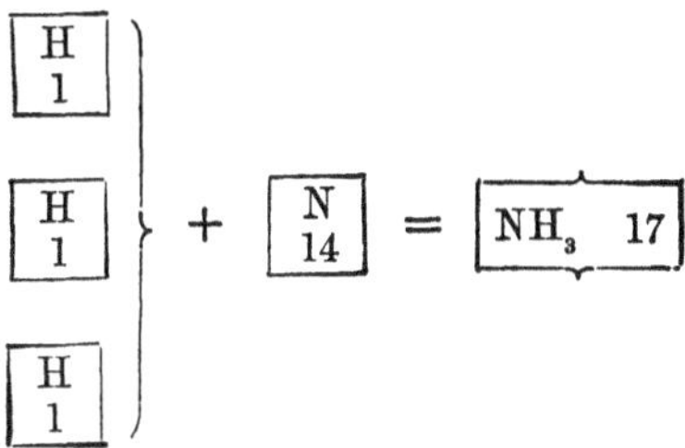

This composition of the gas is verified by its specific gravity as determined by experiment, namely, 8·62.

Three volumes of H weigh	3
One volume of N weighs . . .	14
Two volumes of NH_3 should weigh . .	17
One volume of NH_3 should weigh . .	8·5

a number sufficiently near the result of direct experiment.

90. The knowledge, thus acquired, of the composition of ammonia will enable us to recur with advantage to some of the ex-

periments performed in the first part of this chapter. Ammonia gas, when dissolved in water (as in the Liquor Ammoniæ), must be considered to be in combination with one molecule of water, in the form of the compound $NH_3.H_2O$ or NH_5O. This compound may be supposed to be dissolved in the water present in excess of what is necessary to form the compound. When this water of ammonia combines with nitric acid, as in Exp. 33, to form the compound we have called nitrate of ammonium, a reaction occurs precisely similar to that which takes place where caustic soda combines with nitric acid (p. 75); but in order to bring out the resemblance, the elements of the compound of ammonia and water must be so arranged as to exhibit its analogy with caustic soda, whose formula is NaHO. For that purpose its formula must be written $(NH_4)HO$, so that the group of elements NH_4 shall stand in the formula of water of ammonia where the element sodium stands in the formula of caustic soda. The combination of water of ammonia with nitric acid may then be represented by the equation

$(NH_4)HO$	+	HNO_3	=	$(NH_4)NO_3$	+	H_2O
Ammonia-water.		*Nitric acid.*		*Nitrate of ammonium.*		*Water.*

just like

$$NaHO + HNO_3 = NaNO_3 + H_2O.$$

The student may write both of these reactions in the typical manner; the simple type water, $\left.\begin{matrix}H\\H\end{matrix}\right\}O$, is the only one needed to represent all these substances, whether before or after the reactions which take place between them.

91. Ammonia-water combines with nearly all the acids with which soda is capable of combining, forming a series of compounds in which the group of atoms NH_4 plays the same part which the single atom Na plays in the corresponding compounds of sodium. For this reason it has been found convenient to give to this group of atoms a name bearing some resemblance to the names of metals, and it has therefore been called *ammonium*. Ammonium is known only in its compounds; many attempts have been made to obtain it in a free state, but hitherto in vain; as soon as the group of atoms escapes from combination, it is resolved into ammonia and hydrogen.

The important compounds into which ammonium enters, commonly called the *salts* of ammonium, will be studied hereafter in immediate connexion with the analogous salts of sodium and potassium. Already, however, it will be possible to verify the statement of § 66, to the effect that nitrous oxide contains "all the elements, besides those of water, which enter into the composition of nitrate of ammonium, and therefore of its constituents, nitric acid and ammonia-water." The reaction last given shows that a molecule of nitrate of ammonium contains all the elements of a molecule of nitric acid and a molecule of ammonia-water, less one molecule of water. If the formulæ of ammonia-gas and nitrate of ammonium have been correctly determined, it will be easy to exhibit in an equation the actual result of Exp. 34, as follows:—

$$(NH_4)NO_3 = N_2O + 2H_2O.$$

We thus link together several distinct experiments, and confirm the determinations previously made of the composition of nitric acid, nitrous oxide, ammonia, and water, by showing that the representative formulæ of these substances exhibit with perfect precision reactions already well known to us, but other than those from which the formulæ in question were originally derived.

92. Ammonia exists in very minute quantity in the atmosphere, and hence in rain-water, fog, and dew. The proportion of ammonia in rain-water has been variously given by different observers, from 3·49 parts to 0·744 part of ammonia in 1,000,000 parts of water. The water of fog and dew contains a larger proportional quantity of ammonia; on account of the high solubility of the gas, the proportion of ammonia in water derived from the atmosphere is greater, the smaller the fall of water. Ammonia is given off by putrefying animal and vegetable substances containing nitrogen; and almost every process of slow oxidation in the presence of air and moisture is attended with the formation of ammonia or ammonia salts. Moistened iron-filings, if exposed to the air, become rusty; and this rust is found to contain a small quantity of ammonia. When some of the metals, by preference tin, zinc, or iron, are dissolved in dilute nitric acid, the oxidation of the metal is frequently accompanied, to a greater or less extent, by the production of ammonia. But the chief source of ammonia and its compounds is the decomposition,

either by putrefaction or destructive distillation, of nitrogenous organic matter. The distillation of bones and animal refuse, for the purpose of making bone-black, yields a large amount of ammoniacal liquor, which was formerly the principal source of the compounds of ammonia; the horns of deer used to be thus distilled, whence the name "hartshorn." At present, the destructive distillation of coal in gas-works furnishes the great bulk of ammonia compounds used in the arts. The ammoniacal liquor of the gas-works is water contaminated with tarry matters, and holding in solution a small proportion of very volatile ammonium salts. These volatile salts are distilled off, by application of heat, into dilute sulphuric or muriatic acid, and the sulphate, or chloride, of ammonium thus formed is obtained from the dilute liquid by evaporation and crystallization. These salts will be studied in detail hereafter.

93. The solution of ammonia-gas in water is a reagent continually required, as a test, in the laboratory, and much used in the arts. The solution is colorless, intensely alkaline, has a caustic taste, and, when concentrated, blisters the skin. The solution is lighter than water, and so much the lighter in proportion to the amount of ammonia it contains; for this reason its strength may be accurately determined by its specific gravity. Tables of the relation of strength to specific gravity may be found in chemical dictionaries. The applications of ammonia-water are numerous and various; it is an ingredient in many pharmaceutical preparations; applied to the skin, it is an irritant, and, when very strong, even a caustic; its pungent odor is reviving and stimulating; in veterinary practice it is a useful medicament; on account of its alkalinity it is used in removing grease from cloth, and in restoring colors which have been changed by acids. The solution is ordinarily prepared from a mixture of chloride, or sulphate of ammonium, with slaked lime.

Exp. 48.—Mix intimately 25 grms. of chloride of ammonium, a substance generally sold under the name of *sal-ammoniac*, with about the same weight of cold freshly slaked lime. Introduce the mixture into a flask of 500 c. c. capacity, and place the flask on a sand-bath over the gas-lamp. If a smaller flask be used, the quantities of the materials must, of course, be proportionally diminished. Close the mouth of the flask with a good cork, provided with a delivery-tube so bent as to connect conveniently, by means of a caoutchouc connector, with the first of the series of small three-necked bottles (Woulfe's bottles) represented in Fig. 29.

Fig. 29.

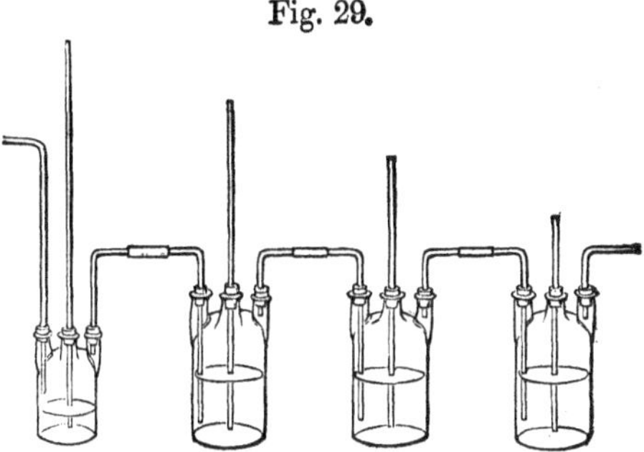

The first of this series of bottles is smaller than the rest, and is not filled so full of water as the others; it should be kept cool by immersion in cold water; the delivery-tube coming from the flask into this bottle must *not* dip into the water at all, so that it will be impossible for any water to suck back into the flask, should the gas suddenly cease to come off from the dry mixture. The construction of the apparatus is easily to be understood from the figure; the open tube which dips beneath the water in each bottle is a safety-tube, which, by admitting air into any bottle in which a partial vacuum may happen to be created by rapid absorption, prevents the contents of the succeeding bottle from flowing back into it. In order to show the action of the safety-tubes, the open tube in the first bottle may be closed for a moment with the finger and the bottle shaken very gently. Water will immediately be forced back from the second bottle through the connecting-tube to fill the vacuum caused by the absorption of the ammoniacal gas; but the moment the finger is removed from the safety-tube air will enter through the latter to fill the vacuum, and the water in the connecting-tube will fall back into the second bottle.

The ammonia-gas cannot avoid four separate contacts with water as it passes through the apparatus, so that all the gas is sure to be absorbed; the contents of the first bottle will not be as pure as those of the succeeding. This apparatus, or modifications of it, is used on the large scale as well as the small, in operations which involve the absorption of a gas by a liquid capable of dissolving it. When a large quantity of gas is continuously delivered from the generating vessel, the absorption can be made equally continuous by successively removing the bottle nearest the flask as soon as the liquid in it is saturated, and adding a fresh bottle at the other end of the series. On heating the flask, the ammonia-gas will be set free from the mixture, and in this experiment will be mostly absorbed in the first and second Woulfe-bottles. It is evident that the Woulfe-bottles in this experiment might be replaced by common bottles with mouths wide enough to admit corks pierced with three holes.

The reaction between the chloride of ammonium and the slaked lime is represented by the following equations:—

$2NH_4Cl$	+	CaH_2O_2	=	$2NH_3$	+	$CaCl_2$	+	$2H_2O$
Chloride of ammonium.		*Slaked lime.*		*Ammonia.*		*Chloride of calcium.*		*Water.*

Chloride of ammonium is a compound which may be obtained by bringing together dry ammonia, NH_3, and dry muriatic acid gas HCl (see Exp. 65).

$$NH_3 + HCl = NH_3HCl = NH_4Cl.$$

It may obviously be regarded as a compound of the group called ammonium, NH_4, with the element chlorine; from this view is derived the name chloride of ammonium. Slaked lime is prepared from water and quicklime, a substance which is chemically the oxide of the metal calcium, -

$$CaO + H_2O = CaO,H_2O = CaH_2O_2.$$

94. In each of the last two chemical equations, two expressions are given for a single substance; in the first case, chloride of ammonium is formulated in two different ways, and in the second, slaked lime. We may now inquire into the meaning of this diversity of expression for one and the same substance. A formula which simply represents the number of atoms of each element in one molecule of any substance, as determined by its analysis, is called an *empirical* formula. The truth of such a formula depends solely upon the correct performance of the analytical process, and upon the accuracy with which the atomic weights have been determined. Concerning such formulæ there is little room for difference of opinion; they express all that we actually *know* of the elementary composition of any compound body. But chemists have endeavoured to contrive formulæ which should express something more than the mere elementary com-

position by weight—which should recall the materials from which the formulated substance was made, and prophesy the products of its decomposition—which should not only name and number the atoms of the substance, but should also suggest such a grouping or arrangement of those atoms as might serve to interpret its known reactions. Such formulæ are called *rational* formulæ. Thus NaHO is the *empirical* formula of caustic soda, while Na_2O,H_2O is a rational formula of the same substance, which recalls the facts that it may be made from anhydrous oxide of sodium and water, and that it enters into many reactions in which the Na_2O goes one way and the H_2O another.

Another rational formula of the same substance is $\left.\begin{matrix}Na\\H\end{matrix}\right\}O$, which suggests many reactions in which caustic soda is involved either as product or ingredient—for example, the decomposition of water by metallic sodium (see Exp. 14), which may be thus written:—

$$\left.\begin{matrix}H\\H\end{matrix}\right\}O + Na = \left.\begin{matrix}Na\\H\end{matrix}\right\}O + H.$$

We shall hereafter meet with many such reactions, in which an atom of sodium replaces an atom of hydrogen; the formula $\left.\begin{matrix}Na\\H\end{matrix}\right\}O$ suggests this large class of reactions by implying that caustic soda is itself constituted as water in which an atom of sodium has replaced an atom of hydrogen. Such formulæ as H_2O,N_2O_5, PbO,N_2O_5, Na_2O,N_2O_5, and Na_2O,H_2O, are called *dualistic*, because they represent these bodies as of a dual nature —as being made up of two oxides which were distinct before they were brought together to form the compound, and will be distinct when separately extracted from it; in a dualistic formula these two distinct parts are conventionally represented as having some separate existence within the compound itself. The supposition is not unnatural; thus, for example, common plaster of Paris is a substance containing the metal calcium and the elements sulphur and oxygen in the proportions by weight which are correctly expressed by the formula $CaSO_4$; but this substance may be made by methods which suggest another formula. If we put together quicklime CaO, and anhydrous sulphuric acid SO_3 in due proportions, under suitable conditions, plaster of Paris, or, as its chemical name is, sulphate of calcium, results—

$$CaO + SO_3 = CaO,SO_3 ;$$

or if we mix slaked lime CaO,H_2O with strong sulphuric acid H_2O,SO_3, in proper proportions, at a suitable temperature, we shall again obtain sulphate of calcium, and water will be eliminated:—

$$CaO,H_2O + H_2O,SO_3 = CaO,SO_3 + 2H_2O.$$

Accordingly we find that the great majority of chemists have hitherto written the formulæ of sulphate of calcium, hydrated oxide of calcium, and hydrated sulphuric acid, in conformity with the suggestion of these reactions, CaO,SO_3, CaO,H_2O, and H_2O,SO_3 respectively. Of positive knowledge concerning the actual grouping of the imaginary atoms which are supposed to make up the hypothetical molecule of a compound body, we have absolutely none, and it must never be forgotten that a rational formula is merely a suggestion of *some* of the chemical processes in which the substance formulated is capable of taking part. In some cases the rational formula may point to the majority of all known transformations of the substance; but generally it suggests only a few. of the possible changes. Since a rational formula never represents a fact, but only an hypothesis or opinion, it is to be expected that a great diversity of rational formulæ should be in use among chemists; and this is really the case.

The dualistic view, above illustrated, has long been, and still is, the prevailing view of the *proximate* composition of inorganic compounds; but in the chemistry of the very numerous compounds which the element carbon forms with oxygen, hydrogen, and nitrogen, a different view, called the doctrine of *types*, widely obtains, and has been adopted by not a few chemists as affording the best theoretical representation of *all* chemical combinations, whether in the inorganic or organic kingdom. According to this doctrine every possible chemical combination may be imagined to be built upon the plan, or framed upon the type or model, of some one of the four substances, chlorhydric acid, water, ammonia, and marsh-gas. These will all shortly be to us well-known substances; but the most important of these types is water, a body with whose composition we are already familiar; we are therefore competent to write upon the type of water the formulæ of several substances with which we have already dealt, and which are classified under this type:—

Water$=\left.\begin{matrix}H\\H\end{matrix}\right\}O$=one molecule; $\left.\begin{matrix}H_2\\H_2\end{matrix}\right\}O_2$=two molecules.

Caustic Soda$=\left.\begin{matrix}Na\\H\end{matrix}\right\}O$=one molecule.

Hydrated Oxide of Calcium, slaked lime,$=\left.\begin{matrix}Ca\\H_2\end{matrix}\right\}O_2=1$ **molecule.**

Monohydrated Nitric acid$=\left.\begin{matrix}NO_2\\H\end{matrix}\right\}O$=one molecule.

Nitrate of Potassium $\left.\begin{matrix}NO_2\\K\end{matrix}\right\}O$=one molecule.

Nitrate of Lead$=\left.\begin{matrix}2\,(NO_2)\\Pb\end{matrix}\right\}O_2$=one molecule.

Sulphuric acid$=\left.\begin{matrix}SO_2\\H_2\end{matrix}\right\}O_2$=one molecule.

Sulphate of Calcium $\left.\begin{matrix}SO_2\\Ca\end{matrix}\right\}O_2$=one molecule.

In these formulæ it is to be observed that K, Na, and NO_2 replace one atom of hydrogen in one molecule of water, while Ca, Pb, and SO_2 replace two atoms of hydrogen in two molecules of water. Facts of this class will accumulate as we advance, and will be the subject of future discussion. The typical notation is doubtless capable of expressing, in a logical and consistent system, the greater part of the reactions of inorganic as well as of organic chemistry; but at present it finds its best application in the chemistry of the compounds of carbon, and has gained but little foothold in the great departments of mineral and industrial chemistry.

The need of rational formulæ is much more urgently felt in that department of chemistry, called *organic*, which treats of the chemistry of carbon, than in the wider field of mineral and inorganic chemistry. Among the very numerous compounds of carbon there are many cases in which one empirical formula represents not one compound, but several; hence it becomes of consequence to determine, or to guess, how the atoms of a compound are arranged, as well as to know what and how many the atoms are. The diversity of opinion concerning this arrangement of atoms is so great, and the possible modes of grouping the numerous atoms which often enter into organic compounds are so many, that the number of rational formulæ proposed for any organic substance is commonly large in proportion to the thoroughness with which the substance has been studied. For acetic acid, for example, one of the best-known of the compounds of carbon with oxygen and hydrogen, no less than nineteen different rational formulæ have been proposed.

Remembering that a rational formula is never to be regarded as the expression of an absolute truth, but only as a guide in classification, an aid to the memory, and a help in instruction, and holding fast to the empirical formula as containing all the results of actual observation and experiment, we shall endeavor to familiarize the student with both the dualistic and typical guesses at the hidden mysteries of chemical processes and the unknowable structure of chemical compounds, giving the preference rather to the dualistic view, as being that which at the present moment prevails in the great bulk of chemical literature, and has become incorporated into the language of the chemical arts.

Lest any doubt should suggest itself to the student's mind as to the value of symbolic formulæ, let it be observed that they express the elementary composition of a compound much more tersely than words can, that they are written and read more rapidly than the sentences of the same signification would be, and that by their brevity, clearness, and precision they greatly facilitate the comparative study and comprehensive classification of chemical compounds. Again, the chemical *equations*, of whose construction we have already had several examples, enable us to set forth with precision the changes which accompany complicated, as well as simple, reactions. Thus the somewhat complex decomposition of nitric acid by copper takes definite form in the appropriate equation which has been given above (p. 74), and the very simple reaction by which nitric oxide yields red fumes of hyponitric acid in contact with air or oxygen is concisely stated by the simple equation $NO+O=NO_2$.

The chemistry of the analysis of nitric oxide by potassium (§ 70) is all condensed into the equation $NO+K_2=K_2O+N$.

When a little ice-cold water is added to liquid hyponitric acid (Exp. 43), the reaction which occurs is very concisely set forth in the equation ;—

Empirical:	$2NO_2$	+	H_2O	=	HNO_2	+	HNO_3
	Hyponitric acid.		*Water.*		*Hydrated nitrous acid.*		*Nitric acid.*

Dualistic: $4NO_2 + 2H_2O = H_2O, N_2O_3 + H_2O, N_2O_5$.

But besides having all the advantages of a *short hand,* chemical symbols are susceptible of another application of hardly less importance; they often direct the chemist beforehand to the most perfect experiment among many similar, or point out in anticipation the possibility of certain methods of research, and the inevitable fruitlessness of others. Thus the equation

$$N_2O_5 + 5Cu = N_2 + 5CuO$$

actually directs the chemist to the due proportion of copper for the exact decomposition of anhydrous nitric acid (§ 73); neither four, nor six, nor any other number than five, parts of copper, would give a perfect reaction without excess of either ingredient. Practically, an excess of copper does no harm, and is always used to make sure of the decomposition.

The student should endeavor, from the beginning, to familiarize himself with the use of chemical symbols and equations, and to this end he should invariably write the formula of every reaction described or actually witnessed in the execution of an experiment.

CHAPTER VII.

CHLORHYDRIC ACID.

95. Muriatic (sea-salt) acid, called in modern nomenclature chlorhydric acid, is a liquid which has been known for centuries, and is to-day an article of commerce, largely employed in the useful arts. The pure acid is a gas, as ammonia is; the liquid muriatic acid of commerce is only an aqueous solution of this gas, and gives it up when heated, precisely as ammonia-water yields ammonia-gas.

This operation may be conveniently performed in the apparatus shown in Fig. 30. About 250 c. c. of the commercial acid is poured into the flask, which is then moderately heated; the gas disengaged is charged with aqueous vapor, which needs to be removed before the gas is collected. For this purpose the delivery-tube is carried to the bottom of a bottle filled with pieces of pumice-stone saturated with strong sulphuric acid; the moisture of the gas is greedily absorbed by the large surface of acid with which the gas comes in contact, as it is forced upward through the acid-soaked stone. The dry, colorless, transparent gas must be collected over mercury, for it is extremely soluble in water.

Fig. 30.

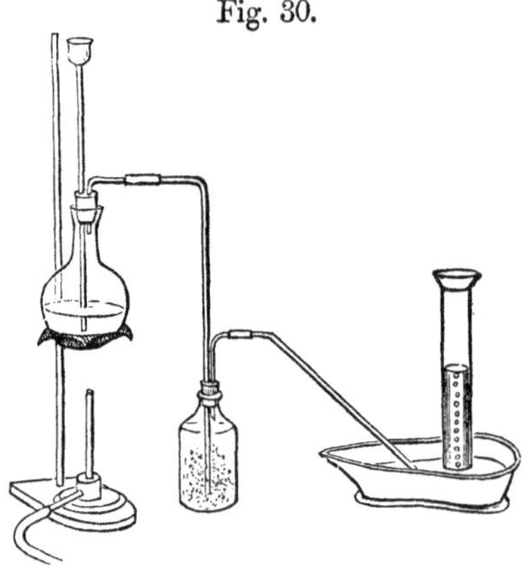

96. The gas is strongly acid in taste and reaction on vegetable colors, provokes violent coughing, and is wholly irrespirable. It is neither combustible nor will it support combustion. The gas is somewhat heavier than air; its specific gravity referred to hydrogen, as determined by experiment, is 18·12, its theoretical density being 18·25; it is possible, though not convenient, to collect it by downward displacement. It forms opaque, white fumes in the air, owing to its union with, and condensation of, atmospheric moisture. Under a pressure of 40 atmospheres, at a temperature of 10°, chlorhydric acid gas is condensed into a colorless liquid. Its great solubility in water would lead us to expect that it could be readily reduced to the liquid state; but, on the contrary, it can be condensed only with difficulty. At 0°, one volume of water dissolves about 500 volumes of chlorhydric acid gas; at common temperatures, something more than 400. The specific gravity of the solution is greater than that of water, and the more concentrated the solution the higher the specific

gravity; so that the strength of any sample of the commercial acid may be ascertained by taking its specific gravity. Tables for this use will be found in chemical dictionaries.

The avidity of water for chlorhydric acid gas may be neatly shown by thrusting a bit of ice into a small cylinder of the dry gas standing over mercury; the ice instantly melts, and the gas as quickly disappears. A solution of the acid containing 20·2 per cent. of the gas, and having a specific gravity of 1·104, distils unchanged at a temperature of about 111°; stronger solutions than this, on being heated under the ordinary atmospheric pressure, lose gas until reduced to this strength; weaker solutions lose water until raised to this degree of concentration. This stable solution, which distils unchanged, is supposed to be a definite compound of the dry gas and water, whose composition the formula $HCl + 8H_2O$ would correctly represent.

97. We propose to answer the question—of what is chlorhydric acid composed—by a partial analysis and a complete synthesis.

One of the elements of this gas can be isolated by a method which we have already applied to the analysis of ammonia. It is only necessary to remove the delivery-tube from the apparatus already used to generate the dry gas (Fig. 30), and to fix in its place a bulb-tube of hard glass, containing a piece of potassium. As soon as the acid gas reaches the potassium, the metal becomes covered with a white incrustation; and if the bulb be now very gently heated (Fig. 31), the potassium fuses, and taking fire, burns with a violet light. During the reaction the chlorhydric acid is decomposed, an inflammable gas, easily recognized as hydrogen, is evolved, and may be lighted at the end of the tube.

Fig. 31.

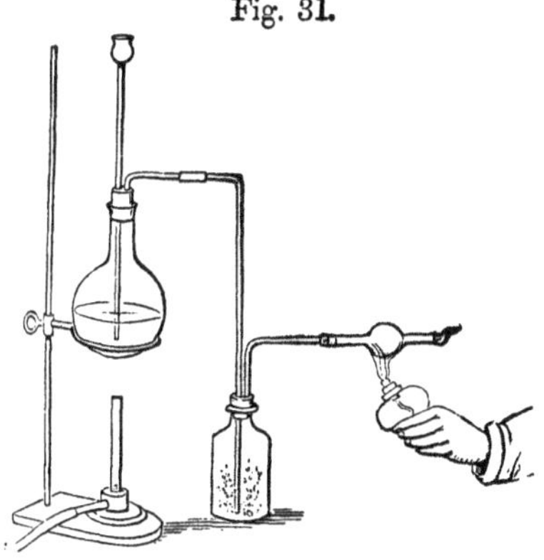

The metal sodium produces similar results, but at a much higher temperature. A solution of sodium in mercury, known amongst chemists as sodium-amalgam, will however bring about the decomposition of the acid at the ordinary temperature. This solution is best prepared by very gently heating some mercury in a glass flask, and gradually adding the sodium, cut into fragments not bigger than a grain of wheat; the fragments dissolve with evolution of light and heat. Why the sodium-amalgam should act at a lower temperature than the sodium itself, is not clear, unless it be that the minute subdivision of the sodium in the mercury gives the gas a freer contact with the metal.

Instead of potassium or sodium, as above described, metallic iron could be employed for analyzing chlorhydric acid. To this end iron turnings should be heated to redness in a glass tube such as is shown in Fig. 8; a wide delivery-tube should be used.

Hydrogen is, then, one ingredient of chlorhydric acid; the other, or others, have combined with the potassium or sodium-amalgam. The isolation of these unknown ingredients may be accomplished by means of the V-tube already used for the analysis of ammonia. Into this tube liquid chlorhydric acid of specific gravity 1·1, colored with indigo solution, is introduced, so as to fill the whole length of the sealed, and about half the length of the open, limb; the negative pole of the battery is connected with the wire of the sealed limb, while the positive pole is inserted into the open limb. Gas rapidly collects at the negative pole in the closed limb, but at the positive pole the disengagement of gas is so slight that it would hardly attract attention but for its intensely disagreeable odor and powerful bleaching action upon the blue liquid. The gas in the sealed limb has no such bleaching power. When enough gas for examination has collected in the sealed limb, it is transferred to the open limb by the manipulation previously described (§ 86); the gas is inflammable, and is, in short, hydrogen.

Fig. 32.

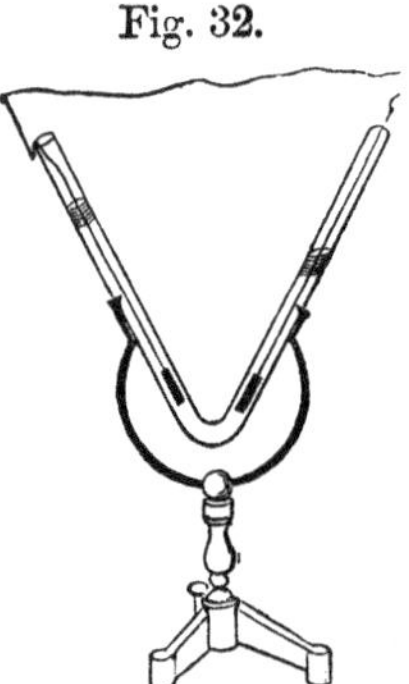

The poles are now reversed, and immediately hydrogen escapes in abundance from the open mouth of the tube, while the liquid in the closed limb becomes decolorized. In the course of fifteen minutes the bleached liquid in the sealed limb begins to assume a yellowish-green color, and the evolution of gas becomes gradually more and more copious, so that in three-quarters of an hour the greater portion of the tube is filled with a transparent, yellowish-green gas. As the gas is transferred to the open limb of the tube for examination, it manifests its powerful bleaching property by decolorizing, as it passes, the portion of the acid which had retained the blue color of the indigo. The tube is no sooner opened to admit a burning taper than the suffocating odor of the gas becomes offensively perceptible; the gas proves to be uninflammable, and it supports combustion but imperfectly, as is evidenced by the sooty cloud which is produced.

This peculiar gas, so different in properties from any gas heretofore studied, is an element; it has been named, on account of its color, Chlorine, from the Greek word for yellowish-green. This element is the subject of the next chapter, where it will be fully studied. As will there be seen (Exp. 51), chlorhydric acid, when heated with a substance called black oxide of manganese, yields chlorine in abundance, with great facility; in fact, this acid is the source of chlorine whenever large quantities of this gas are required. Chlorine is soluble in about one-third of its volume of cold water,—a property which explains its apparently slow evolution at the outset of the foregoing experiment, and the more rapid disengagement of the gas when the liquid has become saturated therewith. Chlorine is heavier than air, and consequently very much heavier than hydrogen; the best experimental determination of its specific gravity has given the number 35·66: but there can be no doubt that the true specific gravity of the gas is 35·5, or in other words that it is $35\frac{1}{2}$ times as heavy as hydrogen.

98. We have thus learned that the electric current sets free from chlorhydric acid two essentially different gases—hydrogen and chlorine,—and that each of these gases may be separately evolved from muriatic acid—the hydrogen by potassium, and the chlorine by oxide of manganese. It remains to prove that chlorhydric acid contains no other than these two constituents, and to demonstrate the proportions in which they are united. To this end the first step shall be to make a partial quantitative analysis of chlorhydric acid gas.

Fig. 33.

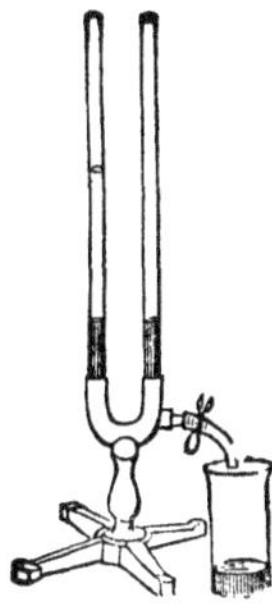

The instrument employed is a glass U-tube, about 50 c.m. long by 1·5 in diameter, having one sealed and one open limb; communicating with the latter is a small outlet-tube which may be closed by a spring-clip on a piece of caoutchouc tubing. The apparatus, mounted on a convenient stand, is represented in Fig. 33. The U-tube is first filled with mercury, and then, the spring-clip being open, the delivery-tube of the apparatus used to generate dry chlorhydric acid gas is passed down the open limb to the bend of the tube in such a manner that the gas bubbles up through the mercury into the sealed limb, from which the mercury escapes as the gas enters. When the closed limb is two-thirds full, the outlet-tube is closed, the gas delivery-tube withdrawn, and mercury is poured into the apparatus until it stands at the same level in both limbs. The space occupied by gas in the tube is then marked by a caoutchouc ring slipped over the tube. That portion of the open limb which is not occupied by mercury is then filled with sodium-amalgam. By closing the orifice of the tube with the thumb, and inclining the tube, the gas may be transferred from the sealed limb to the other, there shaken up with the amalgam, and re-transferred to the sealed limb. This thorough contact with the sodium decomposes the gas. On removing the thumb from the mouth of the open limb, the mercury therein falls a little, and must be further lowered by opening the spring-clip until the mercury stands at one level in the two limbs. When this is the case, it will be observed that the gas is reduced to half its original volume. The gas which remains, is, of course, hydrogen. The experiment proves that any given bulk of chlorhydric acid contains half that bulk of hydrogen.

By availing ourselves of the known specific gravities, or like-volume weights, of chlorhydric acid and chlorine, referred to hydrogen, we may establish a strong presumption in favor of the supposition, that that half of any bulk of chlorhydric acid which is not hydrogen is chlorine without admixture of any other substance.

From the relative weight of any volume of chlorhydric acid gas	18·12
Subtract the relative weight of half that volume of hydrogen..	·50
And the remainder	17·62

is very nearly equal to 17·83, the relative weight of half the same volume of chlorine, according to the best experimental determinations. If we assume, for the moment, that any volume of chlorhydric gas is really composed of half that volume of hydrogen and half of chlorine, and if we use the theoretical specific gravities, which are doubtless the true ones, instead of the above approximate determinations, which are the best which experiment has hitherto furnished, the numerical statement will be as follows :—

From the relative weight of any volume of chlorhydric acid gas	18·25
Subtract the relative weight of half that volume of hydrogen..	·50
And the remainder	17·75

is the relative weight of half the same volume of chlorine $17{\cdot}75=\frac{35{\cdot}5}{2}$. The very close coincidence between the first numerical statement, which is based wholly upon experiment, with the second, which is based on the theory that chlorhydric acid is, by volume, half hydrogen, half chlorine, is evidence enough, when taken in connexion with the preceding experimental isolation of chlorine from chlorhydric acid, to convince us that in any two volumes of chlorhydric acid gas, one volume of hydrogen and one volume of chlorine are united without condensation.

The formula of the molecule of the acid will therefore be HCl, in which Cl is the symbol of chlorine. The following diagram represents the composition of this important compound, both by volume and weight:—

H 1	+	Cl 35·5	=	HCl 36·5

99. The atomic weight of the new element, chlorine, is hereby determined. Hydrogen and chlorine unite by equal volumes to form this single stable compound, chlorhydric acid; and the proportions in which the two elements unite by weight are directly deducible from the proportions in which they unite by volume and the known specific gravities of the two gases. Indeed it also admits of direct proof, by appropriate experiment, that 36·5 parts by weight of chlorhydric acid gas invariably yield 35·5 parts by weight of chlorine and 1 part by weight of hydrogen; and, since it matters not what the absolute weight of these *parts* may be, millionths or millions of grammes, the *molecule* of chlorhydric acid, the least proportional weight in which it is conceived to exist uncombined, must be composed, like any other quantity of the acid, of 35·5 parts by weight of chlorine to 1 of hydrogen. But we conceive of this molecule as consisting of one atom of chlorine and one atom of hydrogen; the chlorine atom, therefore, weighs 35·5 times as much as the hydrogen atom.

100. If it were entirely inconceivable that another substance,

not identical with chlorine, should have precisely the same specific gravity as chlorine, the reasoning by which we have just arrived at the composition of chlorhydric acid would be not only convincing, it would be entirely conclusive; it would do more than establish a very strong presumption, it would furnish a complete demonstration. But it is not inconceivable, though in the highest degree improbable, that there should exist a body, different from chlorine, yet possessing the same specific gravity, and not to be detected by the qualitative tests to which we subjected the acid; and we therefore welcome the perfect demonstration with which the synthesis of chlorhydric acid supplies us.

This synthesis is readily effected; but as the experiment involves the preparation of chlorine, the actual performance of the experiment will be best postponed until chlorine has been prepared and studied in the next chapter. The method is as follows:—Into one of two glass cylinders, standing full of mercury upon the mercury trough, introduce a certain volume of dry hydrogen, not so large as to fill more than half the cylinder, and into the other cylinder bring precisely the same volume of dry chlorine. Cover one of the cylinders from the light with a towel, and deliver the contents of the other cylinder into the protected one. The mixture of equal volumes of the two gases having been thus effected, withdraw the towel, and leave the cylinder for several hours in diffused and not too bright daylight, but sheltered from the direct rays of the sun, which would cause an explosive union of the two elements. Under the influence of the light the two gases gradually combine; when the yellowish tint of the mixture has nearly disappeared, the cylinder may be exposed to the direct influence of the solar rays, in order to complete the reaction. Throughout the experiment there is no change in the volume of the enclosed gas; if the temperature were constant the mercury would neither rise nor fall in the cylinder. The chemical union of one volume of hydrogen with one volume of chlorine is attended neither by condensation nor expansion.

That there has been chemical action, resulting in the disappearance of the properties of the original materials, is evident from the fact that the contents of the cylinder will no longer take fire, or bleach vegetable colors. In contact with air the new gas forms white clouds; blue litmus paper it turns red; and if a little water be passed up into the cylinder, the gas is rapidly absorbed; the taste and smell also, and, in short, all the properties of this gas are those of chlorhydric acid.

By the synthetical method we therefore prove that chlorine and hydrogen are the *only* constituents of chlorhydric acid. On this fact is based the chemical name of this compound. Summing up our previous and present results, we now possess a complete demonstration that chlorhydric acid is composed solely of hydrogen and chlorine, united in equal volumes without condensation.

101. The muriatic acid of commerce is made from the most abundant and cheapest of all the natural compounds of chlorine, common salt, whose chemical name is chloride of sodium, and formula $NaCl$. This substance supplies the chlorine; the necessary hydrogen is obtained from common sulphuric acid (oil of vitriol), whose composition, as expressed in its formula H_2SO_4, we have already become familiar with.

The commercial acid is obtained by heating common salt with sulphuric acid in iron pans or cylinders, and absorbing the evolved gas in water contained in a series of stone-ware Woulfe-bottles, or some similar apparatus. The reaction is somewhat various, according to the proportion of sulphuric acid employed; it may be either of the reactions expressed in the following equations, or may lie between them :—

$NaCl$	+	H_2SO_4	=	HCl	+	$NaHSO_4$
Chloride of sodium.		*Sulphuric acid.*		*Chlorhydric acid.*		*Acid sulphate of sodium.*

$2NaCl$	+	H_2SO_4	=	$2HCl$	+	Na_2SO_4
						Sulphate of sodium.

In the first reaction, only one-half of the hydrogen in each molecule of sulphuric acid is replaced by sodium; in the second, both atoms of hydrogen are replaced. The first reaction requires more sulphuric acid than the second, in proportion to the amount of the product, but is accomplished with less wear of the apparatus, because a less heat suffices for the first than for the second reaction.

We may illustrate the practical importance of the atomic weights, by taking an actual example of each of these reactions. Starting with 100 kilos. of salt in each case, what quantities of sulphuric acid should be employed, and what will be the weights of the products in each reaction? (See § 81.)

The molecular weight of	NaCl	is $23+35{\cdot}5$	$= 58{\cdot}5$
" " "	H_2SO_4	is $2+32+4\times16$	$= 98$
" " "	$NaHSO_4$	is $23+1+32+4\times16$	$= 120$
" " "	Na_2SO_4	is $46+32+4\times16$	$= 142$
" " "	HCl	is $1+35{\cdot}5$	$= 36{\cdot}5$

The weight of sulphuric acid needed in the two cases is ascertained by solving the following proportions :—

First reaction	58·5 *Mol. wt. of* NaCl.	:	98 *Mol. wt. of* H_2SO_4.	=	100 k. *Quantity of* NaCl *used.*	:	x (=167·52 k.) *Quantity of* H_2SO_4 *required.*
Second reaction	117 *Mol. wt. of* 2NaCl.	:	98 *Mol. wt. of* H_2SO_4.	=	100 k. *Quantity of* NaCl *used.*	:	x (=83·76 k.) *Quantity of* H_2SO_4 *required.*

The weight of chlorhydric acid gas produced in the two cases will be precisely the same; it is deduced from the proportions :—

First reaction	58·5 *Mol. wt. of* NaCl.	:	36·5 *Mol. wt. of* HCl.	=	100 k. *Quantity of* NaCl *used.*	:	x (=62·39 k.) *Quantity of* HCl *produced.*
Second reaction	117 *Mol. wt. of* 2NaCl.	:	73 *Mol. wt. of* 2HCl.	=	100 k. *Quantity of* NaCl *used.*	:	x (=62·39 k.) *Quantity of* HCl *produced.*

The weights of the residual sodium-salts in the two cases are deduced from the proportions :—

First reaction	58·5 *Mol. wt. of* NaCl.	:	120 *Mol. wt. of* $NaHSO_4$.	=	100 k. *Quantity of* NaCl *used.*	:	x (=205·128 k.) *Quantity of* $NaHSO_4$ *produced.*
Second reaction	117 *Mol. wt. of* 2NaCl.	:	142 *Mol. wt. of* Na_2SO_4.	=	100 k. *Quantity of* NaCl *used.*	:	x (=121·367 k.) *Quantity of* Na_2SO_4 *produced.*

In each case the sum of the weights of the materials employed is, of course, equal to the sum of the weights of the products. If the questions suggest themselves—how much water will these 62·39 k. of chlorhydric acid gas saturate, and what will be the bulk of the concentrated solution so obtained ?—the answers can be easily deduced from the following data:—The strongest chlorhydric acid has a specific gravity of about 1·20, and contains about 40 per cent. by weight of the gas. 40 k. of chlorhydric acid gas will then saturate 60 k. of water, and it follows that 62·39 k. of the gas will saturate 93·58 k. of water. The weight of the solution of chlorhydric acid produced will therefore be 62·39 + 93·58 = 155·97 k.; and since 1 litre of the solution weighs 1·20 k., the total bulk of concentrated aqueous acid produced will be very nearly 130 litres.

102. If the question suggest itself—why not get the hydrogen wanted from water, H_2O, a much simpler and cheaper substance than sulphuric acid?—the only answer is, that experience has taught that water has no action upon salt except to dissolve it, while sulphuric acid has power to part the two elements of salt, and, giving hydrogen to the chlorine of the salt, to accept the detached sodium of the salt in the place of its own lost hydrogen. Of the nature of the play of forces by which this new adjustment in definite proportions of the atoms of five elements is brought about, we have no distinct conception. All that we know has been said when it is stated that water works no chemical change on salt, while sulphuric acid (and a few other substances of analogous composition) does bring about a very essential change.

In the hope of rendering these and similar facts more intelligible, many chemists have assumed that an element like chlorine, or a group of elements like sulphuric acid, may possess a superior chemical attraction, or a greater *affinity*, for some elements or groups than for others. They would explain the reaction between salt and sulphuric acid by saying that chlorine has a greater affinity for hydrogen than for sodium, while a part of the sulphuric acid has a stronger attraction for sodium than for hydrogen; and, in like manner, they would account for the absence of action between water and salt by saying that the affinity of oxygen for sodium is no stronger than that of chlorine for sodium. If the second of the equations above given be written after the dualistic theory, as follows,

$$2NaCl + H_2O, SO_3 = 2HCl + Na_2O, SO_3,$$

we shall perceive the basis of a still more ample explanation, often given, of such reactions. The reaction above written is said to be determined, or caused, by three affinities:—1. The affinity of the metal for oxygen; 2. The affinity of the hydrogen for chlorine; 3. The affinity of the oxide of sodium for sulphuric acid. It will be at once perceived that the contact of water with salt gives opportunity for the play of the first two affinities; it is, therefore, the third affinity, superadded to the other two, which in this view actually determines the decomposition of salt by sulphuric acid.

Such speculations as these have not been altogether fruitless in the development of chemistry, and to some minds they seem to render the actual phenomenon more intelligible; the term *affinity* is also sometimes convenient in expressing the varying intensity with which one element grapples and holds other elements, or groups of elements; the student must not fail to distinguish, however, between the matters of fact and the matters of speculation, in whatever stands written in chemical literature touching affinities and their play. The best use of the ill-chosen term affinity is as a synonyme for chemical force. Phrases in which the term is used in this sense may contain simple statements of fact; but very frequently, especially when the word "elective" is coupled with it, the term is used in connexion with unprofitable hypotheses. What we actually know of the reaction between salt and sulphuric acid is comprehended in the statement that the hydrogen of the acid and the sodium of the salt change places in the definite proportions by weight which are expressed in the atomic weights of the two elements.

Commercial chlorhydric acid is not pure; its commonest impurities are sulphuric acid which gets mechanically mixed with the acid, iron derived from the iron vessels, arsenic supplied by the impure sulphuric acid employed, the salts contained in the water which dissolved the gas, sulphurous acid, and, not unfrequently, free chlorine.

Exp. 49.—Melt a handful of coarse common salt in a Hessian crucible in a coal fire, and pour out the liquid salt upon a brick or stone floor. Weigh out 30 grms. of the fused salt when it has become cold, and place it in a flask of a litre capacity, provided with a delivery-tube which can be conveniently connected by a caoutchouc connector with a series of small Woulfe-bottles, such as is represented in Fig. 34. Pour 50 grammes of strong sulphuric acid upon the salt, and imme-

Fig. 34.

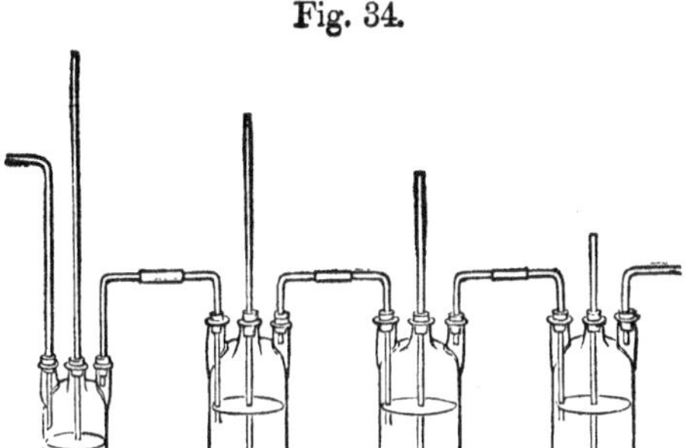

diately cork the flask, place it upon a sand-bath on the iron-stand, and connect the delivery-tube with the Woulfe-bottles. The tubes by which the gas enters the bottles should barely dip beneath the water contained in them, inasmuch as the solution of chlorhydric acid is heavier than water; the bottles should not be more than half full, for the water becomes hot and increases considerably in bulk. As hot water holds less gas in solution than cold water, it is not amiss to place each three-necked bottle in a vessel of cold water. The first Woulfe-bottle should contain but a small quantity of water, and the tube coming from the flask should not dip into this water. The contents of the flask must be very gradually and moderately heated, else a violent frothing is liable to occur, which would spoil the experiment. The process is like that of making ammonia-water, except that the delivery-tube passes to the bottom of each Woulfe-bottle in making ammonia-water, because the solution of ammonia-gas is lighter than water, instead of heavier as is the case with the solution of chlorhydric acid gas. As with the ammonia process, the solution will be purer in the second bottle than in the first, in the third than in the second, and so forth. Reserve the contents of the first bottle to make chlorine from (Exp. 51). The pure acid should be preserved for use in experiments which cannot be performed except with an acid purer than the commercial article.

103. The uses of chlorhydric acid are very numerous. It is employed in making chlorine, chlorate of potassium, and chloride of lime (bleaching-powder), in preparing chloride of ammonium and chloride of tin, in the manufacture of gelatine, for dissolving

metals (either by itself or mixed with nitric acid); it is one of the most useful reagents in the chemical laboratory.

Chlorhydric acid dissolves most metallic oxides, and appears to combine with them; but on evaporating such a solution a compound is obtained which contains neither hydrogen nor oxygen, but only chlorine and the metal. When caustic soda, for example, combines with chlorhydric acid, chloride of sodium and water are the products, as exhibited by the equation

$$NaHO + HCl = NaCl + H_2O.$$

When the black oxide of copper is dissolved in chlorhydric acid, the green liquid produced is an aqueous solution of chloride of copper;

$$CuO + 2HCl = CuCl_2 + H_2O.$$

But though the metal may exist in solution in the form of chloride, it is quite possible to precipitate it as oxide, if it have an insoluble oxide, by adding to the solution of the chloride a soluble oxide of another metal capable of displacing the first. Thus, if to a boiling solution of chloride of copper a hot solution of caustic soda be added, the sodium and the copper change places, and the insoluble black oxide of copper is precipitated.

$$CuCl_2 + 2NaHO = 2NaCl + H_2O + CuO.$$

Chlorhydric acid is, in fact, the chloride of hydrogen, strictly analogous in composition to the chloride of a metal like sodium, and it takes part in double decompositions like any other chloride.

104. *Aqua Regia* (*Royal Water*).—This name was given by the alchemists to a mixture of chlorhydric and nitric acids, because of its power to dissolve gold, the "king of metals."

Exp. 50.—Place two square centimetres of genuine gold-leaf at the bottom of a test-tube, and pour upon the gold six or eight drops of strong chlorhydric acid; put a similar piece of gold-leaf in a second test-tube, and pour upon it two or three drops of nitric acid; neither acid attacks the gold, which remains undissolved. If the contents of the two test-tubes be mixed together in either tube, the gold-leaf will almost immediately dissolve.

Platinum, which like gold resists the action of both chlorhydric and nitric acids singly applied, yields at once to the mixture of the two acids. Both these precious metals are converted by aqua regia into chlorides soluble in water. Strong chlorhydric acid is oxidized by strong nitric acid; chlorine, water, oxides of nitrogen, and unstable compounds containing chlorine, oxygen, and nitrogen, are the products. The decomposition is complex, but may be roughly represented by the equation

$$HCl + HNO_3 = Cl + H_2O + NO_2.$$

The presence of nascent (§ 88) chlorine explains the energetic conversion of metals into chlorides by aqua regia; and the strong oxidizing effect of the liquid is further explained by the presence of the unstable oxygen compounds which result from the reaction. Aqua regia has indeed a very strong oxidizing power; it can change sulphur into sulphuric acid, arsenic into arsenic acid, and effect many other similar oxidations.

This powerful solvent is made by simply mixing the two acids, though in various proportions, according to the use to be made of it; the commonest mixture is composed of one part of nitric acid and three parts of chlorhydric acid.

CHAPTER VIII.

CHLORINE.

105. Chlorine can readily be prepared from chlorhydric acid by removing the hydrogen of that acid by chemical means.

Exp. 51.—In a flask of about 500 c. c. capacity furnished with a suitable delivery-tube, place 8 or 10 grms. of coarsely powdered black oxide of manganese; pour upon it 20 or 30 grms. of common muriatic acid, and gently heat the mixture. Chlorine will soon be disengaged, and may be recognized by its peculiar color. Being very heavy the gas may best be collected by displacement in dry bottles, placed in the open air, or in a case or box provided with an efficient draft. It may also be collected over warm water or brine in the water-pan. It cannot be well collected over water at the ordinary temperature, since it is rather easily soluble therein—though the difficulty may be obviated in part by evolving the gas rapidly, or by passing the delivery-tube to the top of the bottle in which the gas is collected. It must not be left standing over water, since it would soon be entirely absorbed. In experimenting with chlorine, care must always be taken not to inhale it.

The reaction which occurs in this experiment may be thus formulated:—

$$MnO_2 + 4HCl = 2H_2O + MnCl_2 + 2Cl.$$

Black oxide of manganese is a substance rich in oxygen, which, under certain conditions, it readily yields up to other elements. In the case before us, the oxygen of the oxide of manganese unites with the hydrogen of the chlorhydric acid to form water. The chlorine of the chlorhydric acid unites in part with the manganese, and is in part left free.

In place of the black oxide of manganese in this experiment, several other substances which readily give up oxygen may be employed; and instead of the free chlorhydric acid of the foregoing experiment, the mixture of common salt and sulphuric acid, which generates chlorhydric acid (Exp. 49), is often used. The latter method has the advantage of eliminating the whole of the chlorine from the chlorine compound used, whereas, in the decomposition of the oxide of manganese by chlorhydric acid alone, half the chlorine remains combined with the manganese. Moreover, when present in excess, the sulphuric acid has the effect of drying the chlorine. The reaction may be expressed as follows:—

$$2NaCl + 2H_2SO_4 + MnO_2 = Na_2SO_4 + MnSO_4 + 2H_2O + 2Cl.$$

Another method, which has been carried out in practice upon the large scale, is to heat a mixture of common salt and nitrate of sodium with an excess of sulphuric acid. Chlorhydric and nitric acids are evolved, and, reacting upon one another, generate chlorine, hyponitric acid, and water:—

$$HCl + HNO_3 = Cl + NO_2 + H_2O.$$

The hyponitric acid is absorbed by sulphuric acid, and subsequently employed in the manufacture of sulphuric acid, while the chlorine is collected apart and employed in such manner as may be desired.

106. Chlorine is an abundant element, and very widely distributed in nature. It exists chiefly in combination with sodium as a chloride of sodium, which is called rock-salt or sea-salt, accord-

ingly as it is found in beds in the earth, or dissolved in the water of the ocean. Since the atomic weight of chlorine is 35·5 (§ 99), and that of the metal sodium is 23, each molecule, each gramme, or each kilogramme of chloride of sodium contains $\frac{35·5}{58·5}$, or 60·684 per cent. of chlorine. Accordingly in a gramme of chloride of sodium there exists something more than 0·6 grm. of chlorine; and a kilogramme of common salt should yield 606·84 grms of chlorine. Every litre of sea-water will yield about five litres of chlorine gas. Besides chloride of sodium, sea-water contains small quantities of the chlorides of several other metals; there are numerous minerals, also, which contain chlorine.

107. At the ordinary temperature chlorine is a gas of yellowish-green color, 2·5 times as heavy as atmospheric air. Its specific gravity and atomic weight are 35·5. It is excessively irritating and suffocating, even when inhaled in exceedingly small quantities. Any attempt to breathe the undiluted gas would undoubtedly be fatal. Under a pressure of 4 atmospheres at 15° it is condensed to a yellow mobile liquid, having a sp. gr. of 1·33; this liquid has never yet been solidified. It is soluble to a considerable extent in water at the ordinary temperature, 1 volume of it being dissolved by half a volume of water at 15°. This solution, which exhibits the color, odor, and general chemical properties of the gas, is called chlorine-water. At low temperatures, water dissolves a still greater proportion of chlorine; and at 0° a definite hydrate of chlorine, $Cl,5H_2O$, crystallizes out.

Exp. 52.—Fill with water the body of a retort of the capacity of 500 c. c., and without tubulature. Invert the retort and set it upon a ring, or upon a bed of sand, with the neck pointed upwards in such manner that no air shall enter the body. From a flask in which chlorine is being generated pass a long delivery-tube down the neck of the retort to the water, so that the chlorine may slowly bubble through the water. The absorption of the gas may be promoted by gently shaking the retort from time to time. As soon as the water becomes saturated with chlorine, so much gas will collect in the retort that the liquid will be pressed out of the

Fig. 35.

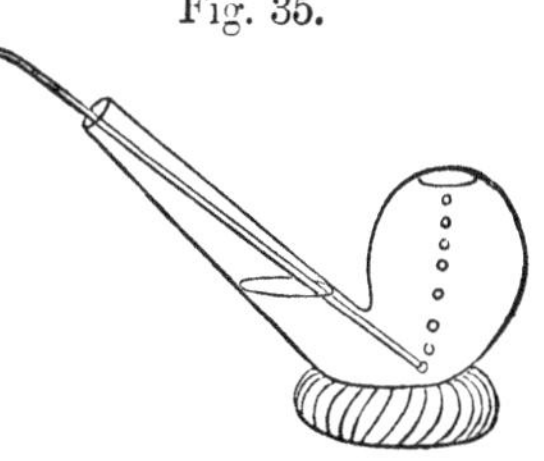

body and will flow over from the neck; when this occurs the operation may be stopped.

At the beginning of the experiment, before all the atmospheric air has been expelled from the flask in which the chlorine is generated, it is well not to push the gas delivery-tube completely to the bottom of the neck of the retort, but to simply immerse it in the edge of the water, so that none of the escaping bubbles of gas shall enter the body of the retort until it has become evident that nothing but pure chlorine is coming over; the tube may then be immersed more deeply.

The water saturated with chlorine should be transferred to a bottle and preserved for future use. It may be employed, more conveniently than the gas, to illustrate many of the properties of the element.

In sunlight, or even in ordinary daylight, chlorine-water suffers decomposition (see § 113), but in the dark it undergoes no change. It should be kept, therefore, either in a cellar or tight closet, or in a stoneware bottle, or in a bottle of black, red, or yellow glass, or in one covered with black paper. Through the blackened glass no light can penetrate to the chlorine-water, and through red or yellow glass few, if any, of the so-called chemical or actinic rays can pass. The violet rays of the spectrum are those which exhibit actinic power, and these are stopped by red or yellow glass, which is red or yellow because it permits the passage of only these colored rays.

108. Chlorine is a powerful chemical agent. It combines with hydrogen with explosive violence upon being heated, or even on being exposed to sunlight.

Exp. 53.—In a soda-water bottle, which must be screened from strong light by wrapping it in a towel, unless direct and reflected sunlight be excluded from the room, mix equal volumes of chlorine and hydrogen, then remove the cork and hold the mouth of the bottle in the flame of a lamp. A sharp explosion will ensue. Or the mixture may be made in a phial of white glass rolled up in a thick towel and filled in a darkened chamber. The explosion can then be brought about by carefully rolling the phial out of its envelope into a ray of sunlight, in a place where the fragments of glass can do no harm. In this last modification of the experiment the phial is, of course, left corked. The operator should stand behind a window-shutter or other suitable screen.

Still another method is to place the bottle in a shady place, and by means of a looking-glass reflect upon it a ray of sunlight. The moment the beam touches it, the bottle will explode.

A mixture of the two gases may be kept in the dark for any

length of time without change; in diffused daylight they usually unite only slowly and gradually; but in direct sunlight the union is so instantaneous as to be attended with explosion.

109. Chlorine combines also very readily with many of the metals, the combination being in several instances attended with evolution of light.

Exp. 54.—Fill a bottle of at least half a litre capacity with dry chlorine gas, by displacement; the gas should be dried by passing it through a tube filled with chloride of calcium, as described in the Appendix, § 15. Gradually sift a gramme or two of very finely powdered metallic antimony into the bottle. The metal will instantly take fire and fall in a glowing state to the bottom of the bottle. This fire attends the formation of a compound of chlorine and antimony, a portion of which will be seen pervading the bottle as a white smoke.

This experiment, and indeed all experiments with chlorine, should be performed only in places where there is a current of air sufficiently powerful to carry away from the operator the volatile products of the reaction, together with any chlorine which may escape from the bottle.

As in the case of the union of sulphur with copper (Exp. 1), so here it will be seen that burning, as commonly understood, is in no wise peculiar to the union of oxygen with the other elements. In the act of chemical combination heat is always evolved, and, of course, light as well, if particles of solid matter be present and become hot enough to be luminous.

Since oxygen is very abundant, we are more accustomed to witness exhibitions of its chemical action than those of any other element; but we must not therefore lose sight of the fact that among the elements there are several which possess chemical power as great when brought into play, though not as frequently exhibited as that of oxygen.

Exp. 55.—Into a small dry bottle throw loosely several leaves of the so-called Dutch-metal (an imitation gold-leaf made from an alloy of the metals copper and zinc), and invert over it a bottle of dry chlorine. As the heavy gas falls into the lower bottle, the chlorine attacks the metal, which becomes red-hot for a moment, shrivels up, and is converted into a mixture of chloride of copper and chloride of zinc. Both these compounds are readily soluble, the chloride of copper imparting to the water a peculiar green tinge. The term *chloride* is used to denote the combination of chlorine with another element, just as the term oxide denotes a compound of oxygen.

110. A burning jet of hydrogen, on being introduced into a jar of chlorine, will continue to burn with a peculiar green light, the two gases uniting to form chlorhydric acid.

Exp. 56.—From a gas-holder containing hydrogen, carry a glass tube, No. 6, outwards horizontally a few centimetres, then downwards to reach the bottom of a wide-mouthed litre bottle filled with dry chlorine; bend the end of the tube, previously drawn to a point, sharply upward, so that the jet of hydrogen may stream upwards through the chlorine. Light the hydrogen jet, and insert it into the bottle of chlorine.

By reversing the experiment, chlorine may just as well be burned in an atmosphere of hydrogen.

Exp. 57.—In a small flask of 75 or 100 c. c. capacity, provided with a small chloride-of-calcium tube prolonged into an upright delivery-tube which is drawn out to a fine point at the top, generate a free supply of chlorine. Inflame a jar of hydrogen, held mouth downwards, and press it slowly down upon the chlorine flask so that the orifice from which the chlorine is issuing may be at the centre of the hydrogen bottle, in the midst of the gas. In passing through the burning hydrogen at the bottom of the jar, the chlorine will be heated to the temperature necessary for its own inflammation, and it will continue to burn in the hydrogen in the same way that oxygen burns in hydrogen under similar circumstances.

111. The heat evolved during the combustion of hydrogen in chlorine is less intense than that produced by its union with oxygen. When one gramme of hydrogen is burned to chlorhydric acid, there are disengaged 23783 units of heat, while 34462 units of heat are evolved when it burns to water.

112. As has been seen, chlorine is both combustible and a supporter of combustion so far as hydrogen is concerned, and it exhibits a strong affinity for many of the metals; but it does not unite directly with either oxygen or carbon.

Exp. 58.—If a burning taper, or a bit of flaming wood or paper, be thrust into a bottle of chlorine gas, the flame will become murky, and after struggling for a moment will go out. Much smoke is at the same time given off.

Exp. 59.—A bit of paper, attached to a wire, dipped in hot oil of turpentine and then quickly plunged into a bottle of chlorine, will usually take fire spontaneously, and burn with evolution of dense black fumes. On account of the volatility and ready inflammability of oil of turpentine, it should be carefully heated upon a water-bath (Appendix, § 17) in a porcelain dish. If by any chance the turpentine take fire in the dish, it can be instantly extinguished by covering the dish.

Exp. 60.—Place an inverted tall bottle full of water upon the shelf of the water-pan and fill it two-thirds full of chlorine; then displace the rest of the water with ordinary illuminating gas. Cover the mouth of the bottle with a glass plate, and, removing it from the water-pan, place it in an upright position upon the table. Remove the cover, and touch a lighted match to the gas; fire will be propagated from above downwards, while clouds of smoke are evolved. Hold a piece of moistened blue litmus paper in the smoke; it will be reddened by the chlorhydric acid which has been formed.

The wax, wood, paper, turpentine, and gas of the foregoing experiments, and indeed most of the substances ordinarily used as combustibles, contain hydrogen and carbon. The hydrogen of these things will burn in chlorine, will unite chemically with the chlorine to form chlorhydric acid; but the carbon will not thus unite with chlorine. Hence it is that, in the experiments in question, the combustion is at the expense of the hydrogen; the hydrogen of the candle, turpentine, and so forth alone unites with chlorine, while the carbon is set free as lampblack or smoke.

113. Chlorine can even decompose water under certain conditions, taking away its hydrogen, while the oxygen is left free. This occurs, for example, when a mixture of chlorine and aqueous vapor is passed through a red-hot glass or porcelain tube filled with fragments of the same material. So, too, when an aqueous solution of chlorine is exposed to light, the water is gradually decomposed, as has been stated in § 107, oxygen being set free, and chlorhydric acid formed.

$$2Cl + H_2O = 2HCl + O.$$

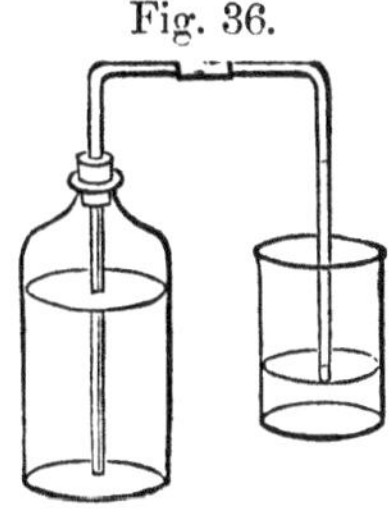
Fig. 36.

Exp. 61.—Fill a narrow-mouthed bottle, of the capacity of at least half a litre, with water which has been saturated with chlorine at a comparatively low temperature—such as is readily obtained by immersing the receiver in ice-water during the absorption of the gas. By means of a perforated cork, or better, a caoutchouc stopper, fit tightly to the bottle a glass tube, No. 6, bent twice at right angles, one branch of which shall be long enough to reach to the bottom of the bottle, while the other arm, made much shorter than the first, dips into an open beaker glass half full of water.

Place the apparatus in such a position that it shall be exposed to as much direct sunlight as possible. After a time oxygen gas will begin to collect at the top of the bottle, and in the course of several hours, or days, so much will have collected that it can be tested by removing the cork from the bottle and thrusting in a glowing splinter. The liquid displaced by the oxygen flows over, through the tube, into the beaker glass. The chlorhydric acid of course remains dissolved in the water of the bottle.

114. The applications of chlorine in the arts depend upon that readiness to combine with hydrogen which has just been exemplified. By virtue of this affinity for hydrogen, chlorine acts indirectly as a powerful oxidizing agent. It acts as a purveyor of *nascent* oxygen, and is hence a much more efficient agent than free oxygen, such as exists in the air. Its chief uses are for bleaching cotton goods, paper stock, and so forth, and for destroying foul and unhealthy emanations.

Exp. 62.—Pour into a test-glass a quantity of chlorine-water (Exp. 52), drop into it a small quantity of a solution of indigo, and stir the mixture with a glass rod. The blue color of the indigo will be immediately destroyed.

In the same way the color of litmus, cochineal, aniline-purple, or of flowers, calico, and the like, can be readily destroyed by immersion in chlorine-water or in moist chlorine gas. The presence of water is essential; perfectly dry chlorine will not bleach.

Exp. 63.—Fill a glass tube, No. 1, about 20 c.m. long, with scraps of coloured calico and bits of paper which have been written upon with ink. Take care that the tube and its contents are perfectly dry, and that the tube is closed at either end with a cork, through which passes a short piece of tubing, No. 6. Place the tube in a vertical position, and pass into it, from below, chlorine gas which has been thoroughly dried by means of chloride of calcium (Appendix, § 15). The coloring-matters will not be destroyed so long as they remain dry; but if, after the dry chlorine has been allowed to act for a few minutes, a little water be poured in at the top of the tube, so that its contents may be wetted, they will be bleached at once.

115. Those coloring-matters which are of vegetable or animal origin are for the most part complex compounds of carbon, hy-

drogen, nitrogen, and oxygen. When moist chlorine is brought into contact with them, a somewhat complicated reaction occurs; a portion of their hydrogen is no doubt taken out by the chlorine, but at the same time some of the water which is present is decomposed, and its oxygen assists the disorganization of the compound which is to be destroyed.

Of the hydrogenized or carburetted compounds exposed to the action of the nascent oxygen in the foregoing experiment, those which are most complex, and of which the elements are held together least firmly, will of course be acted upon, burned up, and destroyed. As a rule, the coloring-matters are far more easily oxidized than the cotton cloth; hence they can readily be removed by the action of chlorine, without injury to the cloth. But if the action of the chlorine were to be continued after the coloring-matter had been destroyed, the cloth itself would gradually be burned up.

In actual practice, where the duration of the exposure of the cloth to the chlorine is carefully regulated, and the portions of bleaching liquor which at first remain adhering to the cloth are completely removed by washing and by chemical treatment, the process is perfectly safe and trustworthy as regards cotton or even linen; but the animal fibres, such as wool and silk, are of more complex composition than cotton and linen; they cannot be bleached by chlorine, since this gas would attack and disorganize them.

116. In destroying noxious effluvia, chlorine either acts upon them as upon coloring-matters, or it simply takes away hydrogen, as in the case of sulphuretted hydrogen hereafter to be studied. Putrid animal matter may be rendered comparatively odorless, by sprinkling it copiously with chlorine-water; hence a solution of chlorine finds some application in inquests and judicial investigations.

The energy with which chlorine seizes upon hydrogen may be further illustrated by causing chlorine to act upon ammonia-water.

Exp. 64.—Into a glass tube, No. 1, about a metre long, pour enough chlorine-water to fill it nine-tenths full, and then ammonia-water enough to fill the remaining space. Close the tube with the thumb, invert it and place it in an upright position upon the water-pan. The ammonia-water, being specifically lighter than the solution of chlorine, will flow upwards and become mixed with the latter; a reaction will immediately ensue; some of the chlorine will unite with the hydrogen of a portion of the ammonia, to form chlorhydric acid, and nitrogen will be set free. Numberless little bubbles of this gas will escape from the liquor and collect at the top of the tube, and may be subsequently tested with a burning match. The chlorhydric acid formed unites with the remainder of the ammonia to form chloride of ammonium:—

$$4NH_3 + 3Cl = N + 3NH_4Cl.$$

By modifying the apparatus employed in the foregoing experiment, so that a current of chlorine can be passed into a vessel containing ammonia-water, the evolution of nitrogen can readily be made continuous, and large quantities of the gas may be collected. It would be an excellent and easy method of preparing nitrogen for use in the laboratory, were it not that care must be taken that the ammonia shall always be present in considerable excess. If this precaution were neglected, there might be formed, by the action of the chlorine upon the chloride of ammonium, a very dangerous compound called chloride of nitrogen. As prepared by this method, the nitrogen is always contaminated with a certain amount of oxygen.

In the foregoing experiment, the chloride of ammonium which is produced remains dissolved in the water. It may be recovered by evaporating the water, or a new portion of it may be prepared by mixing chlorhydric acid with ammonia.

Exp. 65.—Fill one half-litre bottle with dry ammonia-gas, and another with dry chlorhydric acid gas. Invert the latter, and place it over the former, so that the mouth of the upper bottle shall rest upon that of the lower. The gases will immediately unite to form solid chloride of ammonium, a dense white cloud of which will fill the bottles:—

$$NH_3 + HCl = NH_4Cl.$$

One volume of ammonia unites with one volume of chlorhydric acid, and the gases are completely condensed to a white solid.

117. Chloride of nitrogen, the dangerous compound of chlorine and nitrogen which has been alluded to above, is formed when chlorine is brought into contact with a weak solution of chloride or nitrate of ammonium at the temperature of 15° or 20°. As the chlorine is gradually absorbed, yellow oily drops of chloride of nitrogen form upon the surface of the liquid, and soon fall to the bottom:—

$$NH_4Cl + 6Cl = 4HCl + NCl_3.$$

Chloride of nitrogen is a volatile yellow oil, of peculiar, penetrating odor; it is insoluble in water, and does not congeal when exposed to cold. Its specific gravity is 1·653. It decomposes very easily. Upon being heated to nearly 100°, or touched with any fat or oil, with turpentine, or with various other substances, it explodes with extreme violence; indeed it often explodes spontaneously, without any apparent cause. A single drop of it, exploded upon a glass or porcelain dish, shatters the vessel to atoms. The preparation and handling of this body require the greatest caution; it should never be prepared by the novice in chemistry.

118. We have heretofore adduced experimental proof of every proposition and statement so far as was possible at such a stage of the student's progress. The chemical properties of the four elements, oxygen, nitrogen, hydrogen, and chlorine, have been exhibited by experiment, the composition of many of their most important compounds has been demonstrated by analysis or by synthesis, or by both these methods, and the chemical properties of these compounds have been illustrated by actual experiment. Several objects have been thus attained:—first, experimental methods of research have been illustrated by tangible examples; secondly, the foundation, scope, and application of important laws of chemical combination have been explained; thirdly, four leading elements have been minutely studied—hydrogen (the standard atom and the unity of specific gravity for gases), oxygen, nitrogen, and chlorine (three widely diffused elements, each of which is the first member and prototype of an important group of elements, many of whose properties we shall hereafter find we have already become acquainted with in studying the prototypes); fourthly, three compounds of these elements have been carefully studied—chlorhydric acid, water, and ammonia—compounds which are not only interesting in themselves, but of great significance as types, or models, of three large groups of compounds whose properties we have really been studying while we studied their types.

From this point forward the student will be asked to accept on trust many facts, drawn from the accumulated stores of the science and resting on satisfactory evidence, the full exposition of which would be both tedious and inappropriate. The subject-matter of chemistry is so vast and various that it will be necessary to select from the great mass of material only the most valuable portions, and to dwell only on those elements and compounds which are of practical importance in the useful arts, or which are of interest in connexion with instructive theories or recognized laws of the science.

119. *Compounds of Chlorine and Oxygen.*—Free chlorine does ot combine directly with free oxygen. But by resorting to indirect methods several compounds of the two elements can be obtained. As many as five different oxides of chlorine, enumerated below, have been described, though as yet some of them are known only in combination with water or other substances, and not in the free condition:—

NAMES.	COMPOSITION.					FORMULÆ.
	By volume.		Condensed to vols.	By weight.		
	Chlorine.	Oxygen.		Chlorine.	Oxygen.	
Hypochlorous acid	2 vols.	+1 vol.	2	35·5×2=71	16	Cl_2O
Chlorous acid	2 vols.	+3 vols.	?	35·5×2=71	16×3= 48	Cl_2O_3
Hypochloric acid	1 vol.	+2 vols.	2	35·5	16×2= 32	ClO_2
Chloric acid	2 vols.	+5 vols.	?	35·5×2=71	16×5= 80	Cl_2O_5
Perchloric acid	2 vols.	+7 vols.	?	35·5×2=71	16×7=112	Cl_2O_7

120. *Hypochlorous Acid* (Cl_2O).—If a small quantity of slaked lime (hydrate of calcium) be thrown into a bottle of chlorine gas, and the mixture be then left to itself during several hours, the chlorine will be completely absorbed, and there will be formed two compounds, one of which will be found to be hypochlorite of calcium, the other chloride of calcium. The reaction may be thus formulated:—

$$2(CaO,H_2O) + 4Cl = CaO,Cl_2O + CaCl_2 + 2H_2O.$$

This mixture is a substance much used in the arts under the technical names "chloride of lime," and "bleaching-powder;" it will be again referred to hereafter.

121. Hydrated hypochlorous acid may be prepared from "bleaching-powder;" the solution has a yellowish color, an acrid taste, and a peculiar sweet odor. When concentrated it decomposes rapidly, even if kept upon ice. Dilute solutions are

more stable, but they decompose slowly upon being boiled. Hypochlorous acid is a powerful oxidizing and bleaching agent. Its solution produces at once effects which are only slowly obtained when chlorine-water is employed.

Anhydrous hypochlorous acid, which may be obtained by removing the water from the aqueous solution, is a gas of pale yellow color and offensive odor, somewhat resembling that of chlorine. It decomposes very easily into 2 volumes of chlorine and 1 volume of oxygen; even the warmth of the hand is sufficient to decompose it; and it is difficult to preserve it unchanged even for a few hours. At low temperatures, such as are produced by a mixture of ice and salt, the gas condenses to a dark orange-colored liquid, heavier than water and very explosive.

122. *Chlorous Acid*, Cl_2O_3, may be obtained by deoxidizing chloric acid by means of nitrous acid. When in the anhydrous condition it is a gas of a yellowish-green color, liquefiable by extreme cold. It is a dangerous compound to prepare, since at temperatures above 57° it decomposes, with explosion, into chlorine and oxygen. It is readily soluble in water, and the solution possesses strong bleaching and oxidizing properties. It is a weaker acid than chloric acid, § 124, but resembles it in many respects. With metallic oxides it unites to form compounds called chlorites.

123. *Hypochloric Acid*, ClO_2.—This very explosive compound may be prepared by gently heating a mixture of chlorate of potassium and concentrated sulphuric acid. The gas is of a bright yellow color and aromatic odor. Upon being exposed to daylight or to a temperature somewhat below the boiling-point of water, it decomposes into oxygen and chlorine, the decomposition being usually attended with explosion. The preparation of the gas is dangerous, and should never be attempted unless upon a very small scale. At the temperature of a mixture of ice and salt, the gas condenses to a yellow, highly explosive liquid.

124. *Chloric Acid*, Cl_2O_5.—In the present state of science this is the most important of the compounds of oxygen and chlorine. It is not known in the free state, and in the hydrated condition has never been obtained with less than 1 molecule of water, H_2O,Cl_2O_5.

When a current of chlorine is made to flow into a cold dilute aqueous solution of caustic potash, a mixture of hypochlorite and of chloride of potassium is produced:—

$$2(K_2O,H_2O) + 4Cl = K_2O,Cl_2O + 2KCl + 2H_2O,\text{—}$$

the reaction being analogous to that between lime and chlorine, described in § 120. But if the conditions as to the concentration and temperature of the solution of potash be changed—if, instead of using a dilute solution, chlorine be passed into a moderately strong hot solution of caustic potash, or of carbonate of potassium, hypochlorous acid will no longer be formed, but instead of it chloric acid. The reaction may be expressed as follows:—

$$6(K_2O,H_2O) + 12Cl = K_2O,Cl_2O_5 + 10KCl + 6H_2O.$$

Chloride of potassium is formed as before, but the remainder of the chlorine is now more highly oxidized. Chlorate of potassium is less soluble than chloride of potassium; it separates in flat tabular crystals after the liquid has been concentrated by evaporation and cooled. It is the substance which was employed for making oxygen in Exp. 7.

Chloric acid could be prepared directly from chlorate of potassium by boiling a solution of this substance with fluosilicic acid. An almost insoluble fluosilicate of potassium would be formed, and chloric acid set free. But an easier method is to first convert the chlorate of potassium into chlorate of barium, and to liberate the chloric acid from this salt by means of sulphuric acid, with which barium forms a remarkably insoluble compound:—

$$BaO,Cl_2O_5 + H_2O,SO_3 = BaO,SO_3 + H_2O,Cl_2O_5.$$

The solution of chloric acid is separated from the insoluble sulphate of barium by filtration, and concentrated by evaporation over sulphuric acid in the exhausted receiver of an air-pump. By cautious evaporation the acid may be brought to a syrupy consistence, but is then rather easily decomposed, especially if it be heated or exposed to light. At the temperature of boiling it is rapidly converted into perchloric acid, water, chlorine, and oxygen. It is a strong acid, and a powerful oxidizing and bleaching agent.

125. *Perchloric Acid*, Cl_2O_7, is formed, as above stated, when an aqueous solution of chloric acid is boiled; being volatile it may be distilled off and collected. A compound of this acid and potassium, perchlorate of potassium, can be obtained by heating chlorate of potassium to a certain temperature. Perchloric acid is a more stable compound than either of the other oxides of chlorine. The dilute aqueous solution may be concentrated by

evaporation over fire, even at high temperatures. The hydrated acid H_2O,Cl_2O_7 is a colorless, oily liquid, which boils at 203°, and has a specific gravity of 1·782. It is a powerful oxidizing agent.

The student will do well to compare this series of oxides of chlorine with that of the oxides of nitrogen, and to note the points in which the two series resemble and those in which they differ from each other.

CHAPTER IX.

BROMINE.

126. Bromine is an element closely allied to chlorine. It is found in small quantities in sea-water, and in the water of many saline springs. 1 litre of sea-water contains from 0·0143 to 0·1005 grm. of it. As it exists in nature it is combined with metals, bromide of magnesium being the compound most commonly met with. Bromide of magnesium is a constituent of the uncrystallizable residue, called bittern, which remains after the chloride of sodium has been crystallized out from the natural brines; at several saline springs this bittern contains so large a proportion of the bromide that bromine can be profitably extracted from it. Most of the bromine of commerce is thus obtained.

In order to obtain bromine from the bittern, the latter is mixed with black oxide of manganese and chlorhydric acid, and heated in a retort. Chlorine is of course evolved from these materials in the midst of the liquid; it reacts upon the bromide of magnesium and sets free bromine, which distils over into the receiver as a dark-red, very heavy liquid:—

$$MgBr_2 + 2Cl = MgCl_2 + 2Br.$$

All the metallic bromides are readily decomposed by chlorine, bromine being, as a rule, a less energetic chemical agent than chlorine.

127. At the ordinary temperature bromine is a liquid of dark brown-red color, about three times as heavy as water, and highly poisonous. Its odor is irritating and disagreeable, whence the name bromine, derived from a Greek word and signifying a

stench. It boils at about 60°, but is very volatile even at the ordinary temperature of the air.

Exp. 66.—By means of a small pipette, throw into a flask or bottle, of the capacity of 1 or 2 litres, 3 or 4 drops of bromine. Cover the bottle loosely, and leave it standing. In a short time it will be filled with a red vapor, which is bromine gas. This vapor is very heavy, more than five times as heavy as air, and eighty times as heavy as hydrogen.

At about 7° bromine crystallizes in brittle plates. It dissolves sparingly in water, but is soluble in alcohol, and in all proportions in ether.

In its chemical behavior, as well as in many of its physical properties, bromine closely resembles chlorine. Its affinity for hydrogen, though weaker than that of chlorine, is still powerful. Like chlorine, it is an energetic bleaching and disinfecting agent, and it decomposes the vapor of water when passed with it through a tube heated to bright redness, bromhydric acid and oxygen being the products of the reaction. A lighted taper burns for an instant in bromine vapor and is then extinguished. Phosphorus, antimony, potassium, and the like, take fire on being thrown into bromine, in the same way as in chlorine, a bromide of the other element being produced.

Exp. 67.—Fit a thin cork to a large, wide-mouthed bottle; perforate the cork, and through the hole pass a tube of thin glass (No. 2) closed at one end. The tube should reach nearly to the bottom of the bottle, and should project two or three inches above the cork. Within the tube place a few drops of bromine; throw in upon this a very small quantity of finely powdered antimony, and instantly cover the mouth of the tube with an inverted crucible or wide-mouthed phial, in order that nothing may be thrown out of the tube by the violent action which attends the combination. If the tube be broken, its fragments will be retained within the bottle.

Bromine is used to a certain extent in medicine, and largely in photography. In the chemical laboratory it is often employed, not only for its own sake, but as a substitute for chlorine; for, though less energetic, it is more manageable than the latter, especially in those cases where a liquid is desirable.

128. *Bromhydric Acid* (HBr).—In spite of the strong affinity of bromine for hydrogen, these elements cannot readily be made

to unite directly. A mixture of equal volumes of hydrogen and bromine vapor cannot be made to combine with explosion by exposure to the sun's rays, or by the introduction of a burning lamp, though a certain amount of combination occurs in the immediate neighborhood of the flame. But by immersing in the mixture a platinum wire, kept red-hot by a galvanic current, the two elements may be made to unite slowly; and a similar result is obtained by passing the mixed gases through a red-hot tube. Bromhydric acid gas can, however, be readily prepared by decomposing bromide of potassium with sulphuric acid, or, better, with a concentrated solution of phosphoric acid. The reaction is analogous to that in which chlorhydric acid is obtained from chloride of sodium:—

$$2KBr + H_2O,SO_3 = K_2O,SO_3 + 2HBr.$$

If sulphuric acid be employed, a secondary reaction occurs; a small part of the bromhydric acid suffers decomposition, and the product is slightly contaminated with free bromine and with sulphurous acid:—

$$2HBr + H_2O,SO_3 = 2H_2O + 2Br + SO_2.$$

Since phosphoric acid does not thus decompose bromhydric acid, the latter can be obtained in a state of purity by distilling a mixture of bromide of potassium and phosphoric acid.

129. Bromhydric acid is a colorless, irritating gas, which, on coming in contact with the moisture of the air, fumes even more strongly than chlorhydric acid. By powerful pressure it can be reduced to the liquid condition, and upon being exposed to intense cold it may be obtained in the form of a crystalline solid.

It is readily soluble in water, forming a strongly acid solution which resembles chlorhydric acid in many respects, and, like it, fumes in the air. A ready method of preparing the solution is to decompose a strong solution of bromide of barium with sulphuric acid diluted with its own weight of water. The solution of the free acid may then be separated from the insoluble sulphate of barium by filtration or by distillation.

When the solution of this acid is mixed with nitric acid, there is obtained another aqua regia capable of dissolving gold and platinum, like the mixture of chlorhydric and nitric acids, though less readily.

130. The gas undergoes no change when passed through a red-hot tube; but it is readily decomposed by metals like potassium at the ordinary temperature, and by tin gently heated. A bromide of the metal is formed in either case, and there remains a volume of hydrogen equal to half that of the original gas. Observation has shown that the specific gravity of the gas is very nearly 40·5, or half the sum of the specific gravities of bromine vapor and hydrogen. From this fact and the above decomposition of the gas by metals, it follows that bromhydric acid is composed of equal volumes of bromine and hydrogen united without condensation.

131. *Bromic Acid*, H_2O,Br_2O_5.—Only one oxide of bromine has been studied, and even this has never been obtained free from water. It is bromic acid, a substance corresponding to chloric acid in composition and properties. Its compounds, also, known as bromates, generally resemble very closely the corresponding chlorates.

Bromate of potassium can be obtained by the action of bromine upon potash-lye, in the same way that chlorate of potassium is obtained by the action of chlorine:—

$$6(K_2O,H_2O) + 12Br = K_2O,Br_2O_5 + 10KBr + 6H_2O.$$

The bromate, which is less soluble than the bromide, can subsequently be separated by crystallization. In order to obtain the hydrated acid, bromate of barium may be decomposed with dilute sulphuric acid:—

$$BaO,Br_2O_5 + H_2O,SO_3 = BaO,SO_3 + H_2O,Br_2O_5.$$

The insoluble sulphate of barium is separated by filtration. The acid solution can be concentrated to a certain extent by evaporation at a gentle heat, but cannot readily be brought to a syrupy consistency without decomposition. It decomposes, also, on being heated to 100°, and in general gives up oxygen on being brought into contact with substances which readily combine with that element.

132. *Hypobromous Acid*, H_2O,Br_2O.—There can be no doubt of the existence of an oxide of bromine corresponding to hypochlorous acid. When bromine is added to an excess of a solution of nitrate of silver, a liquid of strong bleaching-properties is

obtained, from which a yellowish, acid fluid may be distilled. This distillate bleaches strongly, and yields, on analysis, numbers corresponding with the above formula. When cold dilute alkaline solutions are mixed with bromine they acquire a power of bleaching, and in general behave like the alkaline hypochlorites, which are formed under similar conditions. So too when bromine-water is treated with red oxide of mercury; a sparingly soluble oxybromide of mercury is formed, together with a bleaching liquor supposed to contain hypobromous acid.

The analogies which subsist between chlorine and bromine are, however, everywhere so clearly defined that there is good reason to believe that other oxides of bromine, corresponding to those of chlorine, will be sooner or later discovered.

133. *Chloride of Bromine.*—Liquid bromine absorbs a large quantity of chlorine-gas when the two elements are brought together, and there is formed a very volatile liquid called chloride of bromine. It exhales a pungent, irritating odor, and is soluble in water; the solution possesses considerable bleaching-power.

134. *Bromide of Nitrogen* is an explosive compound analogous to chloride of nitrogen, from which it may be prepared by means of bromide of potassium.

CHAPTER X.

IODINE.

135. In its chemical properties iodine bears a striking resemblance to bromine, and consequently to chlorine also. It exists in sea-water and in the water of many saline and mineral springs. The proportion of iodine in sea-water is exceedingly small, being even smaller than that of bromine. But iodine is obtained more readily than bromine; for iodine is absorbed from sea-water by various marine plants, which, during their growth, collect and concentrate the minute quantities of iodine which the sea-water

contains, to such an extent that it can be extracted from them with profit.

Upon the coasts of Scotland, Ireland, and France, where labor is cheap, the sea-weeds which contain iodine are collected, dried, and burned, at a low heat, in shallow pits. The half-fused ashes thus obtained, called kelp or varec, contains, among other things, iodine in the form of iodide of sodium and iodide of potassium. It is lixiviated with boiling water, and the solution is then evaporated and set aside to crystallize. The iodides, being much more soluble than the other constituents of the ash, remain dissolved in the mother-liquor after most of the other salts have crystallized out. If this mother-liquor, or *iodine-lye*, be now mixed with a small quantity of sulphuric acid and left to itself for a day or two, it may be freed from a further portion of impurity; it is then transferred to a leaden retort, mixed with a suitable quantity of powdered black oxide of manganese, and gently heated. Iodine is set free, just as chlorine would be from chloride of sodium under similar circumstances, and may be collected in appropriate receivers:—

$$2NaI + 2(H_2O,SO_3) + MnO_2 = Na_2O,SO_3 + MnO,SO_3 + 2H_2O + 2I.$$

136. At the ordinary temperature iodine is a soft, heavy, crystalline solid of bluish-black color and metallic lustre. Its specific gravity is 4·948. It melts at a temperature (107°) a little above that at which water boils, and boils at a somewhat higher temperature (178°–180°); but in spite of this high boiling-point it evaporates rather freely at the ordinary temperature of the air, and the more rapidly when it is in a moist condition. It may be entirely volatilized by heating it upon writing-paper. Its odor is peculiar, somewhat resembling that of chlorine, but weaker, and easily distinguished from it. Its atomic weight is 127.

The vapor of iodine is of a magnificent purple color, whence the name iodine, derived from a Greek word signifying violet-colored; it is very heavy, indeed the heaviest of all known gases; it is nearly nine times as heavy as air. The specific gravity of the vapor is 127.

Exp. 68.—Hold a dry test-tube in the gas-lamp by means of the wooden nippers, and warm it along its entire length, in so far as this is practicable. Drop into the hot tube a small fragment of iodine, and observe the vapor as it rises in the tube. If only a small portion of the tube were heated, the vapor would be deposited as solid iodine upon the cold part of its walls.

137. Solid iodine is never met with in the amorphous, shapeless state in which glass, resin, coal and many other substances occur. No matter how obtained, its particles always exhibit a definite crystalline structure. If the iodine be melted and then allowed to cool, or if it be converted into vapor and this vapor be subsequently condensed, crystals will be formed in either case. Perfect crystals can be still more readily obtained by dissolving iodine in an aqueous solution of iodohydric acid, and exposing this solution to the air in a narrow-necked or loosely stoppered bottle; the iodohydric acid will be slowly decomposed by the action of the atmospheric oxygen, and, as it decomposes, well-defined crystals of iodine will be deposited.

In whichever way prepared, the crystals of iodine are octahedrons with a rhombic base, belonging to the crystalline system called trimetric. As commonly seen, the crystals are thin, flattened tables, distorted by excessive elongation in one direction.

138. Iodine is scarcely at all soluble in water, though enough dissolves tô impart a brown color to the water; but it dissolves readily in alcohol and ether. These solutions are much used in medicine, particularly the alcoholic solution, which is called tincture of iodine. When swallowed in the solid state, iodine acts as an energetic corrosive poison; but several of its compounds, and the element itself when taken in small doses, are highly prized as medicaments. It is also largely employed in photography, and is a useful reagent in the chemical laboratory.

139. As has been already stated, iodine, in its chemical behavior, resembles chlorine and bromine, only its affinities are more feeble. It enters into combination with less energy than either of these elements, and is displaced by them from most of its combinations. Like them, it unites directly with the metals and with several other elements. It gradually corrodes organic tissues, and destroys coloring-matters, though but slowly. No oxygen is given off from the aqueous solution when this is exposed to sunlight; but the color of the solution slowly disappears, and a mixture of iodohydric and iodic acids is formed in it.

A singular property of iodine is its power of forming a blue compound with starch.

Exp. 69.—Prepare a quantity of thin starch-paste by boiling 30 c. c. of water in a porcelain dish, and stirring into it 0·5 grm. of starch which has previously been reduced to the consistence of cream by rubbing it in a mortar with a few drops of water.

Pour 3 or 4 drops of the paste into 10 c. c. of water in a test-tube and shake the mixture so that the paste may be equably diffused through the water; then add a drop of an aqueous solution of iodine, and observe the beautiful blue color which the solution assumes. If the solution be heated the blue coloration will disappear, but it reappears when the liquid is allowed to cool.

Dip a strip of white paper in the starch-paste and suspend it, while still moist, in a large bottle, into the bottom of which two or three crystals of iodine have been thrown. As the vapor of iodine slowly diffuses through the air of the bottle it will at last come in contact with the starch, and after some minutes the paper will be colored blue.

This reaction furnishes a very delicate test for iodine. By its means it has been proved that iodine, though nowhere very abundant, is very widely distributed in nature; traces of it have been detected in land plants, and in many well, river, and spring waters, also in rain-water, and even in the air; indeed it would be difficult to say where iodine is not.

In order that it may be detected by this test, the iodine must be free or uncombined. But, as has been stated, chlorine readily expels iodine from most of its combinations. In case, then, we have reason to suspect the presence of a compound of iodine (iodide of potassium, for example) in any substance, a small quantity of chlorine-water, or of some other agent capable of expelling iodine, must be added to this substance. Once displaced from its combination, the iodine may be at once detected by means of starch.

Exp. 70.—Place in a test-tube 10 c. c. of water, a drop of concentrated aqueous solution of iodide of potassium, and 3 or 4 drops of the starch-paste of Exp. 69. If the iodide of potassium be pure, no coloration will occur. Add now 2 or 3 drops of chlorine-water, and shake the tube. The characteristic blue coloration at once appears.

In order to illustrate the extreme delicacy of this reaction, dissolve 0·14 grm. of iodide of potassium in 1 litre of water, and to this solution, which contains 1 part of iodine in 10,000 parts of water, add some of the starch-paste and several drops of red fuming nitric acid, a re-agent on some accounts better fitted than chlorine to disengage iodine in this experiment (see § 150). After a time the solution will exhibit the blue color, though in solutions so dilute as this it sometimes happens that the coloration appears only after the lapse of several hours.

It follows, of course, from the foregoing experiment, that the reaction of iodine upon starch can be used as a test for those substances which, like chlorine or nitric acid, are capable of setting free iodine, as well as for iodine itself. In the chemical laboratory it is customary to keep on hand for this purpose a store of paper upon which has been spread a mixture of starch-paste and iodide of potassium, prepared as follows:—

Exp. 71.—Dissolve 0·5 grm. of pure iodide of potassium (free from iodate) in 100 c. c. of water; boil this solution in a porcelain dish and stir into it 5 grms. of finely powdered starch, taking care not to burn the starch, and stirring until the mass gelatinizes. Remove the lamp, allow the paste to become cold, and by means of a wooden spatula spread it thinly upon one side of white glazed paper. The paper is then dried, cut into strips about 8 c.m. long by 2 wide, and preserved in stoppered bottles kept carefully closed.

Exp. 72.—Place in a test-tube a small quantity of binoxide of manganese, pour upon it 4 or 5 c. c. of chlorhydric acid, heat the mixture, and hold at the top of the tube a moistened strip of the test-paper which was prepared in the preceding experiment. The chlorine evolved by the reaction of the chlorhydric acid upon the binoxide of manganese sets free iodine from the iodide of potassium upon the test-paper, and the starch is thereby coloured blue. The presence of chlorine in chlorhydric acid is thus made apparent. By this test we might discriminate, for example, between dilute nitric and chlorhydric acids.

140. *Iodohydric Acid* (HI).—Hydrogen and iodine do not readily unite together directly. There is here nothing to recall the explosive violence with which chlorine and hydrogen combine. Sunlight has no power to bring about the union of the two elements at the ordinary temperature; but when a mixture of hydrogen gas and iodine vapor is passed through a red-hot tube, iodohydric acid is formed. It has been observed, also, that spongy platinum will cause the union of the two elements even at ordinary temperatures. Even when indirect methods are resorted

to, it is less easy to prepare iodohydric acid than chlorhydric or bromhydric acids.

If iodide of sodium be distilled with sulphuric acid, there will be obtained but little iodohydric acid; for most of that which is produced at first will be subsequently destroyed by the action of sulphuric acid, in the same way as happens to a less extent with bromhydric acid, § 128.

As fast as iodohydric acid is formed in accordance with the reaction

$$2NaI + H_2SO_4 = Na_2SO_4 + 2HI,$$

most of it is decomposed by another portion of sulphuric acid, in a manner which may be thus represented:—

$$2HI + H_2SO_4 = 2H_2O + SO_2 + 2I.$$

Solutions of iodohydric acid can, however, be readily obtained by the action of iodine upon a compound of sulphur and hydrogen, called sulphydric acid. In practice, a current of sulphydric acid gas is made to pass through water in which finely divided iodine is kept suspended by agitation. The sulphydric acid, the formula of which is H_2S, reacts upon 2I, and there is formed 2HI and free sulphur, which is deposited.

A solution of iodohydric acid may also be obtained by distilling a mixture of iodine, phosphorus, and much water, in which case the phosphorus unites with the oxygen of a portion of the water, while the iodine takes the hydrogen. Or it may be prepared by decomposing an aqueous solution of iodide of barium with an equivalent quantity of dilute sulphuric acid, and filtering off the solution from the insoluble sulphate of barium.

141. The dilute acid obtained by either of these methods can be concentrated, by evaporation, to a liquor of 1·7 specific gravity, boiling at 127°, and composed of one molecule of iodohydric acid united with 11 molecules of water. The aqueous solution has a sour, suffocating odor, and pungent acid taste. When concentrated it fumes strongly in the air. It cannot be long preserved when exposed to contact with the air, for the oxygen of the air unites with its hydrogen, and iodine is set free. At first this iodine dissolves in that portion of the iodohydric acid which has not yet been decomposed; but after the acid has become saturated, crystals of iodine are deposited, as has been stated in § 137. The decomposition of iodohydric acid is so rapid that the pure, colorless solution of it becomes red from separation of iodine after a few

hours' exposure to the air, no matter whether it be minute or concentrated. The easy decompositon of this acid shows clearly with how much less force hydrogen holds iodine in combination than it holds either chlorine or bromine.

142. The usual method of preparing anhydrous iodohydric acid is as follows :—

In the bottom of a test-tube place a mixture of 9 parts of iodine and 1 part of phosphorus. Cover the mixture with coarsely powdered glass, and bring about chemical union between the iodine and the phosphorus by gently heating them. Place now a few drops of water in the tube, and connect with it a gas delivery-tube by means of a caoutchouc stopper. Iodohydric acid will be immediately given off, and may be collected by displacement.

Another method is to pack a test-tube with alternate layers of phosphorus, iodine, and moistened glass-powder, and then to gently heat the tube. The operation depends upon the formation of an iodide of phosphorus and the subsequent decomposition of this body by contact with water into iodohydric acid and a compound of phosphorus, oxygen, and water, called hydrated phosphorous acid :—

$$2PI_3 + 6H_2O = 6HI + 3H_2O,P_2O_3.$$

143. Iodohydric acid is a colorless, acid gas, of suffocating odor; it fumes strongly in the air, and is very soluble in water. It can be liquefied rather easily by pressure, and solidified at $-51°$ to a colorless mass like ice. The gas is more than four times as heavy as air, its specific gravity having been found by observation to be 64·11. From this fact, taken in connexion with the striking analogy which the compound bears to bromhydric and chlorhydric acids, it follows that the gas is composed of equal volumes of iodine vapor and hydrogen united without condensation ; for the theoretical density of a gas thus composed would be $(127+1)\div 2=64$, a number with which the observed specific gravity closely agrees. The chemical effect of the small proportion of hydrogen contained in iodohydric acid is most remarkable. Only $\frac{1}{128}$, or less than 1 per cent. of iodohydric acid is hydrogen, yet this very small proportional quantity of hydrogen is competent to impart an entirely new set of properties, both to the iodine and the hydrogen; the acid bears no resemblance to either of its constituents.

144. Iodohydric acid is a compound which decomposes easily.

When a mixture of the gas and oxygen is passed through a red-hot tube, water and free iodine are the products. Chlorine and bromine abstract hydrogen from it, and leave iodine free; and the same effect is produced by many oxygen compounds which readily part with oxygen. With many of the metals it forms iodides, while hydrogen is set free; and it reacts upon most of the metallic oxides, forming water and a metallic iodide.

Though the hydrogen of iodohydric acid is readily removed by means of oxygen in numerous instances, it appears, upon the other hand, that iodine can abstract hydrogen from most of its combinations with the other elements. Only oxygen, chlorine, bromine, and an element (still to be studied) called fluorine, exhibit a stronger tendency than it to unite with hydrogen. Iodine separates hydrogen from its compounds with nitrogen, sulphur, and phosphorus, and from many organic compounds, such as alcohol and ether, iodohydric acid being formed in each case.

145. *Compounds of Iodine and Oxygen.*—Of the compounds of iodine and oxygen, only two have as yet been carefully studied. These correspond respectively to chloric and perchloric acids. Compounds analogous to hypochlorous and hypochloric acids appear to exist, but have not been described with much accuracy.

146. *Iodic acid* (I_2O_5) may be obtained directly by oxidizing powdered iodine with monohydrated nitric acid at a moderate heat. After all the iodine has disappeared, and the excess of nitric acid employed has been evaporated, iodic acid will be left as a white residue.

Iodic acid is readily soluble in water, and crystallizes from an acidulated solution in colorless, six-sided tables, of the formula HIO_3 or H_2O,I_2O_5. It has a peculiar odor, and acid, disagreeable taste. At the temperature of 170°, water is given off and the anhydrous acid remains. This melts upon being heated more strongly, and suffers decomposition.

Iodic acid readily gives up oxygen to many other substances, or, in other words, it is easily decomposed by reducing-agents; for example, when mixed with iodohydric acid it reacts upon it with formation of water and deposition of iodine:—

$$10HI + I_2O_5 = 5H_2O + 12I.$$

All of the metals are oxidized by it, excepting gold and platinum.

With metallic oxides it forms compounds called iodates, which are analogous to the corresponding chlorates and bromates in composition and properties.

147. *Periodic acid* (I_2O_7) may be prepared by passing chlorine gas through a solution of iodate of sodium mixed with caustic soda. Chloride of sodium and basic periodate of sodium will be formed, and the latter, being sparingly soluble in water, will be deposited in crystals :—

$$Na_2O,I_2O_5 + 3(Na_2O,H_2O) + 4Cl = 2Na_2O,I_2O_7 + 4NaCl + 3H_2O.$$

If now the sodium salt be collected and dissolved in water, and the solution be mixed with nitrate of lead, a periodate of lead will be obtained; this may be decomposed by means of dilute sulphuric acid into periodic acid and insoluble sulphate of lead. The latter may then be separated by filtration, and the clear solution of the acid finally concentrated by evaporation.

From the concentrated aqueous solution periodic acid separates in colorless hydrated crystals, which, upon being carefully heated, give off water and yield as a residue the anhydrous acid I_2O_7. At a still higher temperature, the anhydrous acid decomposes and gives off oxygen. It is decomposed also by reducing-agents in the same way as iodic acid.

The other compounds of iodine and oxygen have but little interest for us, except that they serve to increase the number of analogies which subsist between iodine, bromine, and chlorine.

148. *Iodide of Nitrogen* (?).—There appear to be a number of compounds which have hitherto been usually classed under this title. They are produced by the action of ammonia upon iodine, and are mostly of a highly explosive character, though their properties and composition vary to a certain extent according to the mode of their preparation.

Exp. 73.—Place 0·25 grm. of finely powdered iodine in a porcelain capsule, and pour upon it so much concentrated ammonia-water that the iodine shall be somewhat more than covered; allow the mixture to stand during 15 or 20 minutes, when an insoluble dark-brown powder will be found at the bottom of the liquid. This powder is the so-called iodide of nitrogen. It should be collected upon two or three very small filters and well washed with cold water. Remove the filters, together with their contents, from the funnels, pin them upon bits of board, and leave them to dry spontaneously.

As soon as the powder has become thoroughly dry it will explode upon being rubbed, even with a feather, or jarred, as by the shutting of a door, or by a blow upon the wall or table. Though incomparably less dangerous than chloride of nitrogen, and therefore better suited than the chloride to illustrate the explosive character of this obscure class of nitrogen compounds, iodide of nitrogen must nevertheless be handled with great care, and should never be prepared by the student except in very small quantities.

149. *Chlorides of Iodine.*—Iodine combines directly with chlorine in several proportions, a protochloride, ICl, and a terchloride, ICl_3, being the best-known of these compounds.

The protochloride is obtained by passing dry chlorine over dry iodine, the current of chlorine being checked at the moment when all the iodine has become liquid. Or it may be made by distilling iodine with chlorate of potassium, and collecting the product in a cooled receiver.

$$3KClO_3 + 2I = KClO_4 + KIO_3 + KCl + 2O + ICl.$$

Protochloride of iodine is a reddish-brown, oily liquid, volatile, irritating, and of penetrating odor. It decolorizes litmus and indigo, but does not give a blue color with starch.

The terchloride may be produced by treating iodine with an excess of chlorine gas, or by acting upon anhydrous iodic acid with dry chlorhydric acid gas:—

$$I_2O_5 + 10HCl = 2ICl_3 + 5H_2O + 4Cl.$$

It is a yellow crystalline solid, melting at 20°–25°. It acts upon other substances in the same manner as the protochloride; like the protochloride, it decolorizes indigo and does not turn starch blue.

150. A knowledge of the properties of the chlorides of iodine is of some practical importance, since they are liable to be formed incidentally in several chemical processes, which their presence perturbs. Thus, in the manufacture of iodine, as described under § 135, the iodine-lye almost always contains a certain proportion of chloride of sodium. It is evident that if the chlorine in this compound were to be evolved at the same time as the iodine by the action of the black oxide of manganese and sulphuric acid, there would be formed a quantity of the very volatile protochloride of iodine, which would escape condensation. Whatever of iodine was thus combined with chlorine would be lost to the manufacturer. But, as has been repeatedly stated, iodine is an element which

can be much more readily expelled from its combinations than chlorine; and in the case in point it is found that the iodine in the mixture of iodide of sodium and chloride of sodium, which the iodine-lye contains, will all come off before the chlorine, if the distillation be slowly conducted. If, through irregular heating, any portion of the contents of the retort should become hotter than the rest, and so lose all its iodine, chlorine would be disengaged from that portion, and would unite with the vaporized iodine which fills the retort. To ensure the necessary slow and equable heat, the retort is set upon a stove suitable for the maintenance of a slow fire, and is provided with an agitator, by means of which its contents may be continually stirred.

Again, in testing for iodine, as in Exp. 70, chlorine is a far less convenient agent for setting free the iodine from its combinations than fuming nitric acid; for if the slightest excess of chlorine be employed, the iodine will all be converted into chloride of iodine, and the starch will not be colored blue.

151. *Bromides of Iodine.*—There are two compounds of bromine and iodine, and their properties are analogous to those of the chlorides of iodine.

152. Chlorine, bromine, and iodine constitute one of the most remarkable and best-defined natural groups of elements. Whether we regard the uncombined elements or their compounds, it is impossible not to be struck with the close analogies which subsist between them. With hydrogen, all of these elements unite in the proportion of one volume to one volume, without condensation, to form acid compounds extremely soluble in water and possessing throughout analogous properties.

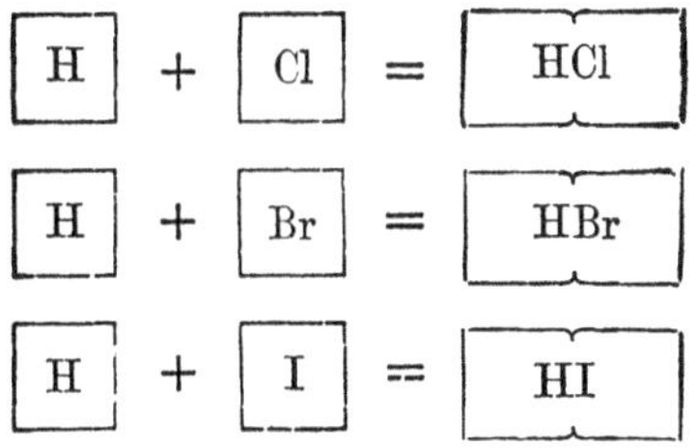

$$\boxed{H} + \boxed{Cl} = \boxed{\quad HCl \quad}$$

$$\boxed{H} + \boxed{Br} = \boxed{\quad HBr \quad}$$

$$\boxed{H} + \boxed{I} = \boxed{\quad HI \quad}$$

With oxygen each of them forms a powerful acid containing five atoms of oxygen, besides divers other compounds of obvious likeness. The compounds furnished by their union with any one metal are always isomorphous (like-formed); the chloride, bromide, and iodide of potassium, for example, all crystallize in cubes. With nitrogen they all form explosive compounds. Many similar analogies will be made manifest as we proceed to study the other elements, and their compounds with this chlorine group.

There is a distinct family resemblance between these three elements as regards their physical as well as their chemical characteristics; but, in all their properties, a distinct progression is observable from chlorine through bromine to iodine. At the ordinary temperature chlorine is a gas, bromine a liquid, and iodine a solid, though at temperatures not widely apart they are all known in the gaseous and liquid states. The specific gravity of bromine vapor is greater than that of chlorine, and that of iodine greater than that of bromine. Chlorine gas is yellow, the vapor of bromine is reddish brown, that of iodine violet. So with all their other properties,—chlorine will be at one end of the scale, iodine at the other, while bromine invariably occupies the intermediate position.

The properties of many of the compounds of chlorine, bromine, and iodine exhibit a similar progression as we pass from the chlorine compounds to those of iodine. For example, the specific gravity of

Chlorhydric acid gas is	18·2
Bromhydric "	40·5
Iodohydric "	64·0

Chlorhydric acid can be liquefied at about $-80°$, and has not yet been solidified. Bromhydric acid liquefies at about $-60°$, and solidifies at about $-92°$. Iodohydric acid liquefies at about $-40°$, and solidifies at about $-50°$.

Chlorhydric acid is a more energetic acid than bromhydric, and bromhydric acid is more powerful than iodohydric. The aqueous solution of chlorhydric acid can be kept without change in contact with air; that of bromhydric acid becomes colored after a while, from separation of bromine; but the solution of iodohydric acid decomposes rapidly, and much iodine is deposited.

As regards the relative chemical power of these elements, it has already been shown that the intensity of this force becomes less as we descend from chlorine to iodine. It is easy, for example, to displace iodine from its compounds by means of bromine,

$$NaI + Br = NaBr + I,$$

and equally easy to displace bromine from its combinations by means of chlorine,

$$NaBr + Cl = NaCl + Br.$$

153. It is an important principle, borne out by most of the other groups of elements, and emphatically true of the natural family now under consideration, that, with kindred elements, the chemical power of each is great, in comparison with that of the related elements, in proportion as its atomic weight is low.

Among the members of a natural chemical group, chemical energy seems to be inversely proportional to atomic weight. Thus the atomic weight of chlorine is 35·5, that of bromine 80, and that of iodine 127, while the chemical energy of these elements follows the opposite order.

154. It is noteworthy that elements of like character almost always occur associated with one another in nature. Bromine and iodine are always found in company with chlorine. That this should be so is in nowise surprising. Those elements which are similar in character and properties must necessarily be similarly acted upon by the natural forces to which they are exposed, and must therefore inevitably tend to be gathered or deposited in like places under like conditions.

CHAPTER XI.

FLUORINE.

155. There is another substance, called fluorine, which is closely analogous to chlorine. This element cannot be readily obtained in the free state, and scarcely anything is known of it in that condition. Special interest attaches to it upon this very account,

and many fruitless efforts to isolate it have been made. Of all the elements, it appears to have the strongest tendency to enter into chemical combination; at all events it is the most difficult to obtain, and to keep, in the free and uncombined condition.

It is not only difficult to expel fluorine from the minerals in which it is found in nature, but on being set free from one compound it immediately attacks whatever substance is nearest at hand, and so enters into a new combination. Hence it is wellnigh impossible to collect it. It destroys at once glass, porcelain, and metal, the materials from which chemical apparatus is usually constructed. Vessels made of the mineral fluor-spar (a compound of fluorine and calcium), are the only ones which have as yet been found capable of withstanding its action. By operating in such vessels, a small quantity of impure fluorine gas appears to have been really obtained; but the process is difficult, expensive, and not uniformly successful. Little or no doubt, however, is entertained as to the general nature of fluorine, since its compounds are closely analogous in many respects to the corresponding compounds of chlorine, bromine, and iodine.

The symbol of fluorine is Fl. Its atomic weight is 19. It occurs tolerably abundantly in nature as fluoride of calcium ($CaFl_2$), in the mineral known as fluor-spar. Small quantities of fluorine are found also in several other minerals, in vegetable and animal substances, particularly in bones; and traces of it occur in sea-water, and in various rocks and soils. It appears to be almost as widely disseminated as iodine, though, from the lack of delicate tests for fluorine, it is far less readily detected. Of late years a considerable mine of a fluorine mineral called cryolite (fluoride of sodium and aluminum) has been worked in Greenland.

156. *Fluorhydric Acid* (HFl).—With hydrogen, fluorine forms a powerful acid corresponding to chlorhydric acid and the other hydrides of the chlorine group. It is a more energetic acid than either of these, but is specially characterized by its corrosive action upon glass. It may be readily prepared by distilling powdered fluor-spar with strong sulphuric acid; the reaction being analogous to that which occurs when common salt is treated with sulphuric acid:—

$$CaFl_2 + H_2SO_4 = CaSO_4 + 2HFl.$$

Since the acid rapidly corrodes glass, the process must be conducted in metallic vessels. Ordinarily, retorts of lead or platinum are employed, and the distillate is collected in receivers made of the same metals, and carefully cooled by means of ice.

157. The product of the distillation is a very volatile, colorless liquid, a little heavier than water. It is strongly acid, emits copious white and highly suffocating fumes in the air, boils at 15°, and remains unfrozen at —20°. On account of its corrosive power, this substance is highly dangerous; if any of it happens to come in contact with the skin, wounds are produced which are very difficult to heal; a single drop of it is sufficient to occasion a deep and painful sore. In preparing the acid, special provision must be made for carrying away from the operator any fumes which may escape condensation.

The acid may be kept in bottles made of lead or silver, or of gutta percha, substances upon which it has no action. It unites with water with great avidity, so much heat being evolved that a hissing noise is produced, as if a bar of red-hot iron had been immersed in the water. In its concentrated form the acid has a specific gravity of 1·061, but on the addition of a certain amount of water the density increases to 1·15, a definite hydrate ($HFl + 2H_2O$) being formed, which boils at 120°, and may be distilled unchanged. The further addition of water to this hydrate is attended with a regular decrease in density.

According to some chemists, the liquid acid obtained as above described is not anhydrous. It is asserted that if it be distilled with an excess of anhydrous phosphoric acid (a substance which has a very strong affinity for water), the anhydride will be set free in the form of a colorless, extremely irritating gas.

158. Upon metals and metallic oxides, fluorhydric acid acts like chlorhydric acid, only more powerfully; but its most striking peculiarity is its action upon silica and the compounds of silica, such as glass or porcelain. If a drop of the concentrated acid be allowed to fall upon a piece of glass, it becomes hot, boils, and partially distils off as a fluoride of silicon, while the glass is corroded and becomes covered with a white powder consisting of compounds of fluorine and various constituents of the glass. If this

powder be washed away a deep impression will be found upon the glass at the point where the acid has acted.

This corrosive power, which is possessed by fluorhydric acid gas as well as its aqueous solution, is made use of for etching glass. The graduations on the glass stems of thermometers and eudiometers may thus be made with great precision and facility; the acid is largely employed also in ornamenting glass with etched patterns.

Exp. 74.—Warm a slip of glass and rub it with beeswax so that it shall be everywhere covered with a thin, uniform layer of the wax. With a needle, or other pointed instrument, write a name, or trace any outline through the wax, so as to expose a portion of the glass. Lay the etching, face downward, upon a bowl or trough of sheet-lead, in which has been placed a teaspoonful of powdered fluor-spar and enough strong sulphuric acid to convert it into a thin paste; if the glass be smaller than the opening of the dish, it may be supported upon wires laid across the latter.

Cover the glass and the top of the dish with a sheet of paper, and then gently heat the leaden vessel for a few moments, taking care not to melt the wax; then set the dish aside in a warm place and leave it at rest during an hour or two. Finally melt the wax and wipe it off the glass with a towel or bit of paper; the glass will be found to be etched and corroded at the places where it was laid bare by the removal of the wax.

This experiment can be performed more rapidly by covering the outside of a watch-glass with wax, tracing characters upon this layer, and then placing the glass upon a small platinum crucible containing a mixture of fluor-spar and sulphuric acid, which is heated over the gas-lamp. The watch-glass is meanwhile kept full of water, in order to prevent the wax from melting. In this way the etching can be effected in the course of a few minutes.

Instead of the gas, a dilute aqueous solution of the acid may be employed in this experiment. The concentrated acid of § 157, diluted with six parts of water, answers a good purpose. In this case the etched surface will appear smooth like the rest of the glass, while in case the gas is employed the etched portion of the glass will be dull and rough.

159. No compounds of fluorine with chlorine, bromine, iodine, nitrogen, or oxygen have yet been discovered, though a sulphur compound has been obtained, as a fuming liquid, by distilling fluoride of lead with sulphur. Fluorine is the only element of

which no oxygen compound is known; this fact, however, will appear less remarkable if it be remembered that, in order to obtain oxygen compounds of chlorine, bromine, and iodine, it is necessary first to isolate these elements, and to have them in the free and uncombined condition. Analogy would therefore teach that a practicable method of preparing free fluorine must be discovered before we can hope to prepare oxides of fluorine.

160. The fact that fluorine forms a powerful acid with hydrogen, connects this element with the three elements (chlorine, bromine, and iodine) which have last been studied. Many of its compounds with the metals are analogous in composition to the compounds of chlorine, bromine, and iodine, and not a few of these compounds are isomorphous with one another. It is customary therefore to study fluorine in connexion with the chlorine group; but the student should remember that in several respects it differs widely from chlorine, and that its connexion therewith is, in any event, less intimate than that of either bromine or iodine.

CHAPTER XII.

OZONE AND ANTOZONE.

161. Besides ordinary oxygen, such as is found in the air and has been prepared in Exps. 5 and 7, two other kinds or forms of this element are known to chemists. These new modifications of oxygen have received special names, and are called ozone and antozone respectively.

162. Several other *elements*, notably sulphur, phosphorus, and carbon, occur, as oxygen does, in very unlike states, or with very different attributes, while the fundamental chemical identity of the substance is preserved. The word *allotropism* is employed to express this capability of some of the elements; it is derived from Greek words signifying *of a different habit, or character.* This word serves merely to bring into one category a considerable number of conspicuous facts, of whose essential nature we have

no knowledge; there is, of course, no virtue in the word itself to explain or account for the phenomena to which it refers.

163. *Ozone* is an exceedingly energetic chemical agent, which resembles chlorine in some respects; it can therefore be advantageously studied in connexion with the chlorine group. Moreover, since ozone and antozone were for a long time confounded with one another, and since they are really intimately related, they should, of course, be studied together. The most natural connexion of these somewhat obscure bodies is with oxygen; but we are better able to appreciate what is known of the properties of ozone and antozone now that we have become acquainted with a number of the elements, and have made ourselves familiar with a considerable variety of chemical processes and reactions, than we were at the very outset, when common oxygen was necessarily studied.

164. It had long been noticed that when an electrical machine was put in operation a peculiar, pungent odor was developed; but it is only at a comparatively recent period that it has been observed that the same odor is manifested during the electrolysis of water (§ 35), and that this odor resembles that evolved by moistened phosphorus when exposed to the air. It has gradually been made out that the odor in each of these cases is due to the presence of a peculiar modification of oxygen, called *ozone* from a Greek word signifying to smell. This modification of oxygen was at one time erroneously supposed by some to be a high oxide of hydrogen, of composition H_2O_2, or H_2O_3; but this view has lately been completely disproved.

Of the methods of obtaining ozone above suggested, that by phosphorus will usually be found most convenient.

Exp. 75.—In a clean bottle, of 1 or 2 litres capacity, place a piece of phosphorus 2 or 3 c.m. long, the surface of which has been scraped clean (under water) with a knife; pour water into the bottle until the phosphorus is half covered; close the bottle with a loose stopper, and set it aside in a place where the temperature is 20° or 30°.

In the course of ten or fifteen minutes a column of fog will be seen to rise from that portion of the phosphorus which projects above the water, the original garlic odor of the phosphorus will soon be lost, and the peculiar odor of ozone will gradually pervade the bottle. After five or six hours, the bottle will be found to contain an abundance of ozone for use in the subsequent experiments.

The chemical changes which occur during this experiment are complicated; it will be enough to say of them that the phosphorus unites with oxygen from the air in the bottle to form an oxide of phosphorus, which will be studied hereafter under the name of phosphorous acid; that during this process of oxidation a portion of the oxygen in the bottle is changed into ozone and antozone, and that some of the ozone remains, even after many hours, diffused in the air of the bottle.

165. It must be distinctly understood that no very large quantity of ozone is obtained in the foregoing experiment. At the best, only a very minute proportion of it will be found in the air of the bottle. But ozone is a substance possessing great chemical power, and but little of it is needed in order to exhibit its characteristic properties.

If it be desired to prepare ozone by passing electric discharges through air or oxygen, either of these gases may be sealed up in narrow glass tubes, through the centres of which are passed platinum wires, welded tightly into the glass, as shown in Fig. 37, and a series of sparks from an electrical machine is thrown through the gas in the tube, during ten or twelve hours. If the experiment be continued longer than this, nothing is gained; for the sparks after this time appear to destroy the ozone previously produced.

To avoid the difficulty last named, a slow current of oxygen may be forced through a tube open at both ends, and electrical discharges may be passed through the gas in its transit; a constant stream of ozonized air will be thus obtained.

Instead of the sparks, the gas within the tube may be subjected to silent discharges of electricity obtained by connecting one of the platinum wires with the ground, the other with the prime conductor of an electrical machine, and slowly turning the crank of the latter. By using a tube having wires near the top, as in Fig. 37, and closing the lower end of the tube by immersing it in a bath filled with an aqueous solution of iodide of potassium, so that the ozone may be absorbed as fast as it is formed, it has been found possible, by some experimenters, to transform and remove all the original oxygen contained in the tube.

Fig. 37.

166. Ozone is produced not only during the slow oxidation of

phosphorus, and by the action of electricity upon air or oxygen; a certain quantity of it appears to be produced also during other processes of oxidation. It is readily formed, for example, during the slow combustion of ether and of various other volatile liquids; it can be at once produced by plunging a heated glass rod or iron wire into a mixture of air and ether vapor.

Into a wide-mouthed bottle, a small quantity of ether is poured; the bottle is shaken for a moment, that the air within it may become charged with the vapor of ether; the liquid ether, if any remain, is then poured away, and a large glass rod, or thick iron wire, heated to about 250°, is thrust into the bottle. The rod must not be too hot, lest the ozone formed be reconverted into ordinary oxygen; if it be insufficiently heated no ozone is produced.

During the slow oxidation of oil of turpentine, oil of cinnamon, oil of lemons, and others of the so-called essential oils, at the ordinary temperature of the air a considerable quantity of ozone is produced. This may be seen in oil of turpentine which has been kept for a long time in half-filled bottles, exposed to sunlight, and frequently opened and shaken. The formation of ozone under these circumstances explains the familiar fact that the corks employed to close bottles containing oil of turpentine and the analogous oils are soon bleached and corroded. At the same time, antozone is also produced in large quantity, as will be explained hereafter.

If quicksilver, to which a little water and a few drops of a solution of indigo have been added, be shaken up violently in a large bottle full of air, the indigo will soon be bleached as if by the action of ozone.

167. One of the best methods of preparing ozone is by treating a compound known as permanganate of potassium with sulphuric acid. It should be observed, however, that in this process, as in all the others, the ozone obtained is mixed with common oxygen; no available method of isolating ozone in a condition of purity has yet been made known.

A small quantity of concentrated sulphuric acid is placed in the bottom of a bottle, and a quantity of pure, dry permanganate of potassium, in fine powder, is added; the proportion of acid to permanganate should be three parts to two, by weight. A strong smell of ozone will be at once perceived, and the pasty mass will continue to give off ozone for a long time.

In this case it is conjectured that a portion of the oxygen of the permanganate of potassium, the empirical formula of which is $K_2Mn_2O_8$.

actually exists in the compound as ozone, and is given off as such when the compound is decomposed.

168. As has been already mentioned, the chemical behavior of ozone is analogous to that of chlorine; it bleaches and destroys vegetable coloring-matters, and is a powerful disinfectant. Like chlorine it instantly decomposes the iodides of the metals; upon this property is based a ready method of testing for its presence.

Exp. 76.—Into the bottle of ozonized air (Exp. 75), thrust a moistened slip of the test-paper, saturated with starch and iodide of potassium, which was prepared in Exp. 71; the paper will instantly acquire a deep blue tint. As in the case where the test-paper was employed for detecting chlorine (Exp. 72), so here, the reaction depends upon the displacement of the chemically feeble iodine by the more powerful ozone:—

$$2KI + O = K_2O + 2I.$$

The ozone here acts as oxygen, in one sense; at all events the oxide of potassium formed is not to be distinguished from oxide of potassium prepared with common oxygen; but this in nowise contradicts the fact that ozone is an extraordinarily active and energetic variety of oxygen, inasmuch as common oxygen will not effect this decomposition.

169. Ozone is an irritating, poisonous gas; air which is highly charged with it is irrespirable, and produces effects on the human subject similar to those produced by chlorine. Its odor, which has been compared to that of weak chlorine, is so powerful that it can be recognized in air containing only one millionth part of the gas. Its oxidizing power is intense. When moisture is present it oxidizes all the metals excepting gold, platinum, and the platinum metals; even silver is oxidized by it at the ordinary temperature, and becomes covered with a brown coating of peroxide of silver. It destroys many hydrogen compounds, such as those of sulphur, phosphorus, and iodine, the hydrogen being oxidized as well as the element with which the hydrogen is associated; iodohydric acid, for instance, is converted into water and iodic acid. In the same way, free iodine is oxidized by ozone, and if test-paper which has become blue by exposure to ozone, as in Exp. 76, be left long in ozonized air, it will become white from oxidation of the iodine. Ozone will even oxidize nitrogen, at the ordinary temperature, when in contact with water and such alkaline oxides as caustic soda, caustic potash,

or caustic lime; thus, if lime-water (a solution of caustic lime in water) be left exposed to ozonized air, a certain quantity of nitrate of calcium will be formed. Ammonia is oxidized by it also, and it converts nitrous and sulphurous into nitric and sulphuric acids.

Many salts of the metals are oxidized by it—for example, the sulphates of iron and of manganese. A valuable test for the presence of ozone is furnished by its behavior towards sulphate of manganese.

Exp. 77.—Dissolve a gramme or two of sulphate of manganese in water; soak in this solution strips of thin white blotting-paper; dry the paper, and preserve it in a bottle. If a slip of this paper be moistened, and then hung in ozonized air (Exp. 75), it will quickly become brown from the formation upon it of black oxide of manganese.

In like manner, most organic substances are quickly oxidized by ozone; when substances such as sawdust, garden-mould, powdered charcoal, milk, or flesh are thrown into a bottle of ozonized air, the odor of ozone instantly disappears; corks and caoutchouc tubes are attacked by it, and must not be used in experimenting with the gas. It destroys the color of indigo, and bleaches litmus without first reddening it. Some organic bodies, on the other hand, become colored when exposed to its action; thus, the cut surface of an apple becomes brown, and fresh surfaces of certain mushrooms become blue. Gum guaiacum also becomes blue. Papers soaked in a dilute alcoholic solution of gum guaiacum, indeed, are often employed as a test for ozone.

Exp. 78.—Dissolve one part of gum guaiacum in thirty parts of ninety per cent. alcohol; add a few drops of this solution to 2 c. c. of ordinary eighty per cent. alcohol; dip in this dilute solution strips of thin white blotting-paper, and dry them in the dark. By exposure to ozonized air this test-paper acquires a bright blue color.

170. By virtue of its strong oxidizing-power, ozone is of great importance as a disinfecting agent. It destroys instantly a multitude of offensive gases, such as arise from decaying animal and vegetable matter, and has been frequently recommended of late as a substance well fitted for the purification of sick-rooms and hospital-wards. Where ozone is employed for purposes of disinfection, it must be borne in mind that the action of the gas

depends solely upon oxidation. A given quantity of ozone can destroy only a certain definite amount of the offensive organic matter; wherever these emanations are incessantly generated, ozone must be as constantly produced in order to destroy them. This disinfecting-power of ozone is interesting in connexion with the observed facts, that ozone is abundant in the air of pine forests, where turpentine abounds, and that pine forests are, as a general rule, remarkably free from malaria. The well-known disinfecting-power of tar is supposed in like manner to be partly due to the formation of ozone during the oxidation of some of its ingredients.

Coal-tar, mixed with plaster-of-Paris, coal-ashes, or dry earth, in quantity sufficient to destroy its stickiness, has been found to be a very efficient disinfectant. The dry powder obtained as above, is simply scattered freely about the offensive locality. The coal-tar, of course, evolves a slight odor, peculiar to itself, which tends to mask or conceal other odors, and also acts as an antiseptic, or arrester of putrefaction; but its chief merit does not appear to depend upon either of these properties; it seems really to destroy the gases which are evolved from putrescent matter, and probably does so by generating ozone.

171. It is supposed that a minute proportion of ozone exists in normal atmospheric air: at all events, there is usually present in air a substance which exhibits the various reactions of ozone, and behaves as ozone would if it were there. This atmospheric ozone, which is supposed to be formed in the processes of oxidation which are always going on in nature, varies in quantity with the locality, the season of the year, the hour of the day, and many other circumstances.

172. Ozone is seldom found in the air of thickly inhabited localities; it often happens that it cannot be detected in the air of cities at the very time when it is abundant in the neighboring country. It is often found to be abundant on the windward side of a city, and altogether absent from the air upon the leeward side, the inference being that it is destroyed by the exhalations which arise from a dense population. Ozone appears to be more abundant in the air in winter than in summer, in cloudy than in clear weather, and by night than by day; it has been observed to be specially abundant at times when dew was falling heavily. As might be expected, comparatively large quantities of it are found

during thunder-storms, and its odor has been recognized in the neighborhood of objects struck by lightning. Ozone is abundant during snow-storms; and it is probable that upon its presence depends the well-known bleaching-power of newly fallen snow.

In searching for ozone in the air, test-paper containing iodide of potassium and starch, such as was prepared in Exp. 71, is usually employed. Dry slips of the prepared paper are exposed, during from six to twenty-four hours, to a free current of air, in a place well sheltered from light and rain. By exposure the dry paper becomes brown, and when wetted acquires shades of color varying from pinkish-white and iron-gray to blue. The shade of color obtained in this way is then compared with a standard chromatic scale, which includes all the shades possible under the circumstances; and the proportion of ozone present in the air is thus roughly estimated.

Although observations of this kind are far from possessing that degree of accuracy and certainty which is desirable, they have nevertheless been considered trustworthy by numerous observers, and have given rise to much speculation concerning the functions of atmospheric ozone, more particularly with regard to its probable influence upon health and disease. If there be ozone in the atmosphere, it will, on the one hand, oxidize and destroy many volatile organic substances which are supposed to be prejudicial to health. Hence many physicians are of opinion that the atmospheric ozone plays an important part in controlling or preventing epidemic diseases through its power of removing infectious matter from the air; and it has been noticed that with the advent of an ozone-bearing wind such diseases have abated or ceased. But, on the other hand, ozone is a highly irritating gas, and in the opinion of some physicians occasions many diseases of the respiratory organs. Numerous statements are upon record to the effect that epidemics of catarrh, colds, sore throat, and influenza have been coincident with the beginning of a spell of ozoniferous wind.

173. Ozone is usually considered to be completely insoluble in water; but it has been recently ascertained that water can take up a small quantity of it, and so acquire some of the properties of ozone. When ozonized air is passed through a solution of caustic soda or caustic potash, a certain amount of ozone is absorbed at first, perhaps by combination with some oxidizable impurity of the solution, but after a little time the ozone will pass through without apparent alteration. Acids do not absorb ozone. It is readily absorbed, however, by aqueous solutions of iodide of potas-

sium and of pyrogallic acid, with the constituents of which it enters into combinations not to be distinguished from those made with oxygen.

174. At moderately high temperatures ozone loses its peculiarities and changes into ordinary oxygen; if ozonized air, such as was obtained in Exp. 75, is made to pass through a narrow glass tube heated to 250°, its peculiar odor, and its power of decomposing iodide of potassium will entirely disappear. The same change occurs gradually if the tube is heated only to 100°,—or instantly if steam be thrown into the ozonized air, so that the whole of it can be heated at once to 100°; hence it may be stated, in general terms, that ozone is converted into ordinary oxygen at temperatures greater than 100°.

175. Ozone is supposed to exist as such in several of the oxides. Black oxide of manganese, for example, is thought to contain it as a constituent; and a method of obtaining it from permanganic acid has been already given, § 167. The oxygen compounds which are supposed to contain ozone are called ozonides. The formulæ of the following compounds, recognized as ozonides, are here given for the sake of reference:—

$$PbO_2 \quad CrO_3 \quad MnO_2 \quad Co_2O_3 \quad N_2O_5$$
$$Ag_2O_2 \quad MnO_3 \quad Mn_2O_7 \quad Ni_2O_3 \quad Bi_2O_5$$

176. *Antozone* (the opponent or opposite of ozone) appears to be produced simultaneously with ozone whenever the latter is formed, whether by electrical action or during processes of oxidation. It may even be that, as some chemists believe, ordinary oxygen is in a certain sense a compound substance, and that when in contact with phosphorus, and in the other circumstances under which ozone is produced, the neutral oxygen is split or decomposed into two opposite and dissimilar modifications—we had almost said elements—one of which is ozone, the other antozone. It is thought that while the greater part of the ozone thus engendered enters into combination with the phosphorus, or other substance, undergoing oxidation, a certain portion of it, together with some of the antozone, becomes mixed with the surrounding air, and so escapes combining with the body which is being oxidized.

Only a comparatively short time has elapsed since antozone

has been recognized as a distinct substance; hence its properties have been less thoroughly studied than those of ozone. Many of its characteristics and properties are still involved in great obscurity, very various and even conflicting statements having been published concerning them.

177. Of the methods devised for preparing antozone, the following deserve notice:—

By passing dry electrized air (§ 165) through a concentrated aqueous solution of iodide of potassium, or of pyrogallic acid, all the ozone contained in the air will be at once absorbed, and the antozone left behind, free from any admixture of ozone.

During the slow oxidation of oil of turpentine and other volatile or essential oils (§ 166), a considerable quantity of antozone is produced, as well as of ozone. While most of the ozone at once combines with the constituents of the oil, to form resins and other products of oxidation, the antozone, which does not oxidize the oil, is dissolved by it. In what state the antozone exists within the oil is still uncertain; but it is, in any event, very loosely held, and is readily given up to other substances.

In the same way that ozone may be prepared, by chemical decomposition, from permanganate of potassium, a compound supposed to contain ozone (§ 167), antozone may be obtained by decomposing certain compounds which are believed to contain this variety of oxygen—such, for example, as peroxide of barium, BaO_2. A little concentrated sulphuric acid is poured into a small bottle, and into this acid are thrown a number of small fragments of peroxide of barium (free from any admixture of nitrate of barium); so soon as an evolution of gas ensues, the air of the bottle will be found charged with antozone. This reaction is sometimes capricious. Usually it occurs at the ordinary temperature of the air; but it is often necessary to place the bottle in a water-bath heated to 50° or 60°, in order to start the evolution of gas; and, on the other hand, the violence of the reaction must sometimes be allayed by immersing the bottle in cold water.

In the preparation of ozone by means of phosphorus in moist air (Exp. 75), or by the electrolysis of water (§ 35), the antozone which is formed at the same time with the ozone, unites with the water present, and must there be sought. (See § 181.)

Antozone has been found in nature in a dark-blue variety of fluor-spar from Wölsendorf, in Bavaria. Upon being rubbed, this mineral emits a peculiar odor, which was formerly thought to be that of chlorine or of hypochlorous acid. More recent investigations have shown that the odor is that of antozone, and that by grinding the mineral with water the antozone can be transferred to the water.

178. Antozone is a gas, the odor of which somewhat resembles that of ozone; there is, however, a decided difference between the two odors, that of antozone being disgusting, while that of ozone is merely pungent and irritating. Antozone changes at once to ordinary oxygen on being heated. Even at the ordinary temperature it reverts to common oxygen very readily—much more readily than ozone. Most of the antozone usually disappears from dry electrized air in the course of an hour, or an hour and a half; and if the air be moist, the change is still more rapid. Ozone, on the cóntrary, is comparatively permanent, under the same conditions; and although when a mixture of ozone and antozone is left in contact with water in a glass-stoppered bottle, some ozone is destroyed during the reversion of the antozone, the larger portion of it will remain almost, if not quite, unaltered for months. Antozone, whether moist or dry, also reverts to the condition of ordinary oxygen on being brought in contact with black oxide of manganese, peroxide of lead, or finely divided platinum.

179. A very remarkable characteristic of antozone is its power of forming fogs and clouds with water. It may even be found, after the matter has been more thoroughly studied, that all the fogs and clouds which occur in nature are dependent for their existence upon the presence of antozone.

If air, charged with antozone, be made to bubble through water, it will emerge from the water in the form of a thick white mist, similar to that formed by the cooling of steam. The same thing occurs when electrized air, or electrized oxygen, issues into a moist atmosphere, though the effect is less marked when ozone is present than when it has been removed by means of iodide of potassium. The mist produced by slowly passing antozonized air through water is heavy; it remains hanging over the surface of the liquid, and may be readily poured from one vessel to another. By conducting it through a tube to the bottom of a dry, tall bottle, it displaces the air, all the while preserving a sharply defined boundary; by gentle agitation it is easily broken up into cloud-like masses.

When a large, dry bottle is nearly filled with this antozone mist, then closed and left to itself, the mist gradually becomes thinner and less opaque, and in the course of half or three-quarters of an hour vanishes altogether. As the cloud thus disappears, water is deposited upon the sides of the bottle, at first as a mere dew, but afterwards accumulating in droplets, which finally flow together to the bottom of the vessel. When the air in the bottle has become clear, no antozone can be detected in it.

It thus appears that antozone has the property of taking up water in such a manner that the water assumes the peculiar physical conditions of a cloud or mist. While the antozone lasts the cloud is permanent; but the antozone is soon transformed into ordinary oxygen, and as fast as this change occurs the water of the cloud is deposited in droplets.

By passing the antozone mist through tubes filled with desiccating substances, such as chloride of calcium (Appendix, § 15), the water may be removed, and transparent antozonized air obtained, capable of again producing a mist on being brought in contact with water. Many strong saline solutions likewise deprive antozone of water; hence the non-appearance of the cloud when electrized air is passed through a strong solution of iodide of potassium; the cloud does appear, however, when the solution is sufficiently dilute.

It has been proved by experiment that electrized air can support or carry nearly twice as much moisture as ordinary air or oxygen at the same temperature, and that this air is much more difficult to dry than the gases with which chemists usually have to deal. This explains how it happened that, before the discovery of the cloud-forming property of antozone, so many observers had been led to consider ozone an oxide of hydrogen. One experimenter would pass recently electrized air through an ordinary drying-tube, such as long experience had shown to be capable of drying common air perfectly, and would then heat the gas; by this treatment both the ozone and the antozone would be changed to ordinary oxygen, and the water which had been carried through the drying-tube by the antozone would be made visible. The remarkable capacity of antozone for moisture being unknown, the water thus obtained was naturally enough supposed to have been derived from some compound of hydrogen and oxygen other than water, and capable of passing unabsorbed through the drying-tube. Other chemists, performing, as they supposed, the same experiment, but in reality operating upon air less recently electrized, and so containing no antozone, were, of course, unable to obtain any water at the point where it had been observed by their predecessors; hence arose a series of controversies which have only recently been composed.

180. As has been already mentioned, antozone, like ozone, is formed in all processes of oxidation and combustion. During combustion most of the ozone produced enters into combination with the substance burned, while the antozone is left free, or enters into combination with water to form peroxide of hydrogen. When the combustion is slow or smouldering, antozone appears in large quantities, and in presence of moisture forms the characteristic mist or cloud. Tobacco-smoke, the gray smoke of chimneys and of gunpowder, and all such smokes are antozone clouds,—facts which support the idea that all clouds, fogs, and mists are caused by the presence of antozone in the atmosphere.

The oxidation of phosphorus affords a ready method of exhibiting the antozone cloud. During the oxidation of phosphorus in moist air, white fumes are formed, which were long a great puzzle to chemists. Whether the phosphorus be allowed to oxidize slowly, as in Exp. 75, or burned rapidly, as in Exp. 13, there is always produced a white mist of very considerable permanence, which remains long after the oxides of phosphorus, which are also formed, have been taken up and removed by the water. This mist is the antozone cloud; it is nothing but water held suspended by antozone.

In the rapid combustion of phosphorus, little or no ozone is left free; all of it seems to unite directly with the phosphorus; but much more antozone is produced when the combustion is rapid than when it is slow. The formation of antozone in this connexion explains the fact already alluded to (Exp. 13), that phosphorus burning with flame, in a confined volume of air, does not wholly exhaust the latter of oxygen. The phosphorus cannot combine with antozone, but only with ozone; hence, when no oxygen other than that in the form of antozone remains, the combustion must cease.

During the burning of a jet of hydrogen under a bell-glass through which a stream of air is drawn, antozone is formed, as is proved by passing the issuing stream through water; the antozone cloud is produced without difficulty, and peroxide of hydrogen appears as a product. The formation of the antozone mist, and of peroxide of hydrogen, may be observed with any other flame if care be taken that the air which streams over the flame be not too strongly heated. A high temperature destroys the antozone as fast as it is formed.

181. Besides its power of forming clouds or mists with water, which is interesting rather as a physical than as a chemical fact, antozone, particularly when newly formed, also unites with water chemically, the substance called peroxide of hydrogen (see § 61), whose composition is expressed by the formula H_2O_2, being the result of the combination.

A simple method of exhibiting the formation of peroxide of hydrogen by the action of antozone upon water, is to place a short, narrow tube, containing concentrated sulphuric acid, within a bottle 2 or 3 c.m. in width, furnished with a ground-glass stopper, and filled with water nearly to the top of the tube. Small portions of peroxide of barium are now added, at intervals, to the sulphuric acid in the tube, elevation of temperature being avoided as far as possible; the stopper should be replaced in the bottle after each addition of the peroxide. Most of the oxygen evolved in this process appears, however, to be in the ordinary inactive state, and the solution of peroxide of hydrogen obtained is consequently extremely dilute. A better method of procedure is to pass a current of carbonic acid gas into a mixture of water and peroxide of barium,

$$BaO_2 + H_2O + CO_2 = BaO,CO_2 + H_2O_2.$$

In this way a highly concentrated solution of the peroxide can be obtained.

Another easy method of preparing peroxide of hydrogen is by the oxidation of amalgams of lead or zinc. In this case also, as in the preceding, the peroxide of hydrogen is probably formed by the union of antozone with water.

One hundred grammes of lead-amalgam, containing so much mercury that it shall be fluid at the ordinary temperature, is shaken in a bottle of the capacity of a litre, together with 200 c. c. of water, acidulated with 2 grms. of sulphuric acid; the water soon becomes milky from separation of sulphate of lead, and in the course of ten or twelve minutes contains enough peroxide of hydrogen to exhibit the characteristic reactions of this substance.

So, too, if pulverulent zinc-amalgam be loosely thrown into a glass funnel, with narrow throat, and a thin stream of water be allowed to flow through it in such manner that the metal may be at the same time acted upon by both air and water, the water will become charged with peroxide of hydrogen. By repeatedly pouring back the dilute solution of the peroxide upon the amalgam, it can be very considerably strengthened. In order to prepare the zinc-amalgam, equal weights of zinc-filings and of mercury are placed in a beaker glass, covered with water acidulated with sulphuric or chlorhydric acid, and thoroughly mixed by stirring with a glass rod; the acid is then poured away, and the last portions of it removed from the amalgam by washing with water.

This power of antozone to oxidize water distinguishes it completely from ozone, which has little or no action upon water.

182. Peroxide of hydrogen, like peroxide of barium, is supposed to contain one atom of oxygen in the form of antozone; the peroxides of potassium, sodium, and strontium also are placed in the same category. They are all called antozonides.

183. Antozone can be distinguished from ozone by the following tests:—

Strips of paper, charged with a solution of sulphate of manganese (Exp. 77), do not become brown when exposed to the action of antozone; on the contrary, manganese papers which have been browned by ozone are bleached by antozone. Guaiacum paper (Exp. 78) does not become blue in antozonized air. The yellow compound called ferrocyanide of potassium, which is converted into red ferricyanide of potassium by the action of ozone, is not changed by antozone. In the absence of acids, antozone has no action upon iodide of potassium.

The chemical behavior of antozone may be conveniently studied by resorting to its compound with water, the antozonide peroxide of hydrogen. If peroxide of hydrogen be brought in contact with an ozonide like peroxide of lead, for example, both of the peroxides will be reduced, and there will result water, protoxide of lead, and free ordinary oxygen. Whenever an antozonide is mixed with an ozonide, a similar reaction occurs; the two active varieties of oxygen disappear, and common oxygen is evolved; hence it has been assumed that ordinary inactive oxygen is a sort of compound, resulting from the union or neutralization of ozone with antozone. Several important tests for antozone are dependent upon this fact of the decomposition of antozonides by ozonides.

If a liquid suspected to contain peroxide of hydrogen be shaken in a test-tube with a small quantity of ether, the ether will dissolve the peroxide, and will finally collect upon the surface of the liquid; on adding to it a small drop of a solution of the ozonide chromic acid, or, what comes to the same thing, a drop of a solution of bichromate of potassium acidulated with sulphuric acid, the ethereal solution will become blue.

If a liquid containing peroxide of hydrogen be added to a dilute red solution of permanganate of potassium, this solution will be decolorized, while common oxygen will be evolved; and in the same way the brown peroxide of lead and the red-colored salts of peroxide of iron are bleached by it.

Another exceedingly delicate and characteristic test for antozone, or rather for peroxide of hydrogen, the rationale of which has not yet been well made out, is the following:—If to a solution containing peroxide of hydrogen there are added a few drops of dilute starch-paste charged with iodide of potassium, and subsequently a very small quantity of a solution of copperas (protosulphate of iron), iodine will be set free, and the starch will become blue. The solution to be tested must be as nearly neutral as possible. The addition of an acid, instead of the copperas solution, will also bring about the same reaction, though less readily.

184. We have thus set forth whatever is best known concerning ozone and antozone, in spite of the details into which so full an exposition has necessarily descended, partly because the subject will evidently be one of primary importance, both theoretical and practical, in the near future, and partly from a desire to show the student how vague and uncertain the prospect is when once the narrow limits of established knowledge are past and the inquirer ventures out into the obscurity which perpetually separates the knowledge of to-day from that which shall be knowledge to-morrow, but also because of the impossibility, with so obscure a subject, of making such a just discrimination between salient and unimportant points as with a well-studied subject is both easy and desirable.

CHAPTER XIII.

SULPHUR.

185. Sulphur occurs somewhat abundantly in nature, both in the free state and in combination with other elements. Many ores of metals, for example, are sulphur compounds. It is a component of several abundant salts, such as the sulphates of calcium, barium, and sodium, and occurs in small proportion in

many animal and vegetable substances. Free sulphur is found chiefly in volcanic districts. Generally it occurs mixed with earthy matters; but it often forms distinct veins, and is sometimes found in the shape of well-defined crystals of considerable size. At the present time about nine-tenths of the sulphur of commerce comes from Sicily.

186. Native sulphur is usually subjected to a rough purification at the place of its occurrence. This purification is sometimes effected by distilling the volcanic earth in retorts or jars of earthenware; the sulphur being volatile, distils over, and is collected in receivers, from which it is drawn off, from time to time, in the liquid state; or if the earth be very rich in sulphur, it is simply heated in large kettles and the melted sulphur dipped off from above, while the earthy impurities settle to the bottom of the kettle. The product thus obtained is known as crude sulphur; it comes to us in irregular lumps of a dirty light-yellow color, and is largely employed for manufacturing-purposes.

This crude sulphur is contaminated with more or less earthy matter. In order to purify it, it is distilled from iron retorts into large chambers constructed of masonry, in which it is deposited either in the form of a light powder, known as flowers of sulphur, or in a liquid state, according to circumstances. At the beginning of the operation, while the chamber is cold, the sulphur vapor condenses as an exceedingly fine, soft, powder (flowers of sulphur) upon the walls of the chamber. But heat is given off as the sulphur vapor condenses, and after a while the walls of the chamber become so hot that sulphur will melt upon them. After this, the incoming sulphur vapor of course condenses only to the liquid state, and a layer of liquid sulphur collects upon the floor of the chamber. This liquid sulphur is drawn off into wooden moulds, and thus cast into the sticks familiarly known as roll-brimstone. It is evident that, by a little management, the sulphur-refiner can obtain, at will, either flowers of sulphur or roll-brimstone, or first the one and then the other.

187. At the ordinary temperature of the air, sulphur is a brittle solid, of a peculiar light-yellow color. It has neither taste nor smell, excepting that when rubbed it exhales a faint and peculiar odor. Most of the odors which in everyday life are

referred to sulphur are really the odors of various compounds of sulphur, and are not evolved by the element itself. It is a bad conductor of heat and electricity. On being rubbed it becomes highly (negatively) electric, and is still employed as a source of electricity in some cases. The symbol of sulphur is S; its atomic weight is 32, being precisely twice as great as the atomic weight of oxygen.

188. Sulphur melts easily at about 112°, a temperature not very far above that at which water boils. A fragment of it may even be melted by heating it on writing-paper over the flame of a candle. It volatilizes freely at temperatures lower than its melting-point, and boils at 440°. Indeed, as is the case with water, it is a substance which can be brought into either of the three states of matter without any difficulty; we can have it as a solid, a liquid, or a gas as we please. It can readily be obtained also in the form of crystals.

Exp. 79.—In a small beaker glass, or porcelain capsule, heat *slowly* 50 to 60 grms. of sulphur until it has entirely melted. Remove the vessel from the lamp, and allow it to cool slowly until about a quarter part of the sulphur has solidified; then pour off, into a basin of water, that portion of the sulphur which is still liquid, breaking through, for this purpose, the crust at the top of the liquid, if any such have formed. The interior of the vessel will be found to be lined with transparent, prismatic crystals.

Exp. 80.—In a test-tube, melt enough sulphur to fill one-quarter of the tube; place the tube in such a position that its contents may cool slowly and quietly, and then watch the formation of crystals as they shoot out from the comparatively cold walls of the tube towards the centre of the liquid.

Exp. 79 represents one general method of obtaining crystals. Crystals of many of the metals, lead and bismuth for example, can be obtained by operating in this way; it is only necessary to melt the metal in a crucible of some refractory material, placed in a furnace. The melted metal having then been allowed to cool until a tolerably firm crust has formed upon its surface, this crust is pierced with an iron rod, and the crucible quickly inverted, so that the portion of the metal which still remains fluid in the interior shall flow out. Upon afterwards breaking the crucible, crystals will be found lining the cavity of the metallic cup which has been formed within it.

189. Exp. 80, besides illustrating the manner in which crystals form, teaches us something of the physical structure of solid bodies. The solid mass of sulphur which is left in the test-tube when it has become cold, is evidently nothing more than a compact bundle of interlaced crystals. If the mass be removed from the tube, and then broken across, it will present a glistening appearance, owing to the reflection of light from the surfaces of the minute crystals of which it is composed. It is said to have a *crystalline structure*. This crystalline structure is apt to render a body brittle; substances which possess it are liable to break "with the grain," or to split in certain directions determined by the shape of the crystals, and called lines of *cleavage*; a stick of roll-brimstone, for example, may be readily broken or cut across, but not so easily in the direction of its length. The same remark applies to many samples of metal. In all cases where tenacity is required, it is important to counteract, or to prevent as much as possible, the tendency towards crystallization. Thus, in manufacturing wrought iron, it is the constant endeavor of the workman to render the metal stringy or fibrous, and not crystalline, and he seeks to accomplish this by appropriate processes of kneading, squeezing, and rolling.

190. Another easy way to crystallize sulphur is by the method of solution and evaporation, such as was employed in the preparation of nitrate of ammonium (Exp. 33). Sulphur is not soluble in water, but it dissolves readily in a liquid compound of sulphur and carbon, known as bisulphide of carbon, which being readily volatile, quickly escapes, on exposure to the air, and so deposits the sulphur.

Exp. 81.—Place in a test-tube a small teaspoonful of flowers of sulphur, pour upon the sulphur 10 or 12 c. c. of bisulphide of carbon, close the tube with a cork, and allow the mixture to stand during half an hour, shaking it occasionally. Decant the clear liquid from the sulphur which still remains undissolved, and pour it into a small porcelain capsule, which place out of doors, or in a draught of air, until the highly offensive bisulphide of carbon has all evaporated. Crystals of sulphur will then be found at the bottom of the dish.

This experiment might be modified by preparing, in the first place, a saturated solution of sulphur in boiling bisulphide of carbon, and then allowing the clear solution to cool slowly. Crystals of sulphur would finally be found beneath the cold liquid. The method by evaporation, as above described, is to be preferred.

It will be noticed that the crystals of Exp. 81 are not shaped like those obtained by the method of fusion in Exp. 79. The two sets of crystals belong in fact to entirely different systems of crystallization.

191. The researches of crystallographers have proved that the crystals of natural minerals and artificial chemical substances may all be included in six general classes of form, called *systems* of crystallization. In every crystal, certain directions may be recognized, with reference to which the bounding planes of the crystal exhibit a more or less symmetrical arrangement. These directions, represented by straight lines drawn through the centre of the crystal, are called *axes*. The thousands of crystal-forms which occur in nature, or are produced by art, have been divided into six systems, or groups, by observation of the number, relative length and mutual inclination of the axes around which they are symmetrically formed. These six systems are defined as follows:—

I. *Monometric (single-measure) or Regular System.*—The axes are three in number, equal in length, and intersect each other at right angles. The cube, regular octahedron, and rhombic dodecahedron, forms of perfect symmetry, belong to this system.

II. *Dimetric (two-measure) System.*—The axes are three in number, and intersect each other at right angles; but one, called the vertical, is either longer or shorter than the two lateral, which are equal. The right square prism and square octahedron are of this system.

III. *Trimetric (three-measure) System.*—The axes are three in number, unequal in length and intersect each other at right angles. The system includes the right rectangular prism, the right rhombic prism, and the rhombic octahedron.

IV. *Monoclinic (single-inclination) System.*—The axes are three in number, and unequal in length; and one, called the vertical, is at right angles with one of the other two axes, which are called lateral, but obliquely inclined to the other; the two lateral axes intersect each other at right angles. The right rhomboidal and oblique rhombic prisms belong to this system.

V. *Triclinic (three-inclination) System.*—The axes are three in number, unequal in length, and all their intersections are oblique. The oblique rhomboidal prism is of this system.

VI. *Hexagonal System.*—The axes are four in number; three, called lateral, lie in one plane, are equal in length, and intersect each other at angles of 60°; the fourth axis, called vertical, is either longer or shorter than the other three, and crosses them at right angles. This system includes the hexagonal prism and the rhombohedron.

Under these systems of crystallization, the variety of possible forms and dimensions is unlimited. Thus, in systems in which the axes are unequal, the inequality may be great or small, through all degrees of discrepancy; in oblique systems the inclination of the axes may vary indefinitely; rhombohedrons may occur of every angle. Thus the actual forms of crystallography become exceedingly numerous, although they all belong to a few simple types.

If the student draws in perspective, upon paper, the axes of the several systems above described, or, better, constructs the different sets of axes out of bits of wood or wire, he will appreciate the fact that forms belonging to different systems are ordinarily so unlike in general appearance as to be readily distinguishable even by those who have no exact knowledge of the mathematical science of crystallography.

192. As a general rule, a substance crystallizes in forms belonging to only one system, and the crystalline form of a substance is something so constant and characteristic as to be one of the chemist's most valued means of recognition and definition. But this general rule is not without exceptions. Sulphur, as has just been proved, may be made to crystallize in forms belonging to two distinct systems of crystallization; and there are other substances, not a few, which when crystallized under different conditions, assume forms of two distinct systems. Substances which are thus capable of assuming crystalline forms belonging to two different systems are said to be *dimorphous* (two-formed). Two such different forms of the same substance often have quite dissimilar physical properties; they are apt to differ from each other in hardness, specific gravity, color, optical properties, and in their relation to heat; the chemical properties, also, of two such different forms are seldom entirely the same.

Fig. 38.

Fig. 39.

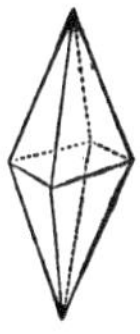

The crystals of sulphur obtained by fusion (Exp. 79) are

elongated oblique rhombic prisms (Fig. 38), and belong to the fourth (or monoclinic) system. The crystals of sulphur which are derived from its solution in bisulphide of carbon (Exp. 81) are rhombic octahedrons (Fig. 39), belonging to the trimetric system. The specific gravity of the octahedral crystals is greater than that of the prismatic in the ratio of 2·07 : 1·91. The specific heat of the octahedral crystals is 0·163, and that of the prismatic somewhat greater. The melting-point of the prismatic crystals is about 120°.

The prismatic crystals of sulphur (Exp. 79) cannot be kept for any great length of time. They soon lose their transparency and characteristic amber color, becoming opaque and light yellow, like ordinary brimstone. If they be examined under the microscope it will be seen that the prisms are now composed of a multitude of little octahedral crystals. The change of color and texture is due to a rearrangement of the particles of the original crystals, though the aggregation of octahedrons which have been formed within the prismatic crystal still retains the shape of the prism. If the prismatic crystals be left at rest, this change of form usually begins in the course of a few hours; but it may be greatly accelerated by scratching the crystals, or shaking them together. Under ordinary circumstances the passage of the sulphur from the one molecular state to the other goes on very slowly, several years being often required for its completion; but the change can be accomplished immediately by moistening the prismatic crystals with bisulphide of carbon. A considerable amount of heat is developed as the prismatic sulphur changes into octahedral; this can readily be appreciated when the conversion is effected by means of bisulphide of carbon.

In the same way that prismatic sulphur slowly changes into the octahedral variety at the ordinary temperature, octahedral sulphur is gradually converted into prismatic sulphur when kept for a long time at a temperature near its melting-point. The change in specific gravity enables us to follow the progress of this conversion.

Sulphur which has been melted and allowed to solidify gradually, is always in the prismatic condition immediately after the solidification. Roll-brimstone, for example, when fresh from the moulds, is translucent, and of a dark amber or brownish-yellow color, like the prismatic crystals of Exp. 79; but in a short time, often in the course of a few hours, the sticks become light-yellow and opaque, as we find them in commerce, and are then composed, at least externally, of a mass of octahedral crystals.

193. There is still a third way of obtaining crystals of sulphur, namely, by sublimation. At slightly elevated temperatures, sulphur is volatile; and if the circumstances be such that the vapor shall condense very slowly, crystals will form. The natural crystals of sulphur found in volcanic countries, which are often very large and of great beauty, have been formed in this way. These native crystals are octahedral, like those obtained by means of bisulphide of carbon (see Exp. 81).

194. As appears from the foregoing, there are three distinct methods of obtaining crystals:—I. By fusion; that is to say, by the slow cooling of molten matter. II. By solution, followed either by removal of the solvent by evaporation or chemical means, or by reduction of its temperature. III. By sublimation.

A familiar instance of the first method is seen in the case of ice, as when a part of the water in any hollow vessel freezes slowly upon the sides of the vessel; of the second, in the manufacture of common salt; and of the third, in the formation of frost upon a window-pane.

There is still a fourth general method of obtaining crystals, which consists in very slowly decomposing some chemical compound of the substance to be crystallized, either by the addition of some other chemical agent, or by means of the galvanic current. Crystals of sulphur may be formed in this way, and are in fact sometimes found in the pipes used to convey illuminating gas through the streets of cities, under such circumstances that it is evident that they have resulted from the decomposition of some one of the sulphur compounds with which coal-gas is always contaminated.

It must not be inferred, from the above enumeration of the ordinary methods of obtaining crystals, that either fusion, solution, or sublimation is a necessary condition of the formation of crystals. Both in nature and in art examples occur of the crystalline arrangement of particles within solid masses, under circumstances which preclude the idea that either fusion, solution, or sublimation, in the ordinary sense of these terms, should have occurred.

195. Sulphur behaves in a very remarkable manner on being heated. When melted at the lowest possible temperature, 110° to 115°, it forms a limpid liquid of a light-yellow color; but if this liquid be heated more strongly, it begins to become viscid and dark-colored at about 150°, and at 170° to 200° it is almost black, and at the same time so thick and tenacious that it cannot be poured from the vessel which holds it, even if the vessel be inverted. At 330° to 340° it regains its fluidity in part, though the liquid is still dark-colored, and finally, at about 440°, it begins to boil, and is converted into an amber-colored vapor. The specific gravity of sulphur vapor, referred to hydrogen, is 32.

196. If melted sulphur, in the viscid state, or, better, that which has regained its mobility, be suddenly cooled, a semisolid modification of sulphur, remarkably different from the ordinary form, will be obtained.

Exp. 82.—Place in a test-tube, of about 30 c. c. capacity, 15 to 20 grms. of coarsely powdered sulphur; melt the sulphur slowly over the gas-lamp, and continue to heat it until it begins to boil, noting, meanwhile, the changes which the sulphur undergoes—as described in § 195. Finally pour the hot sulphur, in a fine stream, into a large dish full of cold water. There will be obtained a soft, elastic, reddish-brown mass, which can be kneaded and moulded like wax, and drawn out into threads like caoutchouc.

This *soft sulphur* cannot be preserved for any great length of time. When left to itself, at the ordinary temperature of the air, it slowly hardens and changes into ordinary brittle yellow sulphur. This change is accelerated by kneading, and is instantaneous at the temperature of 100°. In any event, a certain amount of heat is evolved as the soft sulphur changes into ordinary sulphur. The specific gravity of soft sulphur is somewhat lower than that of the prismatic crystals.

From the foregoing facts it appears that sulphur, like oxygen, is capable of assuming different allotropic states. (See § 162.)

197. In its behavior towards solvents, sulphur presents some curious anomalies. Some specimens of sulphur are freely soluble

in bisulphide of carbon, while of other samples only a comparatively small portion dissolves. We distinguish, therefore, a soluble and an insoluble modification of sulphur.

Octahedral sulphur, the bright-yellow, translucent, native crystals for example, is completely soluble in bisulphide of carbon. But of the soft, elastic sulphur, such as was prepared in Exp. 82, as much as 30 or 40 per cent. is completely insoluble in the bisulphide, whether this liquid be hot or cold.

No method has as yet been discovered of preparing pure insoluble sulphur directly; but it can always be readily obtained by dissolving out the soluble sulphur from a mixture of the two varieties, such as the soft sulphur above mentioned. Flowers of sulphur contain a considerable portion of insoluble sulphur; roll-brimstone much less, though the interior of the sticks contains decidedly more than the outside portions. It may be observed, in this connexion, that flowers of sulphur are prepared by suddenly cooling the vapor of sulphur, while the soft variety is obtained by suddenly cooling melted sulphur.

Insoluble sulphur undergoes no change at the ordinary temperature; but if it be kept for a long time at 100°, or if it be exposed to the vapor of water or alcohol, it is slowly converted into the soluble variety.

198. For some pharmaceutical purposes, sulphur is prepared as a powder finer even than flowers of sulphur. This preparation is known as *milk of sulphur* or *precipitated sulphur.*

Exp. 83.—Place in a small flask as much flowers of sulphur as can be taken up on the point of a penknife; pour into the flask 10 or 15 c. c. of a solution of caustic soda, and boil the mixture for some time. Part of the sulphur will dissolve and color the liquid yellowish brown.

Pour off the clear liquid from the undissolved sulphur, mix it with an equal volume of water and stir in dilute chlorhydric acid, added by small portions, until a drop of the mixture placed upon litmus paper exhibits an acid reaction. As the acid is added, the liquid assumes a milky appearance from the separation of sulphur in the form of an exceedingly fine powder. This powder is so light that, for a long while, it will not subside, but remains suspended in the liquor, imparting to it a milky appearance.

Collect the powder on a small filter, wash it with water, and dry it at a gentle heat. It will now appear as a pale yellowish-grey impalpable powder. If it be heated more strongly, so that it melts, the color will become distinctly yellow, the numberless small particles of the powder being now compacted into a single mass.

It will be remarked that the color of flowers of sulphur is lighter than that of roll-brimstone, while the color of the precipitated sulphur is far lighter than that of the flowers. Such differences as these are common; they depend upon a difference of mechanical condition, upon differences in the state of aggregation of the particles of the substances which exhibit them.

The method of pulverization by precipitation, employed in this experiment, is a general method, applicable to many other substances besides sulphur.

199. Sulphur unites energetically with most of the other elements, such union being, in many cases, attended with evolution of light. Most of the metals, for example, combine with it directly, just as they do with oxygen.

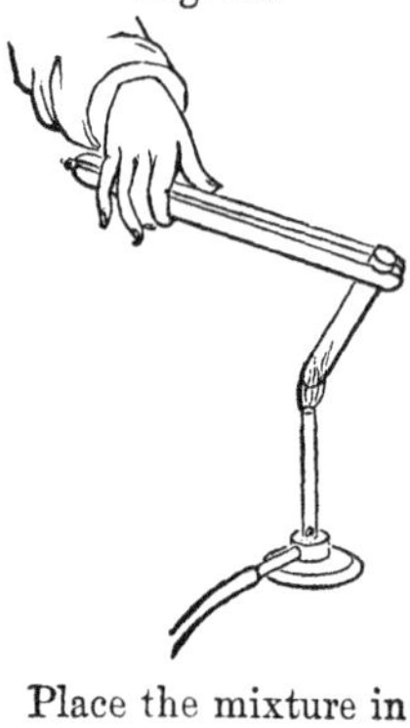

Exp. 84.—Melt in an ignition-tube, 12 to 15 c.m. long, 4 or 5 grms. of sulphur, and heat the liquid until it boils; then throw in small portions of copper filings, or fine turnings, and observe the violent action which ensues.

Fig. 40.

Or a strip of very thin sheet copper or a coil of fine copper-wire may be suspended in the hot sulphur vapor, in the upper part of the ignition-tube; it will glow vividly as it unites with the gaseous sulphur, much in the same way as if it were burning in oxygen gas. The product of the reaction, in either case, is called sulphide of copper.

Exp. 85.—Mix intimately 4 grms. of flowers of sulphur and 7 grms. of the finest iron-filings. Place the mixture in an ignition-tube 10 to 12 c.m. long, and heat the upper part of the tube over the gas-lamp. In a short time the mass will begin to glow, as the sulphur and iron enter into chemical combination, and this ignition will, of itself, pass through the entire length of the tube, even if the lamp be withdrawn. The final product of the reaction is protosulphide of iron.

200. As has been already shown (§§ 2, 109), phenomena of combustion, such as are exhibited in these experiments, are directly referable to chemical union. They are strictly analogous

to the ordinary processes of combustion in which oxygen is involved, though the technical term, combustion, is by custom limited to the act of combination with oxygen.

Sulphur combines with chlorine, bromine, iodine, and phosphorus at the ordinary temperature, and with carbon at a red heat. With oxygen it unites readily at a comparatively low temperature. When heated in the air, it takes fire at about 250°, and burns with a peculiar blue light. This easy inflammability may be readily illustrated by blowing flowers of sulphur into the hot air issuing from the chimney of an Argand gas-lamp; the sulphur takes fire at a considerable height above the flame. The irritating, suffocating gas, which is produced by the union of sulphur and oxygen, will be shortly described under the name of sulphurous acid.

Several important practical applications of sulphur depend upon this property of igniting and continuing to burn at a moderate heat. It is, in fact, largely employed as a kindling material. By means of it, other bodies less readily combustible, can be heated to the temperature at which they continue to burn. Hence its use upon matches and in gunpowder and fireworks.

201. In its chemical properties, sulphur is closely allied to oxygen; like oxygen, it forms a great variety of compounds with a wide range of different elements; and the series of compounds thus obtained is in many respects parallel with, or comparable to, the series of oxygen compounds.

It is an important raw material in the chemical arts, being an ingredient of numerous useful compounds, such as cinnabar, ultramarine, vulcanized caoutchouc, bisulphide of carbon, chloride of sulphur, and the various compounds of sulphur and oxygen, one of which, sulphuric acid, is the most important chemical agent at present employed in manufacturing industry. Sulphur is largely employed in medicine, in the treatment of cutaneous diseases of both men and domesticated animals, and has been of late years extensively used in the vineyards of Europe for destroying a parasitic fungus which infests the vines.

202. *Sulphydric Acid* (H_2S).—When sulphur is sublimed in hydrogen gas, or when hydrogen is passed over melted sulphur, combination takes place between the two elements, though very

slowly and imperfectly, so that only a comparatively small quantity of the compound is obtained, even if the process be continued for a long time. Somewhat larger quantities of it are formed when a mixture of hydrogen gas and sulphur-vapor is passed through a tube filled with fragments of pumice-stone heated to about 500°.

When the two elements meet in the nascent state, they combine readily; thus, when organic bodies containing sulphur putrefy, and when they are subjected to destructive distillation, sulphydric acid is evolved, just as ammonia is under the same circumstances. In either event, the product of the union is a colorless gas of highly offensive odor, like that of rotten eggs.

An easier method of preparing sulphydric acid, or sulphuretted hydrogen, as it is often called, is by acting upon a compound of sulphur and iron with dilute chlorhydric acid.

Exp. 86.—In a gas-bottle (Fig. 41) put 10 or 12 grms. of protosulphide of iron, see Exp. 85; replace the cork in the bottle and introduce the gas-delivery-tube into another small bottle containing cold water, letting the tube dip 5 or 6 c.m. beneath the surface of the water. Through the thistle-tube, pour into the gas-bottle water enough to seal the lower extremity of this tube; then add, through the thistle-tube, as before, 2 or 3 teaspoonfuls of muriatic acid, and observe that bubbles of gas soon begin to pass through the water in the absorption bottle.

Fig. 41.

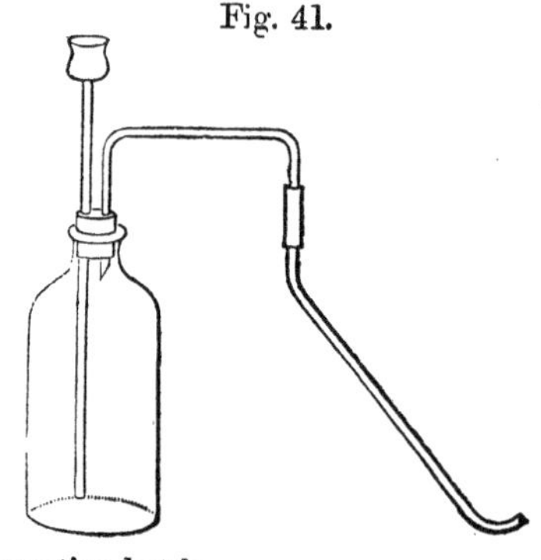

Sulphydric acid is soluble in water to a considerable extent, and is consequently taken up by the water in the absorption bottle. The solution thus obtained, known as sulphuretted hydrogen-water, is much employed as a reagent in chemical laboratories; it will serve us here as a convenient source of sulphydric acid.

When the disengagement of gas slackens, a new portion of muriatic acid may be added through the thistle-tube, and this process continued until the water in the absorption bottle smells strongly of the gas.

This experiment should be performed out of doors, or in a draught of air so arranged that those portions of the gas which escape solution shall be carried away from the operator.

203. The reaction between the sulphide of iron and chlorhydric acid in the foregoing experiment is somewhat analogous to that which occurs in the preparation of hydrogen, § 50. If metallic iron (or zinc) be treated with chlorhydric acid, hydrogen is evolved, according to the equation

$$Fe + 2HCl = FeCl_2 + 2H.$$

But if, instead of simple iron, sulphide of iron, whose formula is FeS, be taken, sulphur will be eliminated, as well as hydrogen, by the action of the acid, and these elements, as they come together in the nascent state, will unite to form sulphydric acid.

$$FeS + 2HCl = FeCl_2 + H_2S.$$

Instead of absorbing the gas evolved in the foregoing experiments in water, it might be collected as such, over a basin holding but a small quantity of water, or, better, filled with warm water or with brine, either of which absorbs less of the gas than cold water. Unless absolutely dry, the gas cannot be collected over mercury, since, when moist, it acts upon this metal.

204. At the ordinary temperature and pressure, sulphydric acid is a gas somewhat heavier than air, its specific gravity being 17, referred to hydrogen; but under a pressure of about 15 atmospheres at 11°, it becomes liquid. The specific gravity of this liquid referred to water is 0·9. At −85° the liquid solidifies to a white crystalline mass.

Fig. 42.

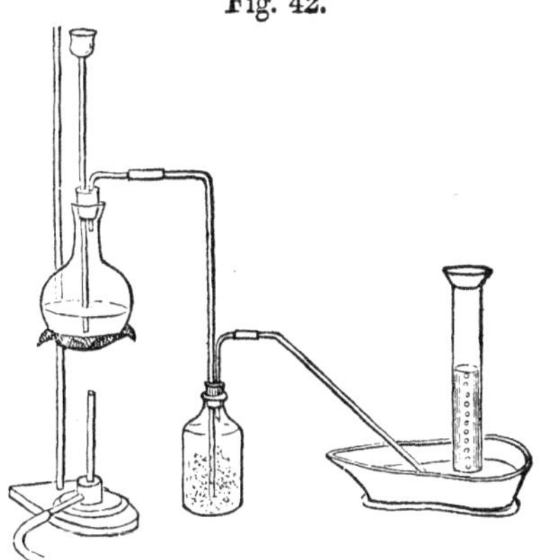

205. Since the sulphide of iron, employed in the preparation of sulphydric acid, is usually mixed with a certain quantity of metallic iron, the gas is liable to be contaminated with free hydrogen. For all ordinary purposes, the gas

thus mixed with hydrogen serves as well as if it were pure; but in some cases a gas free from hydrogen is required. In order to prepare it, sulphide of antimony is substituted for the sulphide of iron.

1 part of powdered sulphide of antimony placed in a thin-bottomed flask is treated with 3 or 4 parts of chlorhydric acid of 1·1 sp. gr. and the mixture gently heated. The apparatus may be arranged as in Fig. 42, in which the bottle into which the gas first enters contains a very small quantity of water; this water serves to remove any particles of the acid or of solid matter which may have been carried over in the current of gas. In case a dry gas be needed, a chloride-of-calcium tube (see Appendix, § 15) must be interposed between the wash-bottle and the mercury-trough.

206. The volumetric composition of sulphydric acid gas is one volume of sulphur-vapor and two volumes of hydrogen, condensed to two volumes. Its molecule, therefore, contains one atom of sulphur and two atoms of hydrogen, and is strictly analogous to the molecule of water.

The composition of sulphydric acid may be determined experimentally by heating metallic tin in a confined volume of the gas. An ignition-tube 20 or 30 c.m. long, bent at an obtuse angle, within 5 to 6 c.m. of the closed extremity, as shown in Fig. 43, should be completely filled with dry sulphydric acid gas over the mercury-trough, and then closed with the thumb and inverted. Some granulated tin should be dropped into the tube and made to lodge in the bent part, the thumb being instantly replaced upon the mouth of the tube the moment the tin has entered.

Fig. 43.

The tube full of gas is now replaced in the mercury-trough, as shown in the figure, and about one-third part of the gas is allowed to escape by inclining the tube so that the gas may bubble out through the mercury. The tube and its contents are left at rest during half an hour, in order that they may acquire the temperature of the surrounding air, a caoutchouc ring is slipped down the tube to mark the height of the gas, and the tin is then heated with the flame of a spirit-lamp. The hot tin combines with the sulphur, and hydrogen is set free. The apparatus is left at rest during another half hour, and the height of the gas in the tube is then noted. If the gas employed was pure, it will be found that its volume has undergone no change. The hydrogen which has been set free occupies precisely the same space as the sulphydric acid did before it was decomposed.

It is evident from this that 1 volume of sulphydric acid gas contains 1 volume of hydrogen, or multiplying these numbers by 2, in order to arrive at the composition indicated by our molecular formula, that 2 volumes of sulphydric acid contain 2 volumes of hydrogen. Now the specific gravity, or unit-volume weight of sulphydric acid has been found, by experiment, to be 17·19, that of hydrogen being 1; and if

From the weight of 2 volumes of sulphydric acid. . . .	34·38
We subtract the weight of two volumes hydrogen . . .	2·00
There will remain	32·38

which is very nearly equal to the unit-volume weight of sulphur-vapor, 31·8, as experimentally determined.

The composition of sulphydric acid, both by volume and weight, may, therefore, be expressed by the diagram.

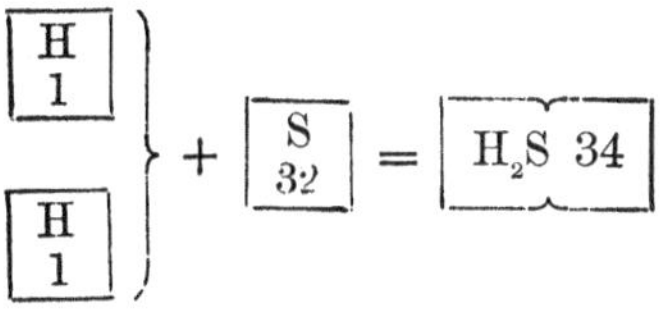

$$\left.\begin{array}{c}\boxed{\begin{array}{c}H\\1\end{array}}\\ \\ \boxed{\begin{array}{c}H\\1\end{array}}\end{array}\right\} + \boxed{\begin{array}{c}S\\32\end{array}} = \boxed{H_2S\ 34}$$

207. The gas is very poisonous; when respired in the pure state it quickly proves fatal, and it is very deleterious, even though largely diluted with atmospheric air. Small birds soon die in air which contains only $\frac{1}{1500}$ of its volume of the gas, dogs in air which contains $\frac{1}{800}$, and horses in air which contains $\frac{1}{200}$ of its volume. Men can support more of it, but in experimenting with it, it is best to do so where there is a free circulation of air. Nausea and headache are often produced when an atmosphere even slightly contaminated with sulphuretted hydrogen has been breathed for any length of time. In case the air of an apartment become contaminated with the gas, the disgusting smell can readily be neutralized by sprinkling the room with chlorine-water, or by evolving a little chlorine gas by adding some dilute acid to a small quantity of bleaching-powder.

The gas exists as a natural constituent of some mineral waters which are thence called sulphurous, such as the Virginia Sulphur Springs, and the mineral springs at Sharon, N. Y. It is also found in the air and water of foul sewers, and wherever animal matter is undergoing putrefaction.

208. Sulphydric acid gas is readily inflammable, and, like hydrogen, extinguishes the flame of a burning candle immersed in it. It burns with a blue flame, producing water and sulphurous acid gas.

$$H_2S + 3O = H_2O + SO_2.$$

In case it be ignited in contact with a quantity of air insufficient to burn the whole of it, the hydrogen will burn first, and a portion of the sulphur will escape combustion.

If a tall glass cylinder be filled with sulphydric acid gas, and the gas be lighted at the top, the flame will pass down the cylinder as the hydrogen is consumed, and a quantity of very finely divided solid sulphur will be deposited upon the walls of the vessel.

It has already been stated that sulphur kindles very easily, and that it has a strong affinity for oxygen; but it appears from this experiment, that hydrogen kindles still more readily, and that its affinity for oxygen is greater than that of sulphur.

When mixed with air, in certain proportions, it is explosive; a fact which should be borne in mind by the experimenter.

209. Water dissolves about three times its own volume of the gas at the ordinary temperature. This solution (see Exp. 86) is transparent and colorless when recently prepared, but, when kept it gradually becomes opalescent and turbid from deposition of sulphur. Oxygen from the air unites with the hydrogen of the sulphydric acid to form water, and sulphur is set free. After the lapse of several weeks or months, it will be found that the solution no longer contains any sulphydric acid; it has lost its nauseous odor, and the bottom of the bottle is covered with sulphur, the result of the decomposition of the dissolved gas.

The aqueous solution of the gas reddens litmus slightly, like the very weak acids. Towards metals and metallic oxides it behaves in a manner somewhat analogous to that of chlorhydric acid and its congeners, while, with regard to metallic sulphides, it stands in the same relation as water to the oxides, as will be explained hereafter.

Exp. 87.—Place a drop of sulphuretted-hydrogen water (Exp. 86) upon a bright piece of copper, lead, or silver. The metal will quickly become black. The sulphur of the sulphydric acid unites with the metal, to form sulphide of copper, sulphide of lead, or sulphide of silver, as the case may be, while the hydrogen escapes.

$$Cu + H_2S = CuS + 2H.$$

Exp. 88.—Place in a test-tube as much litharge (oxide of lead = PbO) as can be held upon the point of a penknife, pour upon it a teaspoonful of sulphuretted hydrogen-water, and observe that the yellow litharge immediately becomes black. Sulphide of lead is formed, as in the preceding experiment, together with a quantity of water.

$$PbO + H_2S = PbS + H_2O.$$

Exp. 89.—In place of the litharge of the last experiment, take a very small crystal of nitrate of lead (Exp. 42); dissolve it in as much water as will half fill the test-tube, and to this solution add a few drops of the sulphuretted-hydrogen water. Black sulphide of lead is thrown down as a precipitate, and nitric acid is set free.

$$PbO,N_2O_5 + H_2S = PbS + H_2O,N_2O_5.$$

210. Since many of the metallic sulphides are, like the sulphides of lead, copper, and silver, insoluble in water and dilute acids, sulphuretted hydrogen is peculiarly well adapted for precipitating the metals from their solutions. After having been thrown down as sulphides, as in the last experiment, they can be readily separated and collected by filtration.

Though many of the metallic sulphides are black, like that of lead, this is not true of all. Several of them exhibit characteristic colors, by which they may be readily recognized; thus the color of sulphide of antimony is orange, that of sulphide of arsenic is yellow, and that of sulphide of zinc white. Upon this fact the application of sulphuretted hydrogen as a test or reagent (that is, as a means of detecting and identifying many metals) is in part based.

211. In the same way that sulphuretted hydrogen can be employed for detecting the presence of metals, so, conversely, solutions of the metals, or, in some cases, the metals themselves, may be used as tests for sulphuretted hydrogen.

Exp. 90.—Prepare a strong aqueous solution of nitrate of lead, or better, of acetate of lead (sugar of lead). Wet strips of white paper 3 to 4 c.m. wide with this solution, and dry them in air which is free from sulphuretted hydrogen. This lead-paper, as it is called, should be kept for use in tightly stoppered bottles.

Moisten a bit of the lead-paper with water and expose it to some source of sulphuretted hydrogen—the mouth of the bottle of sulphydric acid prepared in Exp. 86, for example, or to the fetid air of a sewer. The paper will immediately be blackened from formation of sulphide of lead.

The blackening of silver-ware, of watches, and cards which have been glazed with a preparation of lead, at many mineral springs, and other localities where sulphuretted hydrogen is evolved, is, in like manner, indicative of the presence of this gas. Tests like these, which are continuous and cumulative, are, of course, much more delicate means of detection than the mere odor of the gas.

212. Sulphydric acid is a compound which is very easily decomposed. When simply heated, it breaks up into its components; and it is readily destroyed by various chemical agents.

Exp. 91.—To a gas-bottle such as was employed in Exp. 86, containing sulphide of iron, attach a chloride-of-calcium tube (Appendix, § 15) and a piece of hard glass tubing, No. 4, about 20 c.m. long. To the end of this glass tube, attach another tube bent at right angles and dipping into a bottle of water. Pour chlorhydric acid into the gas-bottle, so that sulphydric acid shall be freely generated, as seen by the flow of bubbles through the final bottle of water. After the lapse of some minutes, when the apparatus has become completely filled with the gas and the last portions of air have been expelled, heat the middle of the tube of hard glass with the flame of the gas-lamp, and observe the ring of sulphur which will collect upon the walls of the cold portion of the tube a short distance in front of the flame.

It will be seen in subsequent chapters that several other of the gaseous compounds of hydrogen are decomposed, like sulphydric acid, upon being passed through hot tubes.

The influence of oxygen, in decomposing the aqueous solution of sulphydric acid, has been already alluded to, § 209; it has been observed, moreover, that air contaminated with sulphydric acid soon becomes odorless of itself, oxygen uniting with hydrogen, as before, and sulphur being set free. All the oxidizing agents (that is, substances which readily give up oxygen) decompose sulphydric acid, water being formed and sulphur deposited.

Exp. 92.—Into a test-tube containing 4 or 5 c. c. of sulphuretted hydrogen-water (Exp. 86), pour half as much concentrated nitric acid. Sulphur will be deposited and nitrous fumes evolved.

Very dilute nitric acid will not thus decompose sulphydric acid.

Chlorine, bromine, and iodine vapor instantly decompose sulphydric acid, uniting with its hydrogen to form chlorhydric, bromhydric, or iodohydric acid, while sulphur is precipitated.

$$H_2S + 2Cl = 2HCl + S.$$

Exp. 93.—In place of the nitric acid of the preceding experiment, pour a few drops of chlorine-water into the solution of sulphydric acid, and observe that the odor of the latter is destroyed.

213. *Persulphide of Hydrogen* (H_2S_2 ?).—This is an exceedingly unstable liquid, the composition of which is not accurately known, though it is supposed to be analogous to that of the peroxide of hydrogen. It can be prepared by adding a solution of persulphide of calcium to diluted chlorhydric acid. The reaction may be conceived to take place in accordance with the following equation:—

$$CaS_5 + 2HCl = CaCl_2 + H_2S_2 + 3S.$$

Exp. 94. Mix 75 or 100 grms. of flowers of sulphur and an equal weight of slaked lime with half a litre of water, place the mixture in a flask and heat it to boiling, taking care to agitate the flask so that the solid matter may not become impacted upon it. Continue to boil for about an hour, then filter off the liquor from the undissolved portions of sulphur and lime. The solution thus obtained is a mixture of several sulphides of calcium, more highly sulphuretted than the protosulphide, but will serve the present purpose as well as if it were the pure quinquisulphide.

Pour the solution of sulphide of calcium into 250 c. c. of a mixture of 2 parts of concentrated chlorhydric acid and 1 part of water. Persulphide of hydrogen will separate in fine oily drops, producing a milky turbidity in the liquid. These drops soon coalesce and settle out beneath the water. A good way of collecting the persulphide is to perform the precipitation in a large glass funnel, provided with a stopper. By carefully opening this stopper, the precipitated oil can nearly all be drawn off without disturbing the water which floats above it.

Persulphide of hydrogen emits a peculiar, disagreeable odor, and is very irritating to the eyes and mucous membrane. It tastes sweet and bitter, but disorganizes the flesh wherever it touches it. Its properties closely resemble those of peroxide of hydrogen; it is very unstable, and is decomposed by the same substances which destroy the oxide. It even decomposes spontaneously when left at rest for a few days; ordinary vegetable colors are quickly bleached by it; it decolorizes also solutions of indigo.

214. In the last section we have used, for the first time, certain technical terms which, perhaps, need brief explanation. As has been stated in § 76, many of the elements are capable of uniting with other elements in several different proportions to form chemical compounds. Sulphur, for example, is specially apt to form more than one compound with a single element. When sulphur unites with a metal, the compound formed is called a sulph*ide*, just as a compound of oxygen and a metal is called an oxide, or one of chlorine and a metal a chloride,—the termination *ide*, which always indicates combination, being added to the first syllable of the word sulphur, or oxygen, or chlorine, and the new word ending in *ide* being then connected with the name of the metal, as in the case of sulphide of copper, Exp. 87. But when, as in the case of calcium, there are several distinct sulphides, it is customary to distinguish one from the other by means of various Latin and Greek prefixes. Thus the compound which contains one atom of sulphur and one atom of calcium is the protosulphide, or simply the sulphide of calcium, the prefix *proto* being derived from the Greek word for first; the compound which contains two atoms of sulphur to one of calcium is the bisulphide of calcium, from the Latin for twice; and in like manner we have a tersulphide, containing three atoms of sulphur to one of calcium, and a quinquisulphide containing five atoms of sulphur. The compound containing the highest proportion of sulphur is often called the *per*sulphide. A good custom is to designate the compounds which contain more sulphur than the protosulphide by prefixes of Latin origin, and to distinguish those which may contain less sulphur than the protosulphide by means of Greek prefixes; thus, if there were a compound of two atoms of calcium and one of sulphur (Ca_2S) it would properly be called a di-sulphide of calcium, the prefix being from the Greek δις. The same prefixes are used in an analogous manner in connexion with the

words oxide, chloride, bromide, iodide, and the similar words ending in *ide*.

215. *Compounds of Sulphur and Oxygen.*—No less than seven different compounds of sulphur and oxygen have been discovered, all of which form acids by union with water. Thus the oxide of sulphur SO_3 forms, by union with the elements of water, common sulphuric acid H_2SO_4; the name sulphuric acid being indiscriminately applied to both bodies, although only that one which contains hydrogen possesses the properties commonly described by the term acid.

Two of these compounds, viz. sulphurous acid (SO_2), and sulphuric acid (SO_3), have long been known, and these are still, comparatively speaking, of most importance, since they are employed upon the large scale in the arts. Subsequently there were found the compounds S_2O_5 (hyposulphuric acid) and S_2O_2 (hyposulphurous acid); and at a still more recent period the compounds S_3O_5, S_4O_5, and S_5O_5.

As long as only two compounds of sulphur and oxygen were known, they were distinguished as sulphur*ous* and sulphur*ic*, in accordance with the rule laid down in § 70; when the two compounds, S_2O_5 and S_2O_2, containing respectively less oxygen than sulphuric and sulphurous acids, were discovered, the prefix *hypo* was resorted to as explained in § 71; lastly, for the later-found acid compounds of sulphur and oxygen, the ordinary rules of chemical nomenclature being inadequate, it was necessary to invent a special set of names. They were all called Thionic acids, from the Greek word for sulphur, and were then distinguished from one another by the prefixes *tri*, *tetra*, and *penta*, in accordance with the number of atoms of sulphur in each. Strictly speaking, the compound S_2O_5, since it contains five atoms of oxygen like the thionic acids, should perhaps follow the new rule and be called dithionic acid, but it is still customary to retain the old name hyposulphuric acid.

The complete list of the names of the compounds of sulphur and oxygen is as follows:—

Sulphurous acid	SO_2
Sulphuric acid	SO_3
Hyposulphurous acid	S_2O_2

Hyposulphuric acid (or Dithionic acid) . .	S_2O_5
Trithionic acid	S_3O_5
Tetrathionic acid	S_4O_5
Pentathionic acid	S_5O_5

Of all these compounds, only sulphurous acid can be readily obtained by the direct union of sulphur and oxygen. The others must be prepared by circuitous methods.

216. *Sulphurous Acid* (SO_2).—This acid is produced when sulphur is burned in the air or in pure oxygen gas.

Exp. 95.—Light a piece of sulphur in a deflagrating spoon and suspend the latter in a half-litre bottle full of air. On examining the contents of the bottle, after the sulphur has ceased to burn, there will be found an irritating, suffocating gas having the peculiar odor which is familiar as that of a burning match. The bottle is now full of sulphurous acid gas, mixed with the nitrogen originally present in the air.

Fig. 44.

217. By burning sulphur in oxygen gas, instead of in air, as in the preceding experiment, a much purer product could, of course, be obtained. But the experiment would be chiefly interesting in enabling us to determine synthetically the composition of sulphurous acid.

If sulphur be burned in a confined volume of dry oxygen gas, it will be found, after the combustion has terminated, and the gas has been allowed to regain its original temperature, that the volume of the sulphurous acid produced is sensibly the same as that of the original oxygen, though its weight is twice as great. Hence 1 volume of sulphurous acid gas contains 1 volume of oxygen. Now, if from the weight

Of 1 unit-volume of sulphurous acid, as determined by experiment	32·256
We subtract the weight of 1 unit-volume of oxygen . . .	15·969
There will remain . . .	16·287

or not far from one-half the number, 32, which represents the real specific gravity or equal volume weight of sulphur-vapor. Consequently 1 volume of sulphurous acid gas contains half a volume of sulphur-vapor, besides 1 volume of oxyen. Or, multiplying these numbers by 2, in order to avoid a fractional volume, it appears that the volumetric composition of sulphurous acid is 1 volume of sulphur vapor and 2 volumes of oxygen condensed to 2 volumes of the compound gas. Or, expressed in the form of a diagram :—

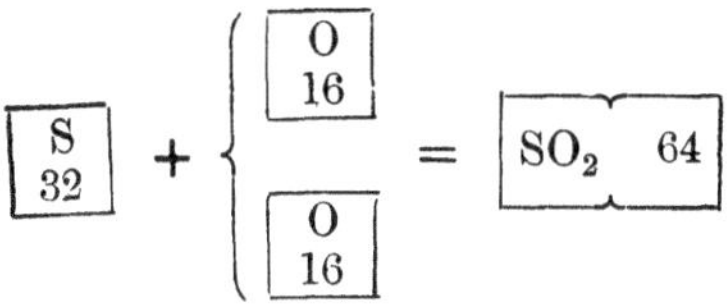

$$\boxed{\begin{matrix} S \\ 32 \end{matrix}} + \left\{ \begin{matrix} \boxed{\begin{matrix} O \\ 16 \end{matrix}} \\ \\ \boxed{\begin{matrix} O \\ 16 \end{matrix}} \end{matrix} \right. = \boxed{SO_2 \quad 64}$$

218. An easier method of preparing pure sulphurous acid is by depriving common sulphuric acid of part of its oxygen. This can be effected by a variety of reducing or deoxidizing agents. For example, when concentrated sulphuric acid is heated with metallic copper or mercury, there are formed a sulphate of the metal, water, and sulphurous acid :—

$$Cu + 2H_2SO_4 = CuSO_4 + 2H_2O + SO_2.$$

Exp. 96.—Into a thin-bottomed glass flask of half a litre capacity, put 14 grms. of copper clippings, or turnings, and 50 grms. of concentrated sulphuric acid. Attach to the flask a delivery-tube and connect this with a series of Woulfe's bottles, such as were employed in the preparation of chlorhydric acid (Exp. 49); heat the flask over the gas-lamp until the acid begins to react upon the copper, then quickly withdraw the lamp for a moment, lest the contents of the flask boil over, and finally regulate the flame so that a steady current of sulphurous acid shall pass through the water in the Woulfe bottles. After the first tumultuous evolution of gas has subsided, the flask can be slowly heated without further trouble. The current of gas should be kept up until the water of the first bottle has become saturated, or, at the least, highly charged with the gas.

Sulphurous acid is freely soluble in water, which, at 15°, takes up something like 44 times its bulk of the gas ; hence the solution obtained as above may be used as a convenient vehicle for sulphurous acid. On account of this easy solubility, the gas cannot be collected over water; but it can be collected over mercury, or by displacement. Since the gas is more than twice as heavy as air, the method by displacement is to be recommended, if an efficient ventilating flue is at command to carry off that portion of the suffocating gas which must escape.

Mercury is, in some respects, better than copper for use in this experiment. It affords a much more regular evolution of gas, and the operation requires less care ; but copper is usually employed on account of its comparatively low cost.

Instead of copper or mercury, as in the foregoing experiment, other reducing agents, such as sulphur or charcoal, may be employed. If 1 part of powdered sulphur be boiled with 12 parts of strong sulphuric acid, sulphurous acid is set free, as exhibited by the following equation:—

$$S + 2H_2SO_4 = 3SO_2 + 2H_2O.$$

The evolution of gas, in this case, though steady and uniform, is comparatively slow, as contrasted with that of the experiment in which copper is employed; hence the process is usually less convenient than that with copper.

If bits of charcoal or dry sawdust are heated with sulphuric acid, a copious evolution of sulphurous acid occurs, though the gas is not, in this case, pure, being mixed with half a volume of carbonic acid.

$$C + 2H_2SO_4 = 2SO_2 + CO_2 + 2H_2O.$$

For many purposes, as in the preparation of the aqueous solution of sulphurous acid, this method with charcoal is to be preferred, on the ground of economy and convenience of application. In the laboratory it is perhaps more frequently employed than either of the others.

Sulphurous acid may also be readily prepared by heating in an ignition-tube a mixture of 4 parts of sulphur and $5\frac{1}{2}$ parts of black oxide of manganese, both in fine powder, and intimately mixed. A mixture of 3 parts of black oxide of copper with 1 part of sulphur answers the same purpose:—

$$2S + MnO_2 = SO_2 + MnS.$$
$$3S + 2CuO = SO_2 + 2CuS.$$

In both these cases, metallic sulphides are left as a residuum in the ignition-tube; but if the sulphur and black oxide of manganese be mixed in the proportion of 1 part sulphur to $5\frac{1}{2}$ parts of the oxide, no sulphide, but only protoxide, of manganese will be formed.

$$S + 2MnO_2 = SO_2 + 2MnO.$$

219. As has been already stated, sulphurous acid gas is transparent and colorless. It is irrespirable and suffocating, and when mixed with air, even in small proportion, occasions violent coughing. It is not inflammable, but, on the contrary, it stops combustion.

The flame of a taper is immediately extinguished on being immersed in sulphurous acid gas, just as it is by nitrogen. A useful application of this property of the gas is in extinguishing burning chimneys. A handful of fragments of sulphur being thrown upon the hot coals in the grate, and the openings of the fire-place being closed in such a manner that no air shall enter the chimney, excepting that which passes through the fire, the chimney will quickly become filled with an atmosphere of sulphurous acid mixed with nitrogen from the air employed in burning the sulphur, and the burning soot upon the walls of the chimney will be immediately extinguished.

It is, of course, essential that the chimney should then be closed at the top, so that air may be excluded and the chimney kept full of the fire-extinguishing atmosphere until its walls shall have cooled to below the kindling temperature of the soot.

The oxygen contained in the sulphurous acid gas is so firmly held that combustibles are powerless to take it away under ordinary circumstances, though at high temperatures this oxygen can be removed by means of hydrogen, carbon, and easily oxidizable metals like potassium.

When hydrogen and sulphurous acid gas are passed together through a red-hot tube, water is formed and sulphur deposited,

$$4H + SO_2 = 2H_2O + S,$$

and when sulphurous acid is passed through a tube containing ignited charcoal, carbonic acid is produced and sulphur deposited, as before.

$$C + SO_2 = CO_2 + S.$$

In case nascent hydrogen come in contact with sulphurous acid, it will decompose it at the ordinary temperature, though in a manner somewhat different from the foregoing. The sulphur, as well as the oxygen, will, in this case, combine with hydrogen, and there will be formed sulphydric acid as well as water.

$$6H + SO_2 = 2H_2O + H_2S.$$

This reaction may be made visible by putting a few drops of a solution of sulphurous acid (Exp. 96) into a gas-bottle from which hydrogen is being evolved (Exp. 19), and testing the hydrogen with a strip of moistened lead-paper (Exp. 90) both before and after the addition of the sulphurous acid.

220. Sulphurous acid can readily be condensed to the liquid state. It is, in fact, one of the most easily liquefiable of the gases. By mere cooling to —10°, under the ordinary pressure of the air, it is converted into a colorless, transparent, limpid liquid.

In preparing small quantities of the liquid, it is sufficient to lead the gas, prepared from copper and sulphuric acid, and dried by passing it through sulphuric acid or over chloride of calcium, into a U-tube which is immersed in a freezing mixture of ice and salt (2 parts of pounded ice and 1 part salt).

Liquid sulphurous acid is a rather heavy liquid, of 1·4911 specific gravity, boiling at about $-10°$ and solidifying at $-76°$, to a colorless crystalline solid. On being exposed to the air at ordinary temperatures, the liquid acid evaporates with great rapidity, and consequently occasions very intense cold. By means of it, mercury may be frozen, and chlorine and ammonia-gas liquefied.

If a quantity of the liquid acid be poured into water, the temperature of which is a few degrees above 0°, a portion of the acid will evaporate at once, another portion will dissolve in the water, and a third portion of the heavy oily liquid will sink to the bottom of the vessel. If the portion, which has thus subsided, be stirred with a glass rod, it will boil at once, and the temperature of the water will be so much reduced that a portion, or even the whole, of the water will be frozen.

221. The specific gravity of the gas, as determined by different observers, is 32·256, or 32·443, or 32·558, instead of 32, as would be indicated by theory. This variation is explained by the fact that sulphurous acid, like all the easily condensible gases, ceases to conform exactly to the law of Mariotte at temperatures near to its point of condensation. Under any given pressure, its volume decreases in somewhat larger proportion than is the case with air and the other permanent gases.

An important property of sulphurous acid is its power of bleaching vegetable colors. It is extensively employed in bleaching articles of straw, wool, silk, &c., which would be injured by chlorine.

Exp. 97.—Into a bottle in which sulphur has been burned (Exp. 95) pour a few teaspoonfuls of a solution of blue litmus, and shake the bottle. The litmus solution will first become red, as it would if any other acid than sulphurous were present, and will then be decolorized.

The same property may be illustrated by holding a red rose in the fumes of burning sulphur, or by immersing the rose in an aqueous solution of sulphurous acid (Exp. 96), and leaving it for a few minutes until it has become white.

In the same way the stains of fruit or wine can be removed from clothing. A bit of sulphur is burned beneath a small open cone of paper, which serves as a chimney, and the stain, having first been slightly moistened with water, is held in the fumes at the top of this chimney. The cloth should finally be carefully washed with water at the place where it has been exposed to the sulphurous acid.

In the arts, the process of bleaching is usually conducted in large chambers, in which the slightly moistened articles are hung while sulphur is burned below. The damp goods absorb the sulphurous acid and gradually become white. The presence of water is essential; perfectly dry sulphurous acid will not bleach.

222. The manner in which sulphurous acid acts as a bleaching agent is not clear. It is remarkable that, as a general rule, it does not actually destroy the coloring-matter; and that upon many coloring-matters it has little or no action. Most of the yellows, and the green coloring-matter of leaves, are in this latter category, and upon litmus, cochineal, and logwood, the acid does not act very readily. In the few instances where it really destroys the color, as in the case of the garden amaranthus, it appears to act as a deoxidizing agent. But in most cases it appears to enter into combination with the coloring-matters and to form colorless compounds. These colorless compounds of sulphurous acid and coloring-matter can be broken up, with restoration of color, by exposing them to the action of various chemical agents capable of expelling sulphurous acid.

Exp. 98.—Bleach a rose, as in Exp. 97, and immerse it in dilute sulphuric acid. Then dry and warm it, so that the volatile sulphurous acid may be driven off. The color of the rose will again appear.

In many cases a solution of caustic soda will restore the color as well as sulphuric acid. A practical illustration of this action of alkaline solutions is seen in the reproduction of the original yellow color of the wool when new flannel is washed with an alkaline soap.

Sulphurous acid is a powerful disinfecting and antiseptic agent. It retards, to a remarkable extent, the processes of putrefaction and fermentation, and is largely employed for this purpose in wine-making; hops and compressed vegetables are charged with it to the same end, and it has been successfully employed for preserving meat. It has often been employed in medicine, in the treatment of skin diseases, as a fumigation.

223. Although sulphur will not take up more than two atoms

of oxygen when burned in the air, or in oxygen gas, it is nevertheless a matter of no very great difficulty to cause it to take up a third atom.

In presence of water it gradually absorbs oxygen from the air, and is converted into sulphuric acid. Hence the aqueous solution of sulphurous acid (Exp. 96) cannot be preserved for any great length of time, unless it be kept in very tight vessels.

224. If a mixture of sulphurous acid gas and oxygen, or air, be brought in contact with hot platinum sponge, the sulphurous acid will unite with oxygen, and sulphuric acid will be formed; the same union occurs when the mixed gases are brought in contact with various other substances, such as pumice-stone, clay, and the oxides of chromium, iron, and copper. Several attempts have been made to put these methods in practice for manufacturing sulphuric acid, but they have been found to be too slow, and in the case of platinum and clay, it has been observed that these substances soon lose their power and cease to convert the mixed gases into sulphuric acid.

Sulphurous acid is, indeed, a deoxidizing agent of very considerable power; and is much employed in the laboratory as a reducing agent. It decomposes iodic acid with separation of iodine, and nitric acid with evolution of hyponitric acid, sulphuric acid being formed in both cases.

$$I_2O_5 + 5H_2O + 5SO_2 = 5(H_2O,SO_3) + 2I.$$
$$H_2O,N_2O_5 + SO_2 = H_2O,SO_3 + 2NO_2.$$

Exp. 99.—Charge a dry bottle, of the capacity of a litre or more, with sulphurous acid gas, by burning in it a bit of sulphur, as shown in Fig. 45. Fasten a shaving, or, better, a tuft of gun-cotton, upon a glass rod or tube bent at one end in the form of a hook; wet the shaving in concentrated nitric acid, and hang it in the bottle of sulphurous acid. Red fumes of hyponitric acid will immediately form about the nitric acid, and will gradually fill the bottle.

Fig. 45.

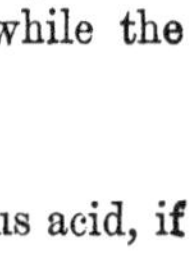

In presence of a mixture of water and chlorine, sulphurous acid takes up an atom of oxygen from the water, while the hydrogen of the water unites with chlorine.

$$SO_2 + 2H_2O + 2Cl = H_2O,SO_3 + 2HCl.$$

A similar reaction occurs between iodine and sulphurous acid, if a very large amount of water be present; in spite of the fact, already mentioned, § 140, that iodohydric acid is readily decomposed by concentrated sulphuric acid with liberation of iodine, sulphurous acid, and water.

225. Sulphurous acid, though a weak acid, forms numerous well-defined salts by uniting with metallic oxides. These salts, called sulphites, are of two classes,—simple or normal sulphites, such as the sulphite of potassium K_2SO_3 (or, dualistic, K_2O,SO_2), and double or acid sulphites, such as the acid sulphite of potassium $KHSO_3$ (or, dualistic, KHO,SO_2). All these salts are decomposed by strong acids, such as chlorhydric, nitric, or sulphuric, sulphurous acid being expelled; but they are not decomposed by carbonic acid. On the contrary, the salts of carbonic acid are decomposed by sulphurous acid; and hence it happens that the impure sulphurous acid gas obtained by heating a mixture of charcoal and sulphuric acid can be used for preparing the sulphites. If, for example, this gas be conducted into an aqueous solution of carbonate of sodium, there will be obtained a solution of sulphite of sodium, and carbonic acid will be set free.

226. Besides the solution of sulphurous acid, such as was prepared in Exp. 96, there is a definite crystalline compound of water and the acid, which can be obtained by passing a current of sulphurous acid gas into ice-water. This compound is very unstable, and is destroyed at temperatures but little above 0°; but by collecting it upon a cooled filter and then pressing the crystals repeatedly between folds of cold blotting-paper, it has been found possible to remove most of the mother-liquor which adheres to them at first, and to obtain the compound in a condition of tolerable purity. The composition of the crystals appears to be SO_2+15H_2O.

227. *Sulphuric Acid.*—The term sulphuric acid is applied somewhat indiscriminately to three or more distinct substances—namely, to a compound of one atom of sulphur and three atoms of oxygen, SO_3, which we shall call anhydrous sulphuric acid, and to certain compounds of sulphur, oxygen, and hydrogen, which have been usually regarded as compounds of the anhydrous sulphuric acid, just mentioned, and water. Of these hydrates the most important are those of the composition H_2SO_4

(dualistic, H_2O, SO_3) [oil of vitriol], and $H_2S_2O_7$ (dualistic, $H_2O,2SO_3$) [Nordhausen or fuming sulphuric acid]. The body, whose formula is H_2SO_4, is one of the most important of chemical substances, and is usually the thing meant when sulphuric acid is spoken of. We will therefore proceed to study its properties before touching upon those of the other substances above-mentioned.

Sulphuric acid is one of the most important products of chemical manufacture, and is made in enormous quantities. In the same way that the metal iron may be said to be the basis of all mechanical industries, sulphuric acid lies at the foundation of the chemical arts. By means of sulphuric acid, the chemist either directly or indirectly prepares almost everything with which he has commonly to deal.

Sulphuric acid might be prepared by passing sulphurous acid gas into boiling nitric acid, until all of the latter had been reduced, and finally distilling off the last traces of the lower oxides of nitrogen which would be formed. Even if sulphur itself were boiled in concentrated nitric acid, it would gradually be oxidized and converted into sulphuric acid. But neither of these processes would be economical. It can be very cheaply prepared, however, by the action of either of the high oxides of nitrogen, nitrous, hyponitric, or nitric acids, upon sulphurous acid, in presence of air and moisture; and this method is the one actually followed in the preparation of sulphuric acid on the large scale. A mixture of the gases above mentioned is effected in enormous chambers constructed of sheet lead, a metal upon which cold sulphuric acid has little or no action.

228. The essential points of the process are, first, that SO_2, when in presence of much moisture, can take oxygen from either N_2O_3, NO_2, or N_2O_5, and reduce them to nitric oxide, NO, while it is itself converted into sulphuric acid, and, secondly, that NO can take oxygen from the air and become NO_2.

In practice, the sulphurous acid is obtained by burning crude sulphur, or more commonly a compound of sulphur and iron, known as iron-pyrites, FeS_2; the gas, together with a large excess of atmospheric air, is then conducted into the first of a series of leaden chambers into which steam is admitted. Nitrous fumes are supplied either by allowing nitric acid to fall in fine streams through the incoming current of sulphurous acid and air, or from the decomposition of a mixture of salt, nitrate of sodium, and sulphuric acid, as described in § 105, or by heating a vessel charged with nitrate of sodium and sulphuric acid, by means of the burning sulphur.

In conformity with the principles above stated, the sulphurous acid, as soon as it comes in contact with the steam, reacts upon the nitrous fumes; there is formed nitric oxide gas and hydrated sulphuric acid, which falls to the floor. But, as there is present in the chamber an excess of air, the nitric oxide immediately unites with a portion of the oxygen therein contained, and is converted into hyponitric acid. This hyponitric acid immediately reacts upon a new portion of sulphurous acid, and the process thus goes on through a whole series of leaden chambers, the very small portion of nitric acid at first taken being sufficient to prepare a large quantity of sulphuric acid. In reality, the oxygen employed in converting the sulphurous into sulphuric acid, all comes from the air, excepting a very little at first; the nitrous fumes serve only as a conveyer of oxygen. The nitric oxide takes oxygen from the air and transfers it to the sulphurous acid, which, as has been stated in § 223, is, by itself and unaided, incapable of combining with oxygen. It will, of course, be understood that, although we trace out these reactions as if they were consecutive, they are really, so far as we know, simultaneous.

Theoretically, a single portion of hyponitric acid would be sufficient to effect the conversion of an unlimited amount of sulphurous into sulphuric acid; but practically this power is qualified by a variety of circumstances. It is found to be impossible, for example, to mix new portions of air with the mixture of sulphurous acid and nitric oxide for an indefinite period; for at a certain point these gases become so loaded down with nitrogen derived from the air already consumed, that they are as good as lost in it. In general the flow of gases is so regulated that all the sulphurous acid shall be oxidized, and that nothing but nitric oxide and waste nitrogen shall pass out of the last leaden chamber.

229. The process of manufacturing sulphuric acid can readily be illustrated upon the small scale.

A large glass balloon, or receiver, of the capacity of several litres, placed in a vertical position, is closed with a cork pierced with five holes, through four of which are passed small glass tubes. All of these glass tubes reach nearly to the bottom of the balloon, and are bent at a right angle above the cork; one of the tubes is connected at the top with a flask containing copper-turnings and sulphuric acid, for the generation of sulphurous acid (see Exp. 96), another with a flask containing copper-turnings and furnished with a thistle-tube, through which nitric acid can be poured, for the generation of nitric oxide (see Exp. 37), and the third with a flask containing water for the evolution of steam; the fourth tube and the fifth hole are both left open.

Everything being in readiness, nitric oxide is generated in the small flask fitted for this purpose; as the gas passes over into the large balloon it unites with oxygen from the air, and red fumes of hyponitric acid are formed. Sulphurous acid is now made to pass into the balloon; this will have no action upon the red fumes, so long as there is no water present, but the moment steam is thrown in from the third small flask, a reaction occurs, the hyponitric acid is reduced, and the sulphurous acid oxidized. By means of bellows, air must, from time to time, be blown into the balloon, through the fourth glass tube, the waste nitrogen passing off through the fifth hole in the cork.

If but little steam be employed in this experiment, a solid compound, formed by the union of nitrous and anhydrous sulphuric acids, is liable to be deposited upon the walls of the balloon; the appearance of this body always indicates that the supply of steam is insufficient; it is never formed when the proper proportion of moisture is present.

230. The sulphuric acid which collects at the bottom of the leaden chambers is necessarily dilute, because of the large amount of water which must be present, in order that the reactions above described may freely occur; moreover it would not be advantageous to allow an acid more concentrated than that of specific gravity 1·4 to form in the chambers, since a stronger acid would absorb and retain a considerable quantity of nitric oxide. To make it fit for the purposes for which sulphuric acid is usually employed, the dilute acid of the chambers must be concentrated by expulsion of the water; to this end, it is run off into shallow leaden pans, and there evaporated until it is of specific gravity 1·71 to 1·75. The concentration cannot safely be carried beyond this point in ordinary leaden vessels, since the strong, hot acid begins to attack the metal, and the temperature at which the liquid boils is so high as to approach the melting-point of lead. This acid of 1·72 specific gravity is somewhat extensively employed, for a variety of purposes, at the factories where it has been prepared, but is still too dilute for transportation. It is therefore transferred from the leaden pans to large glass retorts set in deep sand baths, or to platinum stills, and there evaporated further, until it is nearly of the composition H_2SO_4.

231. The acid thus boiled down is the concentrated sulphuric acid, or oil of vitriol, of commerce ; its specific gravity is usually about 1·83, that of the absolutely pure acid being 1·842. Besides this slight excess of water, it contains also, in solution, a certain quantity of sulphate of lead, and a variety of other impurities. For most purposes, however, it will answer as well as the pure acid. Like the latter, it is a heavy, oily, colorless, and odorless liquid, boiling at about 330°.

Since a comparatively small amount of heat is absorbed in the conversion of the liquid acid to the condition of gas, its vapor can be very easily condensed; in distilling the acid, the receiver need not even be placed in cold water.

From the same cause, combined with the great weight of the liquid, the acid is liable to boil tumultuously, the act of ebullition being irregular and attended with violent blows or shocks. The bubbles of vapor formed at the bottom of the retort condense almost as soon as they are formed, and the heavy liquid above suddenly falls back to fill the vacuum.

In distilling the concentrated acid, it is therefore best to heat only the upper portions of the liquid in the retort; this can be effected either by placing the retort upon a wire-grate so perforated that about half the body of the retort can be sunk below the level of the burning charcoal upon the grate, or by placing a layer of ashes, or of some other bad conductor of heat, beneath the very bottom of the retort, then piling sand around the sides of the retort outside of the ashes, and applying heat beneath the iron pan upon which the whole is supported.

232. The common acid usually freezes at about −34°; but it has been found possible to lower the freezing-point to −80°, by adding a small quantity of water to the commercial acid. When once frozen, it remains solid until the temperature rises to about the freezing-point of water. Crystals of the pure acid melt at about 10°. At the ordinary temperature, sulphuric acid does not vaporize, but, on the contrary, greedily absorbs water from the air and so increases in bulk. In moist weather, its bulk may increase to the extent of a quarter or more, in the course of a single day, and, by longer exposure, a still larger quantity of water will be taken up; the acid must always be kept, therefore, in tightly stoppered bottles.

Exp. 100.—Into a shallow dish of about 200 c. c. capacity, pour about 75 c. c. of concentrated sulphuric acid; place this dish of acid upon one pan of a balance, and upon the other pan put enough small shot, or clean, dry sand, to exactly balance the acid. Preserve the material of the counterpoise, and place the dish of acid uncovered in the open air; from day to day replace it upon the balance, together with the counterpoise, and note the number of grammes or fractions of a gramme that it has increased in weight.

If the acid were allowed to stand for a week or two in a damp place, it might become two or three times as heavy as it was at first. From its power of absorbing aqueous vapor, sulphuric acid is often employed for drying gases. (See Appendix, § 15.)

233. With liquid water sulphuric acid unites with great energy, much heat being evolved at the moment of combination; during the union a certain amount of condensation occurs, the mixture, when cold, occupying less space than was previously occupied by the acid and the water. The water and acid may be mixed in all proportions, being mutually soluble one in the other.

In mixing water and sulphuric acid, the acid should always be poured into the water, in a fine stream, not the water into the acid,—the water being meanwhile stirred. In this way the heavy acid has an opportunity to mix with the water as it sinks down through it.

If, by any accident, water were to fall upon sulphuric acid, it would float on top of it, and great heat would be developed at the point of contact of the two liquids; if the quantities of acid and water were large, sudden bursts of steam would be occasioned and serious damage might arise from the scattering about of portions of the acid.

In mixing water and commercial sulphuric acid as in the following experiment, it will be observed that the solution becomes cloudy, and that a white powder is gradually deposited from it. This precipitate is sulphate of lead, originally derived from the leaden pans in which the acid was concentrated; it is soluble in concentrated, but insoluble in dilute sulphuric acid, and is consequently thrown down when water is added to the commercial acid.

Exp. 101.—Place in a beaker glass of about 250 c. c. capacity, 30 c. c. of water; in accordance with the directions above given, pour into the water 120 grms. of concentrated sulphuric acid, and stir the mixture with a narrow test-tube containing a teaspoonful of water. So much heat will be evolved during the union of the water and the acid that the water in the test-tube will boil.

234. If sulphuric acid be mixed with ice or snow, the latter

will be immediately liquefied. If the proportion of ice in the mixture be small, as compared with that of the sulphuric acid, heat will be evolved much as is the case with liquid water, though to a less extent. But when a large proportion of ice is mixed with a comparatively small quantity of the acid, no heat will be perceived, but, on the contrary, intense cold.

Exp. 102.—Place in a beaker glass of about half a litre capacity 120 grms. of snow, or finely pounded ice; pour upon it 30 grms. of concentrated sulphuric acid, and stir the mixture with a test-tube containing a small quantity of water. The water in the tube will be frozen.

Exp. 103.—Repeat the foregoing experiment, using 30 grms. of snow or ice and 120 grms. of sulphuric acid. A very considerable evolution of heat will occur, as may be seen more clearly by immersing a thermometer in the liquid.

The result of Exp. 102 seems, at first sight, inconsistent with the general fact that heat is always set free during chemical combination; for though chemical union between the acid and water has evidently occurred, no heat, but cold, is manifested. The anomaly is only a seeming one; a certain amount of heat is required, in order that the cohesive force, by which the particles of the ice are held together, shall be overcome; hence the heat which is really produced by the chemical combination is all absorbed, together with much more, taken from the materials and the vessel which contained them, during the liquefaction of the ice.

235. Besides the indefinite mixture or solution above mentioned, several crystalline compounds of anhydrous sulphuric acid and the elements of water, of fixed composition and characteristic form, can be prepared.

If the commercial acid be diluted with water until its specific gravity is reduced to 1·78, and the liquid be then cooled strongly, a substance of composition H_4SO_5 (dualistic, $2H_2O,SO_3$) will crystallize out in the form of large rhombic prisms. These crystals are of sp. gr. 1·785; they melt and solidify at about 8°.

A second hydrate, of composition $3H_2O,SO_3$, can be obtained by evaporating a dilute acid in a vacuum at the temperature of 100°, until it ceases to lose weight; and another of composition $H_2O,2SO_3$, will be described below when we come to speak of fuming sulphuric acid.

236. Sulphuric acid is one of the most powerful acids known. If one drop of it be diluted with a thousand times as much water, it is still capable of reddening blue litmus. It expels most of the other acids from their compounds, in the same way that we have seen it expel nitric acid from nitrate of sodium in Exp. 32. At very low temperatures, however, as at $-80°$, it loses its power of reddening litmus, and has no action upon the carbonates, though it acts violently upon these salts at the ordinary temperature.

It is intensely caustic and corrosive, and quickly chars and destroys most vegetable and animal substances.

Exp. 104.—Into a test-glass pour a tablespoonful of sulphuric acid and immerse in it a splinter of wood. The wood will blacken as if charred by fire, and the acid will become dark-colored. Wood is composed of carbon, hydrogen, and oxygen; and since sulphuric acid unites with compounds of hydrogen and oxygen, rather than with carbon, a portion of the latter is left free; some carbonaceous matter is, however, dissolved by the acid and darkens it. The acid of commerce is often dark-colored, from fragments of straw or other organic matter having accidentally fallen into it. Sulphuric acid which has been colored by organic matter may be rendered colorless by strongly heating it till it becomes fully concentrated.

The action of the acid upon organic matter is more rapid when moisture is present. Thus, if a few drops of oil of vitriol be poured upon dry paper, decomposition will take place only slowly; but if a little water be added to the acid, heat will be developed by the chemical union, and the paper will be at once decomposed by the hot acid.

237. When heated with charcoal or with any organic matter, sulphuric acid gives up oxygen, as has been shown in Exp. 96, and is itself reduced to sulphurous acid; by sulphur, also, and by several of the metals, such as copper and mercury, it is reduced in a similar way. (See Exp. 96.) Towards some metals, such as zinc for example, its behavior is various, according as it is concentrated or dilute. If zinc be treated with cold, dilute sulphuric acid, the zinc simply replaces the hydrogen of the acid, sulphate of zinc is formed, and hydrogen is set free.

$$Zn + H_2SO_4 = ZnSO_4 + 2H.$$

But if zinc be heated with concentrated sulphuric acid, a portion of the latter is reduced, as it would be in presence of copper or mercury, sulphurous acid is evolved, as well as hydrogen, and these gases, reacting upon each other, produce sulphydric acid and a deposit of sulphur, in accordance with the following formulæ:—

$$SO_2 + 6H = H_2S + 2H_2O.$$
$$SO_2 + 4H = S \quad + 2H_2O.$$

As a general rule, concentrated sulphuric acid acts but feebly upon the metals in the cold, though, when boiled upon them, it often behaves as an oxidizing agent.

238. With the oxides of the metals, sulphuric acid unites directly to form the very important salts called sulphates, water being simultaneously eliminated. Oil of vitriol, H_2SO_4, may, in fact, be itself regarded as a salt, in the composition of which, hydrogen fills the same place that sodium does in sulphate of sodium, Na_2SO_4, and it might well be called sulphate of hydrogen, were it not that usage has assigned to it another name. Besides the *normal* sulphates, in which all the hydrogen has been replaced by a metal, as above (or, on the dualistic hypothesis, in which all the water has been replaced by a metallic oxide), there is another class of sulphates, often called *bi* or *acid* sulphates, in which only half of the hydrogen has been thus replaced; as an example of these, the student will recall the acid sulphate of sodium, $NaHSO_4$, mentioned in § 101.

Acids, which, like sulphuric acid, contain two replaceable atoms of hydrogen, and are therefore capable of forming two series of salts, are called bibasic, in contradistinction to the monobasic acids, like nitric acid, which form but one series of salts. There is but one nitrate of sodium, for example, $NaNO_3$. It is for this reason that many chemists object to the doubled formula for nitric acid $H_2N_2O_6$, in spite of its convenience, because this formula suggests, what is not true, that one or both of the atoms of hydrogen might be replaced by any metal which, like sodium, potassium, or silver, replaces hydrogen atom for atom.

239. *Fuming Sulphuric Acid.*—The acid H_2SO_4, above described, has been, for nearly a century, the most important of the several varieties of sulphuric acid; but long previous to the discovery of the process of making it in leaden chambers, there was manufactured another variety, now known as fuming sulphuric acid.

This fuming acid, or Nordhausen acid, as it is often called, from the name of a German town in which large quantities of it were formerly prepared, was at first obtained by distilling in earthen retorts the salt now known as sulphate of iron, formerly called green vitriol. Hence the origin of the name oil of vitriol, which, in England and this country, has come to be applied solely to the common acid H_2SO_4, though it is still used as a synonyme for the fuming acid by German writers.

When dry sulphate of iron is exposed to a full red heat, it suffers decomposition; a considerable quantity of sulphuric acid is given off and can be collected in receivers. The distillate thus obtained, which is a dense fuming liquid of about 1·9 specific gravity, is the acid now in question. Though of far less importance than was formerly the case, considerable quantities of the fuming acid are still prepared for the purpose of dissolving indigo and for other special uses, where an acid stronger than the common acid is needed. It may be regarded as a solution of varying quantities of the anhydrous acid SO_3 in the common acid H_2SO_4; if it be gently heated, all of the anhydrous acid will be expelled, and common sulphuric acid will remain. So, too, if it be exposed to the air, the anhydrous acid will be given off, and, coming in contact with the moisture of the air, will combine therewith to form common sulphuric acid, which, falling as a cloud, occasions the appearance of fumes.

When the fuming acid is cooled to about $-5°$, a crystalline compound of composition $H_2S_2O_7$ (dualistic, $H_2O,2SO_3$) separates out. After having been freed from liquid acid, these crystals melt at 35°. When pure, the fuming acid is colorless; but the commercial article is often brown, from having been in contact with organic matter. It is an excessively corrosive liquid, and destroys most organic matters, even more rapidly than the common acid. On being dropped into water, a noise is emitted as if a red-hot bar of metal had touched the water.

240. *Anhydrous Sulphuric Acid* (SO_3).—As has been mentioned in § 224, this substance can be obtained by passing a mixture of sulphurous acid gas and oxygen over hot, finely divided, metallic platinum, or over various oxides and other porous substances.

Exp. 105.—Prepare a small quantity of platinized asbestos as follows: dissolve about 0·25 grm. of metallic platinum in aqua regia (§ 104), and soak in this solution as much soft, porous asbestos as will form a loose ball of 1·5 c.m. diameter; heat the wet asbestos gently until it has become dry, and then ignite it in the flame of the gas-lamp. The chloride of platinum, which was formed by the solution of the metal, is decomposed by heat, and metallic platinum, in a finely divided condition, is left adhering to the asbestos.

Select a tube of hard glass, No. 3, about 30 c.m. long, and, at a distance of about 10 c.m. from one end, bend it to an obtuse angle, so that when the tube is supported upon a ring of the iron stand above the gas-lamp the shorter bent portion can be thrust into the neck of a receiver; in the centre of the longer portion of the glass tube pack the platinized asbestos loosely; then force into and through the tube a current of mixed sulphurous acid and oxygen; at the same time heat over the gas-lamp that portion of the tube which contains the platinized asbestos, and collect the anhydrous sulphuric acid which is formed in a perfectly dry test-tube or U-tube immersed in ice, or, better, in a freezing-mixture of ice and salt. In the course of the experiment, lift up the receiver for a moment, pour out from it a little of the vapor of anhydrous sulphuric acid, with which it is filled, and observe how rapidly the heavy gas falls through the air, and the cloud which forms as it unites with moisture.

The mixture of sulphurous acid and oxygen may be made before the experiment in a small gas-holder, or, as well, during the progress of the experiment in a bottle behind the asbestos tube. This bottle, which should be of at least half a litre capacity, is fitted with a cork carrying three glass tubes, and is connected with the asbestos tube by one of these tubes, which reaches no lower than the cork; by the other tubes, which pass nearly to the bottom of the bottle, and dip beneath the surface of a layer of common sulphuric acid, which has been placed in it, the bottle is connected with a flask in which sulphurous acid is being generated (Exp. 96), and with a gas-holder containing oxygen. The sulphuric acid in the bottle serves to dry the gases, and the rapidity with which the bubbles of gas pass through the liquid, enables the operator to judge of the proportions in which the gases are being mixed; the flow of oxygen having been fixed at a moderate rate, once for all, the sulphurous acid will alone need attention.

The action of the platinum in this experiment is obscure; it will be treated of under the metal platinum.

Instead of the platinized asbestos, oxide of iron, oxide of copper, or oxide of chromium, or, better, a mixture of the last two can be heated in the tube through which the mixed gases are passing. These processes of preparing sulphuric acid are interesting from a scientific point of view, but, as has been already stated (§ 224), they do not admit of commercial application.

Anhydrous sulphuric acid can readily be prepared by heating the Nordhausen acid (see § 239):—

20 or 30 grms. of fuming sulphuric acid are poured into a perfectly dry, small glass retort; the neck of the retort is thrust into a dry, cold receiver, and the acid is slowly heated until it boils moderately. The vapor of the anhydrous acid will condense and solidify in the receiver.

The anhydrous acid may also be obtained by distilling dry bisulphate of sodium. The bisulphate is prepared by heating a mixture of 3 parts, by weight, of dry sulphate of sodium and 2 parts of common sulphuric acid, until the mixture fuses. All the water of the acid is thus eliminated:—

$$Na_2O,SO_3 + H_2O,SO_3 = Na_2O,2SO_3 + H_2O.$$

The bisulphate of sodium, on being distilled in an eathern retort, will give up one molecule of anhydrous sulphuric acid, and a residue of normal sulphate of sodium will remain in the retort:—

$$Na_2O,2SO_3 = Na_2O,SO_3 + SO_3.$$

241. As thus prepared, anhydrous sulphuric acid is a glistening white solid mass of silky, crystalline fibres, somewhat resembling asbestos; it is tough and ductile, and can be moulded with the fingers like wax. So long as no water is present, it can be handled without danger; when perfectly dry, it is not corrosive, nor does it even react upon blue litmus. It unites with water, however, with great avidity, and is converted into common sulphuric acid. It rapidly absorbs water from the air and deliquesces; at the same time it forms dense fumes; for it is volatile, to a very considerable extent, at the ordinary temperature, and its vapor combines with the moisture of the air. On being brought in contact with a small quantity of water, it combines with it with explosive violence, and much heat is evolved. If a bit of it be thrown into a large quantity of water, the water hisses as if a hot iron had been thrust into it. Owing to its great tendency to deliquesce, the solid acid can only be preserved in dry tubes sealed at the lamp.

The specific gravity of anhydrous sulphuric acid is 1·97. It melts readily upon being heated; but it has been noticed that some samples melt far more easily than others. There appear to be two distinct varieties of the acid; for in some cases a temperature of 18° is sufficient to render the mass fluid, while in others the heat must be carried even to 100°. The easily fusible modification appears to change gradually, by keeping, into that which is more difficultly fusible; and the latter seems to be changed to the former by distillation. The melted acid boils at about 35°, and evolves a colorless and transparent vapor, three times as heavy as air, which, upon coming in contact with the air, unites with moisture and forms dense white fumes. When brought in contact with hot lime or baryta (oxide of calcium and oxide of barium), it unites with them directly; intense heat is evolved, and there is formed sulphate of calcium or sulphate of barium:—

$$CaO + SO_3 = CaSO_4 = CaO,SO_3.$$

242. On being exposed to a strong red heat, the vapor of anhydrous sulphuric acid splits up into oxygen and sulphurous acid—two volumes of it yielding two volumes of sulphurous acid and one volume of oxygen. As has been shown in § 217, two volumes of sulphurous acid gas contain one volume of sulphur vapor and two volumes of oxygen; hence it follows that the volumetric composition of anhydrous sulphuric acid is one volume of sulphur vapor and three volumes of oxygen, the whole condensed to two volumes. The specific gravity of sulphur vapor is 32, that of oxygen is 16, and the proportions, by weight, in which sulphur and oxygen are united in anhydrous sulphuric acid, are consequently 32 and $16 \times 3 = 48$, the combining weight of sulphuric acid being $32 + 48 = 80$. The combining weight of sulphuric acid can also be readily determined in a manner analogous to that employed in the case of nitric acid (§ 73), by saturating with the common acid a known quantity of oxide of lead, evaporating off the water and excess of acid, and then determining the weight of the dry sulphate of lead which is formed. By subtracting from the latter the weight of the original oxide of lead, we obtain the weight of the sulphuric acid which has combined with it. Experiment will show that the weight of this sulphuric acid is to that of the oxide of lead in the ratio of 80 to 223.

The facility with which sulphuric acid is decomposed at a red heat (§ 242) is the basis of a very economical method of preparing oxygen gas in large quantities for manufacturing-purposes. Commercial sulphuric acid is allowed to drop upon fragments of red-hot porcelain, there to be decomposed, in accordance with the formula

$$H_2SO_4 = SO_2 + O + H_2O,$$

and the products of the decomposition are then washed with water, so that the sulphurous acid may be absorbed, the steam condensed, and the oxygen left free. Here again, as in our earlier experiments (§ 10), the oxygen has really been obtained from the air; and if it were desirable, the solution of sulphurous acid obtained in washing this oxygen, might be placed in the leaden chambers and again be converted into sulphuric acid by the addition of oxygen from the air.

243. *Hyposulphurous Acid* (S_2O_2) has never been obtained in the free state, nor is any compound of it with water known; but there are numerous saline compounds of which it makes part, and some of these are of considerable importance in the arts. These salts, called hyposulphites, can be prepared in various ways,—for example, by digesting sulphur in a hot (but not boiling) concentrated solution of an alkaline sulphite. If sulphite of sodium be taken, the reaction can be thus formulated,

$$Na_2O,SO_2 + S = Na_2O,S_2O_2.$$

Another method of preparing the hyposulphites is to pass a current of sulphurous acid gas through the solution of an alkaline sulphide, until no further precipitation of sulphur occurs:—

$$2CaS + 3SO_2 = 2(CaO,S_2O_2) + S.$$

When a hyposulphite is treated with a strong acid, decomposition immediately ensues; S_2O_2 breaks up into SO_2+S; hence our inability to isolate the acid.

Some of the hyposulphites will be more fully described when we come to treat of the metals.

244. *Other Compounds of Sulphur and Oxygen.*—With the exception of hyposulphuric acid, S_2O_5, none of these compounds (see § 215) have been very thoroughly studied; any detailed

description of the methods of preparing them would be out of place in an elementary manual.

245. *Compounds of Sulphur and Chlorine.*—Chlorine and sulphur combine with one another directly and readily, forming several different compounds, whose properties vary in accordance with the varying proportions of chlorine and sulphur which they respectively contain.

246. *Dichloride of Sulphur* (SCl) is the best-known of the compounds of chlorine and sulphur, and is often called simply chloride of sulphur.

It can be prepared by passing a current of dry chlorine through a flask or tubulated retort containing flowers of sulphur. The chlorine is rapidly absorbed by the sulphur, and care must be taken lest the mass become too hot. The reddish-yellow liquid, obtained as the result of the reaction, is a solution of sulphur in dichloride of sulphur; by distilling it the excess of sulphur can be separated.

Dichloride of sulphur is a yellowish-brown liquid of 1·68 specific gravity, and boiling, without decomposition, at 144°. It emits a peculiar odor, which has been likened to that of sea-plants; its vapor excites tears, and its taste is acid, acrid, and bitter. It fumes strongly in the air, being decomposed by the moisture of the air with evolution of chlorhydric acid. It is decomposed by water, but can be mixed with bisulphide of carbon and with benzine. It is remarkable as a powerful solvent of sulphur; 100 parts of dichloride of sulphur can take up about 70 parts of sulphur at the ordinary temperature; on slowly cooling the hot saturated solution, beautiful crystals of sulphur are deposited. Dichloride of sulphur is used in a process of vulcanizing caoutchouc, known as the cold process.

247. *Chloride of Sulphur* (SCl_2).—This compound is formed when sulphur is treated with an excess of dry chlorine, or when a current of chlorine is passed into dichloride of sulphur; the dichloride requires some 278 times its own bulk of chlorine gas, and absorbs it very slowly. Chloride of sulphur is a red liquid, of 1·625 specific gravity. It exhales suffocating and irritating fumes of chlorine and the dichloride, since it slowly decomposes when kept. The decomposition is particularly rapid in a strong light; and so much gas is evolved that a tightly-stoppered bottle

containing chloride of sulphur will explode, after a time, if it be placed in sunlight. On being heated, the liquid gives off so much chlorine at 50° that it seems to boil; but the temperature gradually rises to 64°, which appears to be the real boiling-point of the liquid. It is slowly decomposed by water.

The density of its vapor has been found to be 53; admitting that the gas is composed of one volume of sulphur vapor and two volumes of chlorine, condensed to two volumes of vapor, the calculated specific gravity of its vapor would be 51·5. In view of the instability of the compound, the experimental result is sufficiently near coincidence with the calculated number to make it certain that the composition of the gas is really as above stated.

248. The other compounds of sulphur with chlorine, and with chlorine and oxygen, need not here be discussed; and the same remark applies to the compounds formed by the union of sulphur with iodine, bromine, fluorine, and nitrogen. The compounds of sulphur with carbon, phosphorus, arsenic, and the metals will be treated of hereafter.

CHAPTER XIV.

SELENIUM AND TELLURIUM.

249. These elements are rare, and of little or no industrial importance; but to the chemist they are exceedingly interesting, on account of the close resemblance they bear to sulphur. The three elements, sulphur, selenium, and tellurium, constitute a group which is equally remarkable with that formed by chlorine, bromine, and iodine. (See § 152.)

250. *Selenium*, Se,. is never found in any considerable quantity in any one place. Traces of it occur in many varieties of native sulphur, and in various metallic sulphides. It is now obtained chiefly from the sulphides of iron, copper, and zinc. These sulphides often contain minute traces of selenium, though the quantity is sometimes so small that it can hardly be detected

by the ordinary methods of analysis. When these sulphides are burned for the purpose of manufacturing sulphuric acid, or in metallurgical operations, the selenium goes off with the sulphurous acid produced by the combustion, and is deposited either in the dust-flues of the furnaces or upon the floors of the leaden chambers at the sulphuric-acid works. In this way the selenium from hundreds of tons of the pyritous ores is collected and concentrated to a comparatively small bulk. The deposit taken from the leaden chambers of some sulphuric-acid works contains as much as from 2 to 10 per cent. of selenium. The methods of obtaining pure selenium from these deposits are founded upon the facts that by treatment with nitric acid or aqua regia the selenium can all be oxidized and converted into selenious acid (SeO_2), that selenious acid is soluble in water, and that when a solution of it is treated with sulphurous acid, the selenious acid is reduced and pure selenium deposited.

$$SeO_2 + 2SO_2 = 2SO_3 + Se.$$

251. In its properties and in its chemical behavior, selenium resembles sulphur in many respects, while in others it is like tellurium. Like sulphur and oxygen, it occurs in distinct allotropic modifications (§§ 162, 196). The precipitate obtained by mixing solutions of sulphurous and selenious acids is of a deep red color, almost like that of cinnabar. But, after having been fused and suddenly cooled, selenium appears as a brilliant black mass, amorphous, like glass, and of 4·3 specific gravity. When fused selenium has been slowly cooled, it appears as a dark-grey, very brittle, crumbling mass, of crystalline or granular structure and a metallic lustre like that of lead; the specific gravity of this variety is 4·81. The amorphous or vitreous modification of selenium does not conduct electricity; but the granular or crystalline variety conducts it, and the more readily in proportion as it is hotter. The specific heat of selenium, at the ordinary temperature, is 0·0746, being the same for both the vitreous modification and that with metallic lustre. The vitreous variety is soluble in bisulphide of carbon, but the granular variety is insoluble in that liquid.

Selenium melts readily upon being heated, and the liquid thus obtained boils at about 700°, being converted into a dark yellow vapor, the specific gravity of which has been found to be 82·3. The atomic weight of selenium has been determined to be 79·5. This discrepancy between the vapor-density and the atomic weight is to be ascribed to the imperfection of the experimental determinations. Of itself, selenium has neither taste nor odor. When heated in the flame of a lamp, it burns with a beautiful blue flame and exhales a peculiarly offensive odor, like that of putrid horseradish,—selenious acid, SeO_2, being the chief product of the reaction.

Selenium combines with most of the elements, usually in the same way as sulphur, though not always, since it is a weaker chemical agent than sulphur; its compounds are as a rule somewhat less stable than the corresponding sulphur compounds. With oxygen it forms selenious acid, SeO_2, and selenic acid, SeO_3,—analogous to sulphurous and sulphuric acids respectively. Besides these, there is a lower oxide, SeO (?); it is a colorless gas, having the strong and disagreeable odor like horseradish before mentioned.

252. Both selenious and selenic acids form numerous salts, which closely resemble the corresponding sulphites and sulphates in composition and in many of their properties. Normal seleniate of potassium, for example, K_2SeO_4, cannot be distinguished, by its external appearance, from sulphate of potassium, K_2SO_4,—the crystalline form of the two bodies, as well as their texture, color, and lustre, being identical. If solutions of these two salts be mixed, neither of the salts can subsequently be crystallized out by itself when the solution is evaporated; the crystals obtained will be composed of sulphate of potassium and seleniate of potassium mixed in the most varied proportions. Bodies which are thus capable of crystallizing together in all proportions, without alteration of the crystalline form, are said to be *isomorphous* (like-formed). The formulæ of the two isomorphous salts, just mentioned, differ only in this—that the one contains the atom Se, where the other contains the atom S. It is therefore possible to replace 32 parts by weight of sulphur by 79·5 parts of selenium, or 79·5 of selenium by 32 of sulphur, without changing the crystalline form of the salts; it follows that 32 parts by

weight of solid sulphur, and 79·5 parts of solid selenium, occupy the same space. That this is actually the case may be shown by comparing the quotients obtained by dividing the atomic weights of the two elements by their specific gravities; these quotients will be found to be equal, or as nearly equal as the limits of error of the physical determinations involved will permit. The specific gravity of prismatic sulphur is 1·91, or, in other words, one cubic centimetre of solid sulphur weighs 1·91 gramme; the specific gravity of crystalline selenium is 4·81, or one cubic centimetre of selenium weighs 4·81 grammes; 32 grammes of sulphur will, therefore, occupy $\frac{32}{1 \cdot 91} = 16 \cdot 75$ cubic centimetres; 79·5 grammes of selenium will occupy $\frac{79 \cdot 5}{4 \cdot 81} = 16 \cdot 53$ cubic centimetres. What is true of grammes, is true of any parts by weight, and ultimately of the atoms. This quotient, obtained by dividing the atomic weight of an element by its specific gravity, is called the *atomic volume* of the element; it must be borne in mind that the standard of specific gravity for liquids and solids is water, for gases hydrogen, and, therefore, that the atomic volume of a solid or liquid must not be directly compared with that of a gas. Two elements whose atomic volume is the same can be exchanged in their compounds without alteration of crystalline form, precisely as a brick or stone taken out of a wall can be replaced by another of the same size and shape without changing the form of the wall.

253. With chlorine, selenium forms two compounds, SeCl and $SeCl_4$, the first of which is analogous to dichloride of sulphur. With hydrogen it forms a compound, H_2Se, called selenhydric acid, or seleniuretted hydrogen, which is perfectly analogous to sulphuretted hydrogen, H_2S, but possesses a still more disagreeable odor. In its action upon solutions of the metallic salts, upon metals and metallic oxides, selenhydric acid behaves like sulphydric acid, a selenide of the metal being always formed.

254. *Tellurium* (Te) occurs in nature even more rarely than selenium. Sometimes it is found in the free state, but more generally in combination with the heavy metals, such as gold, silver, lead, copper, and bismuth. It is one of the few elements with regard to which chemists have, at times, been in doubt whether

or not they should be classed as metals. Many of its physical properties are similar to those of the metals, and it particularly resembles the metal antimony; but it is so intimately related to sulphur and selenium in its chemical properties, its crystalline form, and mode of occurrence in nature, that it is now almost always studied as a member of the sulphur group.

255. Tellurium is of a silver-white color and glittering metallic lustre. It is hard and brittle, and crystallizes very easily in rhombohedrons. It is a bad conductor of heat and electricity. Its specific gravity is 6·2; its specific heat is 0·04737, and its atomic weight 128. It melts at a temperature somewhat above the melting-point of lead, and is volatile at a full red heat, the vapor being of a yellow colour, like that of selenium. When heated in the air, it takes fire, and burns with a greenish-blue flame, copious fumes of tellurous acid, TeO_2, being at the same time evolved.

256. The compounds of tellurium and oxygen (tellurous acid, TeO_2, and telluric acid, TeO_3) are analogous to sulphurous and sulphuric acids. By uniting with metallic oxides, they form numerous salts, analogous to, and isomorphous with, the corresponding compounds of sulphur and selenium. So, too, the hydrogen compound, H_2Te, is analogous to sulphuretted and seleniuretted hydrogen, in composition and properties. With the metals it unites directly to form tellurides. There are chlorine compounds also, $TeCl$ and $TeCl_4$.

257. The close relationship which subsists between sulphur and oxygen has been already alluded to, as well as the many points of resemblance between sulphur, selenium, and tellurium; the student is therefore now prepared to recognize the fact that in oxygen, sulphur, selenium, and tellurium we have another group or family of elements, as intimately and naturally related to each other as are the members of the chlorine group. (See § 152.) It will be seen at a glance that in passing from oxygen, at one end of the series, to tellurium, at the other, we meet with the same progression of physical and chemical properties that was so noticeable in passing from chlorine to iodine. The properties of the various compounds formed by the union of the members of the sulphur group with other elements exhibit the same kind

of progression; that these compounds are of analogous composition has been shown in the preceding paragraphs.

With hydrogen the members of the sulphur group unite in the proportion of two atoms of hydrogen to one atom of the other element; thus, H_2O, H_2S, H_2Se, H_2Te. This peculiar relation to hydrogen is an important characteristic of the group.

In this sulphur group, precisely as in the chlorine group, the relative chemical power of each element in the family is great in proportion as its atomic weight is low (§ 153); oxygen is, as a rule, stronger than sulphur, sulphur than selenium, and selenium than tellurium, their atomic weights being respectively:—

$$O = 16, \ S = 32, \ Se = 79{\cdot}5(80\,?), \ Te = 128.$$

CHAPTER XV.

COMBINATION BY VOLUME.

258. A comparison of the formulæ representing the volumetric composition of all the well-defined compound gases and vapors which have been thus far studied, will bring into clear view some of the general facts relating to combination by volume.

It has been established, by experiment, that the following compounds are formed by the chemical union, without condensation, of equal volumes of the two elements which enter into each compound:—

$$\underset{1\text{ vol.}}{\text{Hydrogen}} + \underset{1\text{ vol.}}{\text{Chlorine}} = \underset{2\text{ vols.}}{\text{Chlorhydric Acid}}, \text{ or } \underset{1}{H} + \underset{35{\cdot}5}{Cl} = \underset{36{\cdot}5}{HCl}$$

$$\underset{1\text{ vol.}}{\text{Hydrogen}} + \underset{1\text{ vol.}}{\text{Bromine}} = \underset{2\text{ vols.}}{\text{Bromhydric Acid}}, \text{ or } \underset{1}{H} + \underset{80}{Br} = \underset{81}{HBr}$$

$$\underset{1\text{ vol.}}{\text{Hydrogen}} + \underset{1\text{ vol.}}{\text{Iodine}} = \underset{2\text{ vols.}}{\text{Iodohydric Acid}}, \text{ or } \underset{1}{H} + \underset{127}{I} = \underset{128}{HI}$$

$$\underset{1\text{ vol.}}{\text{Nitrogen}} + \underset{1\text{ vol.}}{\text{Oxygen}} = \underset{2\text{ vols.}}{\text{Nitric Oxide}}, \text{ or } \underset{14}{N} + \underset{16}{O} = \underset{30}{NO}$$

It has further been demonstrated that the following compounds of two elements contain two volumes of one element and one volume of the other, but that these three volumes are condensed, during the act of combination, into two volumes :—

Hydrogen 2 vols.	+	Oxygen 1 vol.	=	Steam 2 vols.	, or	$\frac{H_2}{2}$	+	$\frac{O}{16}$	=	$\frac{H_2O}{18}$
Hydrogen 2 vols.	+	Sulphur 1 vol.	=	Sulphydric Acid 2 vols.	, or	$\frac{H_2}{2}$	+	$\frac{S}{32}$	=	$\frac{H_2S}{34}$
Hydrogen 2 vols.	+	Selenium 1 vol.	=	Selenhydric Acid 2 vols.	, or	$\frac{H_2}{2}$	+	$\frac{Se}{79{\cdot}5}$	=	$\frac{H_2Se}{81{\cdot}5}$
Hydrogen 2 vols.	+	Tellurium 1 vol.	=	Tellurhydric Acid 2 vols.	, or	$\frac{H_2}{2}$	+	$\frac{Te}{128}$	=	$\frac{H_2Te}{130}$
Chlorine 2 vols.	+	Oxygen 1 vol.	=	Hypochlorous Acid 2 vols.	, or	$\frac{Cl_2}{71}$	+	$\frac{O}{16}$	=	$\frac{Cl_2O}{87}$
Chlorine 1 vol.	+	Oxygen 2 vols.	=	Hypochloric Acid 2 vols.	, or	$\frac{Cl}{35{\cdot}5}$	+	$\frac{O_2}{32}$	=	$\frac{ClO_2}{67{\cdot}5}$
Nitrogen 2 vols.	+	Oxygen 1 vol.	=	Nitrous Oxide 2 vols.	, or	$\frac{N_2}{28}$	+	$\frac{O}{16}$	=	$\frac{N_2O}{44}$
Nitrogen 1 vol.	+	Oxygen 2 vols.	=	Hyponitric Acid 2 vols.	, or	$\frac{N}{14}$	+	$\frac{O_2}{32}$	=	$\frac{NO_2}{46}$
Sulphur 1 vol.	+	Oxygen 2 vols.	=	Sulphurous Acid 2 vols.	, or	$\frac{S}{32}$	+	$\frac{O_2}{32}$	=	$\frac{SO_2}{64}$
Selenium 1 vol.	+	Oxygen 2 vols.	=	Selenious Acid 2 vols.	, or	$\frac{Se}{79{\cdot}5}$	+	$\frac{O_2}{32}$	=	$\frac{SeO_2}{111{\cdot}5}$
Tellurium 1 vol.	+	Oxygen 2 vols.	=	Tellurous Acid 2 vols.	, or	$\frac{Te}{128}$	+	$\frac{O_2}{32}$	=	$\frac{TeO_2}{160}$

Lastly, still a third mode of combination by volume with condensation of four volumes to two has been thoroughly studied in the case of ammonia, and has been further illustrated in the composition of anhydrous sulphuric acid :—

Nitrogen 1 vol.	+	Hydrogen 3 vols.	=	Ammonia 2 vols.	, or	$\frac{N}{14}$	+	$\frac{H_3}{3}$	=	$\frac{NH_3}{17}$
Sulphur 1 vol.	+	Oxygen 3 vols.	=	Sulphuric Acid 2 vols.	, or	$\frac{S}{32}$	+	$\frac{O_3}{48}$	=	$\frac{SO_3}{80}$

Throughout these tables the unit-volume is, of course, the same for every element and compound. What the absolute bulk of this unit-volume may be, is not an essential point; for the relations remain the same, whatever the unit of measure. Some chemists have thought that an advantage was gained by using the bulk of one gramme of hydrogen at the ordinary pressure and temperature, viz. 11·2 litres, as the unit-volume, while others prefer to use the litre itself as the unit.

Three condensation-ratios are exhibited in these tables. In the first the condensation is 0 ; in the second it is $\frac{1}{3}$, and in the third it is $\frac{1}{2}$. The typical character of the three compounds, chlorhydric acid, water, and ammonia, is also clearly brought out ; each of these bodies represents a group of compounds which obey the same structural law. The tables also show very clearly the fact that very unequal weights of the compounds tabulated occupy equal spaces, under the same conditions of temperature and pressure. The space occupied by the compound molecule is, in each case, exactly twice the unit-volume.

259. The symbols H, Cl, O, and N represent the relative weights of the same volume of four elements which are gaseous at common temperatures and pressures ; the symbols Br, I, S, and Se, represent the relative unit-volume weights of four other elements which are not gases under the ordinary atmospheric conditions, but which can be converted into gases at a higher temperature. At this higher temperature their unit-volume weights have been experimentally determined, and from these observed volume-weights, the unit-volume weights which they would possess at the ordinary pressure and temperature have been deduced. The symbols of these eight elements, therefore, represent at once the combining weights and the relative weights of equal volumes (specific gravities) of these substances in the gaseous state. In the present state of the science, these eight symbols are the only ones of which this can be affirmed ; tellurium would undoubtedly make a ninth, if the relative size of its combining weight had been experimentally determined, but until this determination has been made, the symbol Te represents only the combining weight of the element, and not its equal-volume weight as well.

The relative sizes of the combining weights of four other elements in the state of vapor, have been experimentally ascertained. These four elements are arsenic, phosphorus, cadmium, and mercury. When we come to study these elements, we shall find that the symbols of arsenic and phosphorus, namely, As and P, represent only the half-volume weights of these two bodies, while the symbols of cadmium and mercury represent the two-volume weights of these volatile metals. Coincidence of the combining weight and the volume-weight has been established for eight elements; discrepancy between the combining weight and the volume-weight has been proved for four elements; of the remaining elements, constituting more than four-fifths of the whole number, the equal-volume weights are wholly unknown, inasmuch as these elements have never been converted into vapor under conditions which permit the experimental determination of the equal-volume weights of their vapors. For example, the symbols Na and K represent the combining weights of these two metals; but they can be held to represent the weights of the unit-volumes of these metals only by pure assumption, or, at best, on the uncertain evidence of analogies, since the unit-volume weights of these metals, when converted by intense heat into gases, have never yet been determined. As the great majority of the known elements cannot be volatilized, or made gaseous, by the highest temperatures as yet at our command, under conditions which permit the chemist to experiment with the gases produced, it is plain that composition by weight is, in the present state of chemistry, of far greater practical importance than composition by volume. The symbols of all the elements represent their combining weights, as determined by ponderal analysis; the symbols of eight elements represent also the equal-volume weights of the substances they stand for. These eight elements, though few in number, are nevertheless the leading elements in inorganic chemistry.

260. The volume of the molecule of every compound gas in the above tables is twice that occupied by the atom of hydrogen. *Two* volumes of compound gas invariably result from the chemical combination of one volume of hydrogen with one volume of chlorine, of two volumes of hydrogen with one of oxygen, of three volumes of hydrogen with one of nitrogen, and these instances are but types of large classes of chemical reactions. In organic chemistry the same law holds good for a great multitude of complicated compounds of carbon; the molecule of every organic compound in the state of vapor occupies a volume twice as large as that occupied by an atom of hydrogen, or, in other words,

twice the unit-volume. This doubled volume is often called the normal or *product*-volume of a compound gas. Since the combining weight of a compound gas or vapor occupies two unit-volumes, it is obvious that the weight of one volume, which is the specific gravity of the gas or vapor, is deduced from the combining weight by dividing the latter by two. The specific gravity of a compound gas or vapor is, therefore, one-half its combining weight.

261. *Molecular condition of elementary gases.*—Bearing in mind our definitions of atom and molecule (§§ 38, 39), let us inquire what inferences concerning the molecular condition of simple gases in a free state can be legitimately drawn from our knowledge of the molecular condition of compound gases. To give definiteness to our conceptions, let us assume the unit-volume of the elements to be one litre; the product-volume of a compound will then be two litres. Two litres, the product-volume, of chlorhydric acid gas are made up of one litre of hydrogen and one litre of chlorine, united without condensation, and each molecule of chlorhydric acid must contain at least one atom of hydrogen and one of chlorine. In these two litres of chlorhydric acid there must be some definite number of molecules; the number is, of course, indeterminable; but let us assign to it some numerical value, say 1000, in order tc give clearness to our reasoning. One litre of chlorhydric acid will then contain 500 molecules, and since equal volumes of *all* gases, whether simple or compound, are assumed to contain, under like conditions, the same numbers of molecules (§ 39), one litre of hydrogen or of chlorine will also contain 500 *molecules.* But the one litre of hydrogen and the one litre of chlorine, which, by uniting, produced 2 litres =1000 molecules of chlorhydric acid, must each have contained 1000 *atoms* of hydrogen and of chlorine respectively, for each molecule of chlorhydric acid demands an atom of hydrogen and an atom of chlorine. The litre of hydrogen, or of chlorine, then, contains 500 molecules, but 1000 atoms,—each molecule of the simple gas being made up of two atoms of the single element, just as each molecule of the compound gas under review is composed of two atoms, one of hydrogen and one of chlorine. It is clear that this train of reasoning is independent of the particular numerical value assumed as the number of molecules in two litres of chlor-

hydric acid. If, therefore, the molecule of chlorhydric acid is represented by the formula HCl, and the diagram

H	+	Cl	=	HCl

there is good reason to assign to *free* hydrogen and *free* chlorine the formulæ HH and ClCl, and to represent the constitution of all uncombined gases by such diagrams as

H	+	H	=	HH	,	Cl	+	Cl	=	ClCl

Upon these models the molecular formulæ of all the elements with which we have become acquainted might readily be written. It is only in a *free* state that the elementary gases and vapors are thus conceived to exist as molecules; when they enter into combination, it is by atoms rather than by molecules. An atom of hydrogen unites with an atom of chlorine; three atoms of hydrogen combine with one of nitrogen.

If this view of the molecular structure of free elementary gases and vapors be correct, perfect consistency would require that no equation should ever be written in such a manner as to represent less than two atoms, or one molecule, of an element in a free state as either entering into or issuing from a chemical reaction. Thus instead of $H_2+O=H_2O$, $N+3H=NH_3$, $HCl+Na=NaCl+\mathbf{H}$, it would be necessary to write

$$2HH + OO = 2H_2O, \qquad NN + 3HH = 2NH_3,$$
$$2HCl + NaNa = 2NaCl + HH.$$

We have not heretofore conformed to this theoretical rule, and do not propose to do so in the succeeding pages, and this for two reasons:—first, because many equations, representing chemical reactions, must be multiplied by two in order to bring them into conformity with this hypothesis concerning molecular structure; the equations are thus rendered unduly complex; secondly, because, in undertaking to make chemical equations express the molecular constitution of elements and compounds, as well as the equality of the atomic weights on each side of the sign of equality, there is imminent danger of taking the student away from the sure ground of fact and experimental demonstration, into an uncertain region of hypotheses based only on definitions and analogies. The symbol Na represents 23 proportional parts by weight of the metal sodium; of the molecular symbol NaNa, the most that can be said is, that some strong analogies justify us in assuming, for the present, in default of any experimental evidence on the subject, that a molecule of free sodium gas, if we could get at it, would be found to consist of two least combining parts by weight of sodium. We know as much, at least, of the molecular structure of sodium as we do of four-fifths of the recognized chemical elements. For the present, the biatomic structure of the molecule of a simple gas or vapor in the free state must take place, in an elementary manual, as an ingenious and philosophical hypothesis, rather than as a general and indubitable fact.

CHAPTER XVI.

PHOSPHORUS.

262. Phosphorus occurs somewhat abundantly and very widely diffused in nature. It is never found in the free state, but almost always in combination with oxygen and some one of the metals. The most abundant of its compounds is phosphate of calcium; small quantities of this mineral are found in most rocks and soils, and in several localities it occurs in large beds. Phosphate of calcium is the chief mineral constituent of the bones of animals; it contains one-fifth of its own weight of phosphorus. The proportion of phosphorus present in most of the ordinary rocks, and in the soils which have resulted from their disintegration, is usually very small, and phosphorus would be an exceedingly costly substance if we were compelled to collect it directly from this source; but it so happens that the phosphorus-compounds are important articles of food for plants and animals, and it is easy to obtain through their intervention the phosphorus which was before widely diffused, but has been by them concentrated.

Growing plants seek out and collect the traces of phosphorus-compounds which exist in the soil; the herbivorous animals in their turn consume the phosphorus which has been accumulated by the plants, and from the bones of animals chemists and manufacturers derive the phosphorus of which they stand in need.

Like oxygen and sulphur, phosphorus occurs in several distinct allotropic modifications. Of these, the best-known are called respectively ordinary phosphorus and red phosphorus.

263. Ordinary phosphorus, when perfectly pure, is a transparent, colorless, wax-like solid of 1·8 specific gravity, which, when freshly cut, emits an odor like garlic, though under ordinary conditions this odor is overpowered by the odor of ozone, which, as has been previously stated (§ 164), is developed when phosphorus is exposed to the air. It unites with oxygen readily, even at the ordinary temperature of the air, and with great energy at somewhat higher temperatures (above 60°); when in contact with air it is all the while undergoing slow combustion.

Exp. 106. Thoroughly wash a piece of phosphorus by rinsing it in successive large quantities of water; place it, for a moment, upon a sheet of filter-paper, in order that a portion of the water adhering to it may be removed, then lay it upon a clean porcelain capsule, and at short intervals press against it a slip of blue litmus paper. In a very few moments the color of the paper will be changed to red; for the products of the oxidation of phosphorus are acid, and they are formed with great rapidity.

If the temperature of the slowly burning phosphorus be slightly increased in any way, the mass will burst into flame and be rapidly consumed. On account of this extreme inflammability, phosphorus must always be kept under water; it is best also to cut it under water, lest it become heated to the kindling-point by the warmth of the hand or by friction against the knife. When wanted for use, the phosphorus is taken from the water and dried by gently pressing it between pieces of blotting-paper.

Phosphorus must always be handled with great caution; for when once on fire, it is exceedingly difficult to extinguish it, and, in case it happens to burn upon the flesh, painful wounds are inflicted, which are exceedingly difficult to heal. Whenever phosphorus is cut or broken, care must likewise to taken that no small fragments of it fall unobserved into cracks of the table or floor, where they might subsequently take fire.

Exp. 107.—Nip a piece of phosphorus, as large as a small pea, between two bits of wood, in such manner that a part of the phosphorus shall project below the wood; rub the phosphorus strongly upon a sheet of coarse paper; it will take fire at the temperature developed by the friction.

264. On account of this easy inflammability by friction, phosphorus is extensively employed for making matches. The matter upon the end of an ordinary friction match usually contains a little phosphorus, together with some substance capable of supplying oxygen, such as red-lead, black oxide of manganese, saltpetre, or chlorate of potassium. The phosphorus and the oxidizing agent are kneaded into a paste made of glue or gum, and the wooden match-sticks, the ends of which have previously been dipped in melted sulphur, are touched to the surface of the phosphorized paste, so that the sulphured points shall receive a coating of it. The sulphur serves merely as a kindling material which, as it were, passes along the fire from the phosphorus to the wood. By rubbing the dried, coated point of the match against a rough surface, heat enough is developed to bring about chemical action between the phosphorus and the oxygen of the other ingredient, combustion ensues, the sulphur becomes hot enough to take on oxygen from the air, and finally the wood is involved in the play of chemical force.

Exp. 108.—Put a piece of phosphorus, as big as a grain of wheat, upon a piece of filter-paper, and sprinkle over it some lampblack or powdered bone-black. The phosphorus will melt after a time, and will finally take fire. As will be more fully explained hereafter, under carbon, the porous, finely divided lampblack has the power of absorbing and condensing within its pores much oxygen from the air; heat is developed by the act of condensation, and, at the same time, oxygen is brought into very intimate contact with the phosphorus, particularly with the vapor of phosphorus which is condensed by the lampblack together with the oxygen, so that chemical action soon results, and ultimately fire. Both the lampblack and the paper are bad conductors of heat; they prevent the phosphorus from losing the heat developed by the condensation and by the slow action of oxygen.

It is remarkable that when dry phosphorus, in very thin slices, is laid upon fine feathers, wool, lint, flannel, dry wood, or other non-conducting substances, it quickly melts, and readily inflames upon the slightest friction, heat enough being produced by the slow combustion of the phosphorus to fuse it, if only this heat can be retained by some bad conductor.

265. At the ordinary temperature of the air, and still more at somewhat higher temperatures, phosphorus shines with a greenish-white light, as may be seen by placing the phosphorus in the dark; hence the name, phosphorus, from Greek words signifying light-bearing. This phosphorescence is seen when an ordinary friction-match is rubbed against any surface in a dark room. Although the phenomena of phosphorescence and of oxidation, or slow combustion, occur simultaneously when phosphorus is exposed to the air, it does not appear that the phosphorescence is a consequence of the oxidation; for phosphorus shines not only in the air, but also when placed in an atmosphere of pure hydrogen, or nitrogen, or carbonic acid, or even in a vacuum, though the light emitted by phosphorus in these inert gases is of different appearance from that developed in presence of oxygen.

266. In warm weather phosphorus is soft and somewhat flexible; it may then be bent without breaking, can be scratched with the nail and cut with a knife like wax; but at 0° it is brittle, and exhibits a crystalline fracture when broken. It melts at 44°, forming a viscid oily liquid, which boils at about 290°, and is converted into colorless vapor. Phosphorus can readily be distilled in a retort filled with some inert gas, like hydrogen, nitrogen, or carbonic acid. The specific gravity of its vapor has been found to be 62·1. Contrary to all our previous experience, however, the density of phosphorus is not identical with its atomic weight, a point which will be discussed when the compounds of phosphorus and hydrogen are treated of.

On being heated to about 230°, out of contact with air, it is converted into red phosphorus. (See Exp. 111.) By exposure to light, also, phosphorus undergoes a certain amount of change; hence it is rarely seen in the perfectly colorless, transparent condition which it exhibits when recently prepared and perfectly pure. The phosphorus of commerce is usually of a light amber-color. When kept for some time under water, phosphorus becomes covered with a white opaque coating, which appears to be a result of the oxidizing action of air held in solution by the water; the surface of the phosphorus is irregularly corroded by this dissolved oxygen, and is thus roughened and made opaque, in much the same way that the transparency of glass is destroyed by grinding one of its surfaces. It is noticed, for that matter, that the water in which phosphorus is kept soon becomes strongly acid; for it dissolves the oxygenated compounds which are produced by the action of the dissolved air upon the phosphorus. The specific heat of solid phosphorus is 0·1788; of liquid phosphorus, 0·2045. It is a non-conductor of electricity, both in the solid and in the liquid state.

Phosphorus is insoluble in water, but is somewhat soluble in ether, petroleum, benzine, oil of turpentine, and other oils; it dissolves abundantly in bisulphide of carbon, in chloride of sulphur, and in sulphide of phosphorus.

Exp. 109.—Pour into a phial of the capacity of 80 or 90 c. c., 10 or 12 c. c. of bisulphide of carbon, and throw into this liquid a bit of phosphorus as large as a pea. Cork the phial, and shake its contents, at intervals, until the phosphorus has dissolved. Preserve the solution for use in subsequent experiments.

267. From the solution in chloride of sulphur and from that in sulphide of phosphorus, crystals of phosphorus, usually in the form either of regular octahedrons or of rhombic dodecahedrons, can be obtained; but, owing to the slowness with which phosphorus passes from the liquid to the solid state, distinct crystals cannot readily be prepared by the method of fusion, unless a comparatively large quantity of phosphorus be operated upon.

268. When a solution of phosphorus in ether, or, better, in bisulphide of carbon, is poured upon the surface of any porous substance and left to evaporate in the air, the volatile solvent will quickly escape, leaving the phosphorus behind in a very finely divided condition. In proportion as a substance is more finely divided, the greater will be the surface which it presents to the oxygen of the air, and the more readily will it combine with this oxygen. In the case before us, the comminuted phosphorus absorbs oxygen very rapidly, and this chemical action is attended with the evolution of so much heat that the phosphorus will take

fire, if the material upon which it has been deposited is a bad conductor of heat.

Exp. 110.—Pour some of the solution of phosphorus obtained in Exp. 109, upon a sheet of filter-paper, and hang the paper upon the iron stand in such manner that the bisulphide of carbon may freely evaporate. The paper will soon burst into flame. It will be noticed that the paper is not completely consumed, but that a very considerable residue of carbon remains unburned. This depends upon the fact that the product of the combustion of the phosphorus, phosphoric acid, quickly covers the paper with a varnish which is not only incombustible in itself, but is quite impervious to air.

In lack of bisulphide of carbon, this experiment can be performed with the ethereal solution of phosphorus, prepared in the manner described in Exp. 109, excepting that common ether is substituted for the bisulphide, and that the mixture is left to digest for a day or two.

269. Ordinary phosphorus is a violent poison, a few decigrammes of it being sufficient to destroy human life. It is the efficient ingredient of many preparations used for poisoning rats, cockroaches, and other vermin. Phosphorus evaporates rather freely at the ordinary temperature of the air; and the vapor has been found to be exceedingly injurious to persons constantly exposed to it. The makers of friction matches are subject to a horrible wasting disease, one of the symptoms of which is the destruction of the bones of the jaws.

270. When phosphorus burns with flame in free air, two atoms of it unite with five atoms of oxygen, and there is formed the compound of P_2O_5; this highest oxide of phosphorus is called phosphoric acid. This compound occurs in bones, and from it phosphorus is prepared. Bone-earth, that portion of bones which remains after all the organic matter has been burnt off in the fire, consists mainly of triphosphate of calcium, $Ca_3P_2O_8$. In order to obtain phosphorus from bone-earth, the calcium and the oxygen must both be removed; the calcium is removed by means of sulphuric acid, the oxygen by means of hot charcoal. Bones are burnt to a white ash (calcined, as the term is), then finely powdered and mixed with a quantity of dilute sulphuric acid. The sulphuric acid removes two of the atoms of calcium and forms sulphate of calcium, while there remains monophosphate of

calcium (superphosphate of lime) in accordance with the following reaction:—

$$Ca_3P_2O_8 + 2H_2SO_4 = CaH_4P_2O_8 + 2CaSO_4.$$

It will be remembered that one atom of calcium replaces **two** atoms of hydrogen. (See p. 89.)

The solution of monophosphate of calcium is then filtered off from the insoluble sulphate of calcium, and evaporated to the consistence of syrup; the syrup is mixed with powdered charcoal, and the mixture dried at a dull red heat; by this means a quantity of water is expelled from the monophosphate:—

$$CaH_4P_2O_8 = CaP_2O_6 + 2H_2O.$$

The porous dry mixture is finally placed in retorts of fireclay and intensely ignited. At high temperatures, charcoal is a powerful deoxidizing agent; it takes away oxygen from the phosphate of calcium, and forms carbonic oxide, which goes off as a gas; phosphorus is thus set free, and distilling over into an appropriate receiver, is condensed under cold water, a quantity of triphosphate of calcium is at the same time reproduced and remains in the retort:—

$$3CaP_2O_6 + 10C = 10CO + 4P + Ca_3P_2O_8.$$

If a quantity of sand (silicic acid) be added to the mixture of charcoal and monophosphate, the whole of the phosphorus can be expelled,—the phosphate of calcium, which would otherwise escape decomposition, being entirely converted into silicate of calcium,

$$2CaP_2O_6 + 2SiO_2 + 10C = 10CO + 4P + 2CaSiO_3.$$

Another proposed method is to pass chlorhydric acid gas over a mixture of bone phosphate and charcoal, maintained at a red heat, in a cylinder of fireclay. By this means all of the phosphorus is set free and chloride of calcium remains:—

$$Ca_3P_2O_8 + 8C + 6HCl = 3CaCl_2 + 8CO + 6H + 2P.$$

The crude phosphorus thus obtained is remelted, and purified by filtration, redistillation, and by chemical treatment with a mixture of bichromate of potassium and sulphuric acid, which oxidizes the principal contaminations. The purified phosphorus is finally remelted and cast into the sticks or cakes in which it is found in commerce.

271. *Red Phosphorus.*—This remarkable allotropic modification of phosphorus is a body as unlike ordinary phosphorus in most respects as could well be conceived. It is of a scarlet-red color, has neither odor nor taste, is not poisonous so far as is known, is not phosphorescent, does not take fire at ordinary temperatures, is insoluble in bisulphide of carbon, and in general behaves altogether differently from the ordinary modification. Yet it is no difficult matter to change one of these modifications into the other. For example, if red phosphorus be heated to about 260°, in an atmosphere of nitrogen, or other inert gas, it will pass into the condition of ordinary phosphorus without undergoing any alteration of weight, or, in other words, without absorbing or disengaging anything.

Exp. 111.—In a narrow glass tube, No. 6, about 30 c.m. long and closed at one end, place a quantity of red phosphorus as large as a small pea; heat the phosphorus gently over the gas-lamp, and note that a sublimate of a light-colored substance is quickly deposited upon the cold walls of the tube a short distance above the heated portion. This light-colored sublimate is ordinary phosphorus, as may be shown by cutting off the tube just below the sublimate, after the glass has been allowed to cool, and then scratching the coating with a piece of wire; the coating will take fire. The air in the narrow tube employed is deprived of its oxygen by the combustion of a small portion of the phosphorus at the moment of its transformation from the red to the ordinary condition; the remaining phosphorus is thus enveloped in nitrogen and so protected from further loss.

272. Red phosphorus is itself neither volatile nor inflammable; it neither rises as vapor nor inflames at temperatures lower than 260°, the point at which it changes into ordinary phosphorus; at 250° it suffers no alteration. As compared with ordinary phosphorus, it may be said that red phosphorus can be handled without danger, and that it may be kept in bottles without special precautions, since it is not liable to take fire by moderate friction; but by powerful friction heat enough may be evolved to convert it into ordinary phosphorus, and if it be even moderately heated, by friction, or in any other way, in contact with oxidizing agents, it instantly bursts into flame

Exp. 112.—In order to observe the comparative difficulty of inflaming red phosphorus, lay an inverted cover of a porcelain crucible upon an iron triangle upon the lamp-stand; place upon the cover, which may be 15 c.m. wide, a small bit of ordinary phosphorus, and, at a distance of 12 c.m., the same quantity of red phosphorus; heat the cover gently and gradually over the gas-lamp. The ordinary phosphorus will soon inflame and burn away; but a considerable space of time will elapse before the red phosphorus takes fire.

By operating in vessels filled with nitrogen, or some other gas which had no chemical action upon phosphorus, the precise temperature at which the red phosphorus ceases to exist can be noted, and the ordinary phosphorus obtained from it can be distilled over and collected.

273. Red phosphorus has been obtained in crystals by dissolving common phosphorus in melted lead, and subjecting the fluid mass to a high temperature for several hours in closed tubes. When the lead cools, the phosphorus separates in thin crystals, which have a metallic lustre and a black color; the crystals, however, transmit a yellowish-red light, and the thinnest of them appear, not black, but red. These crystals of red phosphorus are generally enveloped in the lead; but the lead may be mostly dissolved away by dilute nitric acid, and the phosphorus crystals may thus be obtained in a condition of comparative purity. They are not affected by exposure to the air. These crystals are seen under the microscope to be rhombohedrons; so that phosphorus, like the succeeding members of the family of elements to which it belongs, is dimorphous, presenting forms both of the monometric and hexagonal systems.

The red variety of phosphorus has been not inaptly called *metallic* phosphorus, crystallized in the crystals just described, and amorphous in the usual form of red phosphorus. The crystallized *metallic* phosphorus is less volatile, and has a higher specific gravity than the amorphous. The power of the so-called metallic phosphorus to conduct electricity is small, if compared with that of the common metals, but it is very much greater than the conducting-power of colorless phosphorus, for this latter substance is generally classed with the insulators.

The specific gravity of amorphous red phosphorus is 2·14; its specific heat is 0·1698. When dry, it undergoes no change at the ordinary temperature of the air; but in moist air it oxidizes very slowly. It is easily soluble in nitric acid, which oxidizes it; and since it is much more readily dissolved than ordinary phosphorus, the latter can be purified from any contamination of red phosphorus, by digesting it at a gentle heat in dilute nitric acid.

274. Amorphous red phosphorus is prepared by maintaining ordinary phosphorus, for some time, at a temperature of 230° to 235°, either under water in an air-tight vessel, or in an atmosphere of some gas which has no chemical action upon phosphorus. It is manufactured upon the large scale by heating ordinary phosphorus in a cast-iron vessel provided with a gas delivery-tube dipping into mercury outside the vessel, in such manner that, while the expanded air and some escaping vapors of phosphorus can pass out, no air can enter the vessel. About 200 kilos. of phosphorus are taken for a single charge; this quantity of phosphorus is maintained during ten days or more, as nearly as may be, at the temperature of 240°, care being taken that the heat shall, at no time, much exceed this limit. Under these conditions, the ordinary phosphorus slowly changes into the red variety. After the phosphorus has been exposed during the time which the previous experience of the manufacturer has shown to be most advantageous, the apparatus is allowed to become cold, and the transmuted phosphorus is found adhering to the sides of the vessel, in the shape of a hard, brittle, brick-colored coating, which can be removed by means of hammer and chisel, after covering it with water. It is ground to powder under water, and any particles of ordinary phosphorus which have escaped change are removed from it by means of bisulphide of carbon, or by a solution of caustic soda, which dissolves ordinary phosphorus without acting upon the red modification.

275. Red phosphorus is employed, to a certain extent, as an adjunct to the so-called *safety-matches*. Such matches contain no phosphorus in themselves, and will not take fire readily by friction upon an ordinary rough surface, though they burst into flame at once when rubbed upon a surface specially prepared with red phosphorus. The matter upon the tips of safety-matches is usually a mixture of chlorate of potassium and sulphide of antimony, made into a paste by means of glue; the surface upon which the match is to be rubbed is composed of red phosphorus,

black oxide of manganese, and glue. In favor of the use of red phosphorus for matches are the facts, that, unlike ordinary phosphorus, it is not deleterious to the workmen who have to deal with it, and that it is far less liable to be set on fire by accidental friction. For these reasons, the manufacture of safety-matches has been encouraged by the governments of several European countries, and such matches are now much used in France and upon other parts of the continent, though they are manifestly less convenient, in several respects, than the ordinary matches, which can be ignited by friction upon any rough surface.

276. Phosphorus combines readily with many other elements besides oxygen. The ordinary modification of phosphorus combines violently with sulphur at temperatures near the melting-point of sulphur, the act of combination being attended with vivid combustion and loud explosion. Red phosphorus, on the other hand, does not combine with sulphur at temperatures lower than 230°, and the combustion, though rapid, is not explosive. With chlorine, bromine, and iodine, ordinary phosphorus unites directly at the ordinary temperature of the air, the combination being rapid and attended with inflammation. Red phosphorus also unites with chlorine, bromine, and iodine at the ordinary temperature; and much heat is evolved during the act of combination, though the amount of heat is usually insufficient to produce ignition.

Phosphorus unites directly with most of the metals also; and several of the compounds thus formed closely resemble the so-called *alloys*, or compounds of one metal with another. With hydrogen it forms several interesting compounds, which will be described directly. From the remarkable facility with which it combines with oxygen (§§ 263, 264), it follows necessarily that phosphorus is a powerful reducing agent. Many oxygen compounds can be decomposed by means of it. When immersed in the vapor of anhydrous sulphuric acid, phosphorus takes fire after a time, and combines with the oxygen of the acid, while sulphur is deposited. Monohydrated sulphuric acid is reduced to sulphurous acid, while phosphorous acid is formed:—

$$3H_2SO_4 + 2P = 2H_3PO_3 + 3SO_2.$$

A solution of sulphurous acid, on being heated with phosphorus, yields phosphorous and sulphydric acids, as follows:—

$$SO_2 + 4H_2O + 2P = 2H_3PO_3 + H_2S.$$

When gently heated with chlorate or with nitrate of potassium, or with other highly oxygenated bodies, like the peroxides of lead and manganese, phosphorus combines with their oxygen so rapidly that an explosion ensues; heat enough to bring about the reaction can be developed by gentle friction, as when the phosphorus and the other ingredient are rubbed together upon some hard surface. (Compare § 264.)

Exp. 113.—Provide a bit of ordinary phosphorus, as large as a pin's head, also an equal quantity of red phosphorus; add to each of these portions enough finely powdered chlorate of potassium to cover the phosphorus; fold up each of the mixtures tightly and separately in a small piece of writing-paper; place the parcels, one after the other, upon an anvil and strike them sharply with a hammer. They will explode with violence.

277. *Compounds of Phosphorus and of Hydrogen.*—There are three compounds of phosphorus and hydrogen, one gaseous, PH_3, one liquid, PH_2, and one solid, P_2H, at ordinary temperatures. The gaseous compound, or, rather, the gaseous compound charged with the vapor of the liquid compound, is somewhat interesting, from the fact that it takes fire spontaneously immediately on coming into contact with the air.

Exp. 114.—In a thin-bottomed flask of about 140 c. c. capacity, put 1 gramme of phosphorus and 115 c. c. of potash-lye of 1·27 specific gravity, obtained by dissolving 40 grms. of hydrate of potassium in 110 c. c. of water. Pour two or three drops of ether upon the liquid in the neck of the flask, then close the flask with a cork carrying a long gas-delivery-tube of glass No. 5. Place the flask over the gas-lamp, upon the wire-gauze ring of the iron stand, and immerse the end of the delivery-tube in the water-pan, then gently heat the flask. The ether is added to the contents of the flask in order that the last traces of air may be expelled from the flask by the vapor which arises from this highly volatile liquid so soon as it is warmed.

As the potash-lye becomes hot, small bubbles of gas will be seen to arise from the surface of the phosphorus, and in a short time large bubbles of gas will escape from the delivery-tube; each of these bubbles, as it comes in contact with the air at the surface of the water, will spontaneously burst into flame, and burn with a vivid light and the formation of beautiful rings of white smoke, if the air be not disturbed by draughts. In burning, the phosphuretted hydrogen is converted into phosphoric acid and water, or, rather, into hydrated phosphoric acid; and of this product the white smoke is, of course, composed.

$$2PH_3 + 8O = H_6P_2O_8.$$

Exp. 115.—Place a small inverted bottle full of water over the end of the delivery-tube from which the phosphuretted hydrogen is escaping, as in Exp. 114, and collect 50 or 100 c. c. of this gas. By single bubbles pass the gas thus collected into a litre bottle half-full of oxygen standing inverted upon a shelf in the water-pan. The phosphuretted hydrogen will burn much more vividly in oxygen than in the air. In case, however, several successive bubbles of the gas should fail to inflame on coming into contact with the oxygen, the experiment must be interrupted and the oxygen thrown away, for the introduction of another bubble of phosphuretted hydrogen into this explosive mixture might set fire to it and so shatter the bottle.

278. The reaction which occurs during the preparation of phosphuretted hydrogen is chiefly between water and phosphorus. Phosphorus and water by themselves do not react upon each other, but when in presence of powerful bases, like soda, potash, lime, or baryta, water is decomposed by phosphorus with formation both of oxygenated and hydrogenized phosphorus compounds:—

$$3(K_2O,H_2O) + 8P + 6H_2O = 2PH_3 + 3(K_2O,2H_2O,P_2O).$$

Hypophosphite of Potassium.

Another method of obtaining phosphuretted hydrogen is by decomposing phosphide of calcium with water.

Exp. 116.—Prepare a number of small balls or sticks of quick lime by moulding moistened slaked lime into these forms and then drying and calcining the product. Select a tube of hard glass, No. 2, close it at one end, and place two or three pieces of phosphorus as large as peas at the closed end; fill the tube with the pellets of quick lime, and put it in a sheet-iron trough above a wire-gauze gas-lamp, in the manner depicted in Fig. 8. To prevent the melted phosphorus from flowing against the quick lime, an iron nail may be laid beneath that part of the trough which is farthest from the phosphorus. Heat to redness the portion of the tube which contains the lime, and then cause the vapor of phosphorus to pass over it by cautiously heating the closed end of the tube with an ordinary gas-lamp. To ensure the success of this experiment, that portion of the tube which contains the phosphorus must be heated so slowly that none of the phosphorus can escape uncombined through the lime. After the phosphorus has all been driven forward from the closed end of the tube, the open end of the tube should be stopped with a cork and the lamps should be extinguished; the tube is then left at rest until it has become cold.

When a piece of the impure phosphide of calcium thus obtained is thrown into water, it slowly decomposes with formation of hypophosphite of calcium and disengagement of phosphuretted hydrogen; the bubbles of gas take fire as they reach the surface of the water.

279. Besides the spontaneously inflammable gas, there is another variety of phosphuretted hydrogen which does not take fire of itself in the air. It can be prepared in various ways—for example, by heating hypophosphorous or phosphorous acids (§§ 286, 287), these acids being resolved by heat into phosphoric acid and phosphuretted hydrogen:—

Hypophosphorous Acid. { Empirical: $2H_3PO_2 = PH_3 + H_3PO_4$.
Dualistic: $6H_2O,2P_2O = 2PH_3 + 3H_2O,P_2O_5$.

Phosphorous Acid. { Empirical: $4H_3PO_3 = PH_3 + 3H_3PO_4$.
Dualistic: $4(3H_2O,P_2O_3) = 2PH_3 + 3(3H_2O,P_2O_5)$.

The non-inflammable gas is regarded as pure phosphuretted hydrogen, the property of spontaneously inflaming possessed by the other variety being supposed to depend upon the presence of minute portions of some foreign substance; the vapor of liquid phosphuretted hydrogen, PH_2, produces this effect; on adding to the non-inflammable gas so small a quantity as $\frac{1}{10,000}$th of its bulk of nitric oxide, it acquires the property of inflaming spontaneously.

Pure phosphuretted hydrogen gas is colorless and highly inflammable; its odor is fetid, and has been compared to that of tainted fish; it is slightly soluble in water, and can be liquefied, but has not yet been solidified. Neither the gas nor its solutions have any action on red or blue litmus. It is a powerful deoxidizing agent, and is, in general, easily decomposed. Most of the metals, when heated in the gas, combine with its phosphorus and liberate its hydrogen, just as we have seen the metal potassium set free hydrogen from ammonia. (See p. 77.) This ready decomposition of the gas by hot metals is the basis of the method of determining its composition by weight.

280. Phosphuretted hydrogen is resolved into its two elements, and the proportional weights of the elements which enter into its composition are simultaneously determined, by the following process:—The gas is passed through a hard-glass tube (*A*, Fig. 45), filled with copper turnings and heated to redness; the

Fig. 45.

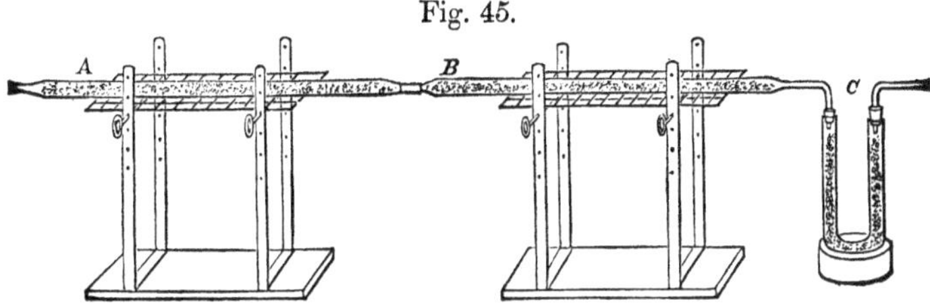

copper retains all the phosphorus, and the hydrogen becomes free. This last gas is carried forward through a second tube, *B*, filled with oxide of copper heated to redness; the hydrogen combines with the oxygen of the oxide of copper, and the steam thus formed is condensed and absorbed in a third tube, *C*, filled with pumice-stone soaked in sulphuric acid. (Appendix, § 15.) The tubes *A* and *C* are weighed both before and after the experiment, and the augmentation of weight gives the phosphorus in *A* and the water in *C*; from the weight of the water is calculated the weight of the hydrogen required to produce it. Care must be taken that the tube *A* be heated so moderately as not to distort it, and that nothing be added to its weight by depositions from the lamp-flames used to heat it. It is also necessary to fill the tubes with nitrogen gas before beginning the actual analysis, and to sweep them out with nitrogen at the end. This operation is easily performed by the aid of a small gas-holder full of nitrogen. It has thus been experimentally proved that any given weight of phosphuretted hydrogen contains 8·57 per cent. of hydrogen and 91·43 per cent. of phosphorus. Now it has been determined, as the result of many experiments and on a careful collation of the formulæ of all known compounds of phosphorus, that the least proportional weight of this element which enters into combination is 31, that of hydrogen being 1. The proportion,

$$91{\cdot}43 : 8{\cdot}57 = 31 : x,$$

gives as the value of x, 2·905. The nearest whole number is 3; and the discrepancy may be attributed to defects of the analytical process, always specially to be feared in cases like the present, where the quantity of one ingredient is many times as large as that of the other. A loss of matter, or error in weighing, which would amount to only 1 per cent. of 90 centigrammes, would cause an error of more than 11 per cent. on 8 centigrammes. The analysis clearly points to the formula PH_3 as representing the composition of phosphuretted hydrogen, inasmuch as for every 31 parts by weight of phosphorus, the gas contains three parts by weight of hydrogen. This result is partially corroborated by volumetric analysis. If the hydrogen liberated from any measured quantity of phosphuretted hydrogen by passing the gas through a tube filled with hot metal, be accurately measured, it will be found that, for every two volumes of the compound gas, three volumes of hydrogen are set free.

Thus far the composition of phosphuretted hydrogen has seemed to be completely analogous to that of ammonia gas; but at this point the analogy fails. In ammonia, three parts by weight of hydrogen are combined with fourteen of nitrogen, and three volumes of hydrogen are united with one volume of nitrogen to form two volumes of the compound gas. If the parallelism between NH_3 and PH_3 were perfect, one volume of phosphorus-vapor ought to be united with the three volumes of hydrogen which two volumes of phosphuretted hydrogen invariably contain. The densities of phosphorus-vapor and of phosphuretted hydrogen, as experimentally determined, prove that this is not the case. The unit-volume being that volume of hydrogen which weighs 1,

From the weight of 2 unit-volumes of PH_3 (sp. gr. = 17·09) . .	34·18
Subtract the weight of 3 unit-volumes of hydrogen	3·00
And there remains for the weight of the phosphorus-vapor . .	31·18

The specific gravity, or relative weight of one unit-volume of phosphorus-vapor, is 62·1, as has been already mentioned. Two volumes of phosphuretted hydrogen, therefore, contain, not one volume, but only half a volume of phosphorus-vapor. The atom of phosphorus weighing 31, combines with the same quantity of hydrogen by weight as the atom of nitrogen weighing 14; but the volume of the phosphorus atom is only one-half the volume of the nitrogen atom. The combining weights and the unit-volume weights of all the elements previously studied have been identical; but the combining weight of phosphorus must be doubled in order to bring it into coincidence with its unit-volume weight. The volumetric and the ponderal composition of phosphuretted hydrogen are both exhibited in the annexed diagram:—

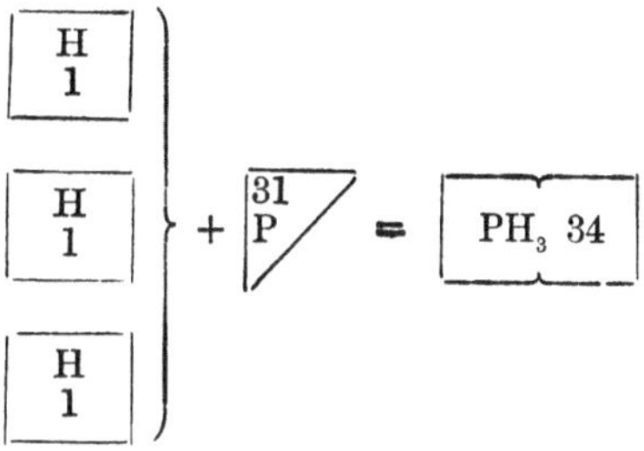

$$\left.\begin{array}{c}\boxed{\begin{array}{c}H\\1\end{array}}\\ \boxed{\begin{array}{c}H\\1\end{array}}\\ \boxed{\begin{array}{c}H\\1\end{array}}\end{array}\right\} + \begin{array}{c}31\\P\end{array} = \boxed{PH_3 \ 34}$$

281. This difference between ammonia and phosphuretted hydrogen is completely outweighed by the essential likeness in composition of these two gases and by the other striking analogies which exist between them. When one or more of the hydrogen atoms in phosphuretted hydrogen are replaced by certain groups of elements, which in organic chemistry play the part of elements, compounds are obtained which, like ammonia, neutralize acids and are strongly alkaline. Phosphuretted hydrogen itself combines with certain of the acids in definite proportions. With bromhydric and iodohydric acids, for example, it forms crystalline compounds whose composition is represented by the formulæ PH_4Br and PH_4I,—formulæ which are evidently comparable with NH_4Br and NH_4I.

282. *Liquid Phosphuretted Hydrogen* (PH_2) may be obtained by passing the spontaneously inflammable gaseous compound obtained in Exp. 114, through a U-tube surrounded by a mixture of ice and salt. Under these conditions, the vapor of the liquid compound, which was diffused in the gas, condenses and separates. Liquid phosphuretted hydrogen is colorless, has a high refracting-power, and is not miscible with water. It does not solidify at $-20°$; when heated to $30°$ or $40°$ it decomposes. It is exceedingly inflammable, and bursts into flame when brought in contact with the air; when a small quantity of its vapor is mingled with combustible gases, such as carbonic oxide, hydrogen, or carburetted hydrogen, these gases acquire the property

of inflaming spontaneously. When exposed to sunlight, it is resolved into gaseous and solid phosphuretted hydrogen:—

$$5PH_2 = P_2H + 3PH_3.$$

283. *Solid Phosphuretted Hydrogen* (P_2H?) is formed by exposing liquid phosphuretted hydrogen to sunshine, or by acting upon the liquid with chlorhydric acid, or by dissolving phosphide of calcium in strong chlorhydric acid. It is a compound insoluble in water or alcohol, but soluble in warm potash-lye with liberation of gaseous phosphuretted hydrogen. It takes fire at about 150°, and is of a yellow color, but becomes red when exposed to light.

284. *Compounds of Phosphorus and of Oxygen.*—Phosphorus unites with oxygen in four different proportions, as follows:—

Oxide of Phosphorus, P_4O.
Hypophosphorous Acid, P_2O.
Phosphorous Acid, P_2O_3.
Phosphoric Acid, P_2O_5.

All of these compounds exhibit a more or less distinct acid character, especially when combined with water, and the one containing most oxygen, phosphoric acid, is a very important acid.

285. *Oxide of Phosphorus* (P_4O).—When ordinary phosphorus is burned in a confined volume of air or oxygen, insufficient for its complete combustion, there will be found mixed with the unconsumed phosphorus, after the chemical action has ceased, a certain quantity of a red powder, which is the oxide of phosphorus now in question.

Exp. 117.—Repeat Exp. 13, and examine the red mass which remains in the porcelain capsule after it has been sunk in the water-pan and thoroughly cooled.

Since the red oxide of phosphorus is insoluble in bisulphide of carbon, it can readily be obtained in a state of purity by dissolving in this liquid the free phosphorus with which it is contaminated.

Although the red oxide is not spontaneously inflammable by itself, a mixture of it with free phosphorus, such as the residue from the preparation of nitrogen (Exp. 13), takes fire with great ease, being even more readily inflammable than phosphorus alone. Such residues must be handled with special care.

Red oxide of phosphorus can be obtained in larger quantities by bringing a stream of oxygen gas into contact with phosphorus melted under hot water.

Exp. 118.—Place about a cubic centimetre of ordinary phosphorus in the bottom of a conical test-glass, or wine-glass, and pour upon it hot water enough to half fill the glass; the phosphorus will melt, but cannot burn, since the water protects it from contact with the air, and since phosphorus by itself is incapable of decomposing water. By means of a narrow gas-delivery-tube of glass, conduct a slow stream of oxygen from a gas-holder to the bottom of the test-glass, so that the oxygen shall come into immediate contact with and bubble through the melted phosphorus. The phosphorus will burn with a vivid light beneath the water; red oxide of phosphorus will be formed, and will float about in the water, from which it may be separated by filtration. In the lack of oxygen, air may be forced down upon the phosphorus; even the impure air blown from the mouth will answer ; but with air the reaction is less intense than with oxygen; hence, when it is employed, the experiment had better be performed in a dark room.

Oxide of phosphorus has neither taste nor smell. On being heated to 350° to 400°, it splits up into phosphoric acid and free phosphorus, the latter, of course, taking fire in case oxygen be present.

286. *Hypophosphorous Acid* ($H_6P_2O_4 = 2H_3PO_2$).—This compound has usually been classed among the oxides of phosphorus, on the supposition that it might be possible to obtain from it an anhydrous oxide, of the composition P_2O; the oxide in question has, however, never yet been obtained.

When ordinary phosphorus is boiled in a solution of caustic potash, soda, lime, or baryta, water is decomposed, a compound of phosphorus and hydrogen (§ 278) is formed, and a hypophosphite of the alkali employed remains in solution, from which it may be separated in crystals by cautious evaporation. If baryta be employed, the reaction may be formulated as follows:—

Empirical: $3BaH_2O_2 + 8P + 6H_2O = 2PH_3 + 3BaH_4P_2O_4$.

Dualistic: $3(BaO,H_2O) + 8P + 6H_2O = 2PH_3 + 3(BaO,2H_2O,P_2O)$.

By cautiously adding sulphuric acid to the solution of the barium salt, sulphate of barium is precipitated and hypophosphorous acid remains in solution:—

$$BaO,2H_2O,P_2O + H_2O,SO_3 = BaO,SO_3 + 3H_2O,P_2O.$$

By evaporating the aqueous solution, after filtration, hypophosphorous acid is left as a viscid, uncrystallizable, acid liquid, which, on being strongly heated, splits up into phosphoric acid and phosphuretted hydrogen. It unites with oxygen readily, and is consequently a powerful reducing agent. Sulphuric acid, for example, is reduced by it, with evolution of sulphurous acid and separation of sulphur.

The hypophosphites are, for the most part, crystallizable salts, soluble in water and often in alcohol also; they can usually be preserved in dry air. Several of them have recently been somewhat extensively employed as medicaments.

287. *Phosphorous Acid* (P_2O_3).—This acid is a product of the slow combustion of phosphorus.

When phosphorus is gently heated in a very slow current of perfectly dry air, it takes on oxygen enough to form phosphorous acid, which, being volatile, condenses upon the cold walls of the tube beyond the phosphorus as a bulky white sublimate. By conducting the operation in a tube drawn out to a fine point at one end and almost completely closed at the other by a perforated cork carrying a narrow tube, and carefully regulating the supply of air which is admitted into the tube, so that just enough oxygen to form phosphorous acid, and no more, shall come in contact with the phosphorus, a tolerably pure product can be obtained. For purposes of illustration, however, a simpler arrangement of the apparatus may be employed, as in the following experiment:—

Exp. 119.—Place a bit of phosphorus, as big as a pea, in the middle of a piece of glass tubing, No. 2, about 30 c.m. long, and open at both extremities; gently heat the phosphorus until it takes fire, and then extinguish the lamp. So long as the tube is held in a horizontal position, the combustion will be so feeble and imperfect that some red oxide of phosphorus will be formed as well as phosphorous acid. On the other hand, if one end of the tube be inclined upwards, so that the products of combustion can pass off and make way for the entrance of fresh air, the combustion will become more vivid, and there will be produced a quantity of the highest oxide of phosphorus, phosphoric acid. If the tube were held perpendicularly, the draught of air, passing through it as through a chimney, would be so powerful that all the phosphorus would be burned completely to phosphoric acid.

It is evident, from the foregoing, that if it were only possible to find out the precise angle at which the tube should be inclined, and, at the same time, to provide means for continually maintaining a suitable temperature within the tube, the phosphorus might all be converted into pure phosphorous acid, instead of the various and mixed products which are actually obtained.

288. Hydrated phosphorous acid, H_3PO_3, or $3H_2O,P_2O_3$, is readily obtained, though in an impure condition, by exposing sticks of phosphorus to moist air.

Exp. 120.—Select a piece of glass tubing, the diameter of which is so much greater than that of an ordinary stick of phosphorus, that the latter can readily be slipped into it; from this tubing prepare three or four short tubes 3 or 4 c.m. long, open above and below, but drawn in at the bottom to such an extent that a stick of phosphorus placed in the upper part of the tube cannot pass the narrowed portion and fall out of the tube. In each of these short tubes put a stick of phosphorus, and place them all in a glass funnel which rests upon a bottle standing in a soup plate full of water; over the funnel and bottle place a tall tubulated bell-jar, from which the stopper has been removed, and allow the apparatus to stand at rest during several days in a cool place where no damage can be done in case the phosphorus take fire.

Under these conditions, the phosphorus will slowly oxidize and waste away (if time enough be allowed it will completely disappear), and the mixture of phosphorous and phosphoric acids which is formed will flow down through the tube of the funnel into the bottle beneath. The mixture thus obtained is often technically termed *phosphatic acid.*

The object of the glass tubes employed to envelope the sticks of phosphorus is, to keep the several pieces of phosphorus from touching one another. If two or three pieces of phosphorus were to be left in contact, in the air, the heat generated during the oxidation of each would be added to that derived from the others, and after a time the mass would become hot enough to take fire spontaneously. But when each stick of phosphorus is placed within a glass tube, the heat generated by its oxidation passes off harmlessly, and a dangerous accumulation of heat is very much less likely to occur than if no such system of isolation were resorted to.

289. The fact that a collection of fragments of phosphorus is thus liable to take fire, so well illustrates the theory of *spontaneous combustion* in general, and the precautionary measures taken in the foregoing experiment to prevent the ignition of the phosphorus point so clearly to the methods which must often be resorted to in order to prevent the spontaneous inflammation of

many highly combustible substances, that a few words may here be appropriately devoted to this important practical subject.

As a rule, all easily oxidizable substances, when finely divided and thrown into heaps, are liable to take fire spontaneously in the air. Many oils, for example, particularly the so-called drying oils, absorb oxygen from the air and enter into combination with it. Wherever chemical combination occurs, heat is developed, and in case the oil be poured upon some porous substance which is both combustible and a non-conductor of heat, like wool or cotton, paper or cloth, the heat developed during the oxidation of the oil may very readily accumulate to the extent necessary to produce inflammation. To prevent this catastrophe, the heap of greasy wool or other matter should be broken up as soon as warmth is perceived in it, and its particles should be scattered about so that air may have free access to them; the heat will then pass off harmlessly from each of these particles as fast as it is generated.

This process of subdivision will prove an effectual protection if the subdivision be carried far enough; but it is a fact not to be lost sight of, that very small parcels of some substances (a hank of oiled twine, for example, or a handful of greasy rags) may take fire when all the conditions are favourable; and it is a matter of the first importance that all such matters should be kept in places where no harm can be done in case they inflame.

A still more familiar instance of the accumulation of heat during chemical action occurs in the ordinary process of hay-making, as when a cock of half-cured hay is left unopened for any length of time; the green hay combines with oxygen from the air, fermentation sets in, and heat is, of course, evolved; but when the hay is scattered about the field, this heat passes off into the air as fast as it is generated, and we cannot perceive it. On the other hand, if, instead of the usual small hay-cocks, the farmer were to throw a large quantity of new-mown hay into one great stack, this stack would undoubtedly take fire if left to itself.

In large heaps of many kinds of bituminous coal also, strong chemical action is induced under very various conditions as regards temperature, moisture, and mechanical subdivision, and the heat evolved becomes at last intense enough to kindle the coal. Protection from the weather, exclusion of moisture, free ventilation, and the avoiding of too large heaps are the most effectual preventives in this case.

290. Anhydrous phosphorous acid is a white amorphous substance, which rapidly absorbs water from the air, and when sprinkled with water, dissolves rapidly with a hissing noise; it is volatile, and may easily be driven from one place to another, in a tube filled with nitrogen (see § 287), by applying a gentle heat. Being a product of the incomplete combustion of phosphorus, it is necessarily combustible; when heated in the air, it undergoes vivid combustion.

Hydrated phosphorous acid is obtained in the form of rectangular prisms, when the aqueous solution is evaporated at temperatures not exceeding 200°. The crystals are deliquescent, and they gradually absorb oxygen from the air; when strongly heated, they are decomposed into phosphoric acid, and phosphuretted hydrogen, and at the same time take fire. The aqueous solution of phosphorous acid exhibits a strong acid reaction; by absorbing oxygen from the air, it is converted into phosphoric acid, quickly in case the solution is dilute, but slowly if it be concentrated. It is a powerful reducing agent; when heated with sulphurous acid, it yields phosphoric and sulphydric acids. Though a very weak acid, it forms salts by combining with those metallic oxides upon which it exerts no reducing action; the phosphites of the alkalies proper are easily soluble in water, but the phosphites of calcium and barium can only be dissolved with difficulty. These salts are more stable than the hypophosphites, but are all decomposed by heat.

291. *Phosphoric Acid* (P_2O_5).—As has been already stated, this highest oxide of phosphorus is the product of the rapid combustion of phosphorus in an excess of air or oxygen.

Exp. 121.—Dry thoroughly a large porcelain plate, a small porcelain capsule, and a wide-mouthed bottle of two litres capacity, by warming them at a fire; place the capsule upon the plate and put in the capsule a bit of dry phosphorus of the weight of half a gramme or thereabouts; light the phosphorus, and cover it at once with the inverted bottle. The phosphoric acid, formed by the combustion of the phosphorus, will be deposited as a white powder, like flakes of snow, upon the sides of the bottle, and much of it will fall down upon the plate below.

The apparatus employed in this experiment can readily be arranged in such manner that fresh portions of phosphorus and of air can be introduced into the bottle as fast as occasion may require; the process will then be continuous, and any desired quantity of phosphoric acid may be prepared by means of it.

The flocculent, amorphous, odorless powder thus obtained unites with water with remarkable facility; if it be left in the air for a few minutes, it deliquesces completely; and upon being thrown into water it dissolves, with a hissing noise and development of much heat. In order to preserve it, it must be placed in a dry tube, and the tube closed by sealing it in the lamp. When touched to the moist tongue, it burns as if it were red-hot metal. On account of this strong affinity for water, it is frequently employed by chemists to withdraw the elements of water from other substances; anhydrous sulphuric acid, for example, can be prepared from oil of vitriol by heating the latter with anhydrous phosphoric acid. On being heated with various organic substances, such as some of the alcohols and essential oils composed of carbon, hydrogen, and oxygen, it decomposes them in such manner that the oxygen of the organic substance, and as much of its hydrogen as is necessary to form water by uniting with this oxygen, combine with the phosphoric acid, while a compound of carbon and hydrogen (technically called a hydrocarbon) is set free.

After the anhydrous acid has once been dissolved in water, it cannot again be completely deprived of water by mere evaporation or ignition. When the aqueous solution is evaporated, there is left, not the anhydrous powder, but a transparent glassy mass, which is a hydrate, of the composition H_2O,P_2O_5. This hydrate is often called glacial phosphoric acid. It is extremely deliquescent, and, at a bright red heat, sublimes in dense white fumes. Besides the monohydrate, there are two other hydrates of phosphoric acid, of the compositions, respectively, $2H_2O,P_2O_5$ and

$3H_2O,P_2O_5$. Of these, the terhydrate is, perhaps, the most important; it is the substance usually meant when phosphoric acid is spoken of.

292. Phosphoric acid can be prepared, also, by the oxidation of phosphorus, or of hypophosphorous or phosphorous acid, or by the decomposition of some one of its salts, such as the phosphate of calcium (bone-earth).

When phosphorus is heated in dilute nitric acid, of 1·2 specific gravity, the nitric acid gives up oxygen to the phosphorus, nitric oxide and phosphoric acid are formed, and the latter goes into solution. When the phosphorus has all disappeared, the solution is evaporated to dryness in order to drive off the nitric acid which was employed in excess, and there is obtained a quantity of the monohydrated glacial acid.

A product less pure than the acid prepared by means of nitric acid, is obtained by neutralizing a solution of monophosphate of calcium, prepared from bones in the manner already described when treating of the preparation of phosphorus (see § 270), with carbonate of ammonium, evaporating the filtered solution of phosphate of ammonium to dryness and heating the residue to low redness. Ammonia is expelled and glacial phosphoric acid remains.

293. It is a remarkable fact that each of the three different hydrates of phosphoric acid possesses properties peculiar to itself, and unlike those of the other two; in fact, each of the hydrates must be regarded as a distinct acid. The monohydrated or glacial acid, H_2O,P_2O_5, is usually called metaphosphoric acid; the bihydrate, $2H_2O,P_2O_5$, is called pyrophosphoric acid; and the terhydrate, $3H_2O,P_2O_5$, is commonly spoken of as ordinary phosphoric acid, or simply as phosphoric acid. The three hydrates are sometimes distinguished as *a* phosphoric acid (meta), *b* phosphoric acid (pyro), and *c* phosphoric acid (ordinary). These different hydrates of the acid retain their peculiar characteristics for a considerable time when dissolved in water, though the mono- and bihydrates change, after a while, to the terhydrate; and in combining with metallic oxides to form salts, they unite with 1, 2, or 3 molecules of the oxide, accordingly as they themselves contain the elements of 1, 2, or 3 molecules of water. There are thus formed three distinct series of salts, each of which corresponds to one of the hydrates, as is seen in the

following formulæ, where M stands for any metal which habitually replaces one atom of hydrogen.

Monohydrated Acid.	*Bihydrated Acid.*	*Terhydrated Acid.*
H_2O,P_2O_5	$2H_2O,P_2O_5$	$3H_2O,P_2O_5$
M_2O,P_2O_5	$2M_2O,P_2O_5$	$3M_2O,P_2O_5$
Metaphosphate of M.	*Pyrophosphate of M.*	*Phosphate of M.*

This behavior is very different from that of the hydrates of nitric or of sulphuric acid; when either of the hydrates of nitric acid, for example, is made to combine with a base, like soda, there is formed always one and the same salt, nitrate of sodium. In each of the three series of salts formed by phosphoric acid, the acid exhibits peculiar properties. A salt of the formula $3M_2O,P_2O_5$ will behave very differently towards many reagents from a salt containing the same metal but in the proportions M_2O,P_2O_5, or $2M_2O,P_2O_5$. As an example of the kind of differences here alluded to, it may be mentioned that, while metaphosphoric acid, on being added to a solution of albumen, will cause the albumen to coagulate, no such coagulation can be brought about by either pyrophosphoric acid or the ordinary terhydrate. Metaphosphoric acid gives a white precipitate when its solution is mixed with a solution of nitrate of silver by either of the other hydrates, unless they are first neutralized with an alkali, in which event a white precipitate is produced by pyrophosphoric acid, and a yellow precipitate by the ordinary acid. These peculiarities will be examined in detail when we come to treat of the phosphates of sodium in the chapter upon sodium and its compounds.

From the formulæ given in the above table, it is apparent that metaphosphoric acid is a monobasic acid, that pyrophosphoric acid is bibasic, and that ordinary phosphoric acid is terbasic. Since each of the atoms of M, in either of the formulæ, can be replaced by an equivalent atom of any other metal, or by hydrogen, it follows that the composition of some of the salts of phosphoric acid is rather complex; thus there is a phosphate of potassium and sodium of the composition $(K_2O,Na_2O,H_2O)P_2O_5$, or $KNaHPO_4$.

Although we have here studied the acids which contain phosphorus, oxygen, and hydrogen, as if they were in reality composed of water and the anhydrous oxide of phosphorus, as the manner of their derivation would suggest, yet it must not be forgotten that we have absolutely no knowledge of the actual structure of the molecules of these compounds, and that the empirical formulæ HPO_3, $H_4P_2O_7$, and H_3PO_4 express all that is absolutely known of their composition.

294. In whichever way prepared, and in all its varieties, phosphoric acid is a very strong acid. Although a less powerful agent, at the ordinary temperature, than sulphuric acid, yet, from being much less volatile than sulphuric acid, it can expel the latter, and most other acids, from their compounds on being heated with them. The behavior of the two acids towards calcium, or its oxide, furnishes an instructive example of the influence of extrinsic or physical circumstances upon the play of the chemical force. When triphosphate of calcium (bone-earth) is treated with dilute sulphuric acid at the ordinary temperature, a quantity of phosphoric acid is set free from the calcium and goes into solution. From this result it might, at first sight, be thought that the calcium was removed from the phosphate of calcium simply by force of the superior chemical power of sulphuric as contrasted with phosphoric acid; but, in reality, the water which is present plays an important part in the reaction. Monophosphate of calcium is readily soluble in water, sulphate of calcium, on the other hand, being well-nigh insoluble. Hence it happens that when triphosphate of calcium is digested in dilute sulphuric acid, monophosphate of calcium goes into solution, while sulphate of calcium is deposited as an insoluble powder. But if the mixture of solid sulphate of calcium and of dissolved monophosphate of calcium, thus obtained, be evaporated to dryness and the residue be strongly heated, all the sulphuric acid will be expelled from the calcium; it will evaporate and pass off into the air, and nothing will finally be left in the vessel but triphosphate of calcium, precisely similar in quality and quantity to that with which the experiment started. In the same way, if a mixture of sulphate of calcium and glacial phosphoric acid be strongly heated, the sulphuric acid, being readily volatile, as compared with phosphoric acid, will all be expelled from its combination with the calcium:—

$$3CaSO_4 + P_2O_5 = Ca_3P_2O_8 + 3SO_3.$$

295. *Chlorides of Phosphorus.*—Phosphorus and chlorine unite readily and directly even at temperatures as low as 0°, the act of combination being attended with evolution of light and heat. If the chlorine be in excess, as regards the phosphorus, there will be formed a solid quinquichloride of phosphorus, while, if an excess of phosphorus be present, a liquid terchloride of phosphorus will be obtained.

296. *Terchloride of Phosphorus* (PCl_3) is a colorless liquid of about 1·5 specific gravity, which boils at about 75°. It fumes in the air, and is decomposed by moist air. When heated in the flame of the gas-lamp, it takes fire and burns with a bright light. When mixed with water it decomposes, yielding clorhydric and phosphorous acids :—

$$2PCl_3+6H_2O=6HCl+3H_2O,P_2O_3, \text{ or } PCl_3+3H_2O=3HCl+H_3PO_3.$$

This reaction is particularly interesting, in view of the fact that by means of it we are enabled to obtain phosphorous acid in a condition of purity. It will be remembered that by the method of direct oxidation (§ 287) it is no easy matter to obtain pure phosphorous acid from phosphorus. But by simply treating terchloride of phosphorus with water and evaporating the solution, so that the chlorhydric acid which results from the reaction may be expelled, hydrated phosphorous acid is obtained as the sole product. The process is a good example of the indirect methods to which chemists are frequently compelled to resort.

Terchloride of phosphorus can be prepared by passing a slow stream of dry chlorine through melted, almost boiling, phosphorus contained in a tubulated retort which has previously been filled with chlorine in the cold, and condensing the chloride in an appropriate receiver as fast as it distils over. The process, like all operations with phosphorus, requires special care.

297. It will be observed that the formula of terchloride of phosphorus is that of phosphuretted hydrogen in which all the hydrogen has been replaced by chlorine. The two substances have a perfectly similar volumetric composition. In phosphuretted hydrogen, three volumes of hydrogen in combination with half a volume of phosphorus, produce two volumes of the compound gas ; if, in terchloride of phosphorus,

Half a unit-volume of phosphorus-vapor, weighing . .	31·00
And 3 unit-volumes of chlorine, weighing (35·5×3) .	106·50
Produce 2 volumes of PCl_3 vapor, weighing	137·50
One unit-volume of PCl_3 vapor should weigh . . .	68·75

The specific gravity of the vapor of terchloride of phosphorus has been found, by experiment, to be 69·12,—a number so nearly identical with the above result of calculation as entirely to confirm the assumption on which the calculation rests, viz. that in terchloride of phosphorus three volumes of chlorine are united with half a volume of phosphorus-vapor.

298. *Quinquichloride of Phosphorus* (PCl_5) is a white or straw-colored crystalline solid, which volatilizes at a temperature below 100° without previously fusing; but when subjected to pressure, it melts at 148°, and boils at a temperature somewhat higher. It burns in the flame of a candle with production of phosphoric acid and evolution of chlorine. It is very deliquescent, and is decomposed by the moisture of the air; by a large excess of water it is immediately resolved into chlorhydric and phosphoric acids:—

$$PCl_5 + 4H_2O = 5HCl + H_3PO_4;$$

with a smaller quantity of water it yields chlorhydric acid and oxychloride of phosphorus:—

$$PCl_5 + H_2O = 2HCl + POCl_3.$$

Sulphydric acid decomposes it in like manner, with production of chlorhydric acid and sulphochloride of phosphorus:—

$$PCl_5 + H_2S = 2HCl + PSCl_3.$$

Quinquichloride of phosphorus reacts upon many organic compounds also, with formation of very interesting products, and is hence an important agent of research in the department of organic chemistry.

299. In order to prepare quinquichloride of phosphorus, a current of dry chlorine may be passed into terchloride of phosphorus until the latter has been completely solidified; the product is then distilled in a current of chlorine. The quinquichloride may be obtained directly from phosphorus in one operation, if a rapid stream of chlorine be conducted into a retort

containing phosphorus, kept so cool that the terchloride of phosphorus at first produced shall not distil over. Again, if powdered red phosphorus is exposed to the action of a rapid stream of chlorine, it will all be quickly converted into the solid quinquichloride.

300. The formula above given for quinquichloride of phosphorus represents the following composition by volume:—

Half a unit-volume of phosphorus-vapor, weighing . .	31·00
And 5 unit-volumes of chlorine, weighing (35·5×5) .	177·50
Should produce 2 vols. of PCl_5 vapor, weighing . . .	208·50
One unit-volume of PCl_5 vapor ought then to weigh .	104·25

The specific gravity of the supposed vapor of quinquichloride of phosphorus, as determined by experiment, does not accord with this calculated result; it is 52·81, almost exactly one-half of the theoretical unit-volume weight above given. If *four* volumes of vapor, instead of *two*, resulted from the union of half a volume of phosphorus-vapor with five volumes of chlorine, the calculated and the actual vapor-density would coincide. But hitherto we have never found a single compound gas in which the product volume was four unit-volumes; two unit-volumes have invariably resulted from the union of the constituent volumes, whatever the character and number of the constituents. It would be necessary to admit this substance as presenting an exceptional volumetric composition, were it not for a well-founded distrust of the experimental determination of the vapor-density of this compound. It not infrequently happens that all attempts to determine the vapor-density of volatile compounds of two or more elements are baffled by their splitting up, at the temperature of vaporization, into their constituent gases or vapors, which, in the act of separating, resume their own proper volumes, however much they may have been condensed during combination. This splitting up of compound vapors, at high temperatures, into less complex compounds, or into the elementary constituents, is termed *dissociation.* Thus, at the elevated temperature necessary to convert quinquichloride of phosphorus into vapor, it is probable that the quinquichloride splits into terchloride of phosphorus and free chlorine, and that it is the specific gravity of this mixture which has been determined, instead of the specific gravity of the real unaltered vapor of the quinquichloride.

Two unit-volumes of PCl_3 weigh	138·24
Two unit-volumes of Cl weigh	71·
Four unit-volumes of the mixture weigh	209·24
One unit-volume of the mixture weighs	52·31

The specific gravity which has been assigned to the quinquichloride is 52·81,—a number very nearly coincident with the above calculated density of the mixture of terchloride-vapor and free chlorine. At the high temperature of vaporization it is therefore probable that quinquichloride of phosphorus undergoes *dissociation* into terchloride of phosphorus and chlorine; but if this be the case, these constituents recombine when the temperature falls, for by lowering the temperature the quinquichloride is recovered.

As we advance, we shall meet with several other examples of the *dissociation* of compound gases and vapors; for the present it will be sufficient to give one more illustration of the meaning of this term. When equal volumes of dry ammonia and dry chlorhydric acid gas are mixed (Exp. 65), the two gases are completely condensed to a white solid, which we are familiar with as chloride of ammonium. Since this ammonium-salt is readily volatilizable, there would be no difficulty in determining the product-volume of the compound of ammonia and chlorhydric acid, were it not for the fact that the vapor of chloride of ammonium undergoes dissociation at the temperature of vaporization. If the real vapor of the compound could be measured, the facts would undoubtedly be correctly represented by the diagram,

$$\boxed{HCl} + \boxed{NH_3} = \boxed{NH_4Cl};$$

but the vapor of the compound is resolved into its constituent gases at the high temperature necessarily employed, so that the following diagram really figures the actual state of things:—

$$\boxed{HCl} + \boxed{NH_3} = \left\{ \begin{matrix} HCl \\ NH_3 \end{matrix} \right\}$$

When the dissociated vapor cools, the parted gases recombine to form solid chloride of ammonium.

It is obvious that the phenomena of dissociation interfere fatally with one of the common methods of arriving at the weight and structure of the molecule of a volatile compound; the indirect method of getting at the volumetric composition of a substance from its ponderal composition and the specific gravity of its vapor becomes impracticable whenever the vapor of the compound under examination is liable to dissociation, inasmuch as experiment cannot determine beyond a doubt the real vapor-density of such a body.

301. *Bromides of Phosphorus.*—When a piece of phosphorus is dropped into bromine, the two elements combine with explosive violence, the burning phosphorus being thrown about in a highly dangerous manner. There are two bromides of phosphorus, PBr_3 and PBr_5, corresponding to the two chlorides. The terbromide is liquid at ordinary temperatures and the quinquibromide solid.

302. *Iodides of Phosphorus.*—Iodine and phosphorus unite directly, when brought in contact with one another, and so much heat is developed by their union, that a portion of the phosphorus will take fire if the mixture be in contact with the air. There are two iodides of phosphorus, both of them solid at the ordinary temperature; their composition is respectively PI_2 and PI_3. It will be noticed that, while the teriodide corresponds to the terchloride and terbromide, the other compound is a biniodide, of which there is known neither a bromine nor a chlorine analogue. The fact is interesting as illustrating the general truth that, when in any group or family of elements we compare the behavior of its several members, analogy ceases to be a sure guide, in proportion as the individuals compared are more widely separated in the natural series. Chlorine and bromine stand next to one another in the family or series of elements to which

they belong; and as we have just seen, their behavior, as regards phosphorus, is well-nigh identical; but iodine, one step further removed from chlorine than bromine is, enters into new combinations not altogether conformable to those of chlorine.

303. *Sulphides of Phosphorus.*—There is a definite sulphide of phosphorus corresponding to each of the oxides, and in addition to these there are two other compounds, which may be represented by the formulæ P_4S_3 and P_2S_{12}. Sulphur and phosphorus may also be melted together in any proportion. The sulphides of phosphorus are exceedingly inflammable, taking fire even more readily than phosphorus itself, and they are all more readily fusible than either of the two elements of which they are composed. They may be prepared by heating sulphur under water in contact with melted phosphorus. The union of the two elements is attended with development of much heat, and sometimes with dangerous explosions. It is well, therefore, to operate only upon small quantities, and to add the sulphur gradually to the phosphorus.

CHAPTER XVII.

ARSENIC.

304. Compounds of arsenic have been known from very early times. The element is sometimes found native, but much more frequently associated with other metals and with oxygen and sulphur. The metals in connexion with which it most commonly occurs are iron, cobalt, nickel, and copper. Ferruginous ores and deposits, in particular, are rarely free from traces of arsenic. In small quantity, arsenic is very widely distributed.

The greater part of the arsenic of commerce is prepared from a native arsenide and sulphide of iron (arsenical pyrites) corresponding to the formula FeAsS, and from the arsenides of nickel and cobalt. Metallic arsenic is obtained directly from the mineral of the formula FeAsS by heating it in earthen tubes laid horizontally in a long furnace; a tube, made by rolling up a piece of thin sheet iron, is inserted in the mouth of each earthen retort, and an earthen receiver is luted on to this iron tube. The arsenic condenses principally in the iron tube, in the form of a compact, whitish, crystalline mass, which is detached, when cold, by unrolling the sheet iron. The metal is also indirectly obtained by reducing the arsenious acid (As_2O_3) which results from *roasting* (heating in a current of air) arsenides, like those of cobalt and nickel; this oxide is heated with charcoal in earthen crucibles covered with conical iron caps, or inverted crucibles, into which the reduced metal sublimes. The metal obtained by the second process is gray and pulverulent, instead of whitish and coherent.

$$FeAsS = FeS + As; \qquad As_2O_3 + 3C = 2As + 3CO.$$

305. Arsenic is a brittle solid, of a steel-gray color and a metallic lustre. Its specific gravity has been variously given at from 5·62 to 5·96. Like the metals, it is a good conductor of electricity. It crystallizes in acute rhombohedrons, and in octahedrons also, thus taking on forms of both the monometric and hexagonal systems, as do phosphorus, the preceding member, and antimony, the succeeding member of this family. At a dull red heat it volatilizes without previous fusion; the vapor is colorless, and possesses a characteristic odor resembling that of garlic. The specific gravity of this vapor is 150, while the atomic weight of the element is 75; arsenic, therefore, resembles phosphorus, and differs from all the other elements heretofore studied, in that its atomic weight is not identical with its unit-volume weight; two combining proportions by weight of arsenic occupy the same volume as one combining proportion of hydrogen; its symbol, As, represents its atomic weight, but only half the weight of the unit-volume of its vapor. At the ordinary temperature the compact metal does not tarnish by exposure to dry air, but a moistened powder of arsenic is slowly converted by the air into a mixture of arsenious acid and metallic arsenic. At a red heat the metal burns with a whitish flame, producing a white smoke of arsenious acid. When thrown, in fine powder, into chlorine gas, it takes fire spontaneously and is converted into

chloride of arsenic ($AsCl_3$). Bromine, iodine, and sulphur also combine readily with arsenic, when aided by a gentle heat. Nitric acid and aqua regia convert the metal into arsenic acid (As_2O_5); chlorhydric acid has little action upon it. Dilute sulphuric acid has no action upon the metal; but the concentrated acid has the same effect upon arsenic as upon phosphorus (§ 276); arsenious acid is formed and sulphurous acid escapes:—

$$3H_2SO_4 + 2As = As_2O_3 + 3SO_2 + 3H_2O.$$

Some fatty oils dissolve arsenic to a slight extent, as they do phosphorus. Metallic arsenic unites by fusion with most metals, forming alloys which the arsenic tends to make hard or brittle. In the manufacture of shot a little arsenic is added to the lead to facilitate the formation of regular globules.

306. *Arsenic and Hydrogen.*—Arsenic forms two combinations with hydrogen; one of these is an unstable, brown solid of uncertain composition; the other is a well-known gas whose constitution is represented by the formula AsH_3, and which is therefore analogous in composition to ammonia (NH_3) and phosphuretted hydrogen (PH_3). The solid hydride is so obscure a substance that nothing need here be said of it, except that it is supposed to contain two atoms of hydrogen and one of arsenic (AsH_2?).

307. *Arseniuretted Hydrogen.*—This very dangerous gas may be prepared in an impure state by decomposing, with sulphuric acid diluted with three parts of water, an arsenide of zinc obtained by fusing together equal weights of powdered arsenic and granulated zinc:—

$$3H_2SO_4 + Zn_3As = 3ZnHSO_4 + AsH_3.$$

As it is not possible to prepare the precise alloy Zn_3As, the arseniuretted hydrogen thus obtained is always mixed with hydrogen. The arsenide of sodium can be decomposed by water, with evolution of arseniuretted hydrogen:—

$$3H_2O + Na_3As = 3NaHO + AsH_3.$$

The same remark, however, applies to this reaction as to the preceding one; the product is contaminated with an indeterminate quantity of free hydrogen. Arseniuretted hydrogen seems also to be formed whenever the oxides of arsenic, or compounds of these oxides, are brought in contact with nascent hydrogen. A mixture of arseniuretted hydrogen and hydrogen may be readily obtained by acting upon zinc by dilute chlorhydric or sulphuric acid in which arsenious acid has been dissolved.

308. Arseniuretted hydrogen is a colorless gas, having a fetid odor ; even when very much diluted with air, it is intensely poisonous, and fatal results have repeatedly followed its accidental inhalation. In experimenting with this deadly gas, the greatest care is required not to inhale the least portion of it. It has been condensed at $-40°$ to a transparent liquid, but it has never been solidified. The gas is soluble in water at the ordinary temperature only to the extent of one-fifth of its volume, and neither the gas nor its aqueous solution has any action upon blue or red litmus-paper. In spite, therefore, of its strong resemblance to ammonia in composition, some of its physical properties are strikingly unlike those of that very soluble and intensely alkaline gas. Arseniuretted hydrogen burns in the air with a whitish flame, forming water and a white smoke of arsenious acid; but if a cold body, like a piece of porcelain, for example, be introduced into a jet of the burning gas, the hydrogen alone will burn, and the arsenic will be deposited in the metallic state upon the porcelain surface, forming a lustrous black spot. This effect is precisely similar to the deposition of soot on a cold body held in the flame of a candle. It is also decomposed when caused to pass through tubes heated to dull redness, metallic arsenic being deposited as a brown or blackish mirror, while hydrogen gas escapes. This decomposition is a good illustration of the dissociation of gases (§ 300). Chlorine in excess reacts violently upon it, forming terchloride of arsenic ($AsCl_3$) and chlorhydric acid:—

$$AsH_3 + 6Cl = AsCl_3 + 3HCl.$$

When, however, chlorine acts on an excess of arseniuretted hydrogen, there are formed chlorhydric acid and metallic arsenic ; flame accompanies this reaction. The reactions of bromine and iodine are similar to, but less violent than, those of chlorine. We recall, in this connexion, the decomposition of ammonia by chlorine, with formation of chlorhydric acid and liberation of nitrogen. Arseniuretted hydrogen decomposes the solutions of the salts of many of the heavy metals, but the products are somewhat various; sometimes a metallic arsenide is precipitated; sometimes the heavy metal is precipitated, while arsenious acid remains in the solution. As we shall shortly see, the chemical properties of this gas are of great importance in the processes used for detecting arsenic in cases of poisoning.

309. Arseniuretted hydrogen may be analyzed by precisely the same method which was used for the analysis of phosphuretted hydrogen (§ 280); but the results can only be approximate, because of the extreme difficulty, not to say impossibility, of obtaining the gas in a state of tolerable purity. The composition of the analogous gases, ammonia and phosphuretted hydrogen, and the specific gravity of the gas, lead us to the following statement of its composition. Two volumes of the gas contain

3 unit-volumes of hydrogen, weighing	3×1	$= 3$
$\frac{1}{2}$ unit-volume of arsenic-vapor, weighing	$\frac{1}{2} \times 150$	$= 75$
2 unit-volumes of arseniuretted hydrogen weigh		78
1 unit-volume of arseniuretted hydrogen weighs		39

The actual specific gravity of arseniuretted hydrogen, as determined by experiment, is as nearly as possible 39,—a fact which makes it certain that two volumes of the gas do not contain one volume of the heavy arsenic-vapor, which is 150 times as heavy as hydrogen, but only half a volume. Herein this gas differs from ammonia, but resembles phosphuretted hydrogen. The weight of the quantity of arsenic which combines with three atoms of hydrogen is 75, just as the weight of the quantity of nitrogen which combines with three atoms of hydrogen is 14; but 75 parts by weight of arsenic-vapor only occupy one half the space which 14 parts of nitrogen fill.

310. *Arsenic and Oxygen.*—Arsenic forms two well-defined oxides, arsenious acid, As_2O_3, and arsenic acid, As_2O_5. The black film which forms on the surface of the metal when exposed to the air is by many supposed to be a suboxide, while others think it is more probably a mixture of metallic arsenic with arsenious acid. The first of the above-mentioned acids corresponds with nitrous and phosphorous acids, the second with

nitric and phosphoric; but arsenious acid is very stable, in comparison with arsenic acid, while the reverse is true of the analogous acids containing nitrogen and phosphorus. The element arsenic possesses many properties which ally it to the metals; but in its compounds its close connexion with nitrogen and phosphorus is clearly exhibited. Its oxides, for example, are both acids, and these acids unite with the oxides of the metals proper to form stable, crystallizable salts, which are in many cases isomorphous with the corresponding salts containing phosphorus.

311. *Arsenious Acid* (As_2O_3).—Arsenious acid, known in commerce as arsenic, or white arsenic, is obtained as a secondary product in the roasting of arsenical ores of nickel, cobalt, and tin, and as a principal product in the roasting of arsenical pyrites. The volatile matters which escape from the roasted ores consist mainly of sulphurous and arsenious acids; the first is allowed to pass off into the atmosphere, the second condenses to the solid state in the chambers and long passages through which the vapors are forced to pass in order that they may deposit their arsenious acid. A second sublimation purifies the raw product. According to the temperature at which the arsenious acid is sublimed and condensed, the product is either in powder or in transparent masses; a low temperature with sudden condensation yields a white powder of minute crystals; a higher temperature with gradual solidification produces a transparent glass.

312. Arsenious acid is a white solid, which occurs not only in two conditions, one amorphous and the other crystalline, but also in two distinct crystalline forms. When the vapor of the acid is cooled so quickly that it solidifies at once, without passing through the semifluid state, each particle of the solid acid assumes more or less perfectly the octahedral form. A hot saturated aqueous solution of the acid also deposits regular octahedral crystals on cooling. The amorphous, glassy variety of the acid changes spontaneously, when kept in contact with the air, into an aggregation of minute octahedral crystals, thereby becoming opaque and porcelain-like in appearance. The other crystalline form of arsenious acid is the right rhombic prism; this form occurs much less frequently than the first, and is con-

verted into the octahedral form by sublimation and by solution in hot water.

The two varieties of arsenious acid, the vitreous and the porcellaneous, differ decidedly in physical and chemical properties, yet they have precisely the same chemical composition; and when either variety changes into the other, no alteration of weight, no addition or subtraction of matter accompanies the change. The two varieties contain the same two elements in precisely the same proportions. When two or more *compounds*, which exhibit essential differences of physical and chemical properties, are, nevertheless, found to be identical in respect to constituent elements and their proportions, the compounds are said to be *isomeric* (equal parts). The term *allotropism* (§ 162) properly applies to the *elements* only, the term *isomerism* to *compounds* only; both terms, however, refer to one and the same unquestionable, though perplexing, truth—namely, that the widest diversity of properties may coexist with absolute identity of ultimate chemical constitution. Two allotropic states of the same element not infrequently present more striking differences than elements recognized as distinct; and among the numerous complex compounds of carbon with which organic chemistry deals, there are many isomeric compounds which are so entirely dissimilar as to lead almost irresistibly to the belief that it is of as much consequence how the atoms of a compound are arranged as what kind of atoms they are. Arsenious acid does not afford a very striking example of isomerism; nevertheless the properties of its two modifications are quite diverse. If it be true that the different arrangement of atoms is the cause of the diversity of isomeric compounds, it is evident that the differences between two varieties of a compound of only two kinds of atoms, united in the simple ratio of 2 to 3, cannot be expected to be so marked as the differences between isomeric compounds which contain four or five elements united in the very complicated proportions which frequently characterize the compounds of carbon. Nevertheless the differences between the two isomeric compounds of arsenic and oxygen are sufficiently distinct. The glassy acid dissolves much more rapidly in water than the porcelain-like variety, being three times as soluble in that liquid. The relation of the two varieties to heat is not the same; for when the vitreous acid changes into the opaque, heat is disengaged. As this change generally takes place slowly, from the surface towards the centre of any fragment of the vitreous variety, the heat evolved is not perceived; but if the change be suddenly accomplished, not only heat, but light also will be disengaged.

Exp. 122.—Dissolve 4 or 5 grms. of the vitreous acid in a hot mixture of 24 grms. of strong chlorhydric acid and 8 c. c. of water, and let the solution cool slowly; the arsenious acid will crystallize in transparent octahedrons, and the formation of the crystals will be accompanied by flashes of light.

The specific gravity of the vitreous acid is 3·738; that of the porcellaneous 3·699. The opaque variety may be changed into the vitreous by long boiling with water. It appears, therefore, that the arrangement of atoms which may be supposed to furnish the vitreous acid is stable only at high temperatures, and that the arrangement of atoms which is peculiar to the opaque acid is stable only at low temperatures.

313. Arsenious acid volatilizes without change when heated with free access of air; if heated in contact with carbon, it gives up its oxygen, and metallic arsenic is liberated. Copper and many other metals reduce arsenious acid.

Exp. 123.—Place a few particles of arsenious acid in an open tube of hard glass (No. 5) about 10 c.m. long, and heat the acid over the lamp, holding the tube in a sloping position; the white solid will be volatilized, but it will immediately be deposited again upon the cold part of the tube. By the aid of a lens, this deposit may be seen to be crystalline.

Fig. 46.

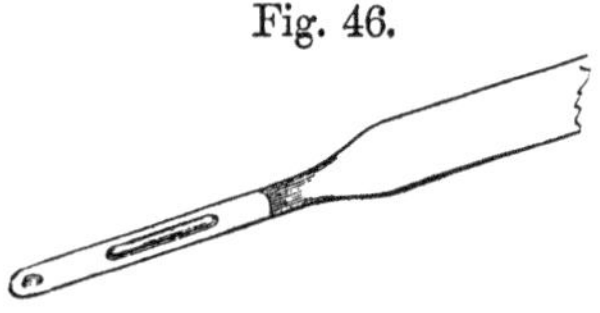

Exp. 124.—Drop into the point of a drawn-out tube of hard glass, No. 5, a morsel of arsenious acid, and above it place a splinter of charcoal (Fig. 46); heat the coal red-hot in the flame of the lamp, and then volatilize the arsenious acid. The acid will give its oxygen to the coal, and the arsenic will be deposited in a ring on the cold part of the tube, presenting a brilliant metallic appearance.

Exp. 125.—Throw a particle of arsenious acid upon a piece of red-hot charcoal; the acid will be partly reduced, and the peculiar garlic odor of the vapor of metallic arsenic will be perceived.

Exp. 126.—Dissolve a few centigrammes of arsenious acid in 5 or 6 c. c. of chlorhydric acid heated in a test-tube; in the hot solution immerse a narrow strip of clean copper; an iron-gray film will be deposited upon the copper. This coating contains metallic arsenic derived from the arsenious acid; it consists of an alloy of arsenic and copper.

314. It is very difficult to say what the solubility of arsenious acid in water really is. The results of different experimenters present very wide discrepancies, due in part to the fact already stated, that the two modifications of arsenious acid are of unlike solubility, and in part also to the circumstance that the acid dissolves with extreme slowness. The difficulty of the determination is increased by the readiness with which either modification passes into the other in consequence of changes of temperature; it is quite possible that both varieties may simultaneously exist in the same solution. A hot aqueous solution usually contains 1 part of the acid in 10 or 12 parts of water; on cooling this solution, a portion of the acid separates, leaving a solution which contains 1 part of acid in 20 to 30 parts of water. The aqueous solution has a feeble acid reaction. No definite hydrate of arsenious acid is known. The acid is much less soluble in alcohol than in water. Hot chlorhydric acid dissolves it with facility, and when cold retains a large proportion in solution; other acids, even some vegetable acids, dissolve it readily when hot, though most of them keep but little in solution when cooled. When the solution of arsenious acid in chlorhydric acid is evaporated, a compound of chlorine and arsenic, the terchloride of arsenic (§ 336), is volatilized, and the solution thus loses a portion of its arsenic. This fact is of importance in examinations for arsenic in cases of suspected poisoning.

315. Solutions of caustic soda and potash readily dissolve the acid, a soluble *arsenite* of sodium or potassium resulting from the reaction. From these arsenites of sodium and potassium the arsenites of other metals are generally obtained by the way of double decomposition. The arsenites are numerous, but they are not very stable and have been but little studied.

316. Arsenious acid is oxidized and converted into arsenic acid by digestion with nitric acid. The same transformation is

brought about, but quicker, by the action of aqua regia, and by chlorine, bromine, and iodine in presence of water. When iodine is added to a solution of arsenious acid mixed with a little starch-paste, the whole of the arsenious acid is converted into arsenic acid before any blue coloration of the starch is produced by the iodine. These facts are turned to account in the volumetric determination of chlorine (§ 565). Sulphuretted hydrogen colors an aqueous solution of arsenious acid yellow, and precipitates a yellow sulphide of arsenic (§ 210) from a solution acidulated with chlorhydric acid.

Arsenious acid is a violent poison, all the more dangerous because it has neither taste nor odor to warn the victim of its presence; two decigrammes of it will cause death. All the soluble salts of arsenious acid are likewise horribly poisonous. The best antidote to the poison is a mixture of moist, freshly precipitated sesquioxide of iron and caustic magnesia.

317. Arsenious acid is largely used for the manufacture of two green paints, an arsenite of copper and a compound of arsenite and acetate of copper; it is applied as an oxidizing agent in the manufacture of glass; it is used for poisoning vermin, and is consumed in considerable quantities for producing the arsenic acid which is used in the dyeing and printing of cloth, and in the manufacture of aniline colors; it is used in very small doses as a remedy for asthma, and in some skin-diseases. Although the acid is so violent a poison, it seems to be possible, by beginning with small doses and gradually increasing them, to accustom the human body to sustain, without injury, doses of 2 to 3 decigrammes, or even more; the arsenic thus taken is said to produce a plump and healthy appearance in those who use it, and especially to increase the power of the respiratory organs. In veterinary practice, it has been found that arsenious acid administered to animals in this manner improves the appearance of the skin.

318. *Arsenic Acid* (As_2O_5).—This compound is produced by oxidizing arsenious acid with nitric acid, aqua regia, hypochlorous acid, or other oxidizing agents.

Exp. 127.—Add 4 grms. of powdered arsenious acid, little by little, to a mixture of 4 grms. of concentrated nitric acid, and 8 grms. of concentrated chlorhydric acid, contained in a small evaporating dish, and gently heated over a lamp in a strong current of air, or beneath a well-ventilated hood. The liquid, which at first gives off red fumes in considerable quantity, must be evaporated until it assumes a syrupy consistency, resembling that of oil of vitriol. This syrupy liquid is arsenic acid.

319. The syrupy solution thus obtained deposits, after standing for some days, at the ordinary temperature, transparent elongated prisms or rhomboidal laminæ. These crystals, heated to 100°, first melt and then yield the terhydrate of arsenic acid ($3H_2O,As_2O_5 = 2H_3AsO_4$) as a crystalline precipitate. The same hydrated acid separates in large prismatic crystals when a concentrated aqueous solution is cooled to a low temperature. There are two other hydrates of the oxide As_2O_5,—a bihydrate, $2H_2O,As_2O_5 = H_4As_2O_7$, and a monohydrate, $H_2O,As_2O_5 = 2HAsO_3$; both these lower hydrates are obtained from the terhydrate by subjecting the latter to the prolonged action of certain temperatures. If either of the hydrates be heated to dull redness, a white amorphous mass remains, which is the anhydrous acid, As_2O_5; this substance has no action upon litmus, and seems to be scarcely soluble in water. After long exposure to moist air, it slowly deliquesces, and if covered with water and soaked for a long time, it at last dissolves, being probably converted into the soluble terhydrate. At a full red heat it is resolved into arsenious acid and oxygen.

320. In spite of the recognized existence of three solid hydrates of arsenic acid, there is but one aqueous solution of this acid, inasmuch as the monohydrate, the bihydrate, and the anhydride, are all immediately converted into the terhydrate when dissolved in water. The solution has a very sour taste and a strong acid reaction on vegetable colors. The concentrated liquid is highly corrosive and produces blisters on the skin. Arsenic acid and its salts are poisonous, but not in so high a degree as arsenious acid and the arsenites.

321. Arsenic acid is a strong acid, capable of expelling all the more volatile acids from their salts at high temperatures. Its three hydrates are strictly comparable with the three hydrates of phosphoric acid.

Hydrates of Phosphoric Acid.	HPO_3 $H_4P_2O_7$ H_3PO_4	$HAsO_3$ $H_4As_2O_7$ H_3AsO_4	*Hydrates of Arsenic Acid.*

Either one, two, or all three of the hydrogen atoms in common arsenic acid, H_3AsO_4, may be replaced by a metal, so that three arseniates of any one metal may exist, as for example,

NaH_2AsO_4	Na_2HAsO_4	Na_3AsO_4
Acid Arseniate of Sodium.	*"Neutral" Arseniate of Sodium.*	*Basic Arseniate of Sodium.*

If an acid arseniate be suitably heated, a meta-arseniate results, as, for example, $NaAsO_3 = NaH_2AsO_4 - H_2O$; if a neutral arseniate be sufficiently heated, a pyro-arseniate results, as, for example, $Na_4As_2O_7 = 2Na_2HAsO_4 - H_2O$; but such meta- and pyro-arseniates, unlike the corresponding meta- and pyro-phosphates, have very little stability, take up again the molecule of water, which the heat expelled, the moment they are brought in contact with water, and are so changed back again into salts of ordinary arsenic acid. The salts of arsenic acid are isomorphous throughout with the corresponding phosphates.

322. Arsenic acid is readily reduced to arsenious acid, and, consequently, acts in some cases as an oxidizing agent. Thus sulphurous acid reduces arsenic acid, and is itself converted into sulphuric acid :—

$$2H_3AsO_4 + 2SO_2 = 2H_2SO_4 + As_2O_3 + H_2O.$$

Sulphydric acid gas, passed through a not too concentrated solution of arsenic acid, slowly precipitates the yellow tersulphide of arsenic, the action being assisted by heat and by the presence of another acid. Charcoal and the metals at a red heat reduce arsenic acid to the metallic state, just as they do arsenious acid.

323. Arsenic acid has been extensively used in calico-printing, in place of the more expensive tartaric acid, for developing white patterns on a colored ground in the chloride-of-lime vat. It is also an excellent preservative of animal substances, and is accordingly used to defend the specimens and preparations of the anatomist and naturalist from decay and from the attack of insects.

324. *Detection of arsenic in cases of poisoning.*—Nearly all compounds of arsenic are poisonous; but arsenious acid is best

known and most easily procured, and is therefore most likely to be met with in cases of poisoning by arsenic, whether accidental or intentional. In criminal trials the solubility of arsenious acid in water has often been much discussed; but this is practically a point of little importance, for the tasteless poison is generally administered in the solid state mixed with soup, gruel, milk, or even with solid food. It thus sometimes happens that small particles of the poison can be found adhering to culinary vessels, cups, plates, or spoons, or even to the coatings of the stomach and intestines after death. If the arsenious acid is too finely divided to be picked out in lumps, it may sometimes be separated by stirring up the mass, under examination, with water, and leaving the heavier particles to settle. Any solid arsenious acid that may be present will be sure to be found in the residue; it may be washed with cold water. It is always very satisfactory thus to obtain the solid poison in the condition in which it was administered, because the examination is, in such cases, very direct and conclusive. It is only necessary to try, with the white powder thus obtained, the experiments already given to illustrate the properties of arsenious acid (Exps. 123–126), together with certain other discriminating tests shortly to be described.

325. It more frequently happens, however, that the arsenic has been dissolved by the acid secretion of the stomach, and has become intimately mixed with the food or excretions, or incorporated into the substance of the organs themselves. The examination then becomes more difficult. The reduction of arsenious acid by copper (Exp. 126) is an available test in such cases. To the suspected matter, if liquid, about one-sixth of its bulk of chlorhydric acid is added, and the mixture is gently boiled. Solid tissues must be cut into small pieces and boiled for some time with dilute chlorhydric acid (1 part acid to 6 parts water) until the whole is disintegrated; this solution is finally clarified by filtration. Strips of copper gauze or foil are then immersed in the boiling liquid; and if any gray deposit is produced, fresh pieces of metal are added so long as the color of the copper is perceptibly changed. They are then removed, washed with water, dried, folded up, placed in a dry tube of hard glass and gently heated. Some of the metallic arsenic in the gray alloy

will be converted into arsenious acid, which collects on the cold part of the tube in the form of a crystalline sublimate. To this sublimate all tests for the identification of arsenious acid can be applied. This mode of operation is known as Reinsch's test. The chlorhydric acid employed must be proved to be free from arsenic.

326. Another method of separating arsenious acid from the organic matters with which it is mixed is that of *dialysis*, a process which depends upon the very different rates at which different substances *diffuse* through water.

Exp. 128.—Select two straight-sided bottles of clear glass about 15 c.m. deep and 8 to 9 c.m. wide. Fill them seven-eighths full of distilled water, or any pure soft water. Dissolve 10 grms. of bichromate of potassium in 100 c. c. of water; suck as much of this solution as will fill the remaining eighth of one of the above-mentioned bottles into a pipette (Appendix, § 22), and carefully convey the colored fluid to the bottom of the bottle by bringing the fine point of the pipette to the bottom of the bottle and then allowing the liquid to flow very slowly out of the pipette. If time enough (5 or 6 minutes) be taken for this process, no sensible intermixture of the two liquids will take place during the delivery. Dissolve 10 grms. of caramel (melted and partially burnt sugar) in 100 c. c. of water, and convey to the bottom of the second bottle, in the same manner as before, enough of the dark-colored solution to fill the bottle.

The two bottles are left at rest for several days in a room where the temperature is nearly constant. Spontaneous diffusion immediately begins; and the very different rates at which the two colored substances diffuse upwards through the water should be from time to time observed.

327. Substances which have a comparatively high diffusive power have generally, though not invariably, the power of crystallizing; their solutions are generally free from viscosity, and always have taste. Such substances are designated by the term *crystalloids*. Among crystalloids there are wide differences of diffusive power; thus caustic potash diffuses twice as fast as sulphate of potassium, and sulphate of potassium twice as fast as sulphate of magnesium.

Substances of very low diffusive power have little, if any, tendency to crystallize, and affect a vitreous structure. Such substances are often very soluble in water; but their solutions have always a certain degree of viscosity when concentrated, and are insipid or wholly tasteless. By combining with water, these substances are apt to form jellies. Gelatine has been taken as the type of this class; and they have hence been called *colloids*, a name derived from Greek words signifying *glue-like*. Among the colloids rank hydrated silicic acid, alumina, starch, gums, caramel, albumen, and animal and vegetable extractive matters.

As we can separate by means of distillation or evaporation two bodies of different volatility, so by the aid of diffusion we can separate one substance more or less completely from another. Jellies and colloid membranes are permeable to crystalloids, but are practically impermeable by colloids like themselves. By means, therefore, of a colloidal diaphragm, or partition, crystalloids can be separated from colloids by diffusion. The most suitable substance for the dialytic diaphragm is parchment paper, prepared by soaking unglazed paper, for a few seconds, in a mixture of 6 parts of strong sulphuric acid and 1 part of water, and immediately washing it, first in water and then in water containing ammonia. The paper subjected to this treatment becomes semitransparent and tough, like parchment.

A *dialyzer*, as the apparatus for effecting separation by diffusion is called, consists of two gutta-percha, or wooden, hoops, one of which should be 5 c.m., and the other 2·5 c.m. deep. The deeper hoop is slightly conical, and the shallower must slip over the small end of the deeper. The hoops may be from 15 to 25 c.m. in diameter. The parchment paper, which is to form the bottom, must be about 8 c.m. wider than the small end of the 5 c.m. hoop. To prepare the dialyzer for use, soak the parchment paper for about a minute in distilled water; stretch it evenly over the small end of the 5 c.m. hoop and strain it on tightly by pushing over it the 2·5 c.m. hoop. The paper must be pressed smoothly up round the outside of the deeper hoop, and the bottom must be flat and even.

There must be no small holes in the paper. To detect such, put distilled water into the dialyser to the depth of 5 m.m., and place the dialyzer on some white blotting-paper. If any wet or dark spots appear, they indicate the existence of small holes. To close such holes, apply to the under surface of the paper about the holes some solution of albumen, put on a small patch of parchment paper, and iron the patch with a hot smoothing-iron. The albumen will coagulate, fix the patch, and close the hole.

Exp. 129.—Into the dialyzer so prepared pour an aqueous solution containing five per cent. of cane-sugar and five per cent. of gum arabic, to the depth of about 1·25 c.m. Then float the dialyzer on distilled water contained in a flat basin. The volume of water in the basin should be from 5 to 10 times as great as the volume of the fluid in the dialyzer. The wider the dialyzer and the greater the quantity of distilled water in the outer basin, the more rapid and effective the diffusion. A dialyzer 15 c.m. in diameter, serves to operate upon 200 to 250 c. c. of liquid; one of 20 c.m. upon 400 to 450 c. c.; one of 25 c.m. upon 600 c. c.

After the lapse of twenty-four hours, the water in the basin should be poured into an evaporating-dish and gently evaporated over a water-bath. Pure sugar will crystallize from the solution. The sugar, a crystalloid, has passed through the diaphragm; the gum, a colloid, has remained in the dialyzer. It should be remarked that gum is wholly uncrystallizable, and that the mixed solution of gum and sugar will not yield crystals, but only an amorphous mass, when evaporated.

328. By means of this dialyzing apparatus, arsenious acid, salts of the metals, strychnine, and other crystallizable poisons, mineral and organic, can be readily separated from organic fluids; and the process has the very great advantage of introducing no metal, chemical reagent, or other foreign substance into the fluids under examination. After twenty-four hours, the crystallizable poison, or a large proportion of it, will have been transferred to the distilled water in the outer basin, and in this solution it may be sought for by the application of the appropriate tests.

Exp. 130.—Dissolve 0·1 grm. of arsenious acid in about 30 c. c. of hot water, and stir the solution into about 200 c. c. of milk, ale, soup, gruel, or other thick organic fluid; place the *poisonous* fluid in a 15 c.m. dialyzer and float the dialyzer on two litres of distilled water in a clean basin. Allow the apparatus to stand at rest in a room where the temperature is tolerably uniform for forty-eight hours. At the expiration of this time, transfer the clear solution in the basin to an evaporating-dish, without losing a drop, rinse the basin carefully with distilled water, and add the rinsings to the contents of the dish; evaporate the solution over a water-bath (see Appendix, § 14) to the bulk of 50 c. c. To one-third of this concentrated solution add a few drops of pure chlorhydric acid, and apply Reinsch's test for arsenic (§ 325), with due regard to the small scale on which the operation must be conducted. About 0·025 grm. of arsenious acid is the quantity which may be expected to respond to the test by copper. Three-quarters of the original decigramme should be transferred by diffusion through the dialyzer in the course of forty-eight hours, and of this solution of this 0·075 grm. of arsenious acid we have taken one-third. The rest of the solution is to be reserved for tests hereafter to be described.

329. When the arsenious acid must be sought in large organs of the body, like the stomach, liver, or intestines, or in considerable quantities of disgusting semifluid materials, it is sometimes necessary to utterly destroy the organic matters by processes which cannot cause the loss of arsenic. Several methods may be employed for this purpose. 1. The organic matter is gently heated in a tubulated retort with strong chlorhydric acid, and strong nitric acid is added from time to time. The organic matter is thus completely destroyed, with the exception of the fat. A cooled receiver is connected with the retort to condense the distillate from the hot mass. The fat is separated from the clear liquid in the retort by decantation, and well washed with water; these washings, together with the distillate in the receiver, are added to the main bulk of the fluid. 2. Chlorate of potassium may be added instead of nitric acid. 3. The organic matter, after being made as fine as possible, is stirred up with water, and chlorine gas is passed through the liquid until the organic substances are partly destroyed and partly deposited in brown flakes.

All these processes, and there are others based on like principles, are processes of combustion; aqua regia, chlorate of potassium, and chlorine are oxidizing agents of great power, as we have already seen; they burn the carbon and hydrogen of the organic materials as literally as the oxygen of the air burns the coal in the grate. The arsenic also is oxidized and converted into its highest oxide, arsenic acid. Whenever chlorhydric acid is used, and heat is applied, there is danger that chloride of arsenic ($AsCl_3$) may be formed; this chloride is a volatile body, against whose loss precautions must be taken, by never allowing the temperature of the fluids to rise much above 100°, and by collecting any distillate which may be formed under circumstances which make it possible for this chloride to be evolved.

330. All these methods of destroying the organic matters in which arsenious acid is to be sought for are liable to one objec-

tion. Considerable quantities, even kilogrammes, of acids must be used, if the quantity of organic substance to be destroyed is large; chlorhydric and sulphuric acids very commonly themselves contain arsenic, and since the liquids, which result from the destruction of the organic tissues, are finally evaporated to a very small bulk, all the arsenic in several kilogrammes of the acids employed, as well as all the arsenic which may have been contained in the bodily organs or fluids submitted to examination, will be concentrated into a small cupful of liquid. It is obviously necessary to demonstrate that the arsenic reactions cannot be obtained from the same quantities of the same acids actually employed, subjected to the same series of operations. The best way is to conduct a parallel examination of normal animal organs or fluids; in this examination the identical processes and the same weights of the same chemical materials must be employed as in the examination of the suspected substances; if arsenic is found in the latter investigation, but not in the parallel examination of the normal animal substances, it will be quite certain that the arsenic was not derived from the chemicals employed in the research.

331. When, by any of the processes above described, a clear arsenical solution, free from organic matter, has been obtained, the identification of the arsenic may be accomplished by many methods, of which the two following will serve as examples :—

1. *By precipitation as sulphide of arsenic.*—If the clear solution contains arsenic acid, it is necessary to reduce this oxide to arsenious acid before the precipitation can be effected. This reduction may be accomplished by passing a slow stream of washed sulphydric acid gas (§ 202) through the solution for several hours, but may be immediately effected by saturating the solution with sulphurous acid gas, the superfluous gas being finally expelled by gentle heating. After the reduction has been effected, a slow stream of washed sulphydric acid gas precipitates the yellow sulphide of arsenic from the liquid.

Exp. 131.—Acidulate one-half of the liquid reserved from Exp. 130 with pure chlorhydric acid, place it in a small beaker glass and pass a slow stream of washed sulphydric acid gas through the solution. The delivery-tube of the gas should be small and the current slow; a piece of unglazed paper should be used as a cover, in order to keep the beaker full of the gas. A yellow precipitate (As_2S_3) will appear, indicating the probable presence of arsenic. When no more precipitate seems to form, stop the current of gas, and let the beaker stand in a warm place till the odor of the gas has nearly disappeared. Collect the precipitate on a small filter (see Appendix, § 14), wash it thoroughly with water, and dry it.

Exp. 132.—Mix intimately the dry precipitate obtained in the last experiment with its bulk of dry carbonate of sodium and its bulk of dry cyanide of potassium, and introduce this mixture into a hard glass tube (No. 5), the end of which has been closed and expanded to a small bulb. If the precipitate stick to the filter-paper, it must be scraped off. Warm the bulb and its contents over the lamp to expel moisture, then wipe the tube out with a tuft of cotton on the end of a wire, and bring the bulb to a red heat. A ring of metallic arsenic, like that of Exp. 124, will be deposited in the tube. Preserve this metallic mirror for subsequent study.

2. *By conversion into arseniuretted hydrogen.*—When an aqueous or acid solution containing arsenious or arsenic acid is added to the contents of a flask in which hydrogen is being generated, the nascent hydrogen reduces the oxide of arsenic, and there is formed a quantity of arseniuretted hydrogen, which mixes with the uncombined hydrogen evolved. (Compare § 307.) This arseniuretted hydrogen is decomposed, with deposition of metallic arsenic, by being passed through a red-hot tube. The undecomposed gas burns with a whitish flame; and if a cold body be held in the flame, a spot of metallic arsenic will be deposited upon it. (See § 308.) Upon these properties and reactions is based the process for detecting arsenic known as *Marsh's test.*

Fig. 47.

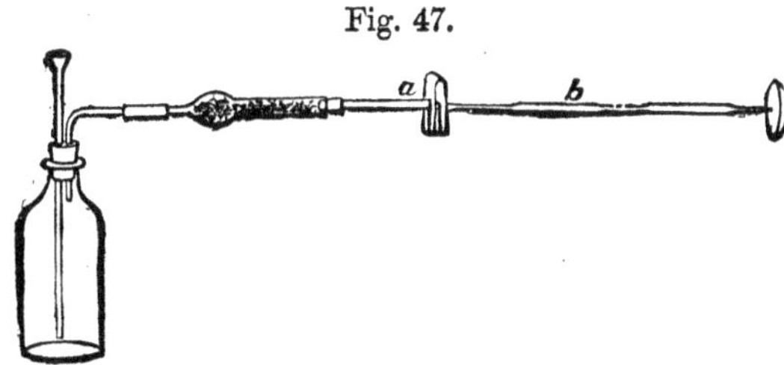

Exp. 133.—To a bottle prepared for generating hydrogen from pure zinc and dilute sulphuric acid, adapt a chloride-of-calcium tube, and with the outer end of this drying-tube connect a tube of hard glass (No. 4) which has been twice drawn to a fine bore and which terminates in a fine open point. (Fig. 47.) Support this long tube at three or four points, in such a manner that the softening of the glass, first at the point *a*, and then at the point *b*, shall not distort the tube. By adding acid through the funnel-tube of the flask, evolve hydrogen, and when the whole apparatus is full of hydrogen, light the gas at the tip of the hard glass tube. By means of an efficient gas-lamp, heat about 2 c.m. of this tube to dull redness at the point *a*; just beyond the hot part of the tube, place a small sheet-iron screen, as shown in the figure, to cut off the heat from the adjoining narrow part of the tube. Maintain the apparatus in this condition for ten minutes, the glass tube red-hot at one point and the hydrogen flowing steadily through the tube and burning with a colorless flame at the point. If no deposit, or only a scarcely perceptible deposit, appears in the fine tube adjoining the heated portion, the zinc and sulphuric acid are pure enough for the experiment, but if a black, shining deposit appears in the fine tube, the materials themselves contain arsenic and are, of course, unsuitable for use in testing for this substance.

If the zinc and sulphuric acid prove to be sufficiently free from arsenic, add to the contents of the flask a few drops of the liquid obtained by dialysis (Exp. 130). In a moment a mirror of arsenic will be deposited in the fine tube adjoining *a*; when this mirror has become large and dense, move the lamp to *b*, transfer the screen, and obtain a similar mirror in the second attenuated portion of the tube; finally, extinguish the lamp and allow the arseniuretted hydrogen to reach the burning jet of gas at the extreme point of the apparatus; the white coloration of the flame will now, for the first time, be seen; introduce into the jet a bit of cold porcelain, and obtain the characteristic black and lustrous spot of metallic arsenic; this experiment may be repeated indefinitely and a large number of spots obtained for subsequent use. Preserve the two mirrors and a number of arsenic spots for future study. In order to prevent the possibility of any arseniuretted hydrogen escaping into the air of the room, the jet of gas must be kept constantly burning, and when the experiments are ended the flask must be washed out promptly and thoroughly.

332. This method is very well adapted for the speedy and certain detection of arsenic in green paints, such as are applied to wall-papers, artificial flowers, lamp-shades, and the like; for in such cases, if any arsenic is present, there is so much as to make any traces of arsenic which may contaminate the zinc and sulphuric acid of no consequence whatever. It is only necessary, in

such examinations, to scrape off some of the green coloring-matter, dissolve it in dilute chlorhydric acid, and add the solution to the hydrogen flask of the apparatus described above. Arsenic greens instantly give enormous mirrors and spots under these conditions.

333. In medico-legal investigations, upon whose results life often depends, it must always be remembered that arsenic is very widely diffused in the mineral kingdom, and that it is a matter of great difficulty to procure reagents absolutely free from it. The substances employed as reagents in Marsh's test are often contaminated with it; and the acids used in destroying organic matter may well contain arsenic enough to become visible after the great concentration of this impurity which inevitably occurs in the evaporation of the liquid which results from the burning of the organic matter. The use of zinc is avoided, and other advantages gained, by obtaining the necessary hydrogen by the electrolysis of acidulated water. (§ 35.) When a solution of arsenious acid, acidulated with chlorhydric or sulphuric acid, is decomposed by the electric current, the greater part of the arsenic eliminated at the negative pole is given off in the form of arseniuretted hydrogen, which may be examined precisely as if it were generated in Marsh's apparatus. This method is very delicate, and seems to possess considerable advantages over Marsh's process, but it has not yet (1870) been actually applied in judicial investigations.

334. To describe the methods by which the analytical chemist purifies his reagents and proves their purity, would involve descending into technical details which are unsuitable for this manual. Zinc and acids, pure enough for illustrative experiments, can be bought of the dealers in pure chemicals. None but expert analysts should ever be intrusted with the chemical investigation in a supposed case of poisoning by arsenic. A difficulty attending such examinations remains to be discussed under the metal antimony, a substance which combines with hydrogen, as arsenic does, to form a gas which is decomposed by heat, as arseniuretted hydrogen is, with deposition of a metallic mirror which cannot be distinguished by mere inspection from that of arsenic. Since preparations of antimony are much em-

ployed as medicines, and particularly since tartar-emetic, a salt containing antimony, is often administered in cases of poisoning, it is essential to find means of distinguishing between compounds of arsenic and the analogous compounds of antimony.

335. *Chloride of Arsenic.*—Only one chloride of arsenic is known, the terchloride ($AsCl_3$), corresponding to the terchloride of phosphorus. No quinquichloride corresponding to the quinquichloride of phosphorus is known. The chloride of arsenic is formed by passing dry chlorine gas over finely divided metallic arsenic placed in a retort. The combination is usually attended with combustion; and the heat developed is sufficient to distil the chloride over into the receiver. It may also be made by distilling a mixture of metallic arsenic and the mercury compound called corrosive sublimate, in accordance with the following equation, in which Hg stands for mercury (Hydrargyrum):—

$$\underset{\textit{Corrosive Sublimate.}}{6\,HgCl_2} + 2As = \underset{\textit{Calomel.}}{3Hg_2Cl_2} + 2\,AsCl_3.$$

Terchloride of arsenic may also be procured by distilling arsenious acid with common salt and sulphuric acid. Small lumps of fused salt should be added from time to time to a mixture of arsenious acid with a large excess of sulphuric acid:—

$$As_2O_3 + 6NaCl + 6H_2SO_4 = 3H_2O + 2AsCl_3 + 6NaHSO_4.$$

336. Terchloride of arsenic is a dense, colorless, oily liquid, whose specific gravity is 2·205. It boils at 132°, producing a vapor whose density is 90·91. It evaporates freely at the ordinary temperature, producing fumes of arsenious acid. It is highly poisonous. The chloride is decomposed by an excess of water into chlorhydric acid and arsenious acid, just as the chloride of phosphorus is resolved by water into chlorhydric and phosphorous acids; this reaction is the basis of the best determination of the atomic weight of arsenic.

$$2AsCl_3 + 3H_2O = 6HCl + As_2O_3.$$

All the chlorine in a known weight of chloride of arsenic is converted by this reaction into chlorhydric acid; the weight of the chlorine contained in this chlorhydric acid can be accurately determined, and the weight of the arsenic with which this quantity of chlorine was originally combined is obtained by simple subtraction. The proportions by weight in which arsenic and chlorine combine are thus determined. In terchloride of arsenic, as in terchloride of phosphorus, three volumes of chlorine unite with only half a volume of arsenic vapor to produce two volumes of the terchloride vapor. Indeed all the volatile compounds of arsenic illustrate the fact, already mentioned (§§ 305, 309), that the unit-volume weight, or specific gravity, of arsenic vapor is the double of its atomic weight.

337. *Bromide and Iodide of Arsenic.* It is enough to say of these two compounds that they are crystallizable solids, obtainable by the direct action of the elements upon each other, and answering to the formulæ $AsBr_3$ and AsI_3 respectively.

338. *Sulphides of Arsenic.*—There are three well-defined sulphides of arsenic, corresponding to the formulæ As_2S_2, As_2S_3, and As_2S_5. The first two occur as natural minerals, *realgar* and *orpiment*, and may also be obtained in the free state by artificial processes; the third is known only in combination.

339. *Bisulphide of Arsenic* (As_2S_2).—The native mineral realgar has this composition. The compound is obtained artificially by melting arsenic with sulphur, or arsenic with orpiment (see the next section), or sulphur with arsenious acid, in such proportions in either case as will bring together those parts by weight of the two elements which the above formula requires. The commercial product is a brownish-red opaque substance of variable composition, generally containing free arsenious acid. Realgar is one of the ingredients of *white Indian fire,* a mixture of 24 parts of nitre, 7 of sulphur, and 2 of realgar, sometimes used as a signal light.

340. *Tersulphide of Arsenic* (As_2S_3).—This sulphide occurs native in translucent rhombic prisms of a yellow color. It is obtained artificially by passing sulphydric acid gas through a solution of arsenious acid, or an arsenite acidulated with chlorhydric acid; the sulphide falls as a bright yellow amorphous powder, insoluble in water and dilute acids. It melts easily, and burns in the air with a pale blue flame; in closed vessels it may be sublimed without change.

Under the name of *orpiment,* this sulphide is used as an orange pigment; a mixture of the sulphide with arsenious acid, called king's yellow, was formerly employed as a yellow pigment. This impure tersulphide was made by subliming 7 parts of arsenious acid with 1 part of sulphur, a proportion of sulphur not sufficient to convert all the acid into tersulphide. If a pattern be printed upon cotton cloth with a preparation containing arsenious acid, and the cloth be then passed through water containing sulphydric acid, orpiment will be deposited in the fibre of the cloth and the pattern will be brought out in orange-yellow. We have already seen (Exp. 132) that the tersulphide of arsenic yields a mirror of metallic arsenic when heated in a closed tube with a mixture of carbonate of sodium and cyanide of potassium. The sulphide is readily dissolved by a cold solution of potash, soda, or ammonia, the oxygen of the alkali converting part of the arsenic in the sulphide into arsenious acid, while the alkali-metal combines with the sulphur liberated; this alkaline sulphide then unites with the undecomposed portion of sulphide of arsenic to form a *sulphur-salt*, whose composition is that of an arsenite in which the oxygen has been replaced by sulphur.

$$4As_2S_3 + 5K_2O = 3(K_2S,As_2S_3) + 2K_2O,As_2O_3.$$

If an acid be added to this solution, no sulphuretted hydrogen is evolved, as is generally the case when an acid is brought in contact with an alkaline sulphide, but the sulphur and arsenic recombine and are separated as tersulphide of arsenic.

$$3(K_2S,As_2S_3) + 2K_2O,As_2O_3 + 10HCl = 10KCl + 5H_2O + 4As_2S_3.$$

341. *Sulpharsenites.*—The tersulphide of arsenic unites with basic metallic sulphides in three different proportions, forming with potassium, for example, the three salts $3K_2S,As_2S_3$, $2K_2S,As_2S_3$, and K_2S,As_2S_3. One mode of preparing a sulpharsenite has been mentioned in the last section; another method is to dissolve arsenious acid in an alkaline sulphydrate, in which case one-half of the alkali is converted into arsenite:—

Dualistic: $2As_2O_3 + 4KHS = K_2S,As_2S_3 + K_2O,As_2O_3 + 2H_2O.$
Empirical: $As_2O_3 + 2KHS = KAsS_2 + KAsO_2 + H_2O.$

The sulpharsenites of the other metals are mostly obtained from the sulpharsenites of sodium and potassium by the method of double decomposition. The sulpharsenites are either yellow or red; they are obscure bodies, of no practical importance at present. They illustrate, however, two points of theoretical interest—namely, the existence of sulphur-salts which bear to sulphides the same relation which oxygen salts bear to oxides, and the parallelism of composition between these two classes of salts. We place beside each other the empirical formulæ of the sulphur-salts of potassium and arsenic, and the corresponding oxygen-salts:—

Sulphur-salts.	*Oxygen-salts.*
K_3AsS_3	K_3AsO_3
$K_4As_2S_5$	$K_4As_2O_5$
$KAsS_2$	$KAsO_2$

342. *Quinquisulphide of Arsenic* (As_2S_5).—A sulphide of arsenic corresponding to anhydrous arsenic acid is not known in the free state. The quinquisulphide is known only in combination with sulphides of the metals in sulphur-salts called *sulpharseniates.* When a solution of sulphide of sodium is digested with some tersulphide of arsenic and sulphur enough to permit the formation of the quinquisulphide, and the solution, after long standing, is concentrated by evaporation and then cooled, large colorless crystals of sulpharseniate of sodium are obtained, which are not changed by exposure to the air. The crystals have the composition indicated by the formula $3Na_2S,As_2S_5 + 15H_2O$. The sulpharseniate, $2Na_2S,As_2S_5$, may be prepared by saturating the aqueous solution of the corresponding oxygen-salt $2Na_2O,As_2O_5$ with sulphydric acid gas. The sulpharseniates of the alkali-metals, and a few others, are soluble in water; but the greater number of sulpharseniates are insoluble. These insoluble salts are prepared by mixing a solution of an alkaline sulpharseniate with a solution of some salt of the metal whose sulpharseniate is desired. The same parallelism is observable between sulpharseniates and arseniates as between sulpharsenites and arsenites.

CHAPTER XVIII.

ANTIMONY.

343. Antimony is found native, both alone and alloyed with other metals, especially with arsenic, nickel, and silver. There exist also a considerable number of minerals which consist of, or contain, large proportions of the compounds of antimony with oxygen and sulphur.

344. All the antimony of commerce is obtained from the mineral tersulphide, Sb_2S_3. The symbol for antimony is Sb, from the Latin name of the substance, *Stibium*. This sulphide is very fusible, melting readily in the flame of a candle; it may therefore be separated from the earthy or rocky gangue in which it occurs by simple fusion at a low temperature. The metal is obtained from the sulphide by several different processes:—1. By adding to the melted sulphide iron nails, filings, or scraps; the iron and the antimony change places.

$$Sb_2S_3 + 3Fe = 3FeS + 2Sb.$$

2. By roasting the sulphide of antimony, reduced to a coarse powder, until the greater part of the sulphur has been burnt off and the antimony converted into the oxide; this residue is then mixed into a paste with water, charcoal powder, and carbonate of sodium, or some equivalent reducing flux, and heated in covered crucibles to full redness; the metal sinks to the bottom of the crucible. 3. By fusing together a mixture of sulphide of antimony, the scales which fall from hot iron when it is hammered (an oxide of iron), carbonate of sodium, and charcoal; this process is a sort of combination in a single operation of the two preceding methods. Since the sulphides and oxides of antimony and the metal itself are somewhat volatile at moderate temperatures, it has thus far been found impossible to avoid a considerable loss of metal during the melting, roasting, and reducing of the ore. From one-fifth to one-half of the metal is lost, according to the skill and care of the workmen.

345. The commonest impurities in commercial antimony are

sulphur, sodium, arsenic, lead, iron, and copper. These impurities injure the antimony for many of its applications in the arts; and the extensive use of antimonial preparations in medicine renders the removal of the arsenic a point of particular importance. The purification may be effected by fusing the powdered metal, first, with a mixture of sulphide of antimony and carbonate of sodium, and, secondly, with a mixture of carbonate of sodium and nitre. These fusions may be several times repeated; the impurities are either oxidized or converted into sulphides, and enter the slag. Lead, however, cannot be got rid of by these processes; this impurity is removed by fusing the antimony with oxide of antimony; the lead changes places with the antimony in the oxide of antimony, and is converted into litharge.

346. Antimony is a brittle metal, having a bluish-white color, a brilliant lustre, and a highly crystalline structure. The cakes of the commercial metal usually present upon their upper surfaces beautiful stellate or fern-like markings. Like phosphorus and arsenic, it is dimorphous, crystallizing both in rhombohedrons and octahedrons. The specific gravity of the metal is from 6·60 to 6·85; its atomic weight is 122. For a metal, it is a poor conductor of heat and electricity. At 450° it melts, gives off vapors at a low red heat, and takes fire at full redness, burning brilliantly, with evolution of white fumes of the teroxide (Sb_2O_3). If the antimony is contaminated with arsenic, as is often the case, a garlic odor, due to the presence of this impurity, may be imparted to the vapors.

Exp. 134.—Melt about 0·5 grm. of antimony by heating it on a piece of charcoal before the blowpipe. (See Chapter XX.) Throw the white, glowing globule into the middle of a large tray made of coarse paper; the globule bursts into a multitude of small beads, which fly over the paper, leaving in their trail a white, powdery oxide.

Exp. 135.—Melt a second small fragment of antimony upon charcoal as before, but, instead of throwing it from the coal, allow it to cool there slowly. The globule will, in this case, become covered with an efflorescence of crystals of the oxide.

The metal is not oxidized by exposure to dry or moist air at ordinary temperatures. Nitric acid oxidizes it easily, but does not dissolve it; the insoluble quinquioxide, or some mixed oxide, is formed, according to the strength of the acid employed.

Powdered antimony takes fire when thrown into chlorine gas, and combines very energetically with bromine and iodine. When finely powdered, it is dissolved by boiling chlorhydric acid, with evolution of hydrogen; if a little nitric acid be added to the chlorhydric, the metal dissolves easily, to form a solution of terchloride of antimony ($SbCl_3$). The metal, when in fine powder, is also dissolved readily by solutions of the higher sulphides of sodium and potassium, with formation of sulphantimonites and sulphantimoniates.

347. In spite of the strong tendency of this metal to crystallize, it can be obtained in an amorphous form by the electrolysis of concentrated antimonial solutions. This amorphous antimony always contains, however, 5 or 6 per cent. of terchloride of antimony and a trace of chlorhydric acid; whether these foreign substances are retained mechanically, or not, within the mass, is not clear. The amorphous metal has a dark steel-color, a smooth surface, a comparatively soft texture, a lustrous amorphous fracture and a specific gravity varying from 5·74 to 5·83. When gently heated or sharply struck, the amorphous antimony suddenly manifests a great heat, the temperature rising from 15° to 230° and upwards, and fumes of terchloride of antimony are evolved. After undergoing this peculiar change, the metal approximates to the crystalline variety in structure, density, and color.

348. Antimony enters into the composition of several very valuable alloys. *Type-metal* is an alloy of lead and antimony, containing about 20 per cent. of antimony. For stereotype plates $\frac{1}{80}$ to $\frac{1}{50}$ of tin is usually added to this alloy. The common white metallic alloys used for cheap teapots, spoons, forks, and like utensils, are variously compounded of brass, tin, lead, bismuth, and antimony; for example, a superior kind of pewter is made of 12 parts tin, 1 part antimony, and a small proportion of copper; *Britannia metal* is sometimes compounded of equal parts of brass, antimony, tin, bismuth, and lead. The value of antimony in these alloys depends upon the hardness which it communicates to the compounds, without rendering them inconveniently brittle.

With zinc, antimony forms two alloys having a definite crystalline character. The alloy containing 43 per cent. of zinc crystallizes in silver-white needle-like prisms; it answers to the formula Sb_2Zn_3. The alloy containing 33 per cent. of zinc crystallizes in broad plates presenting no similarity to the form of the other alloy; it answers to the formula SbZn. These alloys, especially Sb_2Zn_3, decompose boiling water with evolution of hydrogen. The crystals of these two alloys are obtained by the method of fusion (§ 194). In each of these crystallized alloys, the crystalline form may be preserved, although the proportions of the ingredients may vary considerably from the exact atomic proportions indicated by their formulæ. Thus needles may be obtained in which the actual proportion of antimony present varies from 35·77 per cent. to 57·24 per cent., the exact atomic proportion being 55·7 per cent.; and the percentage of antimony in the plates may fall as low as 64·57, or may rise as high as 79·42, although 65·07 per cent. is the true atomic proportion. These interesting crystalline alloys strikingly illustrate, therefore, a principle of wide applicability, namely, that a definite crystalline form is not necessarily a guaranty of an unvarying chemical composition.

349. *Antimony and Hydrogen.*—The composition of the gaseous compound of these two elements is not certainly known, inasmuch as it has never yet been prepared free from admixed hydrogen. When a solution of any salt of antimony is poured into a mixture of zinc and dilute acid which is disengaging hydrogen, the antimony compound is decomposed; one portion of the antimony, and sometimes even the whole of it, is deposited upon the zinc, while another portion usually combines with the hydrogen, and assumes the gaseous state. When this compound gas is passed through a solution of nitrate of silver, a precipitate is produced which has been found to consist of antimonide of silver, $SbAg_3$. Since silver is a metal which replaces hydrogen, atom for atom, it is a natural inference that the gas which has produced this precipitate must have the composition represented by the formula SbH_3. This supposition derives strength from the analogous formulæ of the well-known gases ammonia, NH_3, phosphuretted hydrogen, PH_3, and arseniuretted hyrogen, AsH_3

Antimoniuretted hydrogen is a colorless gas, inodorous when

free from arseniuretted hydrogen, and insoluble in water and alkaline liquids. The gas is decomposed at a red heat into antimony and hydrogen; it burns in the air with a whitish flame, and gives off a white smoke of teroxide of antimony; when a bit of cold porcelain is held against a burning jet of the gas, a sooty spot of metallic antimony is deposited on the porcelain. These reactions resemble those of arseniuretted hydrogen (§ 308).

Exp. 136.—Dissolve 0·5 grm. of tartar-emetic (tartrate of antimony and potassium) in about 30 c. c. of water. Add a few centimetres of the solution thus obtained to the bottle of the apparatus represented in Figure 48, in which hydrogen is already being generated from zinc

Fig. 48.

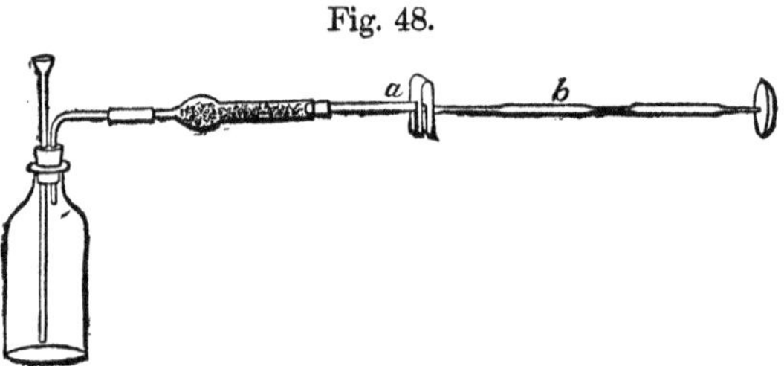

and dilute sulphuric acid. Antimoniuretted hydrogen will be produced, and should be submitted to precisely the same series of operations by which arseniuretted hydrogen was examined. (Exp. 133.) By heating the hard glass tube at *a* and *b* successively, two mirrors of antimony will be obtained; when the gas reaches the jet without decomposition, the white color of the flame will be observable; when a cold piece of porcelain is pressed against the burning jet, spots of antimony will be deposited thereon. Preserve these mirrors and spots.

Exp. 137.—Compare together the spots obtained on porcelain from arseniuretted hydrogen (Exp. 133) and from antimoniuretted hydrogen (Exp. 136). 1. The arsenical spot has a metallic lustre, and a brown color, when thin; the stain of antimony has a feeble lustre, and is smoky-black. 2. The arsenical stain disappears readily on the application of a heat below redness; the stain of antimony is volatile only at a red heat. On account of the comparative want of volatility which characterizes the antimony deposit, the mirrors of antimony obtained in the glass tube (Exp. 136) are always deposited nearer the heated portion of the tube than the arsenic mirrors are. 3. The arsenical stains may be distinguished, moreover, from the antimonial stains by means of a solution of "chloride of soda" (a mixture of hypochlorite of sodium with chloride of sodium, prepared by mixing a solution of chloride of lime with carbonate of sodium in excess, and filtering); this solution, which is analogous to, and indeed may be replaced by, a solution of chloride of lime (§ 120), immediately dissolves arsenical spots, but leaves antimonial spots unaffected for a long time. For the application of this test it is convenient to produce some spots on the interior of a concave bit of porcelain. 4. On warming an arsenic spot with a drop or two of aqua regia, and evaporating to dryness, a slight residue of arsenic acid is left, recognizable by its ready solubility in a drop of water; if to this drop of arsenic acid solution a drop of ammonio-nitrate of silver be added, a brick-red turbidity, due to the formation of arseniate of silver, will be produced. This ammonio-nitrate of silver is prepared by adding exactly ammonia enough to a solution of nitrate of silver to redissolve the precipitate which forms at first. The antimony spot treated in the same way yields no such red precipitate. 5. An antimony stain will dissolve readily in a few drops of a solution of sulphydrate of ammonium which has become yellow by keeping; when such a solution is evaporated to dryness, a bright orange stain remains. The arsenical stain, on the contrary, is not perceptibly affected by the yellow sulphydrate of ammonium solution, unless heat is applied.

Exp. 138.—Connect the tube of hard glass in which two arsenic mirrors were formed, in Exp. 133, with a sulphuretted-hydrogen-generator (Appendix, § 19), interposing between the tube and the generator a suitable drying-tube or bottle filled with chloride of calcium; then transmit through the tube a *very slow* stream of sulphydric acid gas, and heat the mirrors with a small gas-flame, proceeding from the outer to the inner border of the mirrors, in the direction opposite to that of the gas current.

Repeat the same process with the tube containing the antimony mirrors obtained in Exp. 136.

Yellow tersulphide of arsenic is formed in one case, and orange-red or black tersulphide of antimony in the other. When both metals are present in one mirror, the two sulphides appear side by side, the sulphide of arsenic as the more volatile lying invariably beyond the sulphide of antimony.

Exp. 139.—Transmit through the tube which contains the sulphide of arsenic a stream of dry chlorhydric acid gas (§ 95), without applying heat; no alteration will take place in the yellow sulphide.

Transmit the same gas through the tube containing the sulphide of antimony; the sulphide of antimony will immediately disappear. If the gaseous current be then passed through some water, the presence of antimony in the water can be demonstrated by means of sulphydric acid (§ 210).

When both sulphides are present at once, the chlorhydric acid attacks and removes the sulphide of antimony, while the sulphide of arsenic remains behind. A drop or two of ammonia-water, drawn into the tube, will then dissolve the sulphide of arsenic. This solubility in ammonia distinguishes the yellow sulphide from sulphur itself, with which it might otherwise be sometimes confounded.

Antimony and Oxygen.—Antimony forms two well-marked oxides, analogous to the oxides of arsenic, the teroxide or antimonious acid, Sb_2O_3, and the quinquioxide or antimonic acid, Sb_2O_5; a compound of these two oxides $Sb_2O_3, Sb_2O_5 = 2Sb_2O_4$, is sometimes recognized as a distinct oxide under the name of the quadroxide.

350. *Teroxide of Antimony.*—This oxide occurs as a natural mineral, called *White Antimony* or *Antimony Bloom.* Like arsenious acid, it is dimorphous, crystallizing in rhombic prisms belonging to the trimetric system, and also in regular octahedrons. The artificial, as well as the native, teroxide is dimorphous. Antimonious oxide is produced when antimony is burnt in the air, or heated to full redness in imperfectly covered crucibles. The easiest mode of getting it is to heat the tersulphide (Sb_2S_3) with strong chlorhydric acid as long as sulphydric acid continues to escape, and pour the resulting solution of the terchloride ($SbCl_3$) into a boiling solution of carbonate of sodium:—

$$2SbCl_3 + 3Na_2CO_3 = Sb_2O_3 + 6NaCl + 3CO_2.$$

If the solution of carbonate of sodium be cold or only warm instead of boiling, a hydrate of the teroxide is precipitated; $Sb_2O_3,H_2O = 2SbHO_2$.

Antimonious oxide is white or grayish-white at ordinary temperatures, but turns yellow when heated. It melts below a red heat, and sublimes when raised to a higher temperature in a closed vessel. When heated in the air it is partly converted into antimonic acid. It is readily reduced to the metallic state by ignition with hydrogen, charcoal, or potassium. Teroxide of antimony dissolves sparingly in water, but freely in strong chlorhydric acid; it also dissolves in a hot solution of tartaric acid, or of acid tartrate of potassium (cream of tartar). The solution obtained in the latter case contains the tartrate of antimony and potassium ($C_4H_4KSbO_7$), commonly called tartar-emetic. Ordinary nitric acid does not dissolve the teroxide; but fuming nitric acid and fuming sulphuric acid both dissolve it, forming solutions which ultimately deposit shining scales of a nitrate in the one case and a sulphate in the other.

It is obvious, from these facts, that this oxide of antimony differs from all the oxides which we have heretofore studied, in that it is capable of reacting upon strong acids in such wise as to form salts wherein the antimony plays very much the same part which lead plays in nitrate of lead PbN_2O_6 (Exp. 42), or calcium in $CaSO_4$ (p. 88). This truth is expressed in technical language when we say that the teroxide of antimony is capable of acting as a *base*; the oxides heretofore studied have either been acids, like the oxygen acids of the chlorine and sulphur groups, of nitrogen, phosphorus, and arsenic, or they have been indifferent bodies not inclined to form definite, stable compounds by union with other substances.

But if, on the one hand, teroxide of antimony is thus sometimes a base, on the other it also acts as a feeble acid. The artificial teroxide dissolves readily in solutions of caustic potash and soda, forming very unstable *antimonites*, which are decomposed by boiling, or mere evaporation. These antimonites are analogous to the arsenites; but it is to be observed that arsenious acid is not only a stronger acid than antimonious, but that, unlike antimonious oxide, it never plays the part of a base.

351. *Antimoniate of Antimony or Quadroxide of Antimony* (Sb_2O_4).—This oxide occurs as a native mineral. It may be prepared artificially by heating strongly the quinquioxide (Sb_2O_5), or by roasting the teroxide or the tersulphide, or by treating powdered antimony with an excess of nitric acid. As thus prepared, it is white, infusible, and unalterable by heat, slightly soluble in water, more soluble in chlorhydric acid, and easily resolvable, by boiling with a solution of cream of tartar, into antimonious oxide and antimonic acid. $2Sb_2O_4 = Sb_2O_3,Sb_2O_5$. The oxide may therefore be regarded as a compound of the two other oxides of antimony; but it is sometimes considered a distinct oxide on the ground that it yields by fusion with caustic potash,

or carbonate of potassium, an amorphous saline mass whose composition answers to the formula K_2O,Sb_2O_4. This salt itself, however, if such it be, can be regarded as a mixture of an antimonite and an antimoniate:—

$$2(K_2O,Sb_2O_4) = K_2O,Sb_2O_3 + K_2O,Sb_2O_5.$$

352. *Quinquioxide of Antimony or Antimonic Acid* (Sb_2O_5).—This compound is obtained as a hydrate:—1. By treating antimony with nitric acid or aqua regia containing an excess of nitric acid. 2. By decomposing the quinquichloride of antimony, $SbCl_5$ (§ 354), with water:—

$$2SbCl_5 + 5H_2O = Sb_2O_5 + 10HCl.$$

3. By precipitating a solution of antimoniate of potassium ($K_2O,Sb_2O_5 + 5H_2O$) with a strong acid. This antimoniate of potassium is obtained by fusing one part of antimony with four parts of nitre, digesting the fused mass with tepid water to remove nitrate and nitrite of potassium, and boiling the residue for an hour or two with water; the white insoluble mass of anhydrous antimoniate is thereby transformed into a soluble hydrate, and the solution, treated with a strong acid, yields a precipitate of hydrated antimonic acid. The hydrated antimoniate of potassium itself is a gummy, uncrystallizable salt.

The hydrated oxide, obtained by either of these methods, gives off its water at a heat below redness, and yields anhydrous antimonic acid as a yellowish, tasteless powder, insoluble in water and acids. At a red heat it gives off one-fifth of its oxygen, and is converted into the quadroxide. A boiling solution of caustic potash dissolves the oxide.

The hydrated oxide obtained by the first and third of the above methods is not identical with that which results from the second process. The product of the first and third methods is called antimonic acid; the product of the second is called *met-antimonic* acid, a term derived from a Greek adverb which was used in composition to denote a *change* of place, condition, or quality. Antimonic acid forms normal salts of the composition M_2O,Sb_2O_5 and acid salts containing $M_2O,2Sb_2O_5$, while metantimonic acid forms normal salts containing $2M_2O,Sb_2O_5$ and acid salts answering to the formula $2M_2O,2Sb_2O_5$; the acid metantimoniates are isomeric (§ 312) with the normal antimoniates.

The metantimoniates of sodium, potassium, and ammonium are crystalline; the antimoniates of the same bases are gelatinous and uncrystallizable. The antimoniates and metantimoniates of sodium, potassium, and ammonium are the only ones which are readily soluble in water; all other antimoniates and metantimoniates are insoluble or sparingly soluble. Normal antimoniates correspond with normal nitrates:—

$$M_2O,Sb_2O_5 = M_2Sb_2O_6 = 2MSbO_3.$$
$$M_2O,N_2O_5 = M_2N_2O_6 = 2MNO_3.$$

Normal metantimoniates are analogous to pyrophosphates:—

$$2M_2O,Sb_2O_5 = M_4Sb_2O_7.$$
$$2M_2O,P_2O_5 = M_4P_2O_7.$$

Antimony and Chlorine.—Antimony forms two chlorides, a terchloride, $SbCl_3$, and a quinquichloride, $SbCl_5$, both of which have their analogues in the chlorides of phosphorus, already studied; the terchloride is also comparable with the chloride of arsenic. The metal unites directly with chlorine on contact (Exp. 54), and the two chlorides are bodies of considerable stability.

353. *Terchloride of Antimony* ($SbCl_3$).—This chloride is formed when chlorine gas is passed slowly through a tube containing antimony in large excess. It may also be prepared by distilling 3 parts of antimony with 8 parts of corrosive sublimate (chloride of mercury), or 2 parts of the tersulphide of antimony with 4·6 parts of corrosive sublimate:—

$$2Sb + 3HgCl_2 = 2SbCl_3 + 3Hg; \quad Sb_2S_3 + 3HgCl_2 = 2SbCl_3 + 3HgS.$$

The easiest method of preparing this chloride is to dissolve the tersulphide of antimony in strong, hot chlorhydric acid, or metallic antimony in the same acid, to which a little nitric acid has been added; the resulting liquid, in either case, after evaporation to an oily consistency, should be distilled.

At the ordinary temperature, terchloride of antimony is a translucent yellowish substance of fatty consistency, whence its popular name, butter of antimony. It melts at 72° and boils at about 200°, fumes slightly in the air, is deliquescent and highly corrosive. When thrown into water, it is decomposed into chlorhydric acid and teroxide of antimony, which, however, remains united with a portion of the chloride, forming a white powder which contains antimony, chlorine, and oxygen, but is somewhat variable in composition. This white precipitate is redissolved by excess of chlorhydric acid, and the solution thus obtained is the most convenient that can be used for exhibiting the reactions of antimony. The addition of tartaric acid to this solution prevents its decomposition by water.

Exp. 140.—In a flask of about 200 c. c. capacity, heat gently 0·5 grm. of finely powdered antimony with 30 c. c. of strong chlorhydric acid, to which 10 drops of nitric acid have been added. When complete solution has been effected, pour a little of the chloride into water, to demonstrate the decomposition just referred to. Evaporate the rest of the solution to the consistency of a thick syrup; it is the butter of antimony.

The anhydrous terchloride combines with the chlorides of sodium, potassium, and ammonium, and certain other chlorides, to produce crystalline saline compounds, analogous in composition to those oxygen and sulphur compounds to which the term *salt* is commonly applied.

354. *Quinquichloride of Antimony* ($SbCl_5$).—This compound is formed, with brilliant combustion, when finely powdered antimony is thrown into chlorine gas (Exp. 54). It may also be prepared by passing dry chlorine over warm powdered antimony, or over the terchloride.

Exp. 141.—Fill a hard-glass tube, No. 2, 150 c.m. long with coarsely powdered antimony, and fit one end of the tube so charged into a tubulature of a two-necked glass receiver, the other neck of which is connected with a source of dry chlorine. Support the long tube at an angle of 10° or 15° with the table, so that its open end shall be some 20 c.m. higher than the end which enters the receiver. Keeping the tube warm throughout its whole extent, pass chlorine slowly and continuously into the receiver. Combination takes place in the tube and the product flows back into the receiver, where it remains in contact with chlorine; the long layer of antimony prevents the escape of any free chlorine. Preserve the product in a glass-stoppered bottle.

The quinquichloride is a colorless, or yellowish liquid, which is very volatile and emits suffocating fumes. Water in small proportion forms with it white deliquescent crystals, but in large quantity water decomposes the chloride into chlorhydric and antimonic acids.

355. We are familiar with nitric acid (N_2O_5) as an oxidizing agent, as a substance which readily yields some of its oxygen to other bodies with which it is brought in contact; in a perfectly analogous sense, the quinquichloride of antimony and its analogue the quinquichloride of phosphorus, may be said to be chloridizing agents of great power, for they readily impart chlorine to other substances. These two chlorides are much used in organic chemistry for preparing chlorine compounds; thus, for example, the compound of carbon and hydrogen called *ethylene or olefiant gas*, C_2H_4, is converted by passing through boiling quinquichloride of antimony into an oily bichloride, $C_2H_4Cl_2$, known as Dutch liquid. The quinquichloride acts as a carrier of free chlorine, being itself reduced to the terchloride.

Terbromide and Teriodide of Antimony ($SbBr_3$ and SbI_3).—It is enough to mention the existence of these compounds, formed by the direct union of the elements.

Antimony and Sulphur.—Antimony forms two sulphides, Sb_2S_3 and Sb_2S_5, corresponding to antimonious oxide and antimonic acid, and possibly an intermediate sulphide corresponding to the quadroxide.

356. *Tersulphide of Antimony* (Sb_2S_3).—This compound exists in the crystalline and in the amorphous state. Crystallized tersulphide of antimony is a natural mineral called *grey antimony* or *antimony-glance.* It is the source of all the antimony and antimony compounds of commerce. The mineral has a lead-grey color and a metallic lustre; it is friable and very fusible, melting even in the flame of a candle. At a white heat it may be distilled unchanged in closed vessels, but by roasting in the open air it is converted into a fusible mixture of teroxide and tersulphide of antimony. This *oxysulphide,* after it has been fused, constitutes the commercial *glass of antimony,* which contains about 8 parts of the teroxide to 1 part of the tersulphide; the greater the proportion of sulphide, the darker the tint of the glass.

The native tersulphide is seldom pure, being generally contaminated with lead, copper, iron, and arsenic. To obtain pure crystallized tersulphide of antimony, it is best to prepare it artificially by fusing pure metallic antimony with sulphur in the required proportions by weight. The materials must be finely powdered and intimately mixed, and the mixture thrown by small portions into a heated crucible. The reactions of crystallized sulphide of antimony are the same as those of the amorphous sulphide, to be presently described; but they take place less quickly, on account of the greater cohesion of the mass.

Amorphous tersulphide of antimony can be procured by several processes, from which we may select the two simplest:—1. The native grey tersulphide is changed into the amorphous variety by keeping it in the fused state for a considerable time, and then cooling it very suddenly by throwing the vessel in which it has been melted into a large quantity of cold water. The product is an amorphous mass, having a conchoidal fracture, and a less specific gravity, but a greater hardness than that of the crystalline variety. Its color, in thin pieces, is hyacinth-red; in the state of powder, orange-brown. 2. When sulphydric acid gas is passed into an acidulated solution of an antimony-salt (that of tartar-emetic for example), a bright orange-red precipitate of a hydrated tersulphide of antimony is formed, which may be rendered anhydrous at a moderate heat without losing its red color.

Exp. 142.—Dissolve 2 grms. of tartar emetic in 50 c. c. of water and add to the solution a few drops of acetic acid; pass a slow current of sulphuretted hydrogen, from a self-regulating generator (Appendix, § 19), through this solution for ten minutes. The precipitate is the hydrated tersulphide of antimony. Collect this precipitate upon a filter and wash it.

Exp. 143.—Pour a dilute cold solution of caustic soda upon the washed precipitate of the last experiment as it lies upon the filter, and collect the filtrate in a test-tube; if the whole precipitate does not shortly redissolve, pour the filtrate a second time upon the undissolved precipitate in the filter, or use an additional quantity of soda-lye, if necessary. There is produced a mixture of *sulphantimonite* of sodium and teroxide of antimony, which is soluble in the excess of soda-lye.

$$2Sb_2S_3 + 6NaHO = 3Na_2S,Sb_2S_3 + Sb_2O_3 + 3H_2O.$$

Exp. 144.—Pour the clear alkaline solution, obtained in the last experiment, into two or three times its bulk of dilute chlorhydric acid. The whole of the antimony will be thrown down again as tersulphide, without any evolution of sulphuretted hydrogen, because the gas evolved from the sulphantimonite is exactly absorbed by the dissolved teroxide:—

$$3Na_2S,Sb_2S_3 + 6HCl = 6NaCl + Sb_2S_3 + 3H_2S.$$
$$Sb_2O_3 + 3H_2S = Sb_2S_3 + 3H_2O.$$

When hydrated amorphous tersulphide of antimony is boiled with a solution of carbonate of sodium, it is dissolved; the filtered liquid, on cooling, deposits a reddish-brown substance, formerly much used in medicine, and known as *kermes mineral*. This substance is not a definite compound, but is a variable mixture of tersulphide and teroxide of antimony, the latter being combined with a small portion of the alkali. Minute crystals of teroxide of antimony have been recognized in this mixture by microscopic examination. A solution of cream of tartar will dissolve out the teroxide, leaving the tersulphide. On acidulating the cold filtered liquid, after the deposition of the kermes, with chlorhydric acid, a particularly bright orange precipitate of sulphide of antimony, known as the *golden sulphide*, is precipitated. Artificial sulphide of antimony can, indeed, be precipitated of almost any color between a light orange and a blackish brown. A vermilion-red sulphide has found some applications as a paint.

Exp. 145.—Place in a porcelain dish 10 grms. of a solution of chloride of antimony of about 1·35 specific gravity; add to this chloride a cold solution of hyposulphite of sodium made by dissolving 15 grms. of the salt in 30 c. c. of water; heat the dish very slowly, and stir its contents continually so long as any precipitate separates from the liquid. The sulphide of antimony is thrown down of a brilliant red color. The color of the precipitate is darker in proportion as the temperature of the mixture is higher; when, therefore, a fine red is produced, the lamp may be withdrawn, in order to prevent the color from growing brown. The precipitate is collected on a filter, drained thoroughly, and then washed, first with dilute acetic acid and subsequently with water.

Sulphantimonite solutions, similar to those prepared in the wet way, may be obtained by fusing tersulphide of antimony with dry caustic soda or potash, or with the carbonates of sodium or potassium, and boiling the residue with water. During the exposure to air of hot sulphantimonite solutions, a process of oxidation takes place, whereby the sulphur set free from one portion of the salt converts another portion into the state of *sulphantimoniate,* so that on acidulation some quinquisulphide of antimony is precipitated along with the tersulphide.

Like the tersulphide of arsenic, the tersulphide of antimony is a sulphur-acid which unites with basic metallic sulphides to form sulphur-salts. The artificial sulphantimonites of the alkalies have been alluded to above; there are many natural minerals of analogous composition; among such may be mentioned Miargyrite, Ag_2S,Sb_2S_3, Bournonite, $2Pb_2S,Cu_2S,Sb_2S_3$, and Berthierite, $3FeS,2Sb_2S_3$.

357. *Quinquisulphide of Antimony* (Sb_2S_5).—This compound, which is not native, is made by passing sulphuretted hydrogen through quinquichloride of antimony dissolved in tartaric acid. It may also be prepared by acidulating the solution of the sulphantimoniate of sodium, $3Na_2S,Sb_2S_5$:—

$$3Na_2S,Sb_2S_5 + 6HCl = 6NaCl + Sb_2S_5 + 3H_2S.$$

The quinquisulphide is an orange-yellow, anhydrous, amorphous powder, and is chiefly remarkable for the facility with which it unites with the sulphides of the metals to form sulphantimoniates; on this account this sulphide is often called *sulphantimonic acid.* It is readily soluble in the sulphides, sulphydrates, and hydrated oxides of sodium, potassium, and ammonium.

The sulphantimoniates have generally the composition represented by the general formula $3M_2S,Sb_2S_5=2M_3SbS_4$, analogous to that of the tribasic phosphates $3M_2O,P_2O_5=2M_3PO_4$. The sulphantimoniates of sodium, potassium, and ammonium are very soluble in water, and crystallize with facility; those of the heavy metals are insoluble. The latter are precipitated by adding solutions of metallic salts to a solution of the sulphantimoniate of sodium, keeping the latter in excess. The sodium salt may be prepared as follows:—

In a wide-mouthed bottle, or other vessel which can be closed, mix thoroughly 22 grms. of elutriated tersulphide of antimony, 26 grms. of crystallized carbonate of sodium, 2 grms. of flowers of sulphur, 10 grms. of quicklime, slaked after weighing, and 40 c. c. of water. Let the mixture digest at the ordinary temperature for 24 hours, with frequent stirring; then filter the liquid, wash the residue several times with water, and evaporate the filtrate and the wash-water in a porcelain dish or clean iron pan, until a sample yields crystals on cooling. The formation of the salt is accelerated by boiling. The whole is then left to cool; the deposited crystals are washed two or three times with cold water and dried under a bell-jar over a dish of an absorbent like quicklime or oil of vitriol. The salt is sulphantimoniate of sodium $Na_3SbS_4+9H_2O$; it forms transparent, colorless or pale yellow, regular tetrahedrons.

CHAPTER XIX.

BISMUTH—THE NITROGEN GROUP.

358. The metal bismuth is found chiefly in the metallic state, but also occurs in combination with sulphur, oxygen, and tellurium. It is prepared for the arts almost exclusively from native bismuth. The process of extracting the metal from the gneiss and clay-slate in which it generally occurs is very simple, the mineral being merely heated in closed iron tubes, inclined in such a manner that the melted bismuth runs down into earthen pots, which are heated sufficiently to keep the metal in a state of fusion. It is then ladled out and run into moulds. The impure metal, which often contains sulphur, arsenic, copper, nickel, and iron, may be purified by melting it two or three times with about $\frac{1}{10}$ its weight of nitre.

Bismuth is a tolerably hard, brittle metal, of a grayish-white color with a reddish tinge. When pure, it crystallizes more readily than any other metal; by the method of fusion (§ 188) it may be obtained in most beautiful crystals, made highly iridescent by the thin film of oxide which forms on their surfaces while they are still hot; these crystals look like cubes, but are really rhombohedrons. Bismuth, like phosphorus, arsenic, and antimony, is dimorphous, presenting forms both of the monometric and hexagonal systems. The metal melts at 264° and expands about $\frac{1}{32}$ in solidifying; hence its specific gravity is greater in the liquid than in the solid state. When the metal is subjected to strong pressure, its specific gravity, normally 9·83, has been said to become less. At a high temperature bismuth may be distilled. Of all metals it exhibits in the highest degree the phenomena of diamagnetism. Its atomic weight is 210.

Exposed to dry or moist air the metal does not alter; but when exposed to the combined action of air and water, it is superficially oxidized. When heated in the air, it burns with a bluish flame, forming yellow fumes. Strong chlorhydric acid acts on it with difficulty; sulphuric acid attacks it only when hot and concentrated; nitric acid attacks it immediately, and effects complete solution, with formation of nitrate of bismuth and evolution of nitric oxide.

359. Bismuth promotes the fusibility of metals with which it is alloyed to an extraordinary extent. The most remarkable alloy of bismuth is that known as "fusible metal." When composed of 1 part of lead, 1 part of tin, and 2 parts of bismuth, this alloy melts at 93°·75. Solid fusible metal, like liquid water, undergoes an anomalous contraction by heat. It expands regularly from 0° to 35°, then contracts gradually as the temperature rises to 55°, at which point it is less bulky than at 0°, again expands rapidly to 80°, and beyond that temperature continues expanding regularly up to its melting-point. On account of this property of expanding as it cools while still in the soft state, the alloy is much used for taking impressions from dies; the finest and faintest lines are reproduced with great accuracy. It is obvious that an alloy possessing such properties must be something more than a mere mixture of the constituent metals.

No compound of bismuth and hydrogen is as yet known.

Bismuth and Oxygen—Bismuth forms two principal oxides, a teroxide (Bi_2O_3) and an acid oxide (Bi_2O_5); besides these, there is an intermediate oxide (Bi_2O_4) which may be represented as a compound of the other two $Bi_2O_3, Bi_2O_5 = 2Bi_2O_4$.

360. *Teroxide of Bismuth* (Bi_2O_3).—This oxide is formed when the metal is roasted in the air, but is best obtained by gently igniting the nitrate or subnitrate. It is a pale-yellow, insoluble powder, which melts at a red heat, and is easily reduced to the metallic state by heating it with charcoal. A white hydrate of this oxide, $Bi_2O_3,H_2O = 2BiHO_2$, may be precipitated from a salt of bismuth by an excess of ammonia. Teroxide of bismuth com-

bines with acids to form the bismuth salts; in the normal salts one atom of bismuth replaces three atoms of hydrogen, thus:—

$$Bi_2O_3 + 3(H_2N_2O_6) = 3H_2O + Bi_2 3(N_2O_6).$$

Basic salts of bismuth are also known, in which a larger proportion of bismuth is present. Some of the normal salts crystallize well from acid solutions, but they cannot exist in solution unless an excess of acid is present. On diluting solutions of the normal salts with water, insoluble basic salts are precipitated; this reaction recalls the behavior of antimony solutions. The nitrate of bismuth, $Bi_2 3(N_2O_6)+9H_2O$, is the commonest soluble salt of bismuth; it is procured by dissolving the metal in nitric acid. To the basic nitrate, which is precipitated when water is added to the solution of the normal nitrate, the formula $5Bi_2O_3,4N_2O_5 + 9H_2O$ has been assigned. Bismuth salts are heavy compounds, which are colorless unless the acid be colored; they are poisonous in large doses.

361. *Quinquioxide of Bismuth, or Bismuthic Acid* (Bi_2O_5).—When chlorine gas is passed through a concentrated solution of potash holding teroxide of bismuth in suspension, a blood-red solution is obtained, from which there soon separates a red precipitate; this substance is a mixture of hydrated bismuthic acid and teroxide of bismuth. Cold dilute nitric acid dissolves the oxide, but does not attack the acid. The hydrated acid gives up its water at a temperature of 130°, and the anhydrous quinquioxide remains as a brown powder, which, in contact with acids, parts very readily with a portion of its oxygen and falls back to the state of teroxide. The anhydrous quinquioxide may be also converted by a gentle heat into the intermediate oxide Bi_2O_4. Bismuthic acid combines with caustic soda and potash, but the compounds are decomposed by mere washing. The bismuthates are little known, and are of interest only in so far as they go to show the feeble acid character of the quinquioxide.

362. *Terchloride of Bismuth* ($BiCl_3$).—This compound may be obtained by heating bismuth in chlorine, or by mixing the metal in fine powder with twice its weight of corrosive sublimate (chloride of mercury) and distilling. The same substance is produced when bismuth is dissolved in aqua regia, and the excess of acid evaporated. It is a very fusible, volatile, deliquescent body,

which was called *butter of bismuth*, from its resemblance to the *butter of antimony*, long before the relationship now established between bismuth and antimony had been recognized. The terchloride is decomposed by water into chlorhydric acid, which dissolves a portion of the chloride, and a precipitate containing bismuth, chlorine, and oxygen, and called *oxychloride* of bismuth.

$$3BiCl_3 + 3H_2O = 6HCl + Bi_3Cl_3O_3.$$

The same oxychloride is precipitated when a solution of nitrate of bismuth is poured into a solution of common salt. It is used as a pigment, and is known as "pearl-white." It may be distinguished from the analogous oxychloride of antimony by the fact that it is insoluble in tartaric acid and in potash, both of which dissolve the antimony compound. Terchloride of bismuth forms crystallizable double salts with the chlorides of sodium, potassium, and ammonium. These chlorine salts are analogous in composition to, and isomorphous with, the corresponding double chlorides of antimony.

363. *Tersulphide of Bismuth* (Bi_2S_3).—Bismuth glance, a somewhat rare mineral which occurs in acicular prisms isomorphous with the native tersulphide of antimony, is a tersulphide of bismuth. The same compound may be formed artificially by fusing the pulverized metal with one-third its weight of sulphur. Heated in close vessels, the sulphide evolves sulphur; heated with access of air, it forms teroxide of bismuth and sulphurous acid. The tersulphide is also obtained as a brown-black precipitate when sulphuretted hydrogen is passed through a solution of a bismuth salt.

Exp. 146.—Dissolve 0·5 grm. of finely powdered bismuth in aqua regia, in a small flask; the aqua regia should be added by small portions at a time, so as to avoid an unnecessary excess of acid, and the mixture should be gently heated. When complete solution has been effected, pour one or two drops of the acid solution into a tumbler full of water, and observe the precipitation of white oxychloride of bismuth (§ 362). To the remainder of the solution add water, drop by drop, with constant stirring, until a slight cloudiness appears; add a drop of chlorhydric acid to clear the solution, and through the slightly acid liquor, thus obtained, pass a slow stream of sulphuretted hydrogen until the solution has become so thoroughly charged with this gas that it continues to smell of sulphuretted hydrogen even after it has been removed from the source of the gas. Filter the brownish-black precipitate of tersulphide of bismuth and wash it with water upon the filter; scrape off a portion of the precipitate from the filter, with a smooth slip of wood, and place it in a test-tube together with a few drops of a solution of caustic soda; it will not dissolve. Test another portion of the precipitate in the same way with a solution of sulphydrate of ammonium; it will not dissolve. The last two reactions establish distinct differences between the sulphides of bismuth and antimony, in addition to their difference of color. Again filter, and wash the undissolved sulphide and heat a little of it moderately on platinum foil over the gas-lamp; sulphurous acid will be given off, and the oxide of bismuth remains; this oxide readily melts to dark-yellow globules.

364. *The Nitrogen Group of Elements.*—The five elements nitrogen, phosphorus, arsenic, antimony, and bismuth, form a well-marked natural group of elements. In the first place, the elements themselves exhibit a definite gradation of properties; and secondly, the analogy in composition and properties, manifested by the similar compounds of the five elements, is most striking and complete.

Nitrogen is a gas, phosphorus a solid whose specific gravity varies from 1·8 to 2·2, arsenic has the specific gravity of 5·6, antimony of 6·7, while that of bismuth rises to 9·8. The metallic character is most decided in bismuth, is somewhat less marked in antimony, is doubtful in arsenic, and almost vanishes in phosphorus. All four of these elements are dimorphous, presenting forms both of the monometric and hexagonal systems. The series of corresponding hydrides, oxides, chlorides, and sulphides which the elements of this group form are very perfect; they prove the general chemical likeness of the five elements:—

Hydrides.	*Oxides.*	*Oxides.*	*Oxides.*	*Chlorides.*	*Sulphides.*
NH_3	N_2O_3	N_2O_4	N_2O_5	NCl_3 (?)	P_2S_3
PH_3	P_2O_3	Sb_2O_4	P_2O_5	PCl_3	As_2S_3
AsH_3	As_2O_3	Bi_2O_4	As_2O_5	$AsCl_3$	Sb_2S_3
SbH_3	Sb_2O_3		Sb_2O_5	$SbCl_3$	Bi_2S_3
	Bi_2O_3		Bi_2O_5	$BiCl_3$	
					P_2S_5
				PCl_5	As_2S_5
				$SbCl_5$	Sb_2S_5

The first four members of the group form gaseous terhydrides, in which three volumes of hydrogen and one atom of a nitrogen-group element are combined to form two volumes of the compound gas. We have already spoken of the similarity of chemical composition, and the gradation of properties manifested by these four hydrides. Ammonia is a powerful base, and requires a high temperature for its decomposition; phosphuretted hydrogen is a very feeble base, while the basic character is not perceptible in arseniuretted and antimoniuretted hydrogen. Each of the last three hydrides is decomposed by simple exposure to heat, phosphuretted hydrogen requiring the highest temperature, arseniuretted hydrogen decomposing at a lower, and antimoniuretted hydrogen at a still lower degree of heat. The affinity of bismuth for hydrogen is so feeble that it does not appear to form a hydride.

The teroxides also show a gradation of physical and chemical qualities. Teroxide of nitrogen (nitrous acid) is a highly volatile liquid, that of phosphorus (phosphorous acid) a very volatile solid, that of arsenic a less volatile solid, that of antimony a solid volatile only at a full red heat, and that of bismuth a solid which requires an extremely high temperature for its volatilization. The teroxides of nitrogen and phosphorus form with water strongly acid liquids; teroxide of arsenic is but a feeble acid; teroxide of antimony is sometimes an acid and sometimes a base, while teroxide of bismuth is a decided base. Arsenious and antimonious acids are isodimorphous. The series of quinquioxides also shows a very marked gradation of chemical energy, especially when the compounds which they form with the elements of water are considered. Nitric acid is a powerful acid of intense energy; phosphoric acid is still a strong acid, but much less incisive than nitric acid; arsenic acid has the corrosive properties generally attributed to acids, but it is chemically a rather less vigorous compound than phosphoric acid. The phosphates and arseniates are, however, isomorphous, and the two acids are very much alike. In the quinquioxide of antimony the acid character becomes comparatively indistinct; and in the so-called bismuthic acid little remains but the name.

The terchloride of nitrogen has hardly been examined, on

account of its extreme instability. The other four terchlorides are all volatile substances of analogous composition, since three volumes of chlorine and one atom of the nitrogen-group element unite to form two volumes of the compound vapor. The boiling-points of the terchlorides of phosphorus, arsenic, and antimony are 74°, 132°, and 223° respectively, while that of the terchloride of bismuth is considerably higher still. All four terchlorides are decomposed by an excess of water.

The tersulphides of antimony and bismuth are isomorphous.

The elements of this group do not form many combinations among themselves. They combine with hydrogen, metals and compound radicals which replace hydrogen atom for atom, and with the members of the chlorine group, by preference, in the proportion of 1 atom to 3; they also combine with oxygen and the members of the sulphur group, by preference, in the proportions of 2 atoms to 3, or 2 atoms to 5.

When the qualities of the corresponding compounds which the members of the nitrogen group form with other elements are duly taken into account, it will be apparent that the relative chemical power of each element of the group may be inferred from its position in the series of elements :—

N = 14, P = 31, As = 75, Sb = 122, Bi = 210.

The chemical energy of these five elements, broadly considered, follows the opposite order of their atomic weights.

CHAPTER XX.

CARBON.

365. Carbon is an extremely important and a very abundant element. All organic substances, all things which have life, contain it. Large quantities of it occur in the mineral kingdom as well, both in the free state and in combination with oxygen and with other elements. The various forms of coal, graphite, petroleum, asphaltum, and all the different varieties of limestone, chalk, marble, coral, and sea shells contain it in large proportion. It is found also in the atmosphere and in the waters of the globe, and though existing therein in comparatively small proportion, it is an ingredient not less essential than either of their other constituents for the maintenance of the actual balance of organic nature. All vegetable life is directly dependent upon the presence of the compound of carbon (carbonic acid) which exists in the atmosphere.

366. Three distinct allotropic modifications of carbon are distinguished, namely:—1. The diamond; 2. Plumbago or graphite; and 3. Ordinary charcoal or lampblack. There are many subvarieties of the modification last named; but their peculiarities appear to depend chiefly upon physical conditions of aggregation, whereas each of the three principal varieties of carbon above enumerated is really different from the other two in chemical quality or nature.

367. The element carbon, in each of its modifications, is an infusible, non-volatile solid devoid of taste and smell. But the several modifications differ among themselves in color, hardness, lustre, specific gravity, behavior towards chemical agents, power of conducting heat and electricity, and in various other respects. All the varieties of carbon, however, agree in this, that on being strongly heated in presence of oxygen they unite with it and form carbonic acid (CO_2); but in the comparative readiness with which this result is attained great differences are noticeable in the different varieties.

Lampblack and charcoal, as is well known, readily combine with oxygen at the temperature of an ordinary fire; they burn easily in the air. But graphite burns so slowly in air that it is used for making the crucibles in which the most refractory metals are melted. (See Appendix, § 26.) On being heated to a very high temperature, however, in oxygen gas, graphite slowly undergoes combustion; and the same remark is true of the diamond. Both graphite and diamond can be consumed by nascent oxygen, as when heated in the condition of fine powder with a mixture of bichromate of potassium (a substance rich in oxygen) and sulphuric acid. They can be oxidized also by heating them with nitrate or with chlorate of potassium.

368. *Diamond.*—Of this first variety of carbon, little need here be said. The physical properties which render it so valuable, its high refractive power as regards light, and its extreme hardness, are familiar to all. It the hardest known substance, being capable of scratching all other substances; the name diamond is a mere corruption of the word adamant.

Of the chemistry of the diamond very little is known. It consists of pure or nearly pure carbon, crystallized in octahedrons and other forms of the first or regular system; its specific gravity is about 3·55, and its specific heat 0·1469. It conducts electricity and heat but slowly; and yet it conducts heat so much better than glass that this property is sometimes made use of as a test to distinguish false from real diamonds. Its refractive power on light, as compared with that of glass or rock-crystal, is as 2·47 to 1·6.

Chemists are as yet unable to prepare this variety of carbon artificially; the only source of it is the natural mineral. It was thought, at one time, that if there could but be devised means of fusing carbon, crystals of the diamond modification might possibly separate out from the molten liquid as it cooled; but, at present, all the evidence goes to show that at high temperatures, the second modification of carbon (namely, graphite), and not diamond, is produced. If a diamond be heated white-hot between the charcoal points of a powerful galvanic battery, it softens, and swells up, and, after cooling, there is found a hard black brittle mass like the coke obtained by heating bituminous coal. So, too, carbon is soluble in melted iron, and a portion of it crystallizes out as the iron becomes cold, but the crystals thus obtained are crystals of graphite and not of diamond. We can, therefore, only surmise that diamonds crystallize at a low temperature from some unknown solvent of carbon, or, with greater probability, that when carbon is separated by the decomposition of some one of its compounds it is left in the diamond condition.

The diamond is not attacked by the strongest acids or alkalies, not even by fluorhydric acid; nor is it acted upon by any of the non-metallic elements, with the exception of oxygen at high temperatures. At the ordinary temperature of the air, diamond undergoes no appreciable change during the lapse of centuries;

it appears to be well nigh indestructible, in the ordinary sense of the term. Out of contact with the air, or in an atmosphere which has no chemical action upon it, it suffers no alteration at the highest furnace heat.

369. *Graphite or Plumbago*, sometimes called "black-lead," is familiarly known as the material of common "lead pencils." It is found as a mineral in nature in various localities. It occurs both in the form of crystals (six-sided tables belonging to the hexagonal system) and in the amorphous, massive state. In both forms it is always opaque, of a black or lead-gray color and metallic lustre; its specific gravity varies from 1·8 to 2·1; its specific heat is 0·201.

It conducts electricity nearly as well as the metals, and is, on this account, much used for coating surfaces of wood, wax, plaster, and other non-conducting materials so as to render them capable of conducting the galvanic current and so of receiving a metallic film such as is deposited from solutions of the metals when subjected to the action of the galvanic current; it is an important material in the art of electro-metallurgy. The lustre and conducting-power of graphite go far to justify the term which has been often applied to it, *metallic* carbon.

370. Graphite is very friable; when rubbed upon paper it leaves a black shining mark, whence its use for pencils. Amorphous graphite is much more friable than the crystalline variety; it makes a blacker mark upon paper, and is consequently preferred for the manufacture of pencils; it is, in fact, so soft and unctuous to the touch that it is often used as a lubricant for diminishing the friction of machinery. But in spite of all this the particles of which the masses of graphite are composed are extremely hard; they rapidly wear out the saws employed to cut these masses. By powerful pressure the dust of plumbago can be forced into the condition of a coherent solid similar to the native mineral.

In the air, at ordinary temperatures, graphite undergoes no change; hence its use for covering iron articles to prevent their rusting. By virtue of its greasy, adhesive quality, it is easy to cover iron with a thin, lustrous layer or varnish of it; the common stove-polishes, for example, are composed of powdered graphite. Even at very high temperatures it is scarcely at all oxidized by the air; it is, moreover, altogether infusible; hence it is usefully applied in the manufacture of a highly refractory kind of crucible, known as black-lead crucibles or blue pots. (See Appendix, § 26.) An analogous application of graphite is seen in its use as "foundry-facings," a term applied to the infusible dust which the iron-founder sifts over his mould of sand before pouring into it the melted metal; if the hot metal were allowed to come directly into contact with the sand, a quantity of the latter would melt and remain adhering to the cold metal when the casting was taken from the mould. For this purpose coal-dust is an inferior substitute for graphite.

371. Pure plumbago is never met with in nature; when burned in oxygen the mineral leaves from two to five per cent. of ashes, composed mainly of oxide of iron, together with small quantities of silica and alumina. The presence of this impurity is so unvarying that graphite was formerly supposed to be not carbon, but a chemical compound of carbon and iron, a carbide of iron; this view has now been disproved, and it is known that the iron in the native graphite exists there merely as a mechanical admixture.

Soft, fine-grained plumbago, suitable for the manufacture of the best pencils, is rare; but the coarse, foliated crystallized variety is abundant, and this may easily be made soft and unctuous by the action of certain oxidizing agents.

Exp. 147. — Mix 7 grms. of coarsely powdered crystallized graphite with 0·5 grm. of chlorate of potassium in fine powder; add the mixture to 14 grms. of strong sulphuric acid contained in a porcelain dish, and heat the whole over a water-bath as long as yellow vapors of hypochloric acid are evolved. Wash the cooled mass with water, and subsequently dry it on the water-bath.

Ignite a fragment of the dry product upon a piece of platinum foil, and observe the extraordinary intumescence. After the graphite has ceased to swell up, rub a little of it upon a porcelain plate and note the condition of exquisite softness to which it has been reduced, and the ease with which it can be moulded by pressure into any desired form.

372. Regarded from the chemical point of view, graphite resembles the other modifications of carbon inasmuch as it is con-

verted into carbonic acid on being ignited in oxygen, and in that it undergoes no alteration when heated in close vessels, but differs materially from the other varieties of carbon in its behavior towards several of the oxidizing agents. When graphite is repeatedly exposed to the action of a mixture of strong nitric and sulphuric acids, or to that of a mixture of chlorate or bichromate of potassium and sulphuric or nitric acid, it is converted into a peculiar acid, called *graphitic acid.*

This graphitic acid occurs in thin transparent crystals, somewhat soluble in water, but insoluble in water containing acids or salts; on being heated, it decomposes, with explosion and evolution of light. By analysis, it has been found to contain carbon, hydrogen, and oxygen, in proportions corresponding to the complex formula $C_{11}H_4O_5$; but some chemists, who regard this body as an analogue of an acid ($Si_4H_4O_5$) obtained by acting upon one of the modifications of silicon with oxidizing agents, have suggested that the atomic weight of graphite may be different from that of ordinary carbon, and that the composition of graphitic acid could be represented by the simpler formula $Gr_4H_4O_5$, in which the symbol Gr stands for graphon—provided the atomic weight of this graphon were assumed to be 33, instead of 12 (the atomic weight assigned to ordinary carbon).

The graphitic modification of carbon can readily be obtained artificially. When charcoal is added to melted iron, the iron takes up a considerable quantity of it, and if the iron be then left to cool slowly, a portion of the dissolved carbon will crystallize out in the form of graphite; the crystals can readily be isolated by dissolving away their metallic envelope by means of dilute chlorhydric or sulphuric acid.

As has been already remarked, the crystals of graphite are six-sided plates of the hexagonal system, altogether unlike the forms of the regular system which are seen in the diamond. In carbon, then, as in sulphur, we have a striking example of dimorphism. (See § 192.)

373. *Gas-Carbon.*—An interesting sub-variety of carbon somewhat similar to graphite, and standing, as it were, between it and the ordinary modification of carbon, is obtained from the retorts in which common illuminating gas is manufactured. It

is known as "gas-carbon," or "carbon of the gas-retorts," and results from the burning on of drops of tar upon the interior walls of the retort, and the long-continued heating of the crust thus formed.

Gas-carbon is very hard, compact, and dense; it has the metallic lustre, and conducts electricity like a metal; its specific gravity (2·356) and specific heat (0·2036) closely resemble those of graphite. On account of its high conducting-power, it is employed in the manufacture of galvanic batteries and of pencils for the electric lamp; its infusibility and power of resisting chemical agents have led to the employment, in various scientific researches, of crucibles and tubes wrought out of it; it has also been sometimes employed as fuel in experiments where higher degrees of heat are needed than can be obtained from charcoal or coke. The intense heat developed by the combustion of this substance is referable to its high specific gravity; a very considerable weight of carbon can here be burned in a small space. As a fuel, it has the further merit of leaving scarcely any ashes.

374. *Coke* and *Anthracite Coal* are impure sub-varieties of carbon which, from the chemical point of view, may be classed either with graphite or charcoal, or better between the two. They are less like graphite, however, than gas-carbon is. Coke is the residue resulting from the destructive distillation of soft or bituminous coal.

Fig. 49.

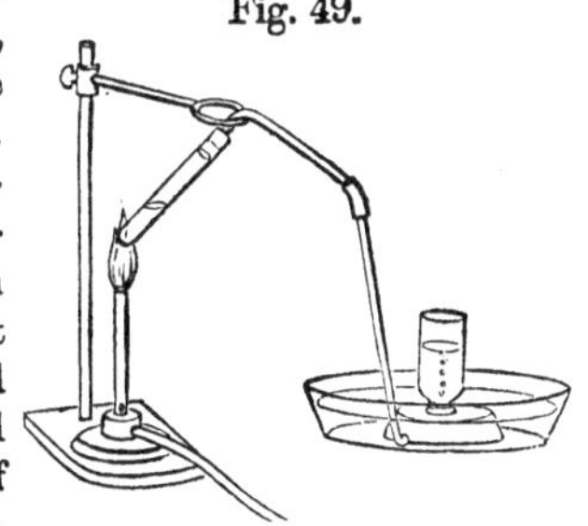

Exp. 148.—Put into a tube of hard glass, No. 1, 12 or 15 c.m. in length, enough bituminous coal, in coarse powder, to fill one-third of the tube. Fit to this ignition-tube a large delivery-tube of glass, No. 4, and support the apparatus upon the iron stand, as shown in the figure. Heat the coal in the ignition-tube, and collect in bottles the gas which will be evolved. This gas is a mixture of several compounds of carbon and hydrogen; for the present, it may be regarded as carburetted hydrogen, it is, in fact, ordinary illuminating gas.

It is inflammable, like hydrogen, but burns with a far more luminous flame. It is very light withal; hence many of the experiments described in the chapter upon hydrogen may be performed with this gas. (See Chapter V.)

As soon as gas ceases to be given off from the coal, take the end of the delivery-tube out of the water and when the ignition-tube has become cold, break it and examine the coke which it contains. The coke used for domestic purposes is obtained as an incidental product in the manufacture of illuminating gas.

In Europe, where anthracite is lacking, immense quantities of coke are prepared for metallurgical uses, the gas resulting from the decomposition of the coal being usually thrown away.

375. Bituminous coal is a substance of vegetable origin, which appears to have been formed from plants by a process of slow decay going on without access of air and under the influence of heat, moisture, and great pressure. Like vegetable matter in general, it is composed of carbon and hydrogen, together with small proportions of oxygen and nitrogen, and a certain quantity of earthy and saline substances, commonly spoken of as inorganic matter. On being heated in the air, it burns away almost completely after a while, leaving nothing but the inorganic component as ashes. But when heated out of contact with the air (that is to say, when subjected to destructive distillation, as in Exp. 148), the volatile hydrogen is all driven off in combination with some carbon, either as gas or as a tarry liquid, and there remains, as a residue, only carbon contaminated with the inorganic matters originally present in the coal.

376. Both coke and anthracite are hard and lustrous. As compared with charcoal, they are rather difficult of combustion; but in suitable furnaces they burn readily, with evolution of intense heat. Both anthracite and coke, the latter in spite of its porosity, conduct heat readily, as compared with charcoal; hence one reason of the difficulty of kindling them. In building a charcoal fire, the heat evolved by the combustion of the kindling material is almost all retained by the portions of charcoal immediately in contact with the kindling agent; but in the case of coke or anthracite a large proportion of this heat is conducted off and diffused throughout the heap of fuel, so that no portion of the fuel can at once become very hot. It follows that in both lighting and feeding fires of coke or anthracite, only small portions

of the fuel should be added at any one time, lest the kindling material, or the existing fire, be unduly cooled. Since coke is usually contaminated with a considerable proportion of inorganic matter, its combustion is apt to be hindered by the accumulation of ashes and consequent exclusion of air, unless special precautions be taken.

377. *Charcoal* or *Lampblack* is commonly taken as the representative of the third or amorphous modification of carbon. This kind of carbon can be obtained in a state of tolerable purity, either by heating in a close vessel sugar, or starch, or some other organic substance which contains no inorganic constituents, or by burning oil of turpentine in a quantity of air insufficient for its complete combustion. A convenient way of obtaining it is to place a vessel filled with ice-water directly in the flame of a lamp fed with oil of turpentine, so that the combustion of the oil shall be impeded, and that soot may be deposited upon the walls of the vessel. In either event, however, the product is liable to be contaminated with traces of hydrogen or of oxygen, or of both these elements, which cannot be expelled even by the application of long-continued and intense heat.

For such illustrations as are required in this manual, charcoal can readily be prepared from wood in the same way that it is made for manufacturing and domestic uses, namely by subjecting the wood to a process of incomplete combustion.

Exp. 149.—Light one end of a splinter of dry wood and slowly push the burning portion into the mouth of a test-tube, as shown in Fig. 50. The portion of the splinter which remains outside the tube and in contact with free air will continue to burn with flame, while that within the tube is either extinguished altogether or barely glimmers as the carbon slowly unites with the small portion of air which can gain access to it.

Fig. 50.

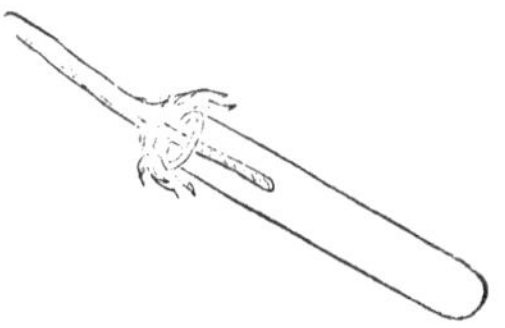

Exp. 150.—Repeat the foregoing experiment, but, instead of the test-tube, provide a cup of sand and slowly thrust the burning splinter into this sand. The flame will be extinguished as fast as the splinter is cut off from the air by immersion in the sand, and a residue of carbon will be thus obtained, as before.

378. Whenever wood, or any other vegetable or animal matter, is not completely consumed, there is left a residue of carbon similar to that obtained in the foregoing experiments. Incomplete combustion in such cases is really a process of destructive distillation, or, rather, in any combustion of wood, or of bituminous coal, there is always destructive distillation at first. When the splinter of wood of Exp. 149, is heated in the lamp, in order to set it on fire, there will distil off from it, in the beginning, certain volatile compounds of hydrogen and carbon; for wood, like coal (§ 375), is composed of carbon, hydrogen, oxygen, nitrogen, and inorganic or earthy matters, and when exposed to strong heat it gives off in the gaseous form the volatile elements hydrogen, oxygen, and nitrogen, together with some carbon. The products of the destructive distillation of the splinter will take fire and burn, and the heat generated by their combustion will be sufficient, not only to distil the contiguous portions of the wood, but also to bring the residual carbon to the temperature at which it unites with oxygen. This kindling-temperature of carbon, it should be remarked, is considerably higher than that at which the volatile distillate composed of carbon and hydrogen takes fire. Now if, as in Experiments 149, 150, the burning splinter be removed from the air as soon as the act of distillation has been completed, but before the combustion of the carbon has set in, the carbon will be preserved, as has been seen. So, too, when burning wood is extinguished by pouring water upon it; the distillatory process has occurred and has been more or less thoroughly completed, but the combustion of the carbon is cut short; for the water not only excludes air but absorbs so much heat that the temperature of the fuel is reduced below the kindling-point. (Compare Exp. 24.)

379. Charcoal can be obtained also by distilling wood in retorts in the same way that we have seen that coke can be procured from bituminous coal. (See Exp. 148.)

Exp. 151.—Provide an ignition-tube and a delivery-tube similar to those employed in Exp. 148. Fill the ignition-tube with shavings or small fragments of wood, arrange the apparatus as in Fig. 51, and light the gas-lamp. Collect in bottles the gas which is given off from the wood, and test it as to its inflammability by applying a lighted match.

After the flow of gas has ceased, remove the end of the delivery-tube from the water, plug it so that no air can enter the ignition-tube, and lay the apparatus aside until it has become cold. Finally remove the cork from the ignition-tube and take out the charcoal which is contained in it. Heat a portion of this charcoal upon platinum foil and observe the manner in which it burns. It will illustrate the fact that solid substances which are incapable of evolving volatile or gaseous matter do not burn with flame ; they merely glow.

Fig. 51.

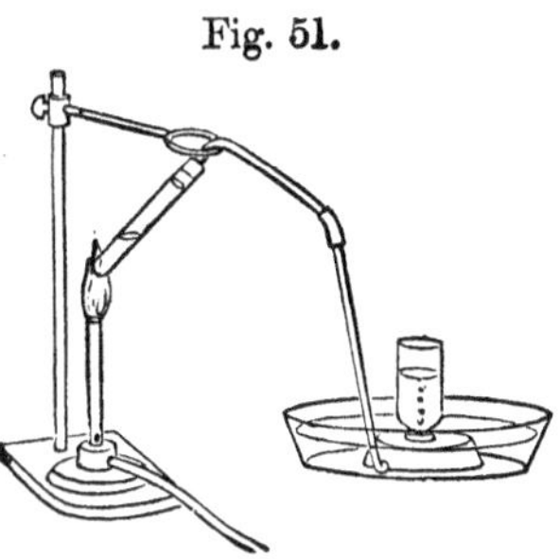

Exp. 152.—Pack an ignition-tube with bits of wood, as in Exp. 151, but, instead of the ordinary delivery-tube, insert in the mouth of this ignition-tube a cork carrying a short piece of glass tubing drawn out to a fine open point. By means of wire, tie the ignition-tube to a ring of the iron stand, and place it over the gas-lamp. The gases evolved from the wood will escape through the narrow tube, and on being kindled will burn with a luminous flame. As has been already stated (§ 57), flame is caused by burning gas.

This experiment, as well as Exps. 148, 151, illustrates the principle of the manufacture of illuminating gas. Upon the large scale, bituminous coal, or sometimes dry wood, is distilled in large iron or clay tubes, called gas-retorts, and the gas evolved is freed from tar and other offensive impurities by processes of cooling and washing with water, and by passing it through layers of lime or oxide of iron; it is then collected in large gas-holders, from which it is pressed through subterranean pipes, it may be for miles.

380. For use in the arts, charcoal is prepared by both the methods above indicated; it is manufactured both by charring or partially burning wood with little access of air, and by methodically distilling wood in actual retorts. The first method is employed in countries where wood is abundant, and is carried on in the forest itself. Logs of wood are piled up into a large mound or stack around a central aperture, which subsequently serves as a temporary chimney, and also for the introduction of burning substances for firing the heap. The finished heap is covered with chips, leaves, sods, and a mixture of moistened earth and charcoal dust, a number of apertures being left open

around the bottom of the heap for the admission of air and the escape of the products of distillation and combustion. The heap is kindled at the centre, and burns during several weeks. When the process is judged to be complete, all the openings are carefully stopped, in order to suffocate the fire, and the heap is then left to itself until cold. Kilns constructed of brick are often used instead of the rude heaps here described. The charcoal retains the form of the wood—the shape of the knots and the annual rings of the wood being still perceptible in it,—but it occupies a much smaller volume than the wood; generally its bulk does not amount to more than three-fourths of that of the wood, and its weight never exceeds one-fourth the weight of the wood.

Where charcoal is prepared by distilling wood in retorts, the liquid products of distillation, namely tar and acetic acid ("pyroligneous acid"), are saved and utilized.

381. *Lampblack.*—Usually when hydrogen is removed from a gaseous compound of carbon and hydrogen, the carbon separates in the form of soft flakes, called lampblack or soot. In Experiment 60, we have already seen that lampblack is formed when hydrogen is removed from carburetted hydrogen by means of chlorine, and we know well that oxygen is capable of producing the same result. Hydrogen is more combustible in oxygen than carbon; hence if carburetted hydrogen be mixed with only enough oxygen to consume its hydrogen, and the mixture be then inflamed, the carbon contained in it will be set free. This is the way in which lampblack is commonly formed; a lamp "smokes" when the supply of air is insufficient to furnish oxygen to both the carbon and the hydrogen of the oil or other combustible.

Upon the large scale, lampblack is manufactured by heating organic matters, such as tar, rosin, or pine knots, which contain volatile ingredients very rich in carbon, until vapors are disengaged, and then burning these vapors in a current of air insufficient for their complete combustion. A dark-red, very smoky flame is thus obtained; a large portion of the carbon of the combustible does not burn, but is deposited as a very fine powder precisely similar to that which constitutes the black portion of common smoke. Lampblack finds important applications in the arts as a pigment and as the chief ingredient of printers' ink.

Exp. 153.—Fill an ordinary spirit-lamp (Appendix, § 5) with oil of turpentine, light the wick and place over it an inverted wide-mouthed bottle of the capacity of a litre or more, one edge of the mouth of the bottle being propped up on a small block of wood, so that some air may enter the bottle. As the supply of air is insufficient for the perfect combustion of the oil of turpentine, a quantity of lampblack will separate and be deposited upon the sides of the bottle.

Hydrogen kindles at a lower temperature than carbon; hence if the flame of a burning compound of carbon and hydrogen be cooled down below the temperature at which carbon takes fire, lampblack will be formed, even if there be present an abundant supply of air.

Exp. 154.—Press down upon the flame of an oil-lamp or candle an iron spoon or a porcelain plate in such manner that the flame shall be almost, but not quite, extinguished. The solid body not only obstructs the draught of air, and thereby interferes with the act of combustion, but it also cools the flame by actually conducting away part of its heat; the temperature is thus reduced to below the kindling-point of carbon, and a quantity of lampblack remains unconsumed and adhering to the spoon or plate. This experiment is, of course, comparable with Exps. 133, 136, in which spots of arsenic and antimony were obtained upon porcelain, as products of the incomplete combustion of their compounds with hydrogen. As has been stated in § 377, lampblack thus prepared usually contains a certain small proportion of hydrogen compounds.

382. Charcoal is altogether insoluble in water; it is odorless and tasteless. Unlike the diamond, it is a good conductor of electricity, but a bad conductor of heat, particularly when in the state of powder. It is a better conductor of heat in proportion as it is denser; when strongly heated out of contact with the air it becomes heavier, harder, and closer in texture than common charcoal, and less easily combustible, but a far better conductor of heat and of electricity than before. The pieces of charcoal, for example, which sometimes fall unconsumed from the bottoms of smelting furnaces, after having passed through a long column of intensely heated fuel, are found to be peculiarly compact and close-grained, and to conduct heat with comparative facility.

As has been seen in § 376, the combustibility of a fuel is diminished in proportion as the fuel is a good conductor of heat;

and since, as has just been stated, charcoal conducts heat the more readily in proportion as it is denser, it follows that, other things being equal, a given sample of charcoal will take fire more quickly if it be light than if heavy. It will appear, moreover, from the foregoing that the combustibility of charcoal should be less, according as the temperature at which the charcoal prepared is higher. If, for example, linen or cotton rags be carbonized at low temperatures, there will be obtained a very light variety of charcoal, called tinder, which takes fire with especial ease. So, too, a very light and easily inflammable charcoal is prepared for the manufacture of gunpowder by distilling light woods, such as willow and alder, at low temperatures. Wood charcoal takes fire easily, as compared with coke; coke as compared with anthracite, and anthracite as compared with gas-carbon. But, inversely, when the charcoal has once been thoroughly lighted, the intensity of the heat obtained from a fire of it will be greater in proportion as the charcoal is more dense. As has been already stated (§ 373), a better fire can be obtained by burning gas-carbon than can be had from coke or charcoal; for in any given space, if the supply of air be ample, more of the dense than of the light fuel can, of course, be burned in a given time. The specific gravity of charcoal is about 1·6; but an ordinary fragment of it readily floats upon water, owing to the air within its pores; if this air be removed, as when the fragment is powdered, the charcoal will sink at once. (See Exp. 159.). It is infusible and non-volatile.

383. In all its varieties, charcoal is a very important chemical agent, chiefly because of the readiness and energy with which it combines with oxygen at high temperatures. Most of the common processes for extracting the useful metals from their ores are based upon the affinity of carbon for oxygen.

Exp. 155.—Mix five grammes of litharge (oxide of lead) with a quarter of a gramme of powdered charcoal; place a portion of the mixture in an ignition-tube made of No. 3 glass, and heat it strongly in the gas-lamp. The charcoal will unite with the oxygen of the oxide of lead, and the compound thus formed will escape in the form of gas, while metallic lead will remain in the tube.

This experiment is analogous to Exp. 124, where arsenious acid was reduced by means of charcoal. Both experiments are typical of the manner in which hot charcoal acts upon metallic oxides. At a white heat it removes oxygen from its combinations with some elements which hold it with great force, such as the oxides of sodium and potassium and phosphoric acid. Even water is decomposed, with liberation of hydrogen, when brought into contact with red-hot charcoal.

Exp. 156.—Fill a piece of iron gas-pipe, about 35 c.m. long and 1 c.m. or more in internal diameter, with fragments of charcoal; adapt to it a delivery-tube, as represented in Fig. 52, and support it upon a ring of the iron stand over one or two wire-gauze gas-lamps. Attach to

Fig. 52.

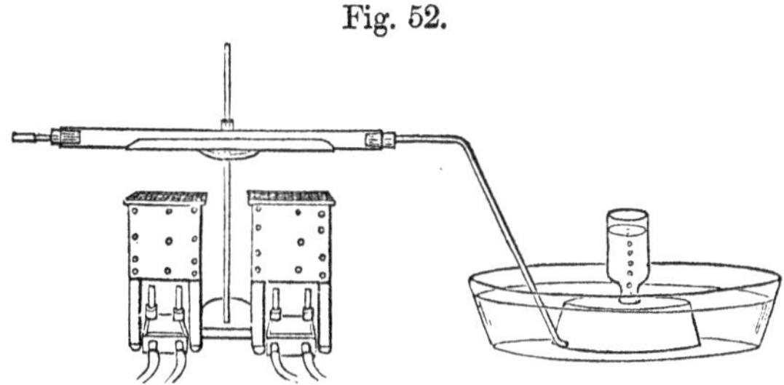

the other end of the tube a thin-bottomed glass flask half full of water and supported upon the ring of a second iron stand. Light the lamps beneath the tube full of charcoal, and wait until it has become red-hot, then heat the water in the flask and cause it to boil slowly. The steam will react upon the hot carbon in a manner which may be formulated as follows:—

$$C + H_2O = CO + 2H,$$

and there will be formed a mixture of a compound of oxygen and carbon called carbonic oxide, and free hydrogen. Collect the mixed gas in bottles upon the water-pan and test it as regards its inflammability by applying a lighted match.

A certain quantity of water may even be decomposed by thrusting pieces of brightly glowing charcoal into water. If the experiment be performed beneath an inverted bottle of water held near the surface of the water-pan, a quantity of gas large enough to be inflamed can readily be obtained.

At a red heat charcoal deoxidizes bodies which are rich in oxygen so readily that there occurs a peculiarly rapid and sparkling combustion known as *deflagration.* Deflagration differs from explosion only in degree; it is less violent than explosion, because the combustion is less rapid.

Exp. 157.—Mix 10 grms. of nitrate of potassium and 5 grms. of recently ignited charcoal, both in fine powder; place the mixture upon a brick, in a current of air, or in any place where the volatile products of the reaction can occasion no inconvenience, and touch it with a lighted stick or red-hot wire. The charcoal will burn violently, with brilliant scintillations, at the expense of the oxygen contained in the nitrate of potassium.

As a modification of this experiment, heat a couple of pieces of charcoal to redness in the fire and sprinkle upon the one a small quantity of powdered nitrate of potassium, and upon the other a little powdered chlorate of potassium.

384. This deoxidizing power of charcoal, above illustrated, is exhibited only at high temperatures. At the ordinary temperature of the air, the chemical energy of charcoal is exceedingly feeble. Charcoal is, in fact, one of the most durable of substances. Specimens of it have been found at Pompeii and upon Egyptian mummies, to all appearance as fresh as if just prepared; the action of the air continued through centuries has exerted no appreciable influence upon it. Experience has proved that many wooden articles which are to be placed in situations peculiarly liable to cause their decay, may be protected by charring them superficially; the carbonization of the interior of casks destined to hold liquids, and of those portions of wooden posts and sills which are to be sunk in the ground, or to remain on or near the surface of the ground, are familiar instances of this custom.

Not only is charcoal unacted upon by air or water at the ordinary temperature, but there are few chemical substances which have any action upon it unless they be hot; neither the neutral solvents, such as alcohol and ether, nor corrosive agents, such as chlorine or fluorhydric and chlorhydric acids, attack it in any way. It is slowly oxidized, however, by nitric acid, and rapidly by perchloric acid, and it dissolves, to a slight extent, in cold concentrated sulphuric acid. At the ordinary temperature, no one of the elements combines with it directly; but at high temperatures it unites directly, not only with oxygen, as we know, but with sulphur as well, forming bisulphide of carbon (CS_2).

At very high temperatures, as when a powerful galvanic battery is discharged from carbon points immersed in an atmosphere of hydrogen, carbon combines directly with hydrogen also, and there is formed a gas called acetylene (C_2H_2). With nitrogen it unites to form cyanogen (CN) when a current of nitrogen gas is passed through a column of ignited charcoal which has previously been charged with carbonate of potassium by soaking it in a solution of this salt; but it will not unite with nitrogen without the intervention of the alkaline carbonate or of some other substance. With chlorine and the other members of the chlorine group it does not unite except indirectly; but with several of the metals it unites directly to form badly defined compounds called carburets or carbides. Upon the chlorides and their analogues carbon has no action, and the same remark is true of some refractory oxides, such as silicic acid and oxide of aluminum, which are not decomposed by charcoal, even at a white heat; but at a strong red heat it reduces most of the sulphates to the condition of sulphides (for example—

$$CaSO_4 + 2C = CaS + 2CO_2),$$

and the nitrates, chlorates, and perchlorates to the state of carbonates—

$$2(K_2O,N_2O_5) + 5C = 2(K_2O,CO_2) + 3CO_2 + 4N.$$

385. A physical property of charcoal, which is of great practical importance, is its power of absorbing and condensing within its pores a great variety of gases and vapors; it absorbs coloring-matters also, and various other substances as well, abstracting them from solutions in which they are contained.

Exp. 158.—Take from the fire a live coal (charcoal) as large as a hen's egg, extinguish it by covering it up tightly in a small metallic vessel or by covering it with sand, and wait until it has become cold: weigh it carefully upon a delicate balance, and record the weight Place the weighed coal in such a situation (in a damp cellar, for example) that it shall be freely exposed to moist air during twenty-four hours again weigh it, and note the increase. Boxwood charcoal has been found to increase in weight 14 per cent. in the course of a single day, and any ordinary charcoal will usually increase in weight from 10 to 12 per cent.

Exp. 159.—Take from the fire, as before, a piece of charcoal which has been heated to full redness for some time; thrust it under water, so that it may be suddenly cooled, and observe that it sinks in the water and that few or no bubbles of gas escape from its pores.

Take another piece of charcoal which has long been exposed to the air and has not recently been heated, attach to it a quantity of sheet lead sufficient to sink it in water, and immerse the whole in a large beaker glass two-thirds full of hot water. The mobile water will immediately enter the pores of the charcoal, and a portion of the air which had previously been absorbed by these pores will be driven out, and can be seen escaping in bubbles through the water, chiefly from the broken ends of the coal.

In a similar way, if a piece of old charcoal be placed in a bottle full of water standing inverted upon the water-pan, a quantity of air will gradually be expelled from it as the water enters its pores, and will collect at the top of the bottle.

The air can at any time be quickly removed by sinking a piece of the charcoal, by means of lead, in water of the ordinary temperature, placing the vessel which contains the water under the receiver of an air-pump and pumping out the air from this receiver; as the pressure of the atmosphere is removed from the surface of the water a multitude of bubbles of air will be seen to issue from the pores of the coal.

To the presence of air and aqueous vapor, which has been thus absorbed, is to be attributed the snapping and crackling of old charcoal when it is thrown upon a hot fire. Charcoal which has been recently made, and which has not yet absorbed air and moisture, does not thus snap and crackle; moreover it kindles much more easily than charcoal which has long been exposed to the air; for from the latter a quantity of gas and vapor must be expanded and driven out before the coal can ignite, and this expansion, being, of course, attended with absorption of heat, keeps down the temperature of the coal below the kindling-point.

It is not necessary that charcoal be freshly made or recently heated, in order that it may kindle readily, if only it has been kept in closed vessels, and thus prevented from absorbing moisture. In the laboratory it is well to throw the remnants of charcoal fires into an iron kettle, furnished with a tightly fitting cover, in order to have always a store of easily inflammable coal for starting new fires in small hand-furnaces.

The absorptive power of charcoal for gases can readily be shown directly by placing recently ignited charcoal in a small confined volume of almost any gas.

Exp. 160.—Fill a glass cylinder of 200 or 300 c. c. capacity with dry sulphurous acid gas (Exp. 96), over the mercury trough; take from the fire a red-hot piece of charcoal as large as can be introduced into the mouth of the cylinder; thrust it beneath the surface of the mercury and hold it there for a moment, in order that it may be extinguished, then pass it up into the jar of sulphurous acid. Mercury will rise in the cylinder in proportion as the gas is absorbed, and will soon completely fill it.

386. As one result of the enormous condensation to which gases are subjected when thus absorbed by coal, heat is necessarily developed; the temperature of the charcoal rises as the gas is condensed in it, and to such an extent that heaps of recently ignited finely divided charcoal often take fire on being exposed to the air.

Different gases are absorbed by charcoal in very different proportions. It has been found, by experiment, that one cubic centimetre of dry, compact charcoal, such as that from boxwood, will absorb in the course of twenty-four hours:—

90 c. c.	of Ammonia gas.	35 c. c.	of Olefiant gas.
85 „	„ Chlorhydric acid gas.	9·4 „	„ Carbonic oxide.
65 „	„ Sulphurous acid gas.	9·3 „	„ Oxygen.
55 „	„ Sulphydric acid gas.	7·5 „	„ Nitrogen.
40 „	„ Nitrous oxide gas.	5·0 „	„ Marsh gas.
35 „	„ Carbonic acid gas.	1·8 „	„ Hydrogen.

As one consequence of this diversity of absorbing-power, it follows that, from a mixture of several gases, charcoal can remove some gases more readily than others. Thus, when recently ignited charcoal, which has been cooled under mercury, as in the preceding experiment, is passed up into a jar of common air, it absorbs the oxygen more rapidly than the nitrogen. But, on the other hand, it is found that, after the charcoal has become saturated with one kind of gas, it can still take up a certain quantity of any other gas which may be presented to it.

387. This power possessed by charcoal of absorbing gases is evidently a particular case of the physical force called adhesion or capillary attraction, whose manifestations are familiar to us in the drinking up of water by a sponge, or of oil by a lamp-wick. If the charcoal be moistened with water (that is to say,

if its pores be clogged by the interposition of a liquid), its absorbing-power for gases will be largely diminished.

This subject is chiefly interesting to chemists because of its intimate connexion with the power of causing various gases to combine with one another, which is possessed by charcoal as well as by finely divided platinum (§§ 224, 240) and several other substances. In the same way that spongy platinum causes hydrogen and oxygen, or sulphurous acid and oxygen, to unite with one another, chemical combination ensues when mixtures of various gases are brought into contact with charcoal. Though the absorbent power of platinum, as regards gases, can hardly be compared with that of charcoal, its power of causing combination is very much greater; platinum appears to possess, moreover, in a marked degree, the faculty of attracting and attaching to itself small quantities of most gases.

Charcoal can be made more efficient as an agent for causing combination, by covering it with a film of platinum. If coarsely powdered charcoal be thoroughly impregnated with bichloride of platinum by boiling it for some time in an aqueous solution of this salt, and if it be then heated to redness in a close crucible, in order that the platinum salt may be decomposed, there will be left a residue of platinum everywhere attached to the sides of the pores of the coal. Such platinized coal is very effective, both as an absorbent of gases and as an agent for producing combination. But even by itself charcoal has a very decided influence upon the combination of gases.

If recently ignited charcoal be allowed to become charged with dry sulphydric acid gas and then introduced into an atmosphere of oxygen, the elements of the sulphydric acid will combine with the oxygen so quickly that an explosion ensues, and both water and sulphurous acid are produced.

If the coal charged with sulphydric acid be brought into contact with air, instead of pure oxygen, combination will occur as before, only more slowly; in this case, however, the hydrogen alone will be consumed, the sulphur of the sulphydric acid being set free in the solid state.

Charcoal is much employed as a disinfecting agent. It is capable of removing many offensive odors from the air—such, for example as the fetid products given off during the putrefaction of animal and vegetable substances. Animal matter in an advanced stage of putrefaction loses all offensive odor when covered with a layer of charcoal; and the flesh of a dead animal buried beneath a thin layer of charcoal will gradually waste away and be consumed without exhaling any unpleasant smell.

Exp. 161.—Place a small quantity of powdered charcoal in a bottle containing sulphydric acid gas and shake the bottle. The odor of the sulphydric acid will quickly disappear. In the same way, an aqueous solution of sulphydric acid (Exp. 86) can be deodorized by filtering it through a layer of charcoal.

Exp. 162.—In a shallow open basket, or in a box through the bottom of which numerous holes have been bored, spread a layer of coarsely powdered boneblack about 5 c.m. thick, placing a sheet of filtering-paper below, if need be, to prevent the powder from sifting through the holes in the box. Place the body of a rat or of some other small animal upon the layer of boneblack, and pour on more boneblack until the rat is covered with a layer of it about 5 c.m. deep. Hang up the basket or box in a warm room, so that air may have free access to it, and leave it at rest. After the lapse of several weeks it will be found, on examination, that all the putrescible portions of the animal have disappeared, and that nothing is left but a mass of hair and bones; but in the interim no odor will have been detected arising from the decomposing animal, excepting a faint odor of ammonia.

In the same way, water can be preserved untainted in casks which have been charred internally; and the quality of some kinds of wine is improved if it be stored in such casks.

In all these cases the use of charcoal as a disinfectant depends not merely upon its mechanical ability to absorb offensive gases, but also and mainly upon the fact that the absorbed gases are chemically destroyed within the pores of the coal by the oxygen which is sucked into these spaces from the air. The purifying action depends upon oxidation, upon the burning up of the offensive gases as fast as they are formed. The charcoal is in no sense an antiseptic, or preservative agent proper to prevent decay; on the contrary, it actually hastens the destruction of putrescible organic matters. Under ordinary circumstances, while in contact with the air, the pores of charcoal are, of course, always charged with oxygen by virtue of their absorptive power. Whenever, therefore, any new gas is dragged in, and forced into intimate contact with this oxygen, it is precisely as if the gas had been carefully collected and then subjected to the action of some corrosive chemical agent. A great merit of charcoal as a disinfectant is, that it constantly draws in to destruction the offensive matters around it; pans of charcoal placed about a room (the wards of a hospital, for example) the air of which is offensive, soon remove the unpleasant smell. Sieves of charcoal, placed across the air-vents of sewers in such manner that the out-going air may be filtered through the charcoal, are found to be most efficient instruments for destroying the noxious effluvia which commonly escape from these openings. In this case, where a current of air is constantly passing through the charcoal filter, the latter will preserve its efficiency for an indefinite length of time, if only it be kept dry; for the action of the coal consists merely in bringing about oxidation and destruction of the offensive gases of the sewer, and as fast as one portion of these is consumed a new portion can be taken in to destruction.

388. Charcoal not only destroys odors, but it removes colors as well; and for this purpose it has long been employed in the purification of sugar and of many chemical and pharmaceutical preparations. Almost any coloring-matter can be removed from a solution by filtering the liquid through a layer of charcoal.

Exp. 163.—Provide four small bottles of the capacity of 100 or 200 c. c., and place in each of them a table-spoonful of boneblack (§ 389); into the first bottle pour a quantity of the blue compound of iodine and starch obtained in Exp. 69, into the second a decoction of cochineal, into the third a diluted portion of the blue liquor obtained by dissolving indigo in Nordhausen sulphuric acid, and into the fourth a solution of permanganate of potassium, enough of the solution being taken in each instance to nearly fill the bottle. Cork the bottles and shake them violently, then pour the contents of each upon a filter (see Appendix, § 14), and observe that the filtrate is in each instance colorless or nearly so. In case the first portions of the filtrate happen to come through colored, they may be poured back upon the filter and allowed to again pass through the coal.

In the purification of brown sugar the coloring-matters are removed in a manner similar to the foregoing, the colored syrup being filtered through layers of boneblack. Many crystallizable organic acids and alkaloids are purified in the same way in the chemical laboratory. But it must never be forgotten that charcoal can absorb many other substances besides coloring-matters; sulphate of quinine, for example, is removed from its solutions, to a very considerable extent, by charcoal; and the same remark applies, with perhaps still more force, to strychnine. The bitter principle of the hop, "lupulin," may be entirely removed from ale by filtering the latter through bone-black. The removal of metals, like gold and silver, from their dilute solutions, by means of charcoal, appears to be a phenomenon of the same order.

In all these cases where coloring-matters, and the like, are removed from solutions, the action of the coal appears to depend in the main directly upon the physical property of adhesion, the subsequent oxidizing action being here far less clearly marked than in the instances previously studied (§ 387) where gases are acted upon. Much of the absorbed color or other matter will usually be found attached to the surfaces of the coal, undecomposed and unaltered. Thus, if the coal which has been charged with a solution of indigo in sulphuric acid, in Exp. 163, be digested in a solution of caustic soda, the latter will dissolve the indigo and remove it from the coal; the alkaloids above mentioned, which have been removed from their aqueous solutions by means of charcoal, can be again recovered by boiling the coal in alcohol; and the metals can be dissolved again by means of strong acids.

When employed to remove coloring-matters, charcoal soon becomes saturated with the color and ceases to absorb any more of it; if the spent coal be then collected and redistilled, it will be found that it has regained but little of its decolorizing-power; for its pores are filled with charcoal resulting from the carbonization of the absorbed coloring-matters and this charcoal is not porous, but, on the contrary, compact and glistening, like the charcoal obtained from sugar or glue, and is almost entirely destitute of decolorizing-power. In order to revivify their coal, the sugar-refiners digest it in chlorhydric acid, allow it to ferment in order that the absorbed matters may be decomposed, and, finally, after washing and drying, subject it to distillation.

389. As obtained from different sources, charcoal exhibits very different degrees of decolorizing-power; but of the varieties commonly met with and to be procured in commerce, boneblack is the most efficient. Boneblack is prepared for the use of sugar-

refiners by subjecting bones to destructive distillation in iron cylinders and carefully cooling the charcoal out of contact with the air.

Exp. 164.—Repeat Exp. 148, p. 293, but instead of bituminous coal, charge the ignition-tube with coarsely-powdered bone. Distil as long as gas is evolved, then remove the delivery-tube from the water, plug it by means of a bit of caoutchouc tube and a glass rod, and wait until the ignition-tube is cold, before opening it to examine the boneblack.

Dry bones contain only about one-third their weight of organic matter, nearly 66 per cent. of them consisting of phosphate of calcium, in the interstices of which the animal matter is distributed; hence it happens that the charcoal obtained by distilling bones is in an exceedingly porous and divided condition. The decolorizing-power of bone-charcoal may, moreover, be increased by digesting it in dilute chlorhydric acid, which dissolves out a portion of the phosphate of calcium and leaves the charcoal even more porous than before.

But because boneblack is most commonly employed for decolorizing, it must not therefore be inferred that it is the most powerful decolorizer of the several varieties of charcoal. A more efficient coal can easily be prepared by mixing nitrogenized organic matters, such as blood, hoofs, horns, or scraps of leather, with carbonate of potassium, distilling the mixture, and finally leaching the product with water. Charcoal of peculiar decolorizing-power may be prepared by igniting a mixture of 4 parts of fresh blood and 1 part of carbonate of potassium in an iron crucible, so long as vapors are evolved, then washing the product with water and boiling it with chlorhydric acid, and again washing, drying, and igniting in a close vessel. A solution of horn-filings in caustic potash, evaporated to dryness and ignited, also yields a very effective charcoal. As with boneblack, the efficiency of this coal prepared with caustic or carbonated potash appears to depend upon the minute subdivision of its particles and the porosity resulting from the mixture of the inorganic matter. The decolorizing-power of the charcoal obtained from vegetable substances can be in like manner increased by mixing these substances with lime or clay before the carbonization. A mixture of 100 parts pipe-clay, stirred up with water to the consistence of a thin paste,

20 parts of tar, and 500 parts of powdered bituminous coal yields a charcoal which decolorizes almost as well as boneblack. It is, however, no very easy matter to determine which, of a number of varieties of charcoal, is the most powerful decolorizer, since the decolorizing-power differs with the nature of the coloring-matter as well as according to the quality of the charcoal. It has been found, for example, that with

EQUAL WEIGHTS OF CHARCOAL PREPARED BY	THE RELATIVE DECOLORIZING-POWER	
	Upon a solution of indigo in sulphuric acid was	*Upon a solution of brown sugar was*
Igniting a mixture of blood and carbonate of potassium	50	20
Igniting a mixture of blood and carbonate of calcium	18	11
Igniting a mixture of blood and phosphate of calcium	12	10
Igniting a mixture of glue and carbonate of potassium	36	16
Igniting acetate of sodium . . .	12	9
Digesting boneblack with chlorhydric acid.	1·9 1·9	1·3 1·6
Ordinary boneblack	1	1

390. *Compounds of Carbon and Hydrogen* are exceedingly numerous. There are many different series of them, each member of either of these series differing but slightly from its next neighbors in composition and properties. But all these substances are commonly classed as organic compounds; they constitute a special branch of chemical science called organic chemistry, and are not discussed in works which treat of all the elements commonly met with, without special reference to any one. This subdivision of the general subject is resorted to merely for convenience; there is no inherent reason why the compounds of hydrogen and carbon should not be studied in this manual as well as the compounds of hydrogen and oxygen. But since carbon is capable of uniting with hydrogen, oxygen, or nitrogen, or with two of these elements, or with all three of them, in the most varied proportions, there are formed so many different compounds, that it has been found advantageous to study them by themselves.

The best definition of the so-called organic chemistry which can be given to-day, is, that it is the Chemistry of the Compounds of Carbon. The department of organic chemistry has grown out of ordinary chemistry solely because of the fact that the compounds of carbon with hydrogen, oxygen, and nitrogen are more numerous, and often of more complex composition, than the compounds formed by any of the other elements. These compounds of carbon with hydrogen, and with the other elements, are all definite chemical compounds conforming to the law of multiple-proportions (§ 76); but they count by thousands, and the mere enumeration of their names and properties would fill a volume.

391. As examples of the series of compounds of carbon and hydrogen, to which allusion was just now made, the following lists may be given:—

	Boils at ° C.		*Boils at ° C.*		*Boils at ° C.*
		CH_2			
CH_4		C_2H_4			
C_2H_6		C_3H_6		C_6H_6	. . 80°
C_3H_8	. . −30°	C_4H_8	. . 5°	C_7H_8	. . 110°
C_4H_{10}	. . 0°	C_5H_{10}	. . 35°	C_8H_{10}	. . 140°
C_5H_{12}	. . 30°	C_6H_{12}	. . 65°	C_9H_{12}	. . 170°
C_6H_{14}	. . 60°	C_7H_{14}	. . 95°		
C_7H_{16}	. . 90°	C_8H_{16}	. . 125°		

The first series of the compounds above formulated is found in petroleum, and in the oil known as coal-oil, obtained by distilling highly bituminous coals and shales at comparatively low temperatures; the petroleum used for lighting contains all these different compounds, together with others of the same class. The second series may generally be obtained by the destructive distillation of various organic compounds; and the members of the third series are found in the most volatile portion of coal-tar —the tar obtained by distilling bituminous coal at high temperatures, as in the manufacture of illuminating gas.

It will be observed that there is a constant difference of one atom of carbon and two atoms of hydrogen (CH_2) between any two contiguous compounds enumerated in the lists. The boiling-points of the several members exhibit a constant difference of 30° for each increment of CH_2, as we go down the lists, so far as has been determined; and so with all the other physical properties, such as specific gravity, mobility, expansibility by heat, and the like; the intensity of these properties increases or decreases regularly, in a constant ratio, as we pass from one member of the series to the members next in order. Such series are called "homologous" series (having the same proportion): they are clearly analogous to the groups, families, or *series* of elements which we have already studied in chlorine, bromine, and iodine (§ 152), in oxygen, sulphur, selenium, and tellurium (§ 257), in nitrogen, phosphorus, arsenic, antimony, and bismuth (§ 364), and are now proceeding to study in carbon, boron, and silicon. Just as in these groups of elements the student has seen a true serial arrangement of the different members, and has observed that the different terms of each series differ from one another in atomic weight and exhibit parallel differences in the intensity of their properties, so here in each of the homologous series of hydrocarbons there have been observed similar constant differences. In the series of hydrocarbons, we know that there is a constant difference of composition of CH_2; and this difference of composition we believe to be at the bottom of the constant difference of physical properties; but to the cause of the constant differences in the homologous series of elements we have, as yet, no clew.

The power of arranging numerous allied compounds into groups or series, like those enumerated upon page 312, has been of great service to chemists in facilitating the study of the compounds of carbon. For every such series a general algebraic expression has been devised which serves as the name of the series. Like any other name, this concise expression brings before the mind of the chemist the general properties of the series. Thus the expression C_nH_{2n+2} is general for the first series above mentioned, C_nH_{2n} for the second, and $C_{n+3}H_{2n}$ for the third.

In this manual we shall describe only one of the compounds of carbon and hydrogen, a compound which occurs ready formed in nature.

392. *Marsh-gas, or Light Carburetted Hydrogen* (CH_4), is a permanent gas which constitutes a large proportion of the ordinary illuminating gas obtained from coal. It is disengaged in large quantities from some sorts of bituminous coal, even at the ordinary temperature of the air, and more rapidly at higher tem-

peratures. In coal mines the gas thus given off is known as "fire-damp;" by mixing with air, in the galleries of badly ventilated mines, it forms explosive mixtures, which frequently occasion frightful accidents when ignited through carelessness.

The gas is evolved also, in large quantities, from the mud at the bottom of stagnant pools; it is one of the products of the putrefaction of vegetable matter under water, where the supply of air is insufficient to oxidize the matter completely to carbonic acid and water; hence the name marsh-gas.

In hot weather marsh-gas can be obtained by thrusting a pole into the mud at the bottom of a pond and collecting the bubbles of gas as they rise, by holding over them an inverted bottle full of water. When the bottle has been filled with gas it should be corked tightly under water, and carried to the laboratory for examination. The gas thus obtained is contaminated with nitrogen and with a large quantity of carbonic acid, these gases being set free, together with marsh-gas, during the process of putrefaction. Before the gas thus collected will burn, it is usually found necessary to remove from it the carbonic acid: this can be done by pouring into the bottle a small quantity of a solution of caustic soda, or some milk of lime, closing and shaking the bottle, and finally removing the stopper under water. The bottle may then be placed upright and a lighted match applied to the gas; it will take fire and burn with a blue flame, the size of which may be increased by pouring water into the bottle so that the gas shall be driven out into the air.

The gas may be prepared artificially as follows:—

Fig. 53.

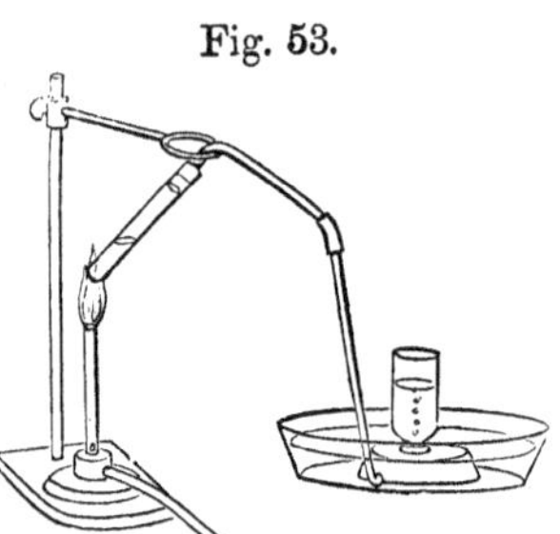

Exp. 165. Mix together 2 grms. of crystallized acetate of sodium, 4 grms. of caustic soda, and 8 grms. of slaked lime. Heat the mixture gently upon an iron plate, until all the water of crystallization of the acetate has been expelled and the mass has become dry and friable. Charge an ignition-tube 20 c.m. long with the dry powder, heat it above the gas-lamp, and collect the gas at the water-pan, as shown in Fig. 53. Carburetted hydrogen is evolved from the mixture at a temperature below redness, and a residue of carbonate of sodium is left in the ignition-tube. The purpose of the lime is to render the mass porous and infusible, or nearly infusible, so that the tube may be heated equably. If caustic soda is heated with the acetate without addition of lime, the tube usually breaks, even when the mixture has been dried beforehand. The reaction may be represented as follows:—

$C_2H_3NaO_2$	+	$NaHO$	=	CH_4	+	Na_2CO_3.
Dry Acetate of Sodium.		*Hydrate of Sodium.*		*Marsh-Gas.*		*Carbonate of Sodium.*

393. Marsh-gas is transparent, colorless, and little more than half as heavy as air. Next to hydrogen it is the lightest known substance, its specific gravity being only 8. It takes fire readily when touched with a lighted match, but is nevertheless more difficult of inflammation than most of the other combustible compounds of hydrogen. While free hydrogen and sulphuretted hydrogen can be lighted by a glass rod which has been heated to dull redness, the rod must be raised to the temperature of bright redness, or even to a white heat, in order that it may kindle marsh-gas. As prepared from acetate of potassium, the gas burns with a pale yellowish-blue flame. It is rather more soluble in water than oxygen; at 0° one volume of water dissolves 0·055 volume of it. It has never been condensed to the liquid condition.

394. We have, thus far, observed three different proportions in which the other elements unite with hydrogen. The members of the chlorine group unite with hydrogen, by preference, in the proportion of one volume to one volume; one volume of any member of the sulphur group combines, by preference, with two volumes of hydrogen; one atom of any member of the nitrogen group unites, by preference, with three atoms of hydrogen. The condensation increases in direct ratio to the increasing proportion of hydrogen, so that, in every case, two volumes only of the resultant compound are produced. We have thus become familiar with the fact that the space occupied by the molecule of a compound gas is always two unit-volumes (§ 258). An examination of the molecule of marsh-gas will reveal a fourth kind of hydrogen-compound—a compound containing in the product-volume (§ 260) four volumes of hydrogen condensed. It is not difficult to prove experimentally that two volumes of marsh-gas yield, on decomposition, four volumes of hydrogen; but, unfor-

tunately, it is not within our power to demonstrate, by experiment, the volume of carbon with which these four volumes of hydrogen are combined; for carbon is a solid incapable of volatilization by the intensest heat at present at our command. We are able to determine the combining volume of each constituent of chlorhydric acid, steam, and ammonia; but we have no positive knowledge whatever concerning the manner in which carbon enters into combination by volume. Its combining proportion by *weight* can be ascertained; but it must be carefully observed that all views respecting the volumetric composition of the very numerous compounds of carbon are purely speculative, so far as the carbon is concerned, until carbon shall have been actually volatilized and its vapor weighed.

That marsh-gas really contains hydrogen and carbon may be readily proved by bringing into play, under appropriate conditions, the strong affinity of chlorine for hydrogen. Chlorine will set free carbon from marsh-gas, just as it liberates nitrogen from ammonia (Exp. 64).

Exp. 166.—Fill a tall bottle of the capacity of 250 or, better, 500 c. c. with water, invert it over the water-pan, and pass marsh-gas into it, until a little more than one-third of the water is displaced; cover the bottle with a thick towel to exclude the light, and then fill the rest of the bottle with chlorine. Cork the bottle tightly, and shake it vigorously, in order to mix the gases together, keeping the bottle always covered with the towel. Finally, open the bottle and apply a light to the mixture. Ignition takes place, chlorhydric acid is produced, while the sides and mouth of the bottle become coated with solid carbon in the form of lampblack, and a cloud of smoke rises into the air. The presence of the acid may be proved by the smell, by its reaction with moistened blue litmus-paper, and by the white fumes which are generated when a rod moistened with ammonia-water is brought into contact with the escaping acid gas.

Carbon and hydrogen are therefore elementary constituents of marsh-gas. We should be glad to add the synthetical to the analytical demonstration, and make marsh-gas out of carbon and hydrogen; but no means are at present known by which these two elements can be directly combined to form marsh-gas. As in the case of ammonia, we are obliged to rely upon the assurance of the balance that the sum of the weights of the two constituents separated from a given quantity of marsh-gas is precisely equal to the weight of the marsh-gas submitted to analysis. The experimental process by which this fact is demonstrated is too complex to be profitably studied at this stage of progress, and the fact must therefore be accepted as the result of experience.

395. It remains to show how the investigation of the composition of the product-volume of marsh-gas leads to the knowledge of the atomic weight, or least combining weight, of the element carbon. Marsh-gas is the compound best suited for ascertaining the atomic weight of carbon, because experience has proved that it contains proportionally less carbon than any other hydride of the element. We determined the atomic weight of oxygen from steam, the hydrogen compound which contains the smallest proportion of oxygen,—of chlorine from chlorhydric acid, the only hydride of chlorine,—of nitrogen from ammonia, the hydride which contains the smallest proportion of nitrogen. So the atomic weight of carbon is the weight of carbon which experiment proves to be contained in two unit-volumes of marsh-gas. By physical determinations, the specific gravity of marsh-gas has been shown to be 8; in other words, one unit-volume of marsh-gas weighs 8; then two unit-volumes, or the product-volume, must weigh 16. How much of this weight of 16 is hydrogen? This question can be answered experimentally by ascertaining how many unit-volumes of hydrogen are locked up in two unit-volumes of marsh-gas.

When a series of electric sparks begin to traverse a measured volume of marsh-gas contained in a U-tube, arranged like the U-tube of figure 11, but without the jacket, *b c*, and its accompaniments, the gas is found to expand, and in a few minutes a light deposit of carbon appears in the vicinity of the points of the platinum wires. The decomposition of the marsh-gas proceeds slowly; so that a considerable time is required for the execution of the experiment. If the mercury in the U-tube be finally brought to a level in the two limbs, it will be seen that the original volume of gas has very nearly doubled. When this point is once attained, the continued transmission of sparks produces no further increase of the volume of the gas. The expanded gas may be shown by the usual tests to be hydrogen.

This experiment is rather difficult to perform, and does not yield perfectly exact results, for a minute portion of marsh-gas escapes decomposition; nevertheless it establishes beyond reasonable doubt the fact that marsh-gas contains twice its volume of hydrogen. Two unit-volumes of marsh-gas, weighing 16, must therefore contain four unit-volumes of hydrogen; but four unit-volumes of hydrogen weigh 4; therefore the quantity of carbon contained in the product-volume of marsh-gas must weigh 12, which number we admit as the atomic weight of carbon.

But it may be said, What means have we of knowing that 12 represents the *least* combining weight of carbon? (for the atom is by definition the *least* quantity of an element which can be conceived to exist in combination). If it were possible to demonstratethat the proportional quantity by weight of carbon which is represented by the number 12 is just the quantity required to fill one unit-volume when converted into vapor, we should have the same reason for believing 12 to be the weight of one atom of carbon that we have for considering 16 to be the weight of one atom of oxygen, or 35·5 the weight of one atom of chlorine. As we cannot convert carbon into vapor at all, it is impossible to ascertain with certainty how many atoms, or how many volumes, 12 proportional parts by weight of carbon really represent; the quantity of carbon which is combined with four atoms of hydrogen in marsh-gas may be four atoms, each weighing 3, or two atoms, each weighing 6, or one atom weighing 12. As it is necessary to assume some number of atoms as represented by the proportional weight 12, that assumption which will conveniently formulate the simplest and most familiar compounds of carbon will be the best. Accordingly it has, of late, been assumed that 12 proportional parts by weight of carbon constitute the least quantity of this element which is conceived to enter into chemical combination, or, in other words, that the atomic weight of carbon is 12. The atom of carbon, thus understood, can then fix, or combine with, four atoms of hydrogen or of any member of the chlorine group, and two atoms of oxygen or of any other member of the sulphur group. The following substances, familiar in common life, or subjects of discussion in this chapter, may be mentioned in illustration of this principle:—

Marsh-gas . . .	CH_4	Carbonic acid . . .	CO_2
Chloroform . . .	$CHCl_3$	Bisulphide of carbon .	CS_2
Chloride of carbon	CCl_4		

If it were assumed that 12 proportional parts by weight, the relative quantity of carbon in each of the above-mentioned compounds, represent two or four atoms or volumes of carbon, instead of one, every one of these formulæ would become less simple. There are a great multitude of compounds which contain a larger proportional quantity of carbon than the compounds cited above; but the greater proportional quantity is always some multiple of 12 proportional parts, or, in other words, is always an integral number of carbon atoms each weighing 12.

In marsh-gas we have thus found a new term in the series of hydrogen compounds. Marsh-gas is an example and type of the hydrides richest in hydrogen; so far as we yet know, hydrogen does not form, with any element whatsoever, any compound whereof the product-volume contains more than four atoms of hydrogen united with one atom of the other element. The following brief table comprehends all the principal types of hydrogen compounds, beginning with chlorhydric acid, the poorest hydride, and passing through water and ammonia, intermediate compounds, to marsh-gas, the hydride richest in hydrogen:—

Chlorhydric acid	$HCl = 2$	volumes.
Water	$H_2O = 2$	"
Ammonia	$H_3N = 2$	"
Marsh-gas.	$H_4C = 2$	"

Each of these types of hydrogen compounds characterizes a group of chemical elements. The first type is characteristic of the chlorine group; the second, of the sulphur group; the third, of the nitrogen group; and the fourth, of the group which we shall soon know as the carbon group.

396. The compounds of carbon and hydrogen are of great practical interest, since the flame of all ordinary lamps and fires results from their combustion. Any allusion to their properties at once suggests the influence which these properties exert in the usual methods of obtaining light and heat, and necessitates a more complete discussion of the subjects of flame and combustion than we have had hitherto. We shall recur to this subject in a

subsequent section, after having studied the oxides of carbon. For the elucidation of the subject of combustion, ordinary illuminating gas, which is a mixture of many hydrides of carbon, will serve as well as any pure hydrocarbon. A brief description of this product may here be given.

About 94-100ths of the volume of purified coal-gas consists of a mixture of marsh-gas, free hydrogen, and carbonic oxide,—the marsh-gas usually amounting to about one-third part of the whole gas. These non-luminous, or very feebly luminous gases, serve as carriers of the six or seven per cent. of real light-producing ingredients which are contained in the gas. This mixture of light-giving ingredients is exceedingly complex. The vapor of benzole is, no doubt, one of the most important of these ingredients. Some of the higher homologues of marsh-gas lend their aid; and a hydrocarbon of composition C_2H_2, called acetylene, is important and very generally present. Sometimes a little olefiant gas (C_2H_4) is present, as well as other compounds of the same homologous series; but the old view, that this substance constitutes the chief luminiferous ingredient of coal-gas, is no longer admitted.

For all practical purposes, we can here regard this mixture of gases as carburetted hydrogen. That it contains hydrogen, can readily be shown by holding a cold, dry bottle over a burning jet of it, and observing that water is a product of the combustion; and that it contains carbon can be seen by holding in the flame a piece of cold porcelain, and noting the deposition of soot. Coal-gas is only about half as heavy as air. In many respects it resembles hydrogen, and most of the experiments which were performed with hydrogen can be equally, or nearly, as well performed with this gas. The student will do well to repeat, as an example, Exp. 24, substituting, for the hydrogen, common gas drawn from the street mains. In the same way he may repeat Exp. 29, mixing 1 volume of coal-gas with from 8 to 12 volumes of air. If, instead of coal-gas, pure light carburetted hydrogen be taken, the explosion will be most violent with 8 to 10 volumes of air; with only 3 or 4 volumes of air, or more than 15 volumes, the mixture is not explosive; either too much or too little air prevents the explosion.

Compounds of Carbon and Oxygen.—There are two of these compounds—carbonic acid (CO_2), and carbonic oxide (CO).

397. *Carbonic Acid* (CO_2) is always formed when carbon or any of its compounds is burned in an excess of air or of oxygen gas, or in contact with substances, gaseous, liquid, or solid, which are rich in oxygen and yield it readily to other bodies.

Exp. 167.—Place a live coal (charcoal) upon a deflagrating-spoon, and thrust it into a bottle full of air, or, better, oxygen gas; cover the bottle closely, and set it aside for examination. Or invert an empty bottle over a burning lamp or candle, so that the products of the combustion of the lamp may be received in it; the bottle will immediately become clouded upon the inside from deposition of water resulting from the combustion (see § 56), and will also be filled with carbonaceous and other gaseous products, simultaneously formed. Cover the bottle and test its contents in the manner described in the succeeding experiment.

Fig. 54.

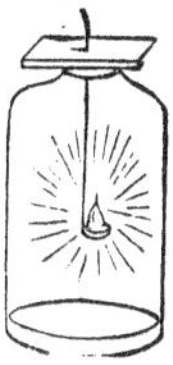

Exp. 168.—Pour some lime-water (a solution of common slaked lime in water) into the bottles filled with gaseous products of combustion in Exp. 167, and shake the bottles. The liquid will become milky and turbid, and, when left at rest, will deposit a white powder (carbonate of calcium). The presence of carbonic acid can readily be detected by means of lime-water, since this insoluble precipitate of carbonate of calcium is formed when the two substances are brought together.

The bottles of gas obtained in Exp. 167, will, of course, contain, besides carbonic acid, a quantity of nitrogen derived from the air which took part in the combustion, unless, indeed, as was suggested, the charcoal be burned in pure oxygen. It is to be observed that, in all ordinary cases of combustion, whether of wood, coal, wax, or oil, there result these same gaseous products —carbonic acid and nitrogen; they ascend, as invisible aërial currents, from every well-regulated flame or fire, and are continually issuing from the chimneys of our houses, though, in the absence of the particles of solid carbon, or of condensed aqueous vapor, which constitute smoke, we can see no product of the combustion.

Exp. 169.—As was just now said, carbonic acid may be produced also by heating carbon in contact with solid bodies which contain oxygen, such, for example, as the red oxide of mercury. Mix 11 grammes of red oxide of mercury with 0·33 grm. of charcoal; place the mixture in an ignition-tube, arranged as in figure 53; heat the tube and collect over water the gas which is evolved. Test the product with lime-water, as in Exp. 168. The reaction between the charcoal and the oxide of mercury may be written as follows:—

$$2HgO + C = CO_2 + 2Hg.$$

The metallic mercury set free condenses in droplets upon the cold upper portions of the ignition-tube. Here, again, as in Exps. 124, 155, the metallic oxide is reduced by the charcoal.

398. Carbonic acid may readily be obtained from certain compounds called carbonates, several of which are abundant minerals. Common chalk, marble, and limestone, for example, are composed of carbonate of calcium; and carbonic acid can readily be obtained by strongly heating them, or by subjecting them to the action of strong acids.

Exp. 170.—Place two or three grammes of coarsely-powdered marble in an ignition-tube provided with a gas delivery-tube bent at a right angle; place the ignition-tube upon the iron stand over the gas-lamp, and dip the outer opening of the delivery-tube into a small bottle containing lime-water; heat the marble strongly, and observe the white precipitate which forms in the lime-water as the carbonic acid gas comes in contact with it. The carbonate of calcium, thus precipitated by bringing together carbonic acid and oxide of calcium, is chemically identical with the chalk or marble from which the acid was expelled.

In actual practice enormous quantities of carbonic acid are expelled from limestone in this way for the sake of the quicklime which is left as a residue; but the carbonic acid, thus expelled by heat, is rarely collected, for a more convenient method of procuring it is to treat the limestone with some acid capable of expelling the carbonic acid.

Fig. 55.

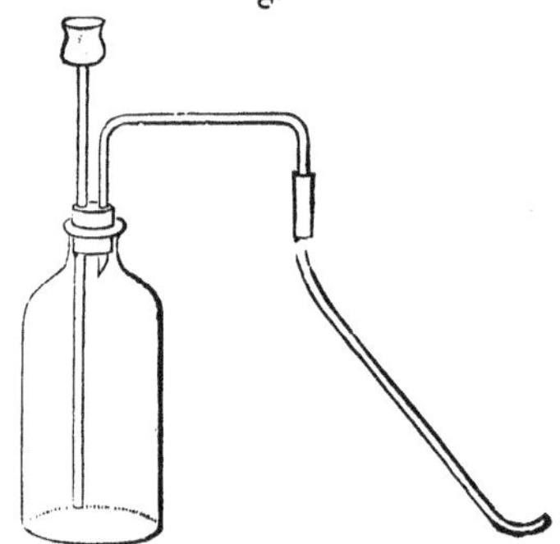

Exp. 171.—In a gas-bottle of 500 or 600 c. c. capacity, arranged precisely as for generating hydrogen (see Exp. 19), place 10 or 12 grms. of chalk or marble in small lumps; cover the chalk with water, and pour in through the thistle-tube concentrated chlorhydric acid, by small portions, in such quantity as shall insure a continuous and equable evolution of gas. Collect several bottles of the gas over water, then replace the anterior portion of the delivery-tube with a straight tube and collect one or two bottles of the gas by displacement; carbonic acid gas is half as heavy again as air. The reaction between the carbonate of calcium and the chlorhydric acid may be thus formulated:—

$$CaCO_3 + 2HCl = CaCl_2 + H_2O + CO_2.$$

When chalk is the material operated upon, sulphuric acid may be substituted with advantage for the chlorhydric acid; for the latter, being rather easily volatile, is liable to be carried forward by the current of carbonic acid, and to contaminate the product. When carbonic acid is prepared for commercial purposes by the action of an acid upon carbonate of calcium, sulphuric acid is almost always the acid employed, and marble-dust the substance acted upon:—

$$CaCO_3 + H_2SO_4 = CaSO_4 + H_2O + CO_2.$$

With a porous material, like chalk, this action occurs readily; but in attempting to operate upon compact varieties of carbonate of calcium such as marble, difficulties are encountered, unless the carbonate be in the state of powder. The sulphate of calcium, which is a product of the reaction, is a rather difficultly soluble substance, and, being deposited upon the surface of the marble, soon covers it with a coating so thick and impermeable that the action of the sulphuric acid upon the marble is well nigh completely arrested.

Carbonate of calcium being cheaper than most other carbonates, is more commonly employed than any other as a source of carbonic acid; but it is sometimes convenient to substitute for it the carbonate of sodium, or of potassium (saleratus), or even of ammonium, since carbonic acid is given off from these compounds very quickly and abundantly. A self-regulating gas-generator, such as is represented in Fig. xxviii. of the Appendix, charged with large solid lumps of the commercial carbonate of ammonium and chlorhydric acid, is, perhaps, the most convenient apparatus which can be employed for preparing the gas upon the lecture-table.

399. At the ordinary atmospheric temperature and pressure, carbonic acid is a transparent, colorless gas, of a slightly acid smell and taste. It is incombustible, being already the product of the complete combustion of carbon, and is, moreover, incapable of supporting the combustion of most other bodies, since the oxygen

contained in it is very firmly held; like nitrogen, it immediately extinguishes burning bodies which are immersed in it.

Exp. 172.—Thrust into a bottle of the gas obtained in Exp. 171, a lighted candle, or, better, a large flame of alcohol burning upon a tuft of cotton; in either case the flame will be instantly extinguished.

The specific heat of gaseous carbonic acid, between 10° and 200°, is 0·2169. Its specific gravity is 22; being thus 1·53 as heavy as air, it can be poured from one vessel to another almost as readily as if it were water.

Exp. 173.—Invert a bottle filled with carbonic acid upon another bottle of equal size filled with air, in such manner that the mouth of the upper inverted bottle shall rest upon the mouth of the lower bottle. After the lapse of several minutes, thrust a burning splinter of wood into each of the bottles; in the upper bottle the splinter will continue to burn, for into this bottle the air from the lower bottle has ascended; while in the lower bottle, now full of carbonic acid, the splinter will be extinguished.

Exp. 174.—From a large bottle full of the gas, pour a quantity of carbonic acid upon the flame of a lamp or candle; that is to say, hold the mouth of the open bottle of carbonic acid obliquely over the candle-flame so that the gas shall fall like water upon it; the flame will immediately be extinguished.

400. Owing to the great weight of carbonic acid, it often fails to rise out of wells and other cavities in the earth, in which it is generated by the decomposition or decay of organic substances. Before allowing workmen to descend into any such place, where there is reason to suspect the presence of carbonic acid, a burning candle should first be lowered; if the candle be extinguished, or even if it burn feebly, the noxious character of the air is indicated, and measures should at once be taken to purify the locality by ventilation or otherwise. One way of removing the carbonic acid is to absorb it by means of some chemical agent, such as slaked lime (hydrate of calcium) or potash-lye. The slaked lime is most efficient as an absorbent when neither very wet nor very dry; it should not be dry enough to be dusty, nor yet noticeably moist. If a quantity of it be suspended in the well, or thrown upon the floor of the cellar containing carbonic acid, this gas will quickly combine with the lime to form carbonate of calcium, as in Exp. 168. Another good method is to lower down a chafing-dish full of

brightly glowing charcoal; if the carbonic acid be present in such quantity that the test-candle has been extinguished by it, the charcoal will, in like manner, be immediately extinguished, and, in cooling, will rapidly absorb this gas (see § 385), together with any nitrogen which may be present; a current of fresh air will, in this way, be induced to flow into the well. If, by the candle test, but little carbonic acid be found in the well, it would, of course, be best to extinguish the charcoal before lowering it into the impure air; this can readily be done by covering it up tightly in an iron kettle.

It should be distinctly understood that the accumulation of carbonic acid in wells and caves is a consequence of the low diffusive power of the gas (see § 53). Carbonic acid mixes with air very slowly; but when once mixed with air it has no further tendency to settle down, or to separate itself in any way.

Exp. 175.—Over a bottle filled with carbonic acid gas, invert another bottle full of air, in such manner that the mouth of the air-bottle shall rest upon that of the upright bottle full of carbonic acid. After some hours, pour lime-water into each of the bottles and shake them; a precipitate of carbonate of calcium will be formed in both cases, for a part of the carbonic acid will, by this time, have ascended out of the lower into the upper bottle. The two gases have, in fact, become intimately mixed or blended; the heavy carbonic acid has diffused upwards into the air, and the lighter atmospheric air has diffused downwards into the carbonic acid, just as, in a previous experiment, we have seen hydrogen and oxygen diffuse into each other. (Fig. 20, p. 43.)

The great importance of this diffusion of gases, in the economy of nature, is well illustrated by the case now under consideration. Whenever, as in the processes of respiration and combustion, oxygen is withdrawn from and carbonic acid thrown into the air, the carbonic acid, and the nitrogen with which it is accompanied, immediately mix with the surrounding air and distribute themselves through the atmosphere. The composition of the atmosphere is thus maintained uniform all over the globe, in spite of the constant removal from it of oxygen in some localities, and the addition of carbonic acid in others. It is only in confined places, where nearly pure carbonic acid is produced more rapidly than it can pass off by diffusion, that the gas accumulates to any appreciable extent.

The singular facility with which gases, and particularly hydrogen, traverse porous bodies is very strikingly illustrated by the following experiment, which our acquaintance with carbonic acid now enables us to perform :—Through a large glass tube, a smaller tube of porous, unglazed earthenware is passed, and the ends of the glass tube are tightly closed by the corks which hold the porous tube. By the tube

Fig. 56.

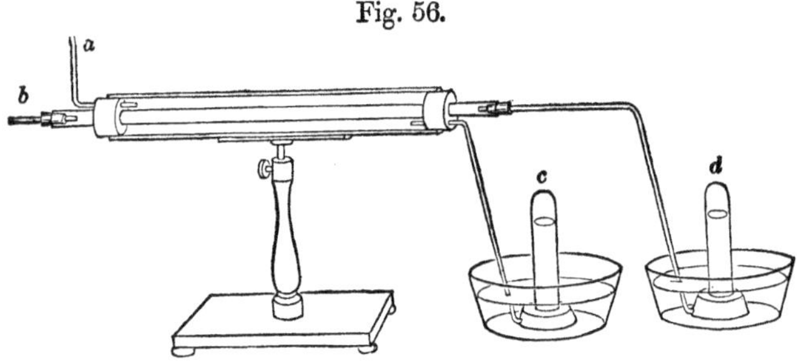

a, a rather rapid stream of carbonic acid is brought from a self-regulating generator into the annular space between the two tubes, while, by the tube *b*, a slower current of hydrogen is introduced into the inner porous tube. It would be expected that hydrogen should be found in the cylinder *d*, and carbonic acid in the cylinder *c*; but, on the contrary, an inflammable gas is found in *c*, and in the cylinder *d* carbonic acid pure enough to extinguish a burning splinter. The hydrogen diffuses almost instantaneously into the annular space, and the carbonic acid enters the inner tube to replace the issuing hydrogen.

401. When pure, or nearly pure, carbonic acid is irrespirable; it produces spasms in the respiratory passages, and is thus prevented from entering the lungs. When so far diluted with air as to admit of being respired, it acts as a narcotic poison; it is, however, far less poisonous than the other oxide of carbon, carbonic oxide, directly to be described.

402. Carbonic acid gas is soluble in water to a considerable extent. One measure of water, at the ordinary temperature and pressure, will dissolve one measure of carbonic acid gas; but its solubility increases if the pressure be increased.

Exp. 176.—Into a long-necked flask or phial filled with carbonic acid, pour a quantity of water, close the bottle with the finger, and shake it; immerse the mouth of the bottle in water, and remove the finger, water will rush into the bottle to supply the place of the gas which has been dissolved. Again place the finger upon the mouth of the bottle, shake the bottle as before, and subsequently open it beneath the surface of the water; a fresh portion of water will flow into the bottle to supply the new vacuum ; in this way, by repeated agitation with water, all of the carbonic acid in the bottle can be absorbed.

Owing to this solubility in water, some carbonic acid is always lost when the gas is collected over water as in Exp. 171; but since considerable time is required to absorb all the gas, there is little objection to collecting it over water.

403. When subjected to increased pressure, carbonic acid gas dissolves in water much more abundantly than at the ordinary pressure of the air; under a pressure of two atmospheres, one measure of water will dissolve two measures of the gas ; under a pressure of three atmospheres, it will dissolve three measures, and so on. Water thus surcharged with carbonic acid has an agreeable, acid, pungent taste, and effervesces briskly when the compression is suddenly removed, as when the liquid is allowed to flow out into the air; such carbonic acid water, or "mineral water," as it is then called, flows from the earth in many localities, as at Seltzer and Saratoga; it is also prepared, artificially, in large quantities, and sold as a beverage under the meaningless name of *soda-water*. Carbonic acid water possesses solvent powers far greater than those of pure water; few minerals are capable of resisting its long-continued action; hence the waters of the springs from which it issues are usually highly charged with saline and mineral ingredients, and are often of medicinal value.

404 The effervescent qualities of fermented liquors, such as cider, champagne, and beer, are in like manner dependent upon the presence of compressed carbonic acid gas. In all these cases the carbonic acid is of value, not only on account of the agreeable pungency which it imparts to the beverage, but also because of the fact that in escaping from solution and assuming the gaseous condition, it absorbs a very considerable amount of heat, and so cools the liquid which contained it.

405. Carbonic acid is widely diffused in nature. Traces of it occur in the air and in water everywhere; and there are many localities, besides the mineral springs before-mentioned, where it issues from the earth in large quantities, notably in several vol-

canic districts. It is produced not only in the actual combustion of all substances which contain carbon, but also during the decay and putrefaction of all animal and vegetable substances. During fermentation it is evolved in large quantities, and it is continually given off during the respiration of animals.

Exp. 177.—Put one or two table-spoonfuls of coarse meal into a bottle of about 250 c. c. capacity; cover the meal with water, and connect the bottle, by means of a cork and glass tube, with a second bottle filled about three c.m. deep with lime-water; the delivery-tube must reach into the lime-water. Out of the top of the second bottle carry a second bent glass tube, whose open end dips into water. The bottle containing the lime-water will, of course, be closed by the cork through which pass the two tubes above described. Keep the apparatus in a warm place; bubbles of gas will pass from the first into the second flask;—they contain carbonic acid, as may be seen from the precipitate of carbonate of calcium which they produce.

Exp. 178.—Dissolve 2 grammes of honey or molasses in 200 c. c. of water; fill a large test-tube with the mixture, and add to it a few drops of bakers' or brewers' yeast; close the open mouth of the test-tube with the thumb, and invert it in a small saucer or porcelain capsule filled with the diluted syrup. Place the saucer and tube, with their contents, in a warm place, having a temperature of about 20° or 30°, and leave them there during 24 hours. In a short time, fermentation sets in, and the sugar of the syrup is gradually converted into alcohol and carbonic acid.

$$\underset{\textit{Sugar.}}{C_6H_{12}O_6} = \underset{\textit{Alcohol.}}{2C_2H_6O} + 2CO_2.$$

The carbonic acid thus formed rises in minute bubbles, causing a gentle effervescence in the liquid, and collects in the upper part of the tube, while the alcohol remains dissolved in the liquid. That the gas obtained in this experiment is really carbonic acid, may be proved by transferring some of it into a clean tube at the water-pan, and then testing with lime-water.

Exp. 179.—Provide two test-glasses or small bottles; place in each 15 or 20 c. c. of lime-water; through a glass tube blow into the lime-water of one of the bottles air coming from the lungs. By means of bellows, to the nozzle of which a gas-delivery tube has been attached, force through the lime-water of the second bottle a quantity of fresh air. The clear liquid of the first bottle will quickly become turbid through deposition of carbonate of calcium, while the lime-water of the second bottle will remain clear for a long while.

This experiment may be modified by constructing an apparatus with valves, in such manner that the air drawn into the lungs shall be made to pass through one bottle of lime-water, while the air expired goes out through another bottle of lime-water. The contents of the first bottle will remain clear, while the liquid in the other immediately becomes turbid.

Ordinary fresh air contains only about one 2000th part of carbonic acid, while air expired from the lungs contains as much as 3 or 4 per cent. of it. In breathing, animals inhale oxygen from the air; this oxygen combines with carbon within their bodies and is exhaled as carbonic acid. A crowd of men consume the oxygen of the air, just as lamps or fires consume it; to carry off this product of animal combustion is one of the objects of systems of ventilation. Air which contains even as little as 1 or 2 per cent. of carbonic acid, exerts a very depressing effect when breathed for any length of time.

406. From the foregoing statements it appears that, in the several processes called "life," "fermentation," "decay," and "combustion," there is involved chemical action; in all of these processes oxygen from the air unites with carbon, while carbonic acid is set free and thrown into the atmosphere. The question now arises, What becomes of all this vast amount of carbonic acid which is constantly coming into our atmosphere from the respiration of animals, from fires, from decaying and fermenting substances, from volcanic fissures, and from various other sources? If this carbonic acid remained in the air, the latter would quickly become unfit to support animal life. But it is not found that the average proportion of carbonic acid in the air does increase; on the contrary, all the evidence goes to show that there was, at certain earlier geological epochs, more of it in the air than now. Many geologists believe that, in the early history of our globe, there was much more carbonic acid in the air than at present; hence immense forests arose whose carbon is now stored away in the form of coal; hence also the formation of enormous beds of limestone covering many parts of the earth's surface,—processes of which the faint continuations are now seen in the formation of peat-bogs and coral-reefs. On the other hand, it is not found that the proportion of oxygen in the atmosphere undergoes any appreciable change, in spite of the enormous volume of it which is absorbed in the processes of breathing, combustion, and decay

above enumerated. For, unlike animals, plants, in breathing, take in carbonic acid and give out oxygen. The leaves of plants are so constructed that they can decompose carbonic acid, fix carbon for the building up of the plant, and set oxygen free. This reciprocal action of plants and animals, tending to maintain unchanged the constitution of the atmosphere, is one of the most wonderful adjustments of nature.

407. Carbonic acid can be liquefied by pressure, and the liquid thus obtained can be solidified by exposure to cold. When the gas is generated in a confined space in a strong vessel, it soon exerts so powerful a pressure that a large portion of it condenses to a transparent, colorless, mobile liquid, somewhat resembling water, though it refracts light less powerfully; or it may be liquefied by mere cooling to —106°, under the atmospheric pressure. A better way of preparing the liquid acid is to pump the gas, by means of a forcing syringe, into a strong wrought-iron vessel surrounded with pounded ice. The pressure can thus be regularly and methodically increased and the receiver finally filled with the liquid. At 0° a pressure of 36 atmospheres is required, in order that the acid shall remain in the liquid state. Liquid carbonic acid does not mix readily with water or with the fixed oils; but with alcohol, ether, petroleum, and similar liquids it is miscible in all proportions. Its specific gravity is 0·83 at 0°; but it expands to such an extent on being heated, that at 30° its specific gravity is only 0·6. It expands to a greater extent, on the application of heat, than any known substance, even to a greater extent than the gases; 20 volumes of it at 0° become 29 volumes when the temperature is raised to 30°; 150 volumes, at 30°, shrink to 100 volumes when the temperature is reduced to —20°. The liquid acid well illustrates the general fact that liquids expand proportionally much more when heated under a high pressure than under a low one.

408. If the stop-cock of a vessel containing liquid carbonic acid be opened, in such manner that a stream of the liquid shall be forced out into the air, a portion of it at once assumes the gaseous state, and in so doing absorbs so much heat from the remainder that the latter solidifies, and is deposited in the form of white flakes like snow. This snow-like substance is slowly con-

verted into gas when exposed at the ordinary pressure of the air, and so disappears. Though its temperature is lower than $-78°$, it may be handled lightly without exciting any special sensation of coldness or pain; for the gas which it is constantly emitting is a bad conductor of heat, and prevents it from coming into intimate contact with the skin. When, however, the solid acid is forcibly pressed between the fingers, it produces painful blisters, as if it were red-hot iron. In order to use the solid acid for producing low temperatures, it is best to mix it with a small quantity of ether; in such a mixture quicksilver can readily be frozen, and many gases can be liquefied or even solidified. If this mixture be placed in the vacuum of an air-pump, a temperature as low as $-100°$ can be obtained; and if a tube containing liquid carbonic acid be then placed in the mixture, the liquid will speedily be frozen to a clear transparent mass like ice.

409. Carbonic acid gas, on being heated from 0° to 100°, does not expand at the same rate as air and the other permanent gases, but increases in volume to a greater extent than any of them. Upon being heated one degree, it expands 0·003688 its volume at 0°. This behavior is in accordance with the general rule, that those gases expand the most which are most readily condensable to the liquid state, while those gases which have resisted all efforts to liquify them scarcely show any appreciable differences in the rate of expansion.

In the same way, carbonic acid, like the other easily condensable gases (see § 221), does not conform precisely to the law of Mariotte. At pressures greater than one-third of the pressure of our atmosphere, its volume diminishes more rapidly, with increasing pressure, than would be the case with air and the other permanent gases under the same conditions.

410. Carbonic acid is one of the compound gases which can be split by heat alone into its proximate constituents; in other words, it exhibits the phenomena of dissociation (§ 300). When the gas is passed through a strongly heated porcelain tube, the gaseous mixture which escapes from the tube contains, besides undecomposed carbonic acid, notable quantities of carbonic oxide (CO) and oxygen.

411. As has been already stated (§ 399), the oxygen in carbonic

acid is so strongly held that it cannot be withdrawn by combustibles under ordinary circumstances; but at high temperatures carbonic acid is decomposed by carbon and by several of the metals—such as iron, zinc, and manganese, besides potassium, sodium, and the other metals of the alkalies and alkaline earths. By the alkali-metals the oxygen is completely removed from carbonic acid, and carbon is set free; but by the other agents above-mentioned, only half the oxygen of the acid is taken away, while carbonic oxide gas is formed:—

$$CO_2 + C = 2CO.$$

Phosphorus, also, at high temperatures and in presence of a fixed alkali, decomposes carbonic acid in the same way, and abstracts part of its oxygen. So, too, if a mixture of equal volumes of hydrogen and carbonic acid be passed into one end of a red-hot tube, steam and carbonic oxide gas will escape at the other:—

$$CO_2 + 2H = CO + H_2O.$$

The decomposing-power of the alkali-metals, above alluded to, furnishes us one means of partially analyzing carbonic acid.

Exp. 180.—To a gas-bottle in which carbonic acid is being steadily evolved, according to Exp. 171, attach a chloride-of-calcium tube, and beyond this drying-tube a short tube of hard glass, from which an exit-tube leads into a small open bottle, as shown in Fig. 57. When

Fig. 57.

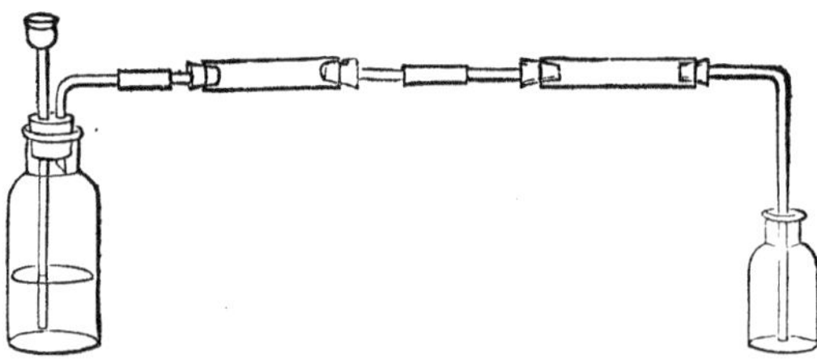

the extinction of a lighted match in the open bottle proves the apparatus to be full of carbonic acid, thrust into the hard-glass tube a bit of potassium as big as a pea, previously dried between folds of blotting-paper, and then gently heat the potassium with a lamp. The potassium will take fire and burn at the expense of the oxygen of the carbonic acid, and black particles of carbon will be deposited upon the walls of the tube. After the reaction has ceased, and the tube has been allowed to become cold, place it in a bottle of water so that the saline mass (carbonate of potassium) may dissolve; the particles of carbon will then be seen more clearly, floating in the liquid; they may be collected upon a filter. The potassium in this experiment may be replaced by sodium, but in this case a somewhat higher temperature is required.

412. The quantitative composition of carbonic acid may be readily ascertained by the method of synthesis. When a quantity of carbon is burned in a confined and measured volume of oxygen, it is found that the volume of carbonic acid gas produced has sensibly the same bulk as the original oxygen. Hence we conclude that the normal or product-volume of the molecule of carbonic acid gas contains two volumes of oxygen.

Now two volumes of carbonic acid weigh 44·152, since the weight of the unit-volume, or the specific gravity of carbonic acid, has been found to be 22·076.

Subtracting from this weight of the product-volume of the gas	44·152
The weight of two unit-volumes of oxygen (15·969 × 2) . .	31·938
There remains as the weight of the carbon in the product-volume of the gas	12·214

The weight of one atom of carbon is 12, as we have seen above (§ 395), and it follows that the formula of carbonic acid is CO_2. From these figures the following percentage composition of carbonic acid may be deduced:—

Carbon	27·66
Oxygen	72·34
	100·00

But these results must be regarded merely as approximations to the truth, since the deviation of carbonic acid from the law of Mariotte (§ 409) renders it well-nigh certain that we have not yet been able to precisely determine, by experiment, the true weight of a unit-volume of this gas.

413. The composition of carbonic acid, however, has been determined with very great accuracy by burning a known weight of pure carbon in a stream of oxygen gas and carefully collecting and weighing the carbonic acid produced.

In order to do this, the product of the combustion of the carbon, together with the excess of oxygen, is made to flow through U-tubes

(Appendix, § 15) filled with fragments of hydrate of potassium, a substance which absorbs only the carbonic acid; the weight of these tubes is determined before the commencement of the experiment and again at its close, the increase of weight during the experiment being, of course, referable to the carbonic acid absorbed. Knowing, then, the weight of the carbon taken and the weight of the carbonic acid into which it has been converted by uniting with oxygen, a very simple calculation, as before, gives us the percentage composition of the acid. Experiments of this kind have yielded the following result:—

Carbon	27·27
Oxygen.	72·73
	100·00

It therefore, appears, that, in uniting to form carbonic acid, the elements carbon and oxygen combine in the proportion of 3 parts by weight of carbon to 8 parts by weight of oxygen, or in the proportion of 12 to 32. If the number 72·73 be divided by 16, the number representing the weight of a unit-volume of oxygen, and if the number 100 be divided by 22, the number which, in accordance with the weight of all the evidence thus far accumulated must be regarded as the true unit-volume weight of carbonic acid, there will be obtained in each case the same quotient, namely, 4·545 volumes; whence we conclude, as before, that any volume of carbonic acid contains the same volume of oxygen.

414. Carbonic acid unites with the protoxides of most of the metals to form well-defined salts called carbonates. The greater number of the best-defined carbonates contain only one molecule of base; but besides the normal carbonates of the general formula M_2O,CO_2 or M_2CO_3, there are sesquicarbonates of the formula $2M_2O,3CO_2$, "bicarbonates" of the formula $M_2O,H_2O,2CO_2$ or $MHCO_3$, and *basic* carbonates containing two, three, or more molecules of the base to one of the acid. In general the term basic salt is applied to all salts in which, as in the carbonates last named, the alkaline constituent or base predominates. The appellation "bicarbonate," though ordinarily applied in the manner indicated above, is a name of very doubtful correctness; strictly speaking, the class of compounds to which it refers should, perhaps, be regarded as double salts of the metal and hydrogen. The normal carbonate, and the so-called bicarbonate of sodium, for

example, differ only in that half the sodium in the normal salt has been replaced by hydrogen in the bicarbonate:—

Normal Carbonate of Sodium.	*Bicarbonate of Sodium.*
Na_2CO_3.	$NaHCO_3$.

Carbonic acid is an exceedingly weak acid; it fails to neutralize (§ 65) completely the causticity of oxides such as those of the alkaline metals; the normal carbonate of sodium, for example, is decidedly alkaline in its reaction and properties. The so-called bicarbonate of sodium is also slightly alkaline; and even the solution of carbonate of calcium in carbonic acid water exhibits an alkaline reaction when tested with turmeric paper. Almost all the carbonates are readily decomposed by acids (even by very weak acids), with an effervescence caused by the escape of carbonic acid. Most of them are decomposed also on being heated; but from the normal salts of sodium and potassium carbonic acid cannot be expelled by heat alone, however intense.

415. *Carbonic Oxide* (CO).—As has been stated in § 411, carbonic oxide can be prepared by acting upon carbonic acid with hot charcoal.

Exp. 181.—In the middle of a tube of hard glass, No. 2 or 2½, about 35 c.m. long, pack a column 15 c.m. in length of coarsely-powdered charcoal. Place the tube upon a sheet-iron trough on a ring of the iron stand above wire-gauze lamps, as shown in the figure. Connect

Fig. 58.

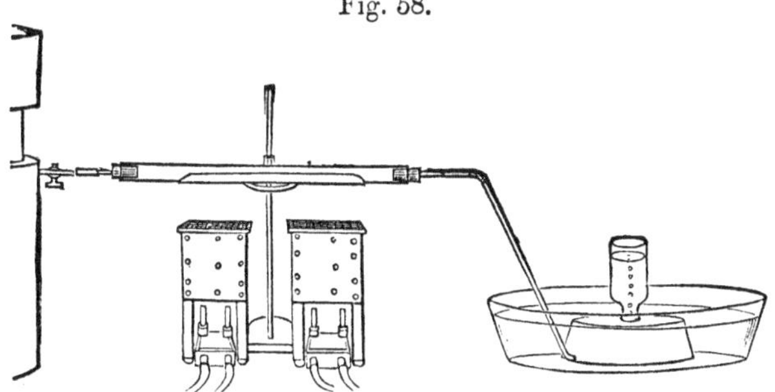

the tube either with a gas-holder containing carbonic acid, or with a bottle in which the gas is being generated. Heat the charcoal intensely, and from time to time test the gas which is delivered at the water-pan, as to its inflammability. Carbonic oxide takes fire on being touched with a lighted match, and burns with a bluish flame. In place of the charcoal, small fragments of iron or of zinc may be employed in this experiment.

Instead of gaseous carbonic acid, the solid compounds called carbonates can be conveniently employed for preparing carbonic oxide.

Exp. 182.—Mix powdered chalk (carbonate of calcium) with an equal weight of iron or zinc filings, place the mixture in an ignition-tube, provided with a gas-delivery tube, and heat it to redness over the gas-lamp. The metal will abstract an atom of oxygen from the carbonate of calcium, oxide of iron or of zinc will be formed, while carbonic oxide will pass off through the delivery-tube to be collected at the water-pan:—

$$CaCO_3 + Fe = FeO + CaO + CO.$$

In the same way, when a mixture of chalk and finely-powdered charcoal is heated to full redness, carbonic oxide gas is given off:—

$$CaCO_3 + C = CaO + 2CO.$$

It should be mentioned, however, that in all these cases the carbonic oxide obtained is more or less contaminated with carbonic acid, portions of which escape reduction by the metal and carbon; the carbonic acid may always be readily removed by causing the gas to pass through a strong solution of caustic soda or through a U-tube filled with bits of pumice-stone saturated with soda-lye.

Carbonic oxide can be obtained also by heating charcoal with other solid oxygen compounds, such as the phosphate of calcium, already mentioned (§ 270), or the oxides of almost any of the metals, provided the charcoal be in excess.

Exp. 183.—Heat in an ignition-tube, as before, a mixture of 1 grm. of finely-powdered charcoal and 8 grms. of red oxide of iron; collect the gas over water, pour into the bottle a little soda-lye, close the mouth of the bottle tightly and shake it, then open the mouth of the bottle under water, and finally test the gas with a lighted match.

416. Another easy method of preparing carbonic oxide is to decompose oxalic acid by means of oil of vitriol; this is the method usually employed in the laboratory. Oxalic acid is a solid vegetable acid, to be procured of the druggists; its composition may be represented by the formula $H_2C_2O_4$. On being heated with concentrated sulphuric acid, it suffers decomposition in a manner which may be formulated as follows:—

$$H_2O,C_2O_3 + H_2O,SO_3 = 2H_2O,SO_3 + CO + CO_2.$$

The elements of water are taken away from the oxalic acid and united with the sulphuric acid, while the remainder of the oxalic acid breaks up into carbonic acid and carbonic oxide.

Exp. 184.—Place in a flask of about 350 c. c. capacity 9 grms of common oxalic acid and 53 grms. of concentrated sulphuric acid; connect the flask with a bottle filled with fragments of pumice-stone saturated with a strong solution of caustic soda, as shown in the figure; heat the contents of the glass gently, and collect the gas which is evolved over water in the usual way. The carbonic acid resulting from the reaction will all be absorbed by the soda-lye, and carbonic oxide will alone be delivered at the water-pan.

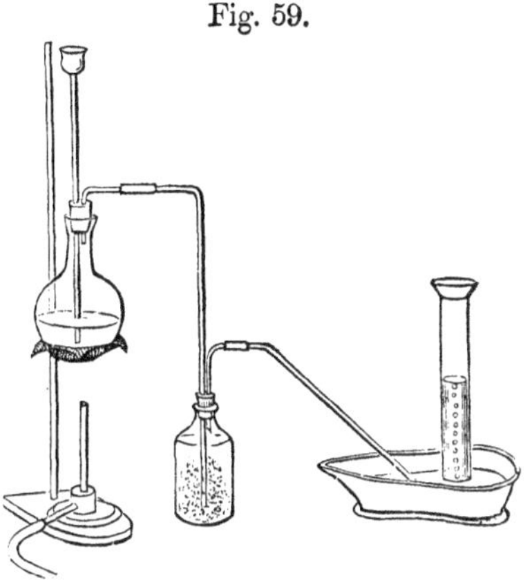
Fig. 59.

None of the methods heretofore given yield pure carbonic oxide directly; in each of the experiments we are compelled to wash out carbonic acid from the gas obtained, if an absolutely pure product is desired; but there are methods by which pure carbonic oxide may be prepared without the need of any process of purification. One of the best of these is as follows:—

Exp. 185.—In a thin-bottomed flask of about 250 c. c. capacity and provided with a suitable gas-delivery tube, heat a mixture of 5 grammes of finely powdered ferrocyanide of potassium (yellow prussiate of potash) and 40 or 50 grms. of strong sulphuric acid. Collect the gas over water, and test it as to its inflammability. Thrust also a lighted splinter into the gas and observe that it will be extinguished. The reactions which occur between the chemicals employed may be expressed as follows:—

Ferrocyanide of Potassium.		*Water.*		*Sulphuric Acid.*
$K_4FeC_6N_6H_6O_3$	+	$3H_2O$	+	$6H_2SO_4$

$$= 6CO + 2K_2SO_4 + FeSO_4 + 3(NH_4)_2SO_4.$$

Carbonic Oxide.	*Sulphate of Potassium.*	*Sulphate of Iron.*	*Sulphate of Ammonium.*

417. Carbonic oxide is a transparent, colorless gas, having little, if any, odor; it has never yet been liquefied. It is somewhat lighter than air, its specific gravity being 14, while that of air is 14·5. It is but little soluble in water, and may be collected and preserved over water without much loss. It extinguishes combustion just as hydrogen does, and destroys animal life. Unlike hydrogen and nitrogen, however, it is a true poison. It destroys life, not negatively by mere suffocation or exclusion of oxygen, but by direct noxious action. Even when largely diluted with air, it is still poisonous, producing giddiness, insensibility, and finally death. It is the presence of this gas which occasions the peculiar sensation of oppression and headache which is experienced in rooms into which the products of combustion have escaped from fires of charcoal or anthracite. Carbonic oxide is very much more poisonous than carbonic acid. Much of the ill repute which attaches to carbonic acid really belongs to carbonic oxide; for since both these gases are produced by burning charcoal, many persons are liable to confound them; but carbonic acid is, comparatively speaking, almost innocuous. Carbonic acid, it is true, is somewhat poisonous; it does not merely suffocate, like water, or nitrogen, or hydrogen; but it is very much less poisonous than carbonic oxide. It has been found, by experiment, that an atmosphere containing only 1-100th of carbonic oxide is as fatal to a bird as one containing 1-25th part of carbonic acid.

Carbonic oxide exhibits neither an acid nor an alkaline reaction when tested with vegetable colors, and, in general, has but little tendency to combine with other substances. With oxygen, however, it combines readily at comparatively low temperatures; an iron wire heated to dull redness is sufficient to inflame it in the air. Unlike most other combustible gases, it contains no hydrogen, and therefore produces no water when burned; nothing but carbonic acid results from its burning.

Exp. 186.—To the apparatus employed for evolving carbonic oxide in Exp. 184, attach a piece of small glass tubing drawn out at the end to a fine point, and bent in such manner that a stream of gas may be delivered upwards from this point. Light the gas as it flows out of the tube, and hold over the pale-blue flame a clean, dry bottle. No moisture will be deposited upon the sides of the bottle. That carbonic acid has been produced by the combustion, may be proved by pouring a little lime-water into the bottle, and shaking it about in the gas therein contained.

418. Carbonic oxide is a very powerful deoxidizing agent. At high temperatures it is capable of taking oxygen away from many of the compounds which contain that element. Hence it plays a very important part in metallurgical operations. Much of the reducing-action which is commonly attributed directly to carbon, is really effected in practice through the mediation of carbonic oxide gas.

Exp. 187.—In the middle of a tube of hard glass, No. 3, about 20 c.m. long, place a gramme of black oxide of copper (CuO); support the tube upon a ring of the iron stand over the gas-lamp, and connect it at one end with a flask in which carbonic oxide is being evolved, as in Exp. 184, and at the other with a tube bent at a right angle and dipping into a bottle which contains lime-water. After the tube, which contains the oxide of copper, has become full of carbonic oxide, heat it and observe that the oxide of copper is reduced, that metallic copper alone remains in the tube, and that the carbonic acid formed has made turbid the lime-water in the bottle.

419. The specific heat of carbonic oxide is considerably greater than that of carbonic acid, being 0·245, while that of carbonic acid is only 0·2103.

When a mixture of carbonic oxide and oxygen, in the proportion of two volumes of the former to one of the latter gas, is lighted, it explodes with about the same degree of violence as a mixture of hydrogen and oxygen (§ 58), a very considerable amount of heat being evolved in the act of combination.

Though considerable heat is evolved during the union of carbonic oxide and oxygen, the amount is much less than that which results from the complete combustion of charcoal to carbonic acid. One gramme of carbonic oxide disengages in burning 2403 units of heat (§ 55), while one gramme of wood charcoal, in burning to carbonic acid, yields 8080 units. The same amount of heat (2403 units) is reabsorbed when the carbonic acid, obtained by burning one gramme of carbonic oxide, is again reduced to the state of carbonic oxide. (Compare Exp. 181.)

The gramme of hydrogen yields, as it unites with oxygen,

34,462 units of heat; but since carbonic oxide is 14 times as heavy as hydrogen, about the same quantity of heat is developed by the complete combustion of a given volume of carbonic oxide as by that of the same volume of hydrogen.

420. Carbonic oxide, unlike carbonic acid, is not decomposed when heated to redness in contact with hydrogen, charcoal, iron, or zinc. Sodium and potassium, however, abstract the oxygen from this gas as they do from carbonic acid.

It unites with chlorine directly, under the influence of sunlight, forming a gaseous compound, the composition of which may be represented by the formula $COCl_2$. When left for some time in contact with caustic potash, at the temperature of 100°, it combines with it, and there is produced a compound known as formiate of potassium:—

$$KHO + CO = CHKO_2.$$

It is absorbed readily by solutions of the salts of dinoxide of copper (Cu_2O) in ammonia-water, and by a solution of dichloride of copper (Cu_2Cl_2) in strong chlorhydric acid, and can thus be separated from a mixture with other gases. Melted metallic potassium also absorbs a certain amount of carbonic oxide and combines with it.

421. Carbonic oxide may be resolved into carbon and oxygen by heat alone; but this dissociation occurs only under very peculiar circumstances.

A porcelain tube is placed in a furnace where it can be raised to a very high temperature; the ends of this tube project beyond the furnace and are closed by corks; through these corks passes, in the axis of the porcelain tube, a very thin brass tube, and each cork carries also a small glass tube; by one of these tubes carbonic oxide enters the porcelain tube, and by the other the products of the reaction escape from the apparatus. Two little screens of porcelain divide internally that part of the porcelain tube which lies in the furnace and is to be heated, from the parts which project beyond the furnace and remain cool. A rapid current of cold water is made to flow through the thin brass tube, in such quantity that in traversing the tube while the furnace is in full action the water shall not be sensibly warmed.

The apparatus being thus disposed, and the porcelain tube heated, a slow and regular current of pure and dry carbonic oxide is passed into the hot tube. The gas, as it issues from the tube, passes immediately through a strong solution of caustic potash, which will absorb the carbonic acid formed, so that the experimenter can weigh the quantity of acid produced, or through lime-water, which will demonstrate the presence of carbonic acid by becoming turbid. Carbonic acid is formed whenever the porcelain tube is bright-red hot. A portion of the carbonic oxide is decomposed into oxygen, which unites with another portion of carbonic oxide to make carbonic acid, and carbon, which is deposited in the condition of lampblack upon the cold brass tube which traverses the hot porcelain tube from end to end. The first action of the heat is to set free particles of carbon from the carbonic oxide, and all such particles which happen to fasten upon the brass tube are instantly chilled down below the temperature at which they will either unite with free oxygen on the one hand, or reduce carbonic acid on the other.

We have repeatedly used the electric spark as a means of decomposing compound gases, such, for example, as ammonia (§ 89) and marsh-gas (§ 395). It is supposed that it is the intense heat of the spark which effects such decomposition, and, in the light of the experiment just described, it seems probable that the efficacy of the spark-current is due to the fact that the few particles of gas which each spark heats intensely are immediately in contact with an atmosphere of gas which is in constant motion and is relatively very cold.

422. The composition of carbonic acid being known, that of carbonic oxide can readily be determined by burning a definite volume of this gas with an excess of oxygen in a eudiometer. If there be introduced into the eudiometer

Fig. 60.

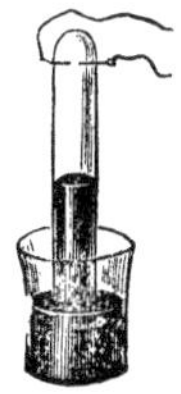

100 volumes of carbonic oxide,
and 100 „ „ oxygen,
and if through the 200 „ „ mixed gases

an electric spark be made to pass, combination will occur, and the gas which remains will occupy only 150 volumes. If a small quantity of a solution of caustic soda be now introduced into the eudiometer, all the carbonic acid which has been formed will be absorbed; there will remain only 50 volumes of gas, which, upon examination, will be found to be pure oxygen.

If only 50 volumes of the original oxygen are thus left free, 50 volumes of oxygen must have been absorbed. It appears, then, that 100 volumes of carbonic oxide have united with 50 volumes of oxygen to form 100 volumes of carbonic acid; and that the original bulk of the carbonic oxide taken has remained unchanged. Since it is admitted, as we have already seen (§ 412), that the product-volume of carbonic acid contains 2 volumes of oxygen, it follows that the double volume of carbonic oxide can only contain 1 volume of oxygen—or, in other words, that 2 volumes of this gas contain 1 volume of carbon-vapor and 1 volume of oxygen united without condensation:—

$$\underset{12}{C} + \boxed{\underset{16}{O}} = \boxed{CO \quad 28}$$

It will be noticed that the addition of a certain quantity of oxygen to a measured quantity of carbonic oxide converts it into carbonic acid without changing the original measured volume of gas. We have often prepared compound gases from elementary gases by this method, and in such cases there is generally a change of volume. We are here, however, converting one compound gas into another compound gas, and the product-volumes of all compound gases are the same.

The specific gravity or unit-volume weight of carbonic oxide has been found, by experiment, to be 13·97.

From the weight of two unit-volumes of carbonic oxide .	27·94
Deduct the weight of one unit-volume of oxygen. . . .	16·
There remains the weight of the atom of carbon	11·94

This result accords very well with the previously given atomic weight of carbon. It will be noticed that the specific gravity of carbonic oxide is the same as that of nitrogen.

423. *Combustion.*—Now that we have become acquainted with carbon, hydrogen, and oxygen, and with some of the more important compounds formed by the union of these elements, the subject of combustion can be more fully discussed than has been possible hitherto. Unlike most of the chemical processes employed by man, which have for their object the preparation of some tangible chemical compound or product, combustion is resorted to merely for the sake of the heat or light which incidentally accompanies the chemical action.

As a general rule, only the compounds of carbon and hydrogen are employed as combustibles—though there are some exceptions to this rule, as when the metal magnesium is burned for light, or the heating of a sulphuretted ore is effected by the combustion of its own sulphur. In the manufacture of sulphuric acid, the sulphur-furnace is often so arranged that the heat from the burning sulphur generates the steam necessary for the operation. In the Bessemer process of making steel from cast iron (§ 633), intense heat is evolved, partly by the combustion of the carbon which cast iron contains, but partly also by the combustion of iron. The carbon compounds are peculiarly well adapted to the purpose, since the products of their combustion are gaseous, and can therefore be readily removed; new portions of the combustible are thus continually laid bare, and a way opened for the admission of fresh air.

424. In almost all cases artificial light results from the intense ignition of particles of solid matter or of dense vapors. When the heat, which is an invariable accompaniment of chemical combination, can play directly upon such solid or semisolid particles with force enough to ignite them, an exhibition of light will accompany the chemical change. The hydrogen-flame affords no light, or as good as none, because in it nothing but a highly attenuated gas is heated. But when a solid body, such as the platinum wire or the piece of lime employed in Exps. 26 and 27, is placed in this non-luminous hydrogen-flame, intense light is radiated from the heated solid.

Exp. 188.—Sprinkle fine iron filings into the flame of an alcohol-lamp, or into the non-luminous flame of the gas-lamp, and observe the light given off by the particles of metal as they become incandescent while passing through the flame. Or rub together two pieces of charcoal above a non-luminous flame, in such manner that charcoal powder shall fall into the flame. Or rub the coat-sleeve beneath a non-luminous flame, or even beneath the luminous flame of an ordinary Argand gas-burner, and observe that the particles of dust detached become incandescent and luminous as they pass upward through the flame. Other things being equal, the hotter the flame the more intense will be the light emitted by the ignited solid.

425. In ordinary luminous flames, such as those of candles, lamps, and illuminating gas, the ignited substance is carbon, or rather a vapor or fog of certain carbon compounds containing more or less hydrogen.

Exp. 189.—By means of caoutchouc tubing, attach to any small gas-burner a piece of hard-glass tubing, No. 4, about 20 c.m. long, the outer end of which has been drawn to a fine open point. Open the cock of the gas-burner, so that gas may flow into and through the glass tube, and light this gas as it escapes. When the last traces of air have been expelled from the tube, heat the middle of the latter intensely with the flame of a lamp. Part of the transparent and colorless carburetted hydrogen, of which the illuminating gas consists (§ 396), will be decomposed by the heat as it passes through the tube, just as sulphuretted hydrogen (Exp. 91), phosphuretted hydrogen, arseniuretted hydrogen (Exp. 133), and antimoniuretted hydrogen (Exp. 136) are decomposed under like conditions, and a ring of carbon will be deposited in the cold portion of the tube a short distance in front of the flame.

In the open channel afforded by the tube it is not easy to heat the whole of the carburetted hydrogen to the temperature necessary for its decomposition; but by lighting the gas as it issues from the tube, heat enough to decompose it can readily be obtained. Precisely as in the combustion of wood (§ 378), after the fire is once started the combustible suffers decomposition; the easily inflammable hydrogen of the gas burns first, and particles of carbon, or, at the first, of hydrocarbons rich in carbon, are set free. These particles are heated to ignition by the burning hydrogen, and as they pass up through the flame emit light; finally they are themselves completely burned to carbonic acid upon the outside of the flame, provided there be present a sufficient supply of air. That there are really particles of free carbon in the flame has already been sufficiently demonstrated in Exp. 154.

If the supply of air furnished to the flame is insufficient to convert all of the components of the gas into carbonic acid and water, then a number of the carbonaceous particles will escape unburned, and a smoky flame will be the result. If, on the other hand, the supply of air is excessive, then all the carbon will be burned at the instant when it is set free, and no light will be afforded. In the gas-lamps commonly employed in chemical laboratories for purposes of heating (see Appendix, § 5), illuminating gas is purposely mixed with a considerable volume of air before it is lighted; there is thus obtained an intensely hot non-luminous flame. Such flames deposit no soot upon the vessels which are heated in them; moreover the heat which would be consumed in heating the particles of carbon, and so producing light, is in such flames utilized for heating-purposes.

Exp. 190.—Unscrew the tube *f* (Fig. 61) from the gas-lamp constructed as described in § 5 of the Appendix, and light the gas as it escapes from the holes in the face of the screw *d*; the flame will be luminous, and, if the holes are large enough to permit a rapid exit of gas, even smoky. Extinguish the burning jet, screw on the tube *f*, and relight the mixture of air and gas at the top; the flame will be nearly colorless. Sometimes, when the gas-cock is too nearly closed, the flame of the mixed gas and air is liable to pass down the tube *f*, and ignite the feeble jet of gas at the apertures in *d*. The lamp then burns with a sickly yellow flame, which is often tinged with green coming from the copper in the heated brass tube *f*. The lamp must be extinguished, and relit at the top of the tube with a freer supply of gas. When the tube *f* is in place, the jets of gas, issuing vertically from the face of the screw *d*, draw in currents of air through the side holes near the bottom of the tube *f*; this air mixes with the gas rising through *f*, and at the top of this tube, where the mixture is inflamed, the carburetted hydrogen is in intimate contact with air enough to burn it at once and completely.

Fig. 61.

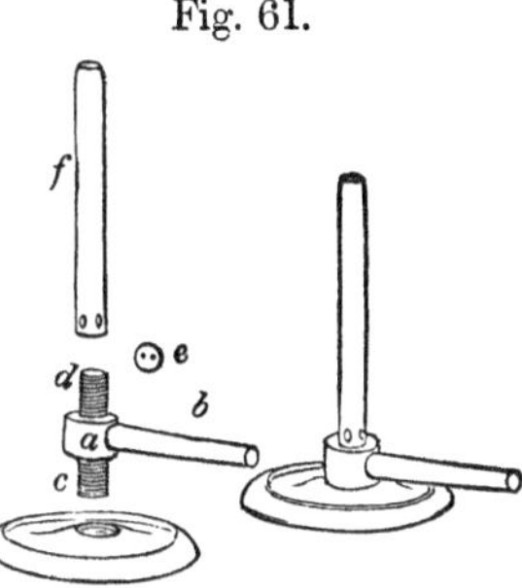

Between the two extremes which a Bunsen burner may be thus made to illustrate, between a smoky flame, on the one hand, and a non-luminous flame, on the other, there are two points which have special significance—the point of *most* light, and the point of most *agreeable* light. The point of most light may always be hit upon by constructing such a burner as will just not allow the gas to smoke.

Exp. 191.—Across the top of the chimney of an Argand gas-burner, which is burning with a shorter flame than usual, place several narrow strips of tin or of sheet iron, so as to obstruct the flow of air through the chimney. The small, low flame with which the experiment began will increase in size as the access of air is diminished, and, at last, the whole interior of the chimney will be filled with a long, smoky flame. The volume of gas burning at any one moment of the experiment is no greater than at another, for the cock which regulates the flow of the gas remains fixed and untouched; but the amount of light afforded by the large smoky flame is manifestly greater than that yielded by the small bright flame with which the experiment started. If any doubt suggests itself as to this point, it will quickly be dissipated by performing the experiment in a darkened room and noting the comparative visibility of the more distant objects therein contained, first with the one flame and then with the other; or the observer may determine at what distances from the two flames fine print can be deciphered.

A murky flame, such as was just now obtained, before actual smoking begins, in which the largest possible number of the particles of carbon are heated, though none of them are heated very hot, yields the largest amount of light which the particular sample of gas under examination is capable of affording. Such flames are called, technically, *quantity flames*; they are better adapted than any others for lighting streets and large halls. In practice, such flames are obtained by burning the gas at a low pressure, that is, under such conditions that it shall be very gently pressed out into the air, so that air shall mix with it and act upon it but slowly.

But besides this point of the maximum amount of light, there is another, of the most agreeable light; and this is something which each individual must determine for himself. Few persons would choose, as a study-lamp, either the murky flame of Exp. 191, or the intense lime-light of Exp. 27; but between these two extremes no one light is likely to suit many people equally well.

If a bright *intensity flame* is required, we have only to arrange matters in such a way that air may come to the gas so quickly and abundantly that a portion of the carbon in the gas, as well as the hydrogen, shall be burned at once in the lower part of the flame, and by the heat of its combustion ignite more intensely the remaining particles of carbon. Among the very great number of gas-burners which have been devised, there may be found those adapted to meet almost any requirement. Each kind of burner brings the gas and the air into contact with one another in some special way, producing a flame of convenient shape, of peculiar economy, or of particular steadiness or brilliancy. It is obvious that the conditions under which gas is most advantageously burned are different for different uses, and that no one burner can be equally available under such varying and, in some sense, antagonistic conditions. The Argand burner may, perhaps, be made to fulfil as many of these conditions as any other; from it there may be obtained, at will, either an intensity or a quantity flame, as has been shown in Exp. 191.

426. The chemical composition of the gas to be burned is, of

course, an important point to be considered in the construction of the burner. A gas rich in carbon requires, for its combustion, far more air than gas which is less carboniferous.

Exp. 192.—Place three small tufts of cotton upon an earthen plate; moisten one with alcohol, another with petroleum, and the third with benzin; touch a lighted match to the vapor which arises from each. In the one case there will be seen hardly any luminous particles of carbon, in the second a bright light; and in the third so much carbon will be set free that, under the conditions of the experiment, a great deal of it cannot find air with which to unite, and consequently escapes as smoke.

The composition of alcohol may be represented by the formula C_2H_6O; it contains a large proportion of hydrogen and some oxygen; hence steam is necessarily produced when it burns; this steam spreads or diffuses the flame, and promotes the prompt union of the alcohol vapor with the oxygen of the air, so that few carbonaceous particles have time to become incandescent before they are consumed. But in benzin, the formula of which may be written C_6H_6, there is no oxygen, and a far larger proportion of carbon than in alcohol; hence the necessity of supplying a large amount of air to the lamps in which its vapor is burned. The best way of consuming benzin is to mix its vapor with air in suitable proportions, and to press this mixture through a gas-burner as if it were ordinary illuminating gas. When thus treated, it burns without smoke, and affords a brilliant white light. Petroleum (C_2H_6), like benzin, contains no oxygen, but it contains far less carbon than benzin, though much more than alcohol; it does not smoke like benzin, and yet it smokes so much that it cannot readily be burned from a simple wick; it is commonly burned in lamps provided with a special draught of air.

427. Ordinary lamps and candles are, strictly speaking, gas-lamps. In all cases their flames are composed of burning gas.

Exp. 193.—Construct a lamp as follows:—To a wide-mouthed bottle of the capacity of about 50 c. c. fit a cork loosely; bore a hole in the cork and place therein a short piece of glass tubing, No. 3, open at both ends; through this glass tube draw a piece of lamp-wicking, or any loose twine, long enough to reach to the bottom of the bottle. It is essential, either that the cork should fit the bottle loosely, or that there should be a hole in the cork, in order that the pressure of the external air may act upon the surface of the alcohol; to this end a very small glass tube may be inserted in the cork at some distance from the tube which carries the wick. Fill the bottle nearly full with alcohol, and, after a few minutes, touch a lighted match to the top of the wick. The fluid alcohol is drawn up out of the bottle by the capillary attraction exercised by the pores of the vegetable fibre of which the wick is composed. When heat is applied to the alcohol at the top of the wick, some of it is converted into vapor; this vapor then takes fire, and, in burning, furnishes heat for the vaporization of new portions of the alcohol. From the top of the wick there is constantly arising a column of gas or vapor, and upon the exterior of this conical column chemical combination is all the while going on between its constituents and the oxygen of the air. The dark central portion of the alcohol-flame is nothing but gas or vapor.

Exp. 194.—Thrust the phosphorus end of an ordinary friction-match directly into the middle of the flame of the alcohol-lamp of Exp. 193. The combustible matter upon the end of the match will not take fire in the atmosphere of carbonaceous gases of which the centre of the flame consists. The wood of the match-stick, of course, takes fire at the point where it is in contact with the outer edge of the flame, for it is there heated in contact with air. In withdrawing the match from the middle of the flame, it is not easy to prevent it from taking fire as it passes through the outer edge of the flame; for the materials on the tip of the match have been so strongly heated by radiation, during their sojourn within the circle of fire, that they are now ready to burst into flame immediately on coming in contact with the air; by a quick jerk, however, the match may often be withdrawn from the flame without taking fire.

Exp. 195.—Hold a thin wire (best of platinum, though iron will answer well enough) or a splinter of wood across the flame of an alcohol-lamp, as shown in Fig. 62. The wire will be heated to redness, and the wood will burn, only at the outer edges of the flame where the gas and air meet; in the interior of the flame the wire will remain dark and the wood unburned; for there is no combustion there, and comparatively little heat. If the wire be successively placed at different heights in the flame, the size and shape of the internal cone of gas can easily be made out; it will appear, moreover, that the hottest part of the flame is just above the top of the interior cone of gas. As a rule, when glass tubing, or the like, is to be heated in a flame, it should never be placed below this point of the greatest heat.

Fig. 62.

428. The flame of an ordinary oil-lamp or of a petroleum-lamp, in the same way as the flame of an alcohol-lamp, is composed of an inner cone of gas, or vapor of hydrocarbons, and an envelope where chemical combination is going on; and a candle-

flame is really the flame of an oil-lamp (Exp. 196). The presence of vapor in the candle-flame can be readily shown (Exps. 197, 198). In the candle-flame, as in that of the alcohol-lamp, there is a cone of unburnt gas surrounded by a shell of burning substances (Exps. 199, 200).

Exp. 196.—Touch a lighted match to the wick of a new candle; the cotton of which the wick is composed takes fire and is at once consumed for the most part; but, in burning, the cotton gives off considerable heat, and some of the wax or tallow of which the candle is composed is thereby melted and converted into oil. The liquid oil ascends the wick by virtue of capillary attraction, and is converted into vapor or gas by the heat of the cotton still burning at the stump of the wick; this gas then burns precisely like the alcohol vapor in Exp. 193 or the illuminating gas in Exp. 189, and by the heat thus disengaged new portions of wax or tallow are continually melted. There is always a little cup of oil at the top of the rod of wax or tallow of which the candle consists, and the apparatus is as truly an oil-lamp as if the oil were held in a vessel of glass or metal.

Exp. 197.—Let a candle, best of tallow, burn until the snuff has become long; blow out the flame, and observe the cloud of vapor which ascends from the hot wick. Touch a lighted match to this column of vapor, and notice that it takes fire at some little distance from the wick. After the flame has been extinguished, the wick retains heat enough for a few moments to distil off a quantity of gas, although there is not heat enough generated to inflame this gas. To the gas or vapor thus evolved is to be referred the disagreeable odor which is observed when a candle is blown out.

Exp. 198.—Draw a glass tube (No. 5 or 6) 10 or 15 c.m. long to a moderately fine open point; with a piece of wire bind this tube in an inclined position to a ring of the iron stand, and place the lower end of the tube in the middle of a candle-flame, just below the centre, so that a portion of the gas of the inner cone of the flame may escape through the tube; light the gas at the point at the top of the glass tube, and observe that it will burn there steadily, if the experiment is performed in a quiet place where there are no draughts of air.

Fig. 63.

Exp. 199.—Press down a piece of white letter-paper, for an instant, upon the flame of a candle until it almost touches the wick, then quickly remove the paper before it takes fire, and observe that its upper surface is charred in the manner shown in Fig. 63. There will be obtained, in fact, burned into the paper, a diagram of the cylindrical column of unburnt gas and of the shell of burning matter which surrounds it. Within the charred ring the paper is unacted upon; for that part of it was in contact only with the unburnt gas in the centre of the flame.

Exp. 200.—Replace the paper of Exp. 199 with a strip of glass, so held that the conical flame of the candle shall be cut across horizontally by the glass as it was by the paper in Exp. 199. Look down from above through the glass into the hollow cylinder of unburnt gas within the circle of combustion.

429. In any flame, which is rendered luminous by incandescent carbonaceous particles, three portions can be distinguished:—1st, the dark interior cone of gas, *a*, Fig. 64; 2nd, the zone of intense chemical action, *b*, where the hydrogen is burning and the carbonaceous particles are heated to whiteness; and, finally, upon the very outside a thin, scarcely perceptible film of burning carbonic oxide, *c*.

Fig. 64.

430. From the study of luminous flames we pass to a consideration of flames employed only as sources of heat. In the experiments (25–27) with the oxyhydrogen blowpipe, it has been already shown that a very intense heat may be obtained by throwing oxygen into the hydrogen-flame, and so localizing the chemical action and the heat with which this action is accompanied. The subject may be here conveniently studied by employing coal-gas and air in place of hydrogen and oxygen.

Exp. 201.—Fill one gas-holder with air, and screw to it a metallic jet, such as is shown in Fig. 65. Fill another gas-holder with ordinary illuminating gas and connect the opening of this gas-holder with

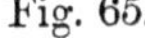

Fig. 65.

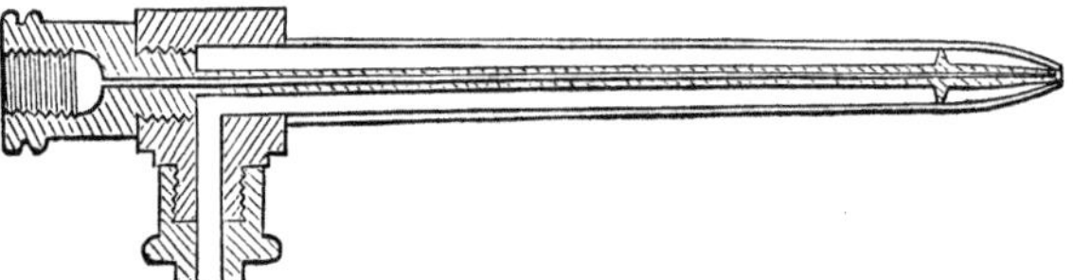

the lower opening of the metallic jet. Open the cock of the holder which contains the coal-gas, and inflame the gas at the point of the metallic jet. There will be thus obtained a long stream of gas burning at the expense of the air which bathes its surface. The chemical action between the oxygen of the air and the constituents of the coal-gas, and the heat resulting from this action, are diffused over the entire surface of this long flame. Without touching the cock of the holder which contains the coal-gas, or in any way altering the amount of gas which flows out of this holder, open the cock of the holder which contains air, so that air may be thrown into the middle of the coal-gas flame. The latter will be immediately shortened down to almost nothing. The constituents of the coal-gas will now all combine with oxygen in a very small space, and the heat of combination, which was diffused before, will be correspondingly concentrated. It is much the same as if the coal-gas and the air had been mixed together beforehand and then lighted. Indeed, in one of the first forms of the oxyhydrogen blow-pipe, a mixture of the two gases, such as we have exploded in Exp. 30, was first prepared, and then forced out of a single gas-holder of peculiar construction, provided with an exceedingly minute orifice, at the mouth of which the mixed gases were burned. This apparatus was inconvenient and dangerous, and has long since been superseded; but it well illustrated the local concentration of heat now under discussion.

431. The principle of the common mouth-blowpipe, of the glass-blower's lamp (Appendix, § 6), and of all blasts and blowers, is identical with that of the oxyhydrogen blowpipe, which, as has been already stated (§ 55), is the simplest of all intense and concentrated combustions. Air, or more strictly speaking, oxygen, is thrown into the combustible gas or fuel, in order that the combustion may go on in a small space.

The mouth-blowpipe may be used with a candle, or with any hand-lamp proper for burning oil, petroleum, or any of the so-called *burning-fluids*, provided that the form of the lamp below the wick-holder is such as to permit the close approach of the object to be heated to the side of the wick. When a lamp is used, a wick about 1·2 c.m. wide and 0·5 c.m. thick is more convenient than a round or narrow wick; a wick of this sort, though hardly so wide, is used in some of the open burning-fluid (naphtha) lamps now in common use. The wick-holder should be filed off on its longer dimension a little obliquely, and the wick cut parallel to the holder, in order that the blowpipe-flame may be directed downwards when necessary (Fig. 67).

The cheapest and best form of mouth-blowpipe for chemical purposes is a tube of tin-plate, about 18 c.m. long, 2 c.m. broad at one end, and tapering to 0·7 c.m. at the other (Fig. 66); the broad end is closed, and a little above this closed end a small cylindrical tube of brass about 5 c.m. long is soldered in at right angles; this brass tube is slightly conical at the end, and carries a small nozzle or tip, which may be made either of brass or platinum. The tip should be drilled out of a solid piece of metal, and should not be fastened upon the brass tube with a screw. A trumpet-shaped mouth-piece of horn or box-wood is a convenient, though by no means essential, addition to this blowpipe.

Fig. 66.

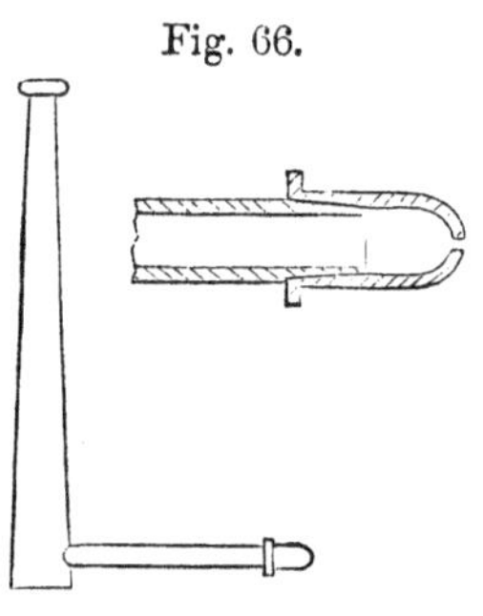

Exp. 202.—To use the mouthblowpipe, place the open end of the tin tube between the lips, or, if the pipe is provided with a mouth-piece, press the trumpet-shaped mouth-piece against the lips; fill the mouth with air till the cheeks are widely distended, and insert the tip in the flame of a lamp or candle; close the communication between the lungs and the mouth, and force a current of air through the tube by squeezing the air in the mouth with the muscles of the cheeks, breathing, in the meantime, regularly and quietly through the nostrils. The knack of blowing a steady stream for several minutes at a time is readily acquired by a little practice. It will be at once observed that the appearance of the flame varies considerably, according to the strength of the blast and the position of the jet with reference to the wick.

When the jet of the blowpipe is inserted into the middle of a candle-flame, or is placed in the lamp-flame in the position shown in Figure 67, and a strong blast is forced through the tube, a blue cone of flame, *a b*, is produced, beyond and outside of which stretches a more or less colored outer cone towards *c*. The point of greatest heat in this flame is at the point of the inner blue cone, because the combustible gases are there supplied with just the quantity of oxygen necessary to consume them; but between this point and the extremity of the flame the combustion is concentrated and intense. The greater part of the flame thus produced is *oxidizing* in its effect; and this flame is technically called the *oxidizing flame*. From the

Fig. 67.

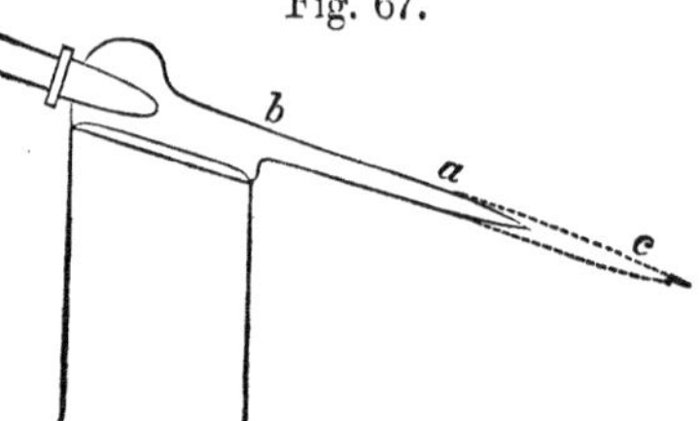

point *a* of the inner blue cone, the heat of the flame diminishes in both directions—towards *b*, on the one hand, and towards *c*, on the other; most substances require the temperature to be found between *a* and *c*. Oxidation takes place most rapidly at, or just beyond, the point *c* of the flame, provided that the temperature at this point is high enough for the special substance to be heated.

A flame of precisely the opposite chemical effect may be produced with the blowpipe. To obtain a good *reducing* flame, it is necessary to place the tip of the blowpipe, not within, but just outside of the flame, and to blow rather over than through the middle of the flame (Fig. 68). In this manner, the flame is less altered in its general character than in the former case, the chief part consisting of a large, luminous cone, containing carbonaceous vapors in a state of intense ignition and just in the condition for taking up oxygen. This flame is therefore *reducing* in its effect, and is technically called the *reducing* flame. The substance which is to be reduced by exposure to this flame should be completely covered up by the luminous cone, so that contact with the air may be entirely avoided. It is to be observed that, whereas to produce an effective oxidizing flame a strong blast of air is desirable, to get a good reducing flame the operator should blow gently, with only enough force to divert the lamp-flame.

Fig. 68.

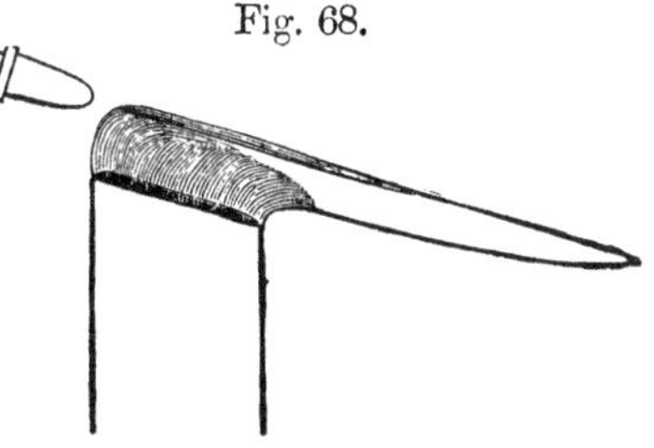

Substances to be heated in the blowpipe-flame, are supported, sometimes on charcoal, sometimes on platinum foil or wire or in platinum spoons or forceps, and sometimes on little capsules made of clay or bone-earth. Charcoal is especially suited for a support in experiments of reduction.

Exp. 203.—The heat of the oxidizing flame may be well shown by melting the extremity of a very fine platinum wire into a little ball. To effect this fusion, bend the wire at right angles at 5 to 6 m.m. from the end, and hold this bent end precisely in the axis of the flame, with the angle outward and the extremity of the wire at the hottest part of the flame. If the wire is kept steadily in position, and the blast is strong and the flame pure, a little knob will soon appear on the end of the wire. The bend in the wire is made in order to keep a certain length of the wire hot, and so to diminish the conduction of heat from the point.

Exp. 204.—Place a kernel of metallic tin, as large as this o, in a little hollow scooped out at one end of a bit of charcoal 8 to 12 c.m. long. Melt this tin in the reducing flame of the blowpipe, and endeavour to preserve the metallic lustre of the fused metal untarnished. A pure reducing flame is necessary for this purpose. A touch of the oxidizing flame upon the metal covers its surface with a white, infusible, incandescent ash.

432. Another method of supplying the burning fuel with air is by means of chimneys. Chimneys, whether of lamps or furnaces, are simply devices for bringing air in abundance, and therefore oxygen, into the fire; that in so doing they, at the same time, carry off the waste products of combustion, is an incidental advantage.

Exp. 205.—Light a piece of a candle 8 or 10 c.m. long and stand it upon a smooth table; over the candle place a rather tall, narrow lamp-chimney of glass, the bottom of the chimney being made to rest upon the table, and observe that the candle-flame will soon be extinguished. No fresh air can enter the chimney from below to maintain the chemical action, and the small quantity of air which can creep down the chimney from above is altogether insufficient to meet the requirements of the case.

Exp. 206.—Relight the candle of Exp. 205, and again place over it the lamp-chimney; but instead of allowing the chimney to rest closely upon the surface of the table, prop it up on two narrow strips of wood, so that air can have free entrance into the chimney from below. The candle will now continue to burn freely; for the heavy cold air outside will continually press into the lower part of the chimney, and push out the warm, light products of combustion, and the candle-flame will all the while be supplied with fresh air.

Fig. 69.

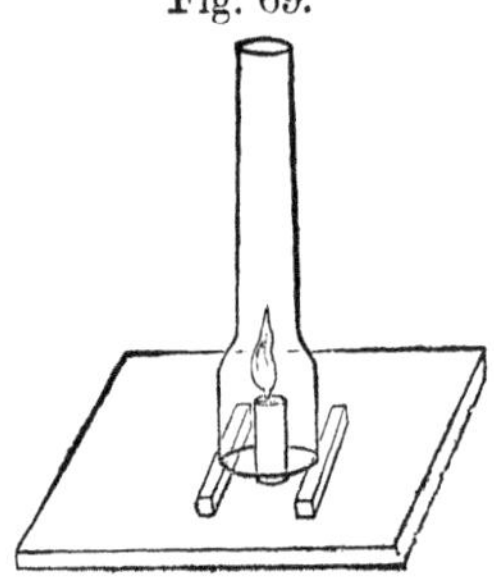

Exp. 207.—Prepare several strips of "nitre-paper" by soaking ordinary brown paper in a strong solution of nitrate of potassium and drying the product. On being lighted, paper thus prepared will burn without flame while emitting clouds of smoke. Light a piece of the nitre-paper and place it at the foot of the chimney arranged as in Fig. 69. The smoke of the burning paper will instantly pass up through the chimney, and so indicate the direction of the invisible air which is all the while entering below and passing out above at the top of the chimney, as fast as it is heated and made lighter by the burning of the candle.

Exp. 208.—Repeat Exp. 206, and when the candle is burning quietly, cover the top of the chimney tightly with a piece of tin or sheet-iron, or with a strip of window glass; the candle will soon cease to burn, precisely as if the chimney were closed at the bottom, for, the escape of the hot products of combustion being prevented, no air can pass into the chimney to reach the candle-flame.

It is by inducing the current of fresh air (Exp. 206), or draught, as it is ordinarily termed, that chimneys are specially useful. They give direction and precision both to the incoming cold air and the outgoing hot gas. Where there is no chimney, the hot air from a lamp goes off at a comparatively slow rate, and vaguely; through the chimney, on the other hand, it flows straight forward and rapidly, and, of course, a correspondingly direct and rapid current of fresh air presses in to supply its place. Owing to this power of rapidly supplying air, chimneys are employed upon lamps burning petroleum and other highly carbonized oils which are liable to smoke; in general they are made use of upon lamps in all cases where intensity flames are required.

Exp. 209.—It is not absolutely necessary that the fresh air should flow into a chimney from below. Divide the upper part of the chimney of Exp. 205, into two channels, by hanging in it a strip of sheet-iron or tin, as a partition at the centre of the chimney. (See Fig. 70.) Place the chimney thus divided over a burning candle, and observe that the candle will continue to burn as if in a strong draught of air, although no air can enter the chimney from below. Hold a piece of burning nitre-paper (Exp. 207) at the top of the divided chimney; the smoke will be drawn down into the chimney on one side of the partition and thrown out again upon the other, as indicated by the arrows in Fig. 70. It appears from this, as well as from the tremulous motion of the flame, that a current of cold air passes down upon one side of the division wall and supplies the required oxygen.

Fig. 70.

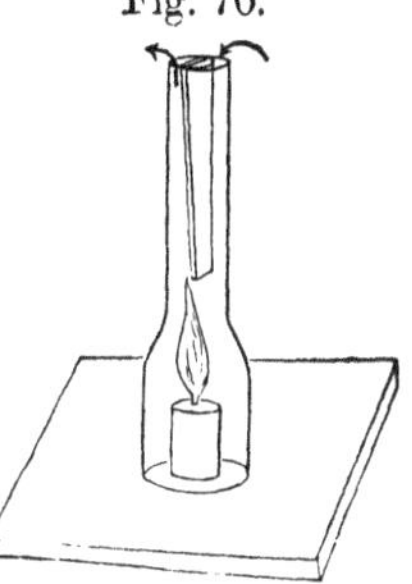

433. One exceedingly important point in chemical philosophy which we have hitherto taken for granted, may now be readily illustrated. As the result of long-continued experimentation and the most rigid scrutiny of all chemical facts, it is admitted, as a

fundamental truth, that matter is indestructible. When substances undergo chemical change, none of their ingredients are really destroyed, not an atom of them is annihilated; nor, upon the other hand, is any new matter created; it is the form only of the old substances which is changed; their weight remains in every case unaltered.

By bringing about chemical combination between two or more bodies, we can entirely change their appearance, their condition, and their properties, but in every case it will be found that the weight of the resulting compounds is precisely equal to the sum of the weights of their components. Thus, when a candle burns in the air and gradually disappears, none of the elements which compose it are either lost or destroyed. Though by uniting with the oxygen of the air, the components of the candle have been converted into compounds which are invisible, it is nevertheless easy to satisfy ourselves of the existence of these compounds. Already, in Exps. 167, 168, it has been demonstrated that carbonic acid and water are products of the combustion of the candle; and we now proceed to show that these products weigh more than the candle.

Exp. 210.—Take a glass tube 2 or 3 c.m. in diameter and 25 or 30 c.m. long, such, for example, as the neck of a broken retort; fit a cork to each extremity of the tube, and at a distance of 6 or 8 c.m. from its upper, wider end, fix across the tube a partition of wire gauze. Through the upper cork insert a glass tube, No. 3 or 4, bent at a right angle, and in the lower cork bore several open holes, besides a central orifice into which a small wax taper is fastened. Fill the space between the upper cork and the shelf of wire gauze with recently slaked lime, in not too fine powder; replace the cork, hang the tube, with its contents, at the end of one arm of a balance, and counterpoise the apparatus precisely by placing shot or sand in the pan upon the other side of the scales. Next make an "aspirating flask," by fitting to the upper orifice of a bottle with a stopcock, such as is depicted in the upper part of Fig. xvii. in the Appendix, a sound cork, carrying a glass tube, No. 3 or 4; the tube should reach to the bottom of the bottle and be bent at a right angle a short distance above the cork.

By means of flexible tubing, connect the upper tube of the aspirating flask with the tube issuing from the upper cork of the apparatus upon the balance, and open the stopcock of the aspirator to such an extent that water shall flow out from it slowly. Remove the lower cork from the apparatus upon the balance, in order to light the taper, then quickly replace it and regulate the flow of water from the aspirator, so that sufficient air shall be drawn in through the holes in the cork which supports the taper to maintain the latter in lively combustion.

After the taper has burned during 4 or 5 minutes, close the aspirator, remove the flexible tube from the candle apparatus, so that the latter may again hang freely from the arm of the balance, and observe that the weight of the apparatus is now greater than it was at the beginning of the experiment, for the lime within the tube has absorbed the carbonic acid and water which were produced by the combination of the ingredients of the candle with the oxygen of the air.

434. In order that any combustible substance shall burn, or, in other words, in order that brisk chemical action shall occur between the combustible and the oxygen of the air, it must first be heated to a certain temperature, and then kept at that heat. The temperature at which any substance takes fire is known as the kindling-temperature of that substance.

Exp. 211.—Place a small bit of phosphorus and another of sulphur, not in contact with the first, upon a fragment of porcelain 6 or 8 c.m. across, and heat them slowly over the gas-lamp. The phosphorus will soon take fire at a temperature of 68°-70°; but the sulphur will not inflame until the temperature of the porcelain support has risen to about 250°, as can be readily ascertained by the thermometer.

As was just now said, the degree of heat necessary to start any fire must be kept up continually, or the fire will go out. Whenever burning bodies are cooled below the kindling-temperature they are extinguished,—the chemical action which occasioned the appearance of heat and light ceases.

Exp. 212.—Pile up upon an iron grate, thick in metal, and supported in such manner that air may enter beneath it, several pieces of red-hot charcoal. The charcoal will go on burning until nearly all of it has been consumed; for the heat generated by the combustion of the portions first burned keeps up the temperature necessary to kindle the subsequent portions.

Upon a cold grate, similar to the one just employed, scatter about several small pieces of red-hot charcoal, taking care that no two pieces of the coal shall come in contact, or be placed so as to heat one another. Each of the pieces of charcoal will soon cease to burn; for the metallic grate is so good a conductor of heat that it removes heat from the isolated pieces of charcoal more rapidly than these can produce it; consequently the temperature of the charcoal is soon reduced to below the kindling-point.

A result similar to the foregoing is obtained when any fire is broken up and scattered about to such an extent that its several portions cannot assist in one another's combustion—though if, instead of being placed upon iron, the separate glowing coals be laid upon ashes or dry earth, they will be extinguished only after a much longer time, for ashes and dry earth are very poor conductors of heat in comparison with iron.

In all ordinary fires the heat evolved by the combustion of the fuel is more than sufficient to maintain a temperature higher than the kindling-point of the fuel—though, generally speaking, the fuel becomes at last so clogged with ashes, that the oxygen of the air cannot get at the remaining combustible matter in sufficient quantity to maintain lively chemical action.

435. Precisely as coals can be extinguished by placing them upon cold metal, so flames may be put out.

Exp. 213.—Upon a ring of the iron stand place a sheet of clean wire gauze about 10 c.m. square; lower the ring so that the gauze shall be pressed down upon the flame of a lamp or candle almost to the wick, as shown in Fig. 71. No flame will be seen above the gauze, but instead of flame a cloud of smoke. The gauze is a mere open sieve; there is nothing about it which can prevent the gas, which was just now burning with flame above the wick of the candle, from passing through. Indeed it may be seen from the smoke that the particles of carbon which, in the original undisturbed flame, were becoming incandescent, and so affording light, do now actually come through the gauze.

Fig. 71.

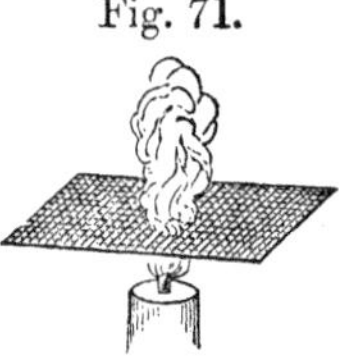

The explanation of the phenomenon is simply that the metallic sieve conducts away so much heat that the temperature of the candle-flame is reduced to below the kindling-point. That this is really so is proved by the fact that, after the gauze has become sufficiently heated by long-continued contact with the flame below—after it has attained the kindling-point of the candle-gas, it will no longer extinguish the flame. In like manner, a candle-flame may be cooled to such an extent that it will go out, by placing over it a small coil of cold copper wire, while if the wire be previously heated the flame will continue to burn.

If the smoke and unburned gas which has passed through the cold wire gauze be touched with a lighted match, and so brought to the kindling temperature, it will burst into flame.

The power of wire gauze to prevent the passage of flame has been usefully applied in several ways, notably for the prevention of explosions in those coal-mines which are liable to accumulations of marsh-gas. For this purpose safety-lamps are constructed by enclosing an ordinary oil-lamp completely in wire gauze, so that the flame within the gauze cannot kindle any combustible or explosive gas into which it may be carried. In case such a lamp be carried into a place filled with explosive gas, the latter will, of course, pass into the lamp through the meshes of the gauze and burn within the cage. This combustion gives warning of the presence of the dangerous gas, and indicates to the workman that he should withdraw from the locality; the gas can then be expelled by appropriate methods of ventilation.

The wire-gauze lamps employed in chemical laboratories (for one form of which see Appendix, Fig. viii.) are simply applications of the same general idea.

Exp. 214.—Beneath the sheet of wire gauze of Exp. 213, place an unlighted ordinary gas-lamp (Bunsen's burner), at such distance that the gauze shall be 3 or 4 c.m. above the top of the lamp; turn on the gas and light it above the wire gauze; it will continue to burn on the top of the gauze for an indefinite period, for the gauze will, in this case, always be kept cool by the cold gas which is continually passing through it. Carefully and gradually lift the ring which carries the gauze, and determine how far it is possible to lift the gauze above the gas-jet without extinguishing the flame.

The student will remember that other experiments illustrating the influence of cooling agencies in extinguishing combustion (Exps. 136 and 154), have already been performed. Compare also Exp. 196 and § 200, as regards kindling.

436. An effect somewhat similar to that produced by wire gauze is often seen in ordinary fires. When a mass of red-hot anthracite, charcoal, or coke is burning freely upon a grate in the open air, there is always a blue flame of carbonic oxide burning above the coal. This gas results from the reduction of carbonic acid by means of hot carbon, precisely as in Exp. 181. Air enters at the bottom of the grate and combines with the hot coal which it finds there to form carbonic acid (CO_2). This carbonic acid, as it rises through the hot coal in the middle of the fire, is deprived by the heated carbon of half its oxygen,

$$CO_2 + C = 2CO,$$

so that two molecules of carbonic oxide gas finally emerge at the top of the coal, instead of the single molecule of carbonic acid which was formed at first. The carbonic oxide being combustible, will at once take fire on coming in contact with the air, provided the temperature at the summit of the fire be equal to the kindling-temperature of carbonic oxide. But if the temperature of the fire is in any way reduced below this point, as, for example, by throwing on too large a quantity of cold fuel, which is, of course, equivalent to covering the fire with a sheet of wire gauze, then the carbonic oxide will be extinguished, and, escaping into the chimney, will produce no useful effect.

Were it not for the formation of carbonic oxide, as above mentioned, neither anthracite coal, nor coke, nor charcoal would burn with flame after having once got well on fire; they would simply glow, as a single live coal, or a bar of metal, glows when taken from the fire.

437. In heating steam-boilers and other large vessels, it is often a point of great importance to obtain from the fuel a large flame, in order that the heat from the fuel may be quickly distributed and brought into contact with the matter to be heated. With anthracite and coke this result is effected by placing beneath the grate, upon which the fuel is burned, a quantity of water. From this water steam gradually rises, as hot ashes and cinders fall into it and as heat radiates down upon it from the fire above. The steam, as it enters the fire, is decomposed by the hot coal (see Exp. 156), in accordance with the following reaction,

$$C + H_2O = CO + 2H,$$

and the combustible gases thus obtained are superadded to the carbonic oxide which is formed in due course from the action of air upon the coal. All these gases burn again to carbonic acid and water above the fire, where air is thrown in to meet them through appropriate orifices. In this use of water as an adjunct to the combustion of coal, the absolute amount of heat given off by the fuel is in nowise increased; but in many instances much heat may undoubtedly be saved by thus equally distributing and applying it by means of flame.

438. On the other hand, furnaces are sometimes seen con-

suming fuel under such conditions that all the carbonic oxide produced within them escapes unburned into the chimney. In such cases, more than two-thirds of the amount of heat which the fuel is capable of yielding must necessarily be lost; for while 1 gramme of charcoal gives off in burning to carbonic acid 8080 units of heat (§ 55), 1 gramme of carbon in burning to carbonic oxide gives off only 2473 units of heat. The number last given is determined as follows. It has been found, by direct experiment, that 1 gramme of carbonic oxide, on being burned to carbonic acid, yields 2403 units of heat; carbonic oxide is composed (§ 422) of one atom of carbon, weighing 12, and one atom of oxygen weighing 16,—the weight of the molecule of carbonic oxide being consequently 28. In one gramme of carbonic oxide, therefore, there can be only $\frac{12}{28} = 0{\cdot}4286$ of a gramme of carbon; but $0{\cdot}4286 : 1 = 2403 : x = 5607$, whence it appears that there is evolved by one gramme of carbon in carbonic oxide 5607 units of heat when this carbon unites with the additional oxygen to form carbonic acid; and the difference between this number (5607) and the number (8080) denoting the amount of heat given off by one gramme of charcoal in burning to carbonic acid will show how much heat is evolved by one gramme of carbon burned to carbonic oxide: $8080 - 5607 = 2473$, as above stated.

In order to thoroughly burn the carbonic oxide in any case, the stove or furnace should be so arranged that a volume of air, as large as that which has already passed through the fire, can be constantly supplied to the carbonic oxide and nitrogen as they emerge from the coal, and be intimately mixed with these gases while they are still hot.

439. The amount of air needed for the complete combustion of coal or other fuel can always be readily calculated. We have only to determine how much oxygen will be needed by the combustible, and then how much air must be taken in order to supply this oxygen. Let it be supposed, for example, that we wish to learn how much air is needed in order to burn one kilogramme of charcoal. Having learned the full significance of the formula CO_2, a moment's consideration of this formula informs us that, for every 12 parts by weight of carbon, 32 parts by weight of

oxygen are needed in order to its complete combustion, or, for one part of carbon, 2·67 parts of oxygen. Air contains 23·1 per cent. of oxygen by weight; hence the proportion $23{\cdot}1 : 100 = 2{\cdot}67 : x = 11{\cdot}558$, from which it appears that to burn 1 kilo. of charcoal, 11·558 kilos. of air are needed. Since the weight of a litre of air, at the ordinary temperature, is only 1·2258 grm., these 11·558 kilos. of air will occupy about 9429 litres, or, in other words, nearly 9½ cubic metres. In round numbers, it may be said that about 12 kilos., or 9¾ cubic metres, of air are required to burn one kilo. of charcoal. For coke and anthracite, corrections must, of course, be made for the ashes which they contain, as well as for a certain portion of hydrogen which may also be present. If a gramme of pure carbon will disengage 8080 units of heat, a gramme of well-burned coke, containing 15 per cent. of ashes, will disengage only 6868 units.

440. *Chloride of Carbon* (CCl_4).—Chlorine does not unite directly with carbon; but several compounds of the two elements can be obtained by subjecting compounds of carbon and hydrogen to the action of chlorine. Of these compounds, only the so-called bichloride (CCl_4) need here be mentioned, the others being usually treated of in works upon organic chemistry. Bichloride of carbon may be obtained by the action of chlorine on marsh-gas, by subjecting chloroform or wood-spirit to the action of an excess of chlorine in sunlight, by passing a mixture of bisulphide-of-carbon vapor and chlorine through a red-hot porcelain tube, or by the action of quinquichloride of antimony upon bisulphide of carbon:—

$$CS_2 + 2SbCl_5 = CCl_4 + 2SbCl_3 + 2S.$$

At the ordinary temperature of the air it is a transparent, colorless liquid, of pungent aromatic odor, boiling at 77°, and having a specific gravity of 1·56. At —23° it solidifies in the form of crystals of pearly lustre. The specific gravity of its vapor has been determined to be 76·96, which would indicate that a molecule of the vapor is composed of 1 atom of carbon and 4 volumes of chlorine, condensed to 2 unit-volumes:—

For, since the weight of one atom of carbon is	12
And the weight of 4 atoms of chlorine is	142
The weight of the two volumes of gas produced would be . .	154

And the weight of 1 volume would be 77, a number with which the experimental determination very nearly agrees.

In composition, this chloride of carbon (CCl_4) is analogous to the hydride CH_4 which we have already studied under the name light carburetted hydrogen, or marsh-gas (§ 392), and in the same way that the hydride may be converted into the chloride by acting upon it with chlorine, so, conversely, the chloride, on being brought into contact with water and sodium-amalgam (§ 97), may be deprived of chlorine and converted back into the hydride,—the hydrogen of the water being substituted for the chlorine, which, like the oxygen of the water, unites with the sodium.

441. *Compounds of Carbon and Nitrogen.*—With nitrogen, carbon forms a number of highly interesting compounds, which may be found treated of in works upon organic chemistry. Prominent among these compounds are cyanogen, CN (§ 384), and cyanhydric acid, HCN. Cyanhydric acid corresponds to chlorhydric acid and the other hydrides of the chlorine group of elements; by its action upon metallic oxides there may be formed a series of cyanides of the metals, corresponding perfectly with the metallic chlorides:—

$$M_2O + 2H(CN) = 2M(CN) + H_2O.$$
$$M_2O + 2HCl = 2MCl + H_2O.$$

The group of atoms (CN) which constitutes cyanogen, acts, in fact, as if it were a single element. In the same way that the group NH_4, called ammonium, is capable of replacing a metal like sodium (see § 91), so the group CN can replace chlorine and the elements allied to chlorine. Groups or knots of atoms, such as these, are often called *compound radicals.* Cyanhydric acid, or prussic acid as it is sometimes called, is notorious as a violent poison. Many of the cyanides are of great importance in the arts, particularly in processes of gilding, plating, electrotyping, and dyeing, as well as in the preparation of pigments. Some of them will be described hereafter under the respective metals.

442. *Bisulphide of Carbon* (CS_2), or sulpho-carbonic acid, as it is often called, is especially interesting from its correspondence with the binoxide of carbon, carbonic acid, in accordance with the

general rule that sulphur compounds are analogous to the compounds of oxygen. With the metallic sulphides this compound forms a series of salts, of the general formula M_2CS_3, or M_2S,CS_2, analogous to those formed by the union of carbonic acid with the metallic oxides, the general formula of which, as we know, is M_2CO_3 or M_2O,CO_2.

Bisulphide of carbon may be obtained by bringing the vapor of sulphur in contact with red-hot charcoal. It is a mobile, colorless liquid of 1·27 specific gravity, which refracts light powerfully. It boils at 45° and evaporates rapidly at the ordinary temperature of the air. The density of its vapor is 38·19. It has a peculiarly fetid odor, and is very easily inflammable, burning with a blue flame to carbonic and sulphurous acids. It is not soluble in water, but is easily soluble in alcohol, and is itself a powerful solvent of fats and various other substances; it has been, of late years, somewhat extensively employed as a solvent of phosphorus (§ 274) and of chloride of sulphur (§ 246) in the cold process of vulcanizing caoutchouc. Mixtures of its vapor with oxygen, air, nitric oxide, and other gaseous oxygen compounds burn, without violent explosion, with a sudden brilliant flash of intensely blue light. This flame is remarkable for its great actinic power; it acts very energetically upon a prepared daguerreotype plate, and causes a mixture of hydrogen and chlorine to combine almost as readily as sunlight would (Exp. 53).

Exp. 215.—Fill a tall bottle, of the capacity of 700 or 800 c. c., with nitric oxide gas, at the water-pan; cover the bottle with a plate of glass and stand it upright upon the table; draw the cover aside far enough to admit of the introduction, from a pipette, of 8 or 10 c. c. of liquid bisulphide of carbon; replace the cover and leave the bottle at rest for a few minutes, in order that the vapor of the bisulphide may have time to diffuse through the nitric oxide. Finally, touch a lighted match to the opened mouth of the bottle and observe the brilliant flame which is produced.

CHAPTER XXI.

BORON.

443. Boron is an element of the same natural family as carbon. It is found in nature, in combination with oxygen, as boracic acid, and in combination with oxygen and some of the metals—notably as a biborate of sodium, commonly called borax, and as a double borate of sodium and of calcium.

In certain volcanic districts in Tuscany, jets of steam mixed with other vapors escape continually from cracks in the soil, and bring to the surface small quantities of boracic acid. Since boracic acid is not volatile, in the ordinary sense of the term, at temperatures so low as 100°, it appears that it is transported mechanically by the steam, much in the same way that dust is carried along by a current of air. The jets of vapor, laden with boracic acid, are made to bubble through water as they escape from the earth; this water retains the acid, and so concentrates it to a very considerable extent. After the water has become as highly charged with the acid as has been found in practice to be desirable, it is run off into pans, lower down upon the hill-side, beneath which hot jets of vapor from the earth are caused to circulate. The excess of water is thus evaporated by heat which the earth supplies, and the solution becomes so concentrated that, on cooling, crystals of boracic acid separate from it. About 120 millions of kilogrammes of water are thus evaporated annually, and 1,300,000 kilogrammes of boracic acid produced without the intervention of any artificial motor or the consumption of any fuel. After having been purified, the boracic acid is sent into commerce as such, or it is treated with a hot solution of carbonate of sodium, and so converted into borax.

Besides this principal source of the boron compounds, a certain quantity of native borax is obtained from the mud and waters of certain lakes in Tartary, Ceylon, Thibet, and California. In Peru, also, a mineral composed of borate of sodium and borate of calcium is found associated with nitrate of sodium.

444. The element boron has been obtained in two distinct allotropic conditions (§§ 162, 366) comparable with those of carbon. It can be had amorphous like charcoal, and crystallized like the diamond.

To prepare the diamond-like modification, fused boracic acid is intensely heated with metallic aluminum; a portion of the aluminum takes oxygen away from the boron, while another portion of the molten metal dissolves this boron as fast as it is formed. As soon as the solution has become saturated (that is to say, when the melted aluminum has dissolved all the boron that it is capable of dissolving), a portion of this boron is deposited in diamond-like crystals. When the crucible and its contents are broken up after cooling, these crystals are found lining cavities within the mass.

The amorphous modification may be prepared by heating together boracic acid and metallic sodium or potassium beneath a layer of fused chloride of sodium. After having been allowed to cool, the products of the reaction are treated with water, which dissolves away the chloride and the borate of sodium, and leaves the boron as an amorphous powder which may be collected upon a filter.

A third variety of boron, similar to graphite, was at one time described; but recent experiments have shown that the so-called graphitoidal boron is really a boride of aluminum (AlB_2).

445. Of the properties of these varieties of boron, little need here be said; they are analogous to, and closely resemble, the corresponding modifications of carbon, which have already been fully described. The diamond modification is transparent, and sometimes colorless, though usually of a yellow or reddish color. It crystallizes in the same forms as the real diamond, and refracts light very powerfully. Its specific gravity is 2·68. It is almost as hard as the real diamond, being capable of scratching the ruby, and even of polishing the diamond. It is only with extreme difficulty that it can be burned in oxygen, since a coating of boracic acid soon forms which protects it from further action of the oxygen. It is remarkable that at the moment of its combustion it swells up as the diamond does when intensely heated. Amorphous boron is an infusible greenish powder, which readily takes fire on being heated in air or oxygen. It is, in fact, necessary, in preparing it, to take care that the filters upon which it has been collected shall be dried at low temperatures, lest the finely divided

boron take fire spontaneously; if exposed to the sun's rays in summer, the filters containing boron will often take fire as soon as they have become dry. Unlike the other modification of boron, it is readily attacked by most chemical agents. Like the corresponding modification of carbon, it is an energetic reducing agent.

No compound of boron and hydrogen has yet been discovered, but it unites readily with chlorine, bromine, and iodine; it can be made to combine also with fluorine, nitrogen, and sulphur.

446. The best-known of the compounds of boron is the oxide B_2O_3, called boracic acid. This oxide occurs ready formed in nature, as has been said, and is the sole product when boron is burned in oxygen. The composition of boracic acid may be determined synthetically by burning amorphous boron in pure oxygen, or by treating it with nitric acid. A certain definite weight of boron being taken in the first instance, the weight of the dry boracic acid obtained is carefully determined, and from these data the percentage composition of the boracic acid is calculated; 100 grammes of boracic acid consist of

Oxygen	68·71 grammes.
Boron	31·29 „

From the specific gravities of the vapors of chloride of boron and of fluoride of boron, as determined by experiment, and from some other rather inconclusive considerations which need not here be dwelt upon, chemists have been led to admit that the atomic composition of boracic acid may be represented by the formula B_2O_3. If boracic acid be really composed of 3 atoms of oxygen and 2 atoms of boron, the weight of the atom of boron will follow from the proportion $68{\cdot}71 : 31{\cdot}29 = (16 \times 3) : x$, in which 16 equals the weight of an atom of oxygen and x the weight of two atoms of boron. Upon this assumption, the weight of one atom of boron will be 10·93.

447. Boracic acid is but a feeble acid at the ordinary temperature, and may be set free from its compounds by almost any of the acids, excepting carbonic acid.

Exp. 216.—Dissolve 4 grms. of borax in 10 grms. of boiling water, in a beaker glass or porcelain capsule of 30 or 40 c. c. capacity, and add to the solution 2·5 grms. of concentrated chlorhydric acid. After the lapse of some time, hydrated boracic acid will be deposited from the solution in the form of glistening, colorless plates or scales. These crystals contain as much as 43·6 per cent. of water; their formula is H_3BO_3, or, dualistic, $3H_2O, B_2O_3$.

On being heated in a clean iron spoon, the crystals will first dissolve in the water which they contain, or, as the fact is usually stated, they will "melt in their water of crystallization;" if the heat be continued, the mass will become pasty, and will swell up as the water is expelled. After all the water has been driven off by strong heat, the anhydrous acid is left as a clear, viscous liquid, from which long threads of the solid acid may be drawn out by touching the surface of the liquid with the end of a stick or glass rod, and then gently pulling away the stick with the matter which has adhered to it.

If the fused acid be allowed to cool, it will solidify to a hard, transparent glass, which soon cracks in every direction and splits up into fragments.

Anhydrous boracic acid is of about 1·8 specific gravity; it is odorless and destitute of corrosive power; it has a slightly bitter, but not sour, taste. It is much more soluble in hot than in cold water, and more soluble in alcohol than in water. It imparts to the flame of burning alcohol a peculiar green tint, which is quite characteristic, and affords a valuable test by which the presence of the acid may be detected. Upon litmus and turmeric, boracic acid acts somewhat differently from other acids.

Exp. 217.—Dissolve a little of the crystallized boracic acid of Exp. 216, in a teaspoonful of alcohol in a small porcelain capsule. Set fire to the alcohol and stir the burning solution with a rod, or agitate it by jarring the dish. Or moisten a tuft of cotton with alcohol, strew upon it some powdered boracic acid, and light the alchohol. In either case the flame of the alcohol will be of a fine green color.

Exp. 218.—Pour into a test-glass 20 or 30 c. c. of a solution of blue litmus; in a small quantity of water, contained in another test-glass or tube, dissolve a little of the boracic acid of Exp. 216; add the solution of boracic acid to the litmus, and observe that the color of the latter changes to a brownish wine-red, decidedly different from the bright clear red which is obtained by the action of other acids upon litmus. If a large quantity of boracic acid, however, be added to a small portion of the litmus solution, the latter will be colored strongly, as if by a powerful acid.

Exp. 219.—Dip into a solution of boracic acid a slip of yellow turmeric paper, and observe that the yellow color is changed to brown, as it would be by ammonia-water or by any other alkaline solution. None of the other acids produce a like effect.

448. When an aqueous solution of boracic acid is boiled, an appreciable quantity of the acid goes off with the vapor of water; but the dry acid, when heated by itself, is nevertheless one of the least volatile of all the acids. It does slowly sublime, however, at a white heat, and may be completely evaporated, if left for a long time in the hottest part of a porcelain furnace. As a consequence of this fixity, or lack of volatility, it follows that boracic acid is a comparatively powerful acid at temperatures high enough to volatilize the ordinary acids. On being heated with nitrates, or sulphates, for example, it quickly expels nitric or sulphuric acid, and unites with the other ingredients of the salt, though either of these acids would at once decompose the borate thus formed, if they were collected and added to it at the ordinary temperature. Even phosphoric acid is expelled by it from the phosphates.

449. *Chloride of Boron* (BCl_3) is a colorless, mobile liquid, of 1·35 specific gravity, which boils at 17°. The specific gravity of its vapor has been found to be 58·78, a result which points directly to the formula BCl_3 as representing the true composition of the compound.

From the weight of 2 vols. of chloride of boron (58·78×2) .	117·56
Subtract the weight of 3 vols. of chlorine (35·5×3) . . .	106·50
and the remainder will be	11·06

a number almost precisely equal to the weight of one atom of boron as previously determined (§ 446).

Upon being mixed with water, chloride of boron decomposes, with formation of boracic and chlorhydric acids:—

$$2BCl_3 + 3H_2O = B_2O_3 + 6HCl.$$

Chloride of boron may be prepared by slightly heating amorphous boron in an atmosphere of chlorine, or more readily by passing a current of chlorine over a mixture of anhydrous boracic acid and charcoal, heated to redness in a porcelain tube. In presence of the hot charcoal which stands ready to take oxygen from the boracic acid, chlorine can take boron away from boracic acid; and, conversely, in presence of the chlorine, ready to combine with the boron, carbon can take away oxygen:—

$$B_2O_3 + 3C + 6Cl = 2BCl_3 + 3CO.$$

The method here described, of converting an oxide into a chloride through the intervention of carbon, is a method of very general applicability, and is often employed for the preparation of chlorides of the metals.

450. *Fluoride of Boron* (BFl_3) is a colorless gas of 34·19 specific gravity, as determined by experiment. Upon the assumption that an atom of boron weighs 11, its specific gravity would be $(11+3\times19)\div2=34$. It fumes strongly in damp air, and, by pressure, may be readily condensed to a colorless and very mobile liquid. The gas is exceedingly caustic and corrosive; it carbonizes and destroys wood and other organic substances, in the same way as concentrated sulphuric acid. As with sulphuric acid (Exp. 104), so here, the fluoride of boron unites with the elements of water which are contained in the organic matter, and the integrity of the latter is destroyed.

451. Fluoride of boron is absorbed by water rapidly and in large quantity, 1 volume of water being capable of dissolving 700 or 800 volumes of the gas; but in the act of solution decomposition occurs as well, and there is obtained, not a simple solution of fluoride of boron in water, but a mixture, or rather a compound, of fluorhydric and boracic acids:—

$$2BFl_3 + 3H_2O = B_2O_3,6HFl.$$

The reaction is interesting in all its stages, inasmuch as it well illustrates the vagueness and indefiniteness of a considerable class of chemical reactions. When water dissolves fluoride of boron, it increases in bulk to a considerable extent, and in density also, its specific gravity rising as high as 1·77. Upon warming the saturated solution, some fluoride of boron is again disengaged, perhaps as much as one-fifth of all that had been absorbed; but on continuing to heat the solution, it distils over unchanged, and the condensed liquid presents the appearance of oil of vitriol. In it the elements of boracic and fluorhydric acids are undoubtedly held together in a loose condition of chemical combination. By many chemists the compound is called fluoboric acid, though, in order to avoid confusion, it would, perhaps, be better if the name fluorhydrate of boracic acid were allotted to it; for when this compound is largely diluted with water, boracic acid is deposited, and another acid compound is left in solution, the composition of which may be represented by the formula HFl,BFl_3. This new acid is called indifferently fluoboric acid or fluorborhydric acid, and the salts formed by its union with metals are called fluoborates. It is remarkable that the first-named compound, the fluorhydrate of boracic acid, upon being neutralized with alkalies, yields, not mixtures of a borate and a fluoride of the alkali employed, but true chemical compounds, double salts, of the general formula M_2O,B_2O_3; $6MFl$, whence the name fluoboric acid has arisen. The best way of preparing the fluorhydrate of boracic acid is to dissolve boracic acid by small portions in fluorhydric acid.

Fluoride of boron may be itself prepared by heating in a glass flask a mixture of 1 part of fused boracic acid, 2 parts of powdered fluorspar, and 10 or 12 parts of concentrated sulphuric acid:—

$$3CaFl_2 + B_2O_3 + 3H_2SO_4 = 3CaSO_4 + 3H_2O + 2BFl_3.$$

On account of its easy solubility in water, the gas must be collected over mercury. Fluoride of boron may be employed as a test to determine whether a given sample of any gas is completely dry; if a few bubbles of it are added to the gas to be tested, the slightest trace of moisture in the gas will be made manifest by the appearance of white fumes of the fluorhydrate of boracic acid above described.

452. *Sulphide of Boron* (B_2S_3) is a white, crystalline solid, decomposable by water, in accordance with the following formula:—

$$B_2S_3 + 3H_2O = B_2O_3 + 3H_2S.$$

It may be prepared by passing a current of sulphide-of-carbon vapor over a mixture of boracic acid and charcoal strongly heated in a porcelain tube.

453. *Nitride of Boron* (BN) is a soft, white, amorphous solid, tasteless, odorless, infusible, and non-volatile. It is, in general, but little acted upon by chemical agents.

454. It will be remarked that while boron is closely analogous to carbon in many respects, it differs from it decidedly in others. Thus, while in their allotropic modifications the two elements are

almost precisely alike in appearance and properties, and while many of the salts of boracic acid are strikingly similar to the corresponding carbonates, the compounds of boron are not comparable as regards their composition with the compounds of carbon. While one atom of carbon unites by preference with four atoms of any member of the chlorine group (as in CCl_4), or with two atoms of any member of the sulphur group (as in CO_2), an atom of boron unites with only three atoms of chlorine (BCl_3), or two atoms of it unite with three atoms of oxygen or sulphur (as in B_2O_3). The exceptional character of the composition of boron compounds will appear still more clearly in the next chapter, where it will be shown that silicon, the third member of the carbon group, resembles carbon as regards the atomic composition of its compounds, as well as in other respects. The atomic weight of boron above given (§ 446) cannot be accepted as established beyond a doubt, and it may happen that further investigation will show that the boron compounds are really formed upon the same type or pattern as those of carbon and silicon; but in face of the experimental evidence now at hand, this view cannot be maintained. In the meantime the student will better understand the physical and chemical properties of boron and its compounds, if this element is studied in company with carbon and silicon, which it so closely resembles, than if it were described in connexion with arsenic, antimony, and the other elements of the nitrogen group, which form teroxides and terchlorides indeed, but which present not the least other analogy to boron, either in the simple or the compounded condition.

CHAPTER XXII.

SILICON.

455. Like carbon, silicon may be obtained in three distinct allotropic conditions, which have been designated as amorphous, diamond-like, and graphitoidal. After oxygen, it is the most abundant and widely diffused of all the chemical elements; at

least one quarter of the solid crust of the earth is composed of it. In combination with oxygen, it occurs in silicic acid, which is one of the commonest substances upon the surface of the globe. Common quartz and flint, as well as rock-crystal, agate, and the like, are pure silicic acid. The yellow sand of sea-beaches and of many sterile tracts of country is silicic acid contaminated with a trace of oxide of iron. In like manner, sandstones and a great variety of other rocks are mainly composed of it.

In order to obtain pure silicon, the following methods may be employed:—When metallic potassium is heated with a double compound of fluoride of potassium and fluoride of silicon, known as fluosilicate of potassium, a violent reaction occurs, fluoride of potassium is formed and silicon set free in the amorphous state,

$$2KFl,SiFl_4 + 4K = 6KFl + Si;$$

by washing with water, the silicon may then be readily freed from the fluoride of potassium, which is soluble. The graphitoidal modification of silicon can be prepared either by heating the amorphous variety very strongly, in which event the powder contracts upon itself and becomes much more dense, or by melting together in a crucible a mixture of metallic aluminum, in excess, and fluosilicate of potassium; a fusible double fluoride of aluminum and potassium is formed, and the silicon thus set free dissolves in the melted aluminum. If the aluminum be dissolved away, after the mass has become cold, by means of chlorhydric acid, the silicon will be left in the form of hexagonal scales. Still a third method must be resorted to, in order to obtain the diamond-like modification of silicon; a mixture of dry fluosilicate of potassium, metallic zinc, and metallic sodium is thrown into a hot crucible, the mass is covered with a layer of fluosilicate of potassium, and the crucible covered. A lively reaction ensues, and the mixture within the crucible fuses; the fused mass is then stirred with an iron rod until the zinc begins to escape as vapor. The crucible is then removed from the fire and allowed to cool; within it there will be found a button of metallic zinc filled with long crystals of silicon, which can readily be isolated by dissolving the zinc in chlorhydric acid.

456. The amorphous variety of silicon is a brown powder, which, when touched, soils the fingers. It may be melted at a temperature not far from that at which cast iron becomes liquid. When mixed with common salt and exposed to a degree of heat strong enough to volatilize the salt, amorphous silicon changes to the graphitoidal variety, as has been already remarked. Amor-

phous silicon burns readily in the air and in oxygen. It is not acted upon by acids, excepting fluorhydric acid, which dissolves it with disengagement of hydrogen.

Graphitoidal silicon is very similar to true graphite; it crystallizes in lustrous hexagonal plates of a leaden-gray color, and is an excellent conductor of electricity. It is, however, much harder than graphite. By exposing it to an exceedingly high temperature, it can be transformed to the diamond-like condition. The diamond-like modification of silicon occurs in the form of regular octahedral crystals, exhibiting a decided metallic lustre. The specific gravity of these crystals is 2·49. They are less hard than the corresponding crystals of carbon and boron, and melt at the same temperature at which cast iron melts.

Either graphitoidal or diamond-like silicon may be heated to redness in oxygen gas without burning to any appreciable extent, for a film of silicic acid is formed which protects the remainder of the silicon and prevents it from being consumed; but if the silicon be heated together with a substance capable of furnishing oxygen in presence of a base competent to unite with silicic acid and form a fusible silicate, it can readily be oxidized and converted into a silicate. Thus, when heated to intense redness with carbonate of sodium, it decomposes the latter with evolution of light and heat, and there is produced silicate of sodium, while carbon is set free:—

$$Na_2O,CO_2 + Si = Na_2O,SiO_2 + C.$$

In a similar way, if silicon be heated in very concentrated solutions of caustic potash or soda, water will be decomposed and a silicate of potassium or of sodium formed, while hydrogen is set free:—

$$2NaHO + H_2O + Si = Na_2SiO_3 + 4H.$$

Diamond-like silicon is not attacked at the ordinary temperature by any of the acids, excepting a mixture of fluorhydric and nitric acids, by which it is converted into fluoride of silicon. Hot chlorhydric acid gas, as well as chlorine, attacks it readily.

457. *Silicon and Hydrogen* (SiH_4).—A gaseous compound of these elements, called siliciuretted hydrogen, may be obtained, mixed with free hydrogen, by acting upon silicide of magnesium (Mg_2Si) with chlorhydric acid; other methods of preparing this

gas give a purer product. It is a colorless gas, which takes fire spontaneously on coming in contact with the air, and burns to silicic acid and water. On being heated in a tube, out of contact with the air, it is decomposed into free hydrogen and free silicon, and the latter is deposited upon the walls of the tube as a shining mirror, similar to the mirrors of arsenic and antimony obtained in Exps. 133, 136.

Silicon and Oxygen.—Two compounds of these elements have been discovered, though but one of them is as yet well known.

458. *Oxide of Silicon* is described as a white, amorphous, hydrated substance, so light that it floats upon water; it may be obtained by treating with water at 0° the compound of silicon, hydrogen, and chlorine described in § 472; its composition is probably that represented by the formula $Si_2H_2O_3$. It is scarcely at all acted upon by cold water; but in presence of alkalies it decomposes water rapidly, with evolution of hydrogen and formation of the higher oxide of silicon, silicic acid, directly to be described. At temperatures above 300° it suffers decomposition, breaking up into silicic acid and siliciuretted hydrogen. It burns brilliantly in the air, and still better in oxygen. None of the acids, excepting fluorhydric acid, have any action upon it.

459. *Silicic Acid* (SiO_2) constitutes at least one half of the crust of the earth. Besides being thus abundant, it is one of the most important acids known, regarded merely as a chemical agent. It is found everywhere, sometimes free, as quartz and sand, sometimes in combination with metallic oxides, in the form of salts known as silicates. It occurs more or less abundantly in all soils, plants, and waters, while rocks like granite and earths like clay are largely composed of it. The common minerals feldspar and mica are double silicates of aluminum and potassium, and of aluminum, iron, and potassium respectively. In plants, the silicic acid (or silica, as it is often called) is contained particularly in the outer covering of the stalks and the husks of grain. The cuticle of rattan, for example, contains a large proportion of silica, and the same remark is true of most of the grasses and grains. The value of the plant called horse-tail (*Equisetum*) as a polishing or scouring agent depends upon the large quantity of silica contained in it.

460. Silicic acid occurs in at least two distinct isomeric modifications,—in one of which it is completely insoluble in water and in acids, excepting fluorhydric acid, and is only slowly soluble in boiling potash-lye; while, when in the other modification, it dissolves easily in a solution of potash, and may be retained in solution in considerable quantities both by water and by acids. Rock crystal, as it is found in nature, may be regarded as the type of the first or insoluble modification, and the soluble variety may be prepared artificially by treating the solution of some one of the soluble silicates with chlorhydric or sulphuric acid.

Exp. 220.—In a block of charcoal 12 or 15 c.m. long by 4 or 5 c.m. wide and thick, scoop out a small cup-shaped cavity large enough to hold a pea; place in this cavity a fragment of amorphous quartz or flint, and heat it intensely by means of the blowpipe (Exp. 202) during several minutes. When the quartz has become red-hot, suddenly throw it from the coal into a dish of cold water; it will break up into numerous small fragments or become filled with a multitude of cracks, and can hence be readily pulverized. Grind the broken quartz to fine powder in a wedgwood, or, better, in an iron or agate mortar; weigh out 1 grm. of the powder, also 2 grms. of caustic soda, and to the mixture of these ingredients add 8 or 10 c. c. of water. Boil the mixture in a porcelain dish for an hour or two, taking care to add, from time to time, water enough to supply that lost by evaporation; then pour the solution into a tall, narrow bottle, and leave it at rest until the undissolved portions of silica have settled and the liquid has become clear. The highly alkaline solution thus obtained may be regarded either as an aqueous solution of basic silicate of sodium, or as a solution of normal silicate of sodium in the soda-lye.

Exp. 221.—To one-third part of the solution obtained in Exp. 220 add 10 or 12 times its bulk of water, and to this solution add dilute chlorhydric acid, drop by drop, until the liquor manifests a decided acid reaction (Exp. 33). The liquid will remain clear, and no silicic acid will be deposited. All the silicic acid which has been set free from its combination with the alkali remains dissolved in the acidulated water.

If the clear solution be placed in a dialyzer (§ 327), all the chloride of sodium, together with the free chlorhydric acid, will pass off through the parchment-paper in the course of a few days, and there will be left in the dialyzer nothing but pure silicic acid and water. Aqueous solutions of silicic acid, containing from 5 to 14 per cent. by weight of the acid, may be thus obtained. This pure aqueous solution of silicic acid exhibits a decided acid reaction. It has little or no taste, though when applied to the tongue it occasions an unpleasant sensation. It is by no means so permanent as the solution, above mentioned, acidulated with chlorhydric acid. When left to itself, the aqueous solution deposits silicic acid after a while as an insoluble gelatinous precipitate. In like manner, it has been observed that when chlorhydric or another acid is added to the aqueous solution of basic silicate of sodium in no greater quantity than is sufficient to exactly neutralize the alkali, the liquor, though it remain clear for a considerable space of time, will gradually become cloudy and deposit silicic acid.

Exp. 222.—To one-third part of the solution obtained in Exp. 220 add at once enough concentrated chlorhydric acid to render the solution acid. All the silicic acid will immediately be thrown down as a thick insoluble jelly.

Exp. 223.—Place in a small porcelain dish a portion of the solution of silicic acid in acidulated water obtained in Exp. 221, evaporate it to dryness upon a water-bath, and heat the residue, over the gas-lamp, to a temperature of 180° or 190°. Add water to the cold dry residue, and observe that the silicic acid does not redissolve; it remains as a fine white powder, which may be readily collected upon a filter and there washed clean by pouring upon it several successive portions of water. After having been once thoroughly dried, silicic acid is completely insoluble in water.

461. By adding chlorhydric acid to a sufficient quantity of the water of almost any spring or river, and then evaporating the water to dryness, a small quantity of silicic acid will be found in the residuum. The silica, in this case, may be combined with an alkali, or may be held in solution by the action of an alkaline carbonate. Silicic acid which has been finely powdered, or, better, that which has recently been precipitated, is soluble, to a considerable extent, in aqueous solutions of the alkaline carbonates, or even of the bicarbonates, only a small proportion of the carbonic acid being expelled by it as it dissolves. It is not improbable, therefore, that the silicic acid found in waters may be retained in solution by force of this solvent action of the alkaline carbonates. It is, at all events, a matter of fact that waters highly charged with carbonate of sodium, such as flow from the geysers or boiling springs of Iceland, contain, in solution, very large quantities of silica, much of which is deposited as the

water becomes cold, upon the rocks and other objects with which it comes in contact; siliceous petrifactions are thus formed.

462. Besides the soluble and insoluble modifications above mentioned, a distinction is made between the variety of silicic acid which is found crystallized in nature and that which occurs in the amorphous state. Crystallized silicic acid is anhydrous, has a specific gravity of from 2·6 to 2·66, and is scarcely at all acted upon by alkaline lyes. The amorphous silicic acid found in nature contains a certain amount of water, has a specific gravity of only 2·1 to 2·2, and is soluble in alkaline solutions.

As it exists in the insoluble modification, either as found in nature or as prepared by calcining the artificial product, silicic acid is a white, tasteless solid, incapable of forming a cohesive plastic mass with water. It is infusible, excepting at very high temperatures, such as may be obtained by means of the oxyhydrogen blowpipe, in the flame of which it melts to a colorless glass. When melted, it is tough and viscous, and, like glass, can be drawn out into fine threads which are exceedingly elastic, particularly if they be dipped, while white-hot, into water. Silicic acid is not volatile by itself; but when exposed to a current of gas or vapor at a white heat, portions of it are carried off by the vapor; like boracic acid, it can be transported in considerable quantities by superheated steam.

When in the soluble modification, whether in the gelatinous hydrated condition or in the state of an air-dried powder which has never been exposed to heat, it dissolves readily in solutions of the caustic and carbonated alkalies, particularly if these be heated. Some varieties of native silica, such, for example, as the soft, pulverulent, infusorial earth found at the bottoms of many ponds and swamps, are as readily soluble in the alkalies as that which has been prepared artificially. All the varieties of silica can be dissolved by heating them with alkaline lyes in close boilers under a pressure of 4 or 5 atmospheres; in other words, even crystallized silica is soluble in the alkalies at high temperatures.

463. At the ordinary temperature silicic acid is but a weak acid; almost any of the other acids are capable of decomposing the aqueous solution of a salt of silicic acid and of expelling the

latter from its combination with the metal. But, as is the case with boracic acid also (§ 448), at high temperatures the reverse of all this is true, silicic acid being then capable of expelling all acids which are more volatile than itself. Neither variety of silicic acid is acted upon at any temperature by either carbon, hydrogen, phosphorus, or chlorine, if these elements be taken separately; but when exposed at high temperatures to the combined action of carbon and chlorine, chloride of silicon and carbonic oxide are produced, much in the same way that chloride of boron is formed under similar circumstances, as has been shown in § 449. When silicic acid is heated together with carbon, or other reducing agents, in contact with metals like iron or platinum, some of it is decomposed, and a silicide of the metal is formed. Platinum crucibles are often injured in this way by the nascent silicon which forms when certain minerals are fused in them.

464. Silicic acid combines with many of the metallic oxides to form salts, many of which are of very complex composition. Hundreds of silicates are found in nature as crystallized minerals, and they may, in general, be formed by fusing together silicic acid and the appropriate metallic oxide, or carbonate, in suitable proportions. Most of the silicates are fusible, and the greater number of them, when in the molten state, have the power of dissolving, either an excess of base, or an excess of silicic acid over and above the quantities corresponding to strict atomic proportions; hence it happens that mixtures of silicic acid and of bases may be melted together in the most varied proportions. The study of the silicates thus becomes difficult, since it is often impossible to determine whether a given silicate be really a definite chemical compound or a chemical compound contaminated with an excess of silicic acid or of the base; in very many instances, moreover, it is probably true that the excess of silica, above the normal atomic proportion, or of the base, is really held in combination by virtue of the chemical force, though it be held feebly and indefinitely, as is the case with the ingredients of many solutions and metallic alloys. (See §§ 49, 76.)

As a general rule, the simple silicates containing but a single base, and no excess of either base or acid, crystallize in passing from the liquid to the solid condition; but the double silicates (compounds formed by the union of two or more simple silicates, and therefore containing at least two different metals) usually solidify to a non-crystalline, homogeneous glass. All the common varieties of glass consist essentially of such double salts of silicic acid. In general, those silicates of the alkali-metals and of lead which contain an excess of base are more readily fusible than the normal or acid salts; those silicates which contain easily fusible oxides melt at correspondingly low temperatures; and many mixtures of two or more different silicates fuse at temperatures lower than the melting-point of either of the simple silicates of which the mixture is composed.

The silicates of potassium and of sodium, containing 1 or 2 or more molecules of base to one molecule of the acid, are readily soluble in cold water, and compounds containing as many as 4 molecules of the acid to one molecule of the alkaline base may be completely dissolved in water when boiled therewith for a considerable time. These acid compounds (such, for example, as the sodium-salt of the formula Na_2O, $2–4SiO_2$) are much employed in the arts under the name of *Waterglass* or *Soluble glass*. When, however, the proportion of acid is more than $4\frac{1}{2}$ molecules to 1 molecule of the alkaline base, the silicate may, for all practical purposes, be regarded as insoluble in water. The silicates of the metals not included in the alkali group (§ 544) are, in like manner, insoluble in water, in the ordinary sense of the word, though, in the last analysis, it would be hard to find any silicious minerals, or glasses, which are not susceptible of being decomposed and dissolved, to a greater or less extent, by water. Some metallic silicates may be prepared, not only by the method of fusion, but also by adding the solution of an alkaline silicate to the solution of a salt of the metal whose silicate is desired.

465. The most commonly occurring silicates may be referred to three or four general classes, similar to the classes of carbonates (§ 414):—1st. Normal silicates, of the general formula M_2O,SiO_2, such as the crystallizable silicate of sodium Na_2O,SiO_2, or the normal silicate of calcium, CaO,SiO_2. 2d. Bisilicates, of the formula $M_2O,2SiO_2$, such as the bisilicate of calcium, $CaO,2SiO_2$. 3d. Disilicates (basic silicates) of the general formula $2M_2O,SiO_2$,

such, for example, as the silicates of iron, $2FeO,SiO_2$, of magnesium, $2MgO,SiO_2$, of manganese, $2MnO,SiO_2$, and of zinc, $2ZnO,SiO_2$. Also, sesquisilicates, of the formula $2M_2O,3SiO_2$, to which the mineral meerschaum may, perhaps, be referred, its formula being $2MgO,3SiO_2 + 2H_2O$.

466. Many of the natural silicates may be decomposed by digestion with the strong mineral acids, such as concentrated chlorhydric acid; and this is especially true of those silicates which contain a large proportion of base, and of those containing water of crystallization. But many anhydrous normal or acid silicates are not decomposed by any acid, excepting fluorhydric acid. Fluorhydric acid attacks and dissolves, not only pure silicic acid in all its varieties, but all silicates as well, and, other things being equal, the rapidity of its action is in proportion to the degree of its concentration. Crystallized silica, however, is much less readily acted upon by fluorhydric acid than the amorphous variety; it dissolves slowly and without development of heat, while amorphous silica not only dissolves rapidly, but the act of solution is accompanied with the evolution of considerable heat. From those silicates which are decomposed with difficulty by chlorhydric acid the silicic acid separates out as a soft powder; but from those which are more readily decomposed the silicic acid separates as a hydrated gelatinous mass when the finely powdered mineral is digested with chlorhydric acid; in this case the mineral is said to gelatinize with acids. Some silicious minerals may even be completely dissolved, silica and all, in dilute chlorhydric acid.

467. It is remarkable that, while some of the hydrated silicates (such, for example, as the class of minerals called zeolites) can no longer be readily decomposed by chlorhydric acid after they have been ignited, there are other silicates (such as the minerals garnet, epidote, and idocrase) which, though scarcely at all acted upon by chlorhydric acid when in the natural condition, may be decomposed thereby with gelatinization after they have been melted. Facts like these would seem, at first sight, to indicate that silicic acid exists in combination, sometimes in the insoluble, and at other times in the soluble modification, and that it may be transformed from the one state to the other without being first set free;

but some of them admit of being explained in another way. When any silicate, no matter whether it be decomposable by chlorhydric acid or unacted upon by this agent, is mixed with an excess of one of the fixed alkalies, or alkaline earths, or with a carbonate or nitrate of either of the alkaline or alkaline-earthy metals, and the mixture then heated to intense redness, there will be obtained a melted or agglutinated mass, which decomposes readily on being treated with chlorhydric acid, and yields gelatinous silica. At the high temperature to which the mixture is exposed, the strong alkaline or alkaline-earthy bases combine with the silicic acid, and there are formed silicates of potassium, sodium, calcium, or the like, easily decomposable by acids. Now, in the case of the minerals garnet, epidote, and idocrase, above mentioned (which are attacked by chlorhydric acid after they have been melted), it may be conceived that the strong bases (such as lime, magnesia, and protoxide of iron) which are contained in these minerals have acted upon the silicate of alumina (also contained in them) in much the same way that an extraneous alkali would act if it were fused with the powdered mineral.

468. Sulphuric acid, when concentrated or but slightly diluted, acts much more energetically than chlorhydric acid upon the silicates, apparently because of its comparatively high boiling-point. Common clay, for example (silicate of aluminum), which is but little acted upon by chlorhydric acid, may be completely decomposed when digested with boiling concentrated sulphuric acid, silicic acid being set free and sulphate of aluminum produced. This decomposition, like that of the garnet above mentioned, is in any event very much more readily effected if the clay be first gently roasted or calcined. The mineral feldspar also, when in fine powder, may be decomposed by boiling oil of vitriol.

469. Contrary to the view formerly held by many chemists, the atomic weight of silicon is now thought to be 28, instead of 21, as formerly. This change makes the formula of silicic acid SiO_2, in strict analogy to carbonic acid, instead of SiO_3 which was for a long time the accepted formula, or Si_2O_3.

The percentage composition of silicic acid has been determined by direct experiment (by oxidizing a weighed quantity of silicon to silicic acid) to be 51·92 per cent. of oxygen and 48·08 per cent. of silicon. But since it is difficult in this way to oxidize the whole of the silicon, inasmuch as portions of it become covered with silicic acid, and are so protected from the further action of oxygen, the above result can be regarded only as a rough approximation to the truth. These numbers would correspond to 29·63 as the atomic weight of silicon. For

$$51{\cdot}92 : 48{\cdot}08 :: (32 = 2 \text{ atoms of O}) : (x = 1 \text{ atom of Si} = 29{\cdot}63).$$

More accurate experiments, however, upon another plan, have since shown that the real atomic weight of silicon is 28, or very nearly 28, as will be fully set forth in the next section.

470. *Chloride of Silicon* ($SiCl_4$) is formed when silicon is heated in an atmosphere of dry chlorine, or more conveniently by heating together finely divided silicic acid, charcoal, and chlorine.

The experiment may be conducted as follows:—Mix together intimately equal weights of lampblack and dry pulverulent silicic acid, such as is obtained by decomposing an alkaline silicate with chlorhydric acid and evaporating the residue to dryness, as in Exp. 223. Knead into the mixture enough oil to render it plastic, and from the thick paste thus formed make a number of small balls or pellets; roll these pellets in powdered charcoal, place them in a Hessian crucible, cover the crucible, and heat in a strong fire until all the oil has been distilled off or decomposed. The mixture of silicic acid and charcoal is thus left in an open porous condition, well adapted for the action of the chlorine. Place the dry pellets in a porcelain tube about 2 c.m. in diameter, connect one end of the porcelain tube with a U-tube surrounded with a freezing-mixture of ice and salt, and to the other end attach a flask in which to generate chlorine, taking care to interpose between this flask and the porcelain tube one U-tube filled with pumicestone soaked in oil of vitriol, and another filled with chloride of calcium, in order that the gas may be thoroughly dried. Place the porcelain tube in a suitable stove or furnace; pass into it a slow current of chlorine, in order to expel the air, and then build around it a hot fire of charcoal or coke; finally, when the tube has become heated to intense redness, pass in the chlorine as fast as it is absorbed. The chloride of silicon will be condensed in the U-tube as a liquid which has usually a yellow color, owing to the presence of dissolved chlorine. In order to purify it, the liquid may be shaken in a dry flask with a quantity of quicksilver in which a little potassium has been dissolved;

after having been decanted from the mercury and subjected to redistillation in dry vessels, it will be found to be pure.

It will be noticed that the method here described, of preparing chloride of silicon, is similar to that recommended for preparing chloride of boron (§ 449). Chlorine alone is not capable of decomposing either silicic or boracic acids; but in presence of carbon combination of chlorine and silicon is effected, at the same time that carbon and oxygen unite to form carbonic oxide :—

$$SiO_2 + 2C + 4Cl = SiCl_4 + 2CO.$$

471. Chloride of silicon is a transparent, colorless, very mobile, volatile liquid, of pungent, acid, and irritating odor. It fumes strongly in the air, boils at 59°, and is of 1·52 specific gravity. The specific gravity of its vapor has been found to be 85·74, which accords well with the supposition that a molecule of chloride of silicon contains 4 volumes of chlorine and 1 volume of silicon-vapor condensed to 2 volumes :—

For if to the weight of 4 volumes of chlorine (4 × 35·5) = . .	142
There be added the weight of 1 volume of silicon-vapor = . .	28
The two volumes of gas produced will weigh	170

The weight of one volume of the gas should, consequently, be equal to 170 ÷ 2 = 85.

Chloride of silicon is at once decomposed by water with formation of chlorhydric acid and deposition of gelatinous silicic acid,

$$SiCl_4 + 2H_2O = SiO_2 + 4HCl,$$

no gas of any kind being set free, and no other products being formed. The fact is important, since it shows at once that the composition of chloride of silicon must correspond with that of silicic acid—that in it four atoms of chlorine simply replace the two atoms of oxygen contained in the silicic acid. For knowing, as we do, the composition of chlorhydric acid (§ 98) and of water (§ 36), we are sure that, for every atom of chlorine eliminated from the chloride of silicon, one atom of hydrogen is required, in order that chlorhydric acid may be formed, and that for every two atoms of hydrogen thus taken from the water, one atom of oxygen will be detached. But as no oxygen escapes from the solution, all of that derived from the water must evidently have combined with the silicon to form silicic acid. Now, it is a comparatively easy matter to determine the amount of chlorine in any solution by adding to the solution nitrate of silver, and then collecting and weighing the insoluble chloride of silver which is formed. Hence, if a weighed quantity of chloride of silicon be decomposed with water, and the amount of chlorine contained in the solution be determined, the difference between the weight of chlorine found and the weight of chloride of silicon taken will give us the weight of the silicon with which the chlorine was combined. By careful experiments, it has been proved in this way that chloride of silicon contains in 100 parts by weight 16·52 parts of silicon and 83·48 parts of chlorine. But 83·48 parts of chlorine are equivalent to 18·81 parts of oxygen, for

$$\underbrace{(35{\cdot}5 \times 2 = 71)}_{\textit{Weight of two atoms of Cl.}} : \underbrace{16}_{\textit{Weight of one atom of O.}} :: 83{\cdot}48 : (x = 18{\cdot}81).$$

Hence 18·81 parts of oxygen must have combined with every 16·52 parts of silicon contained in the solution; or, reduced to per cent., every 46·75 parts of silicon must have combined with 53·25 parts of oxygen; and if this proportion of oxygen lead to the formula SiO_2, the proportion of chlorine will, in like manner, require the formula $SiCl_4$. From the composition of chloride of silicon, as thus determined, the atomic weight of silicon may be accurately derived (compare § 469):—

$$\underbrace{\underset{\textit{Per cent. of Oxygen}}{53{\cdot}25} : \underset{\textit{Per cent. of Silicon}}{46{\cdot}75}}_{\textit{in Silicic Acid.}} = \underset{\textit{Weight of two atoms of Oxygen.}}{32} : \underset{\textit{Weight of one atom of Silicon.}}{(x = 28{\cdot}1)}$$

Or, more directly, by the proportion

$$\underbrace{\underset{\textit{Per cent. of Chlorine}}{83{\cdot}48} : \underset{\textit{Per cent. of Silicon}}{16{\cdot}52}}_{\textit{in Chloride of Silicon.}} = \underset{\textit{Weight of four atoms of Chlorine.}}{142} : \underset{\textit{Weight of one atom of Silicon.}}{(x = 28{\cdot}1)}$$

At the ordinary temperature, liquid chloride of silicon does not act upon potassium; but if this metal be heated in its vapor, chloride of potassium is formed and silicon set free; the reaction affords, in fact, a good method of obtaining silicon in the amorphous state. There is a bromide of silicon ($SiBr_4$) analogous to the chloride in composition and properties; it may be prepared in a similar way. No compound of iodine and silicon has yet been discovered.

472. *Compound of Silicon, Hydrogen, and Chlorine.*—By passing a current of dry chlorhydric acid gas over crystallized silicon, heated nearly to redness in a glass tube, and condensing the product in a receiver immersed in a freezing-mixture, there is obtained a colorless, fuming liquid, which boils at 34°; the density of its vapor has been determined at 4·64, the calculated value being 4·69. At a red heat it decomposes, yielding chloride of silicon, chlorhydric acid, and amorphous silicon. Its vapor is very inflammable, and, when mixed with air or oxygen, detonates violently on being lighted; chloride of silicon, together with a smoke of chlorhydric and silicic acids, are the products of this reaction. Water decomposes it instantly, with production of hydrated oxide of silicon, as has been already set forth (§ 458).

473. *Fluoride of Silicon* ($SiFl_4$) is formed whenever dry fluorhydric acid comes in contact with silicic acid, either free or combined. It may be readily prepared by treating a mixture of siliceous sand and fluorspar with oil of vitriol.

In a flask of about 250 c. c. capacity, place an intimate mixture of 9 grms. of finely-powdered quartz-sand, 9 grms. of powdered fluorspar, and 53 grms. of concentrated sulphuric acid. Heat the flask, and collect the gas which is evolved, in tall bottles of 150 c. c. capacity, at the mercury trough. The sand and fluorspar should both be heated before being used, in order that they may be perfectly dry; the flask also and the mercury in the trough should be thoroughly dried; for fluoride of silicon is decomposed by moisture, and the bulky precipitate of silicic acid which would be formed might clog the delivery-tube of the apparatus or coat the glass vessels and render them opaque.

The decomposition of the gas by water can readily be shown by causing some of it to pass out from the delivery-tube into a bottle inverted upon the mercury-trough, one half of which has been filled with water and the other half with quicksilver. The water in the bottle will soon become filled with gelatinous silica. With the quantities of material above given, there will be obtained first a bottle of 150 c. c. capacity of air from the flask, which should be thrown away; a bottle of the same size can then be filled with sufficiently pure fluoride of silicon, and in a third bottle the decomposition with water may be shown.

Instead of sand, coarsely powdered glass may be placed in the flask as the source of silicon. In any event the glass of the flask will be somewhat corroded; it is better to conduct the operation in a platinum retort, if such a vessel can be had.

The reactions which occur among the materials in the flask may be represented by the following equations:—In the first place, fluorhydric acid is formed by the action of sulphuric acid upon the fluoride of calcium,

$$CaFl_2 + H_2SO_4 = CaSO_4 + 2HFl,$$

and the fluorhydric acid then attacks the silica,

$$SiO_2 + 4HFl = 2H_2O + SiFl_4.$$

The water here formed unites with the sulphuric acid which has been employed in excess, and is thus prevented from acting upon the fluoride of silicon. The reaction last named is essentially the same as that which occurs when glass is etched with fluorhydric acid gas (§ 158). When the fluoride of silicon is brought into contact with water, there is produced, together with the silicic acid, fluosilicic acid ($2HFl,SiFl_4$), a compound to be described directly.

474. Fluoride of silicon is a colorless gas which fumes strongly in the air, especially if the air be moist. It has an acid odor and taste, extinguishes combustion though itself uninflammable, has no action upon glass, can support a high degree of heat without being decomposed, and may be condensed by pressure and cold to a colorless, very mobile liquid. Its specific gravity has been determined to be 51·98, or 3·6 as compared with air. It can consequently be readily collected by displacement, if pains be taken to protect it from contact with moist air by means of drying-tubes.

The specific gravity of this gas points very decidedly to the formula $SiFl_4$, and consequently to the number 28 as the atomic weight of silicon, and to the formula SiO_2 for silicic acid. For, if a molecule of gas be composed of

One volume of silicon-vapor = ...	28
And four volumes of fluorine (19 × 4) = ...	76
Condensed to two volumes = ...	104

One volume of the gas will weigh 52, or almost precisely as much as has been indicated by experiment.

The behavior of the gas with water is peculiarly interesting. If water be sprinkled into a bottle filled with fluoride of silicon, or if the gas be conducted into water, decomposition occurs instantly, and silicic acid is deposited in the gelatinous condition. At first sight it might seem as if the reaction were strictly analogous to that which occurs when chloride of silicon is mixed with water (see § 471), and that it could be represented by the formula

$$SiFl_4 + 2H_2O = SiO_2 + 4HFl;$$

but if it be remembered that fluorhydric acid, unlike chlorhydric acid, is capable of dissolving silica, it will be at once evident that the reaction between water and fluoride of silicon cannot be directly comparable with the other reaction, in which chloride of silicon is concerned. The reaction which really occurs between water and fluoride of silicon may be represented by the following equation:—

$$3SiFl_4 + 2H_2O = SiO_2 + 2(2HFl,SiFl_4),$$

—a double compound of fluorhydric acid and fluoride of silicon, known as *fluosilicic acid*, being the other product besides silica.

Fluosilicic acid may be prepared either in the manner indicated in § 473, or more conveniently by conducting fluoride of silicon gas into an upright bottle full of water; in this case, however, the orifice of the tube which delivers the gas should be immersed in a layer of mercury, 3 or 4 c.m. deep, beneath the water, so that the gas may bubble up through the mercury, and the water never come in contact with the mouth of the delivery-tube. In this way the tube may be kept free from the silicic acid which would quickly clog it if it opened directly into the water. As the bubbles of fluoride of silicon escape from the mercury, each of them becomes covered with a thick crust of gelatinous silica, which it carries with it to the surface of the water. Sometimes one of these crusts will remain adhering to the mercury and will be gradually prolonged into the form of a tube extending from the mercury to the surface of the water. Such tubes should, however, be broken up, by stirring the liquor, lest fluoride of silicon escape through them into the air without coming in contact with water.

Another modification of the method of preparing fluosilicic acid is to conduct the gas into a large flask containing but little water, and to agitate this flask so that its sides may be continually moistened with water, although no water can come in contact with the opening of the gas delivery-tube. The following quantities of materials have been found convenient in practice. In a flask of 600 or 700 c. c. capacity place a mixture of 35 grms. of powdered sand, 35 grms. of powdered fluorspar, and 210 grms. of concentrated sulphuric acid. Place the flask upon a sand-bath over the gas-lamp, and by means of a delivery-tube connect it either with the bottom of a tall bottle containing 150 c. c. of water and enough mercury to form the layer above described, or with the middle of a flask of 700 or 800 c. c. capacity, and containing 150 c. c. of water. In the latter case the receiving-flask must be agitated, as aforesaid, so that its walls may be kept moist. When the evolution of gas has ceased, the gelatinous mass in the bottle should be thrown upon a piece of cotton cloth, and the liquid contained in it separated from the solid matter by pressure. The cloudy liquor thus obtained may be rendered clear by filtering it through paper. The clear liquid is a strong aqueous solution of fluosilicic acid, and the solid matter is hydrated silicic acid.

475. Fluosilicic acid is known only in aqueous solution. The saturated solution is a transparent, colorless, fuming, and very acid liquid, which has no action upon glass, and may consequently be kept in glass bottles. It cannot be distilled without suffering decomposition, nor can it be evaporated beyond a certain degree of concentration without breaking up into fluoride of silicon and fluorhydric acid. If the evaporation is conducted in a glass vessel, the silica of the glass will retain the fluorhydric acid, and only fluoride of silicon and water will be set free. The reaction which occurs in this case,

$$SiO_2 + 2(2HFl,SiFl_4) = 2H_2O + 3SiFl_4,$$

is just the reverse of that which takes place when fluoride of silicon is decomposed by water—

$$2H_2O + 3SiFl_4 = SiO_2 + 2(2HFl,SiFl_4).$$

With bases fluosilicic acid unites to form compounds known as fluosilicates, such as the fluosilicate of potassium, $K_2SiFl_6 = 2KFl,SiFl_4$, and fluosilicate of barium, $BaSiFl_6 = BaFl_2,SiFl_4$, if no excess of the base be present; but if an excess of the base be added, silicic acid will be precipitated, and the whole of the fluorine will unite with the metal contained in the base to form a fluoride. In the first case, where no excess of base is employed, the reaction may be thus represented:—

$$2KHO + H_2SiFl_6 = 2H_2O + K_2SiFl_6.$$

In the second case, where an excess of the base is present,

$$6KHO + H_2SiFl_6 = 4H_2O + SiO_2 + 6KFl.$$

Most of the fluosilicates are easily soluble in water; but some of them are so nearly insoluble that the acid is sometimes employed as a precipitant of various oxides. It is remarkable that precisely those bases, namely the alkalies, which form soluble salts with almost all the other acids, should here yield insoluble compounds. Fluosilicic acid is in fact a good reagent for the detection of potassium; and it is valuable as a means of removing potassium from many of its salts when we desire to obtain in a free state the acids which these salts contain. (Compare § 124.)

476. *Sulphide of Silicon* (SiS_2) occurs in white, needle-like crystals, unalterable in dry air. It is volatile at the temperature of redness, and is decomposed at once by water, with deposition of gelatinous silica and evolution of sulphydric acid:—

$$SiS_2 + 2H_2O = SiO_2 + 2H_2S.$$

Like sulphide of boron, it may be obtained by passing the vapor of bisulphide of carbon over a mixture of silicic acid and carbon heated to redness.

477. As has appeared abundantly from the foregoing, the three elements carbon, boron, and silicon constitute a distinct natural family. This family differs in character from each and every one of the other natural groups of elements hitherto described (§§ 152, 257, 364). The occurrence of carbon and silicon in three distinct and extraordinary modifications (as diamond, graphite, and charcoal), and of boron in the diamond and amorphous modifications, distinguishes this group from all others. The eminently refractory nature of the several members, and their fixity as regards heat and solvents, are marked characteristics of the three elements. The vitrifiable character of the oxides of boron and silicon (boracic and silicic acids), and of the borates and silicates of many of the metals, is a noteworthy resemblance between these elements. Remarkable resemblances between the hydrated carbonate, borate, and silicate of sodium often manifest themselves to the chemical manipulator. The corresponding compounds of carbon, boron, and silicon, with the members of the chlorine group, and with the members of the sulphur group, have similar proper-

ties, result from like reactions, and suffer, for the most part, analogous decompositions; but, in the present state of the science, only the compounds of carbon and silicon can be said to have also the same atomic constitution. The three members of this group are not arranged in the order of their atomic weights, as has been the case with all the preceding groups; but if, in the future, boracic acid comes to be written BO_2, the atomic weight of boron will be 14, and therefore higher than that of carbon. The existing anomaly in the arrangement of the group will then disappear. The collocation of these three elements, and the order in which they now stand, is based upon too many natural resemblances to be lightly set aside.

CHAPTER XXIII.

SODIUM.

478. This abundant element is chiefly found in nature in the state of chloride, nitrate, carbonate, borate, and silicate. The most abundant of its compounds is common *salt*, which is the combination of sodium with chlorine ($NaCl$). Sea-water contains two and a half per cent. of salt, and enormous deposits of the same substance are found in the solid crust of the earth. There are also many natural salt springs, whose waters yield on evaporation the chloride of sodium which they hold in solution. Sodium also occurs, in the condition of silicate, in very many common minerals and rocks. From the soil which has resulted from the disintegration and decomposition of these minerals and rocks, and from its soluble compounds, like common salt, sodium enters into plants, and thence into animals. Its chloride is one of the essential mineral constituents of the food of man and other animals. On account of the inexhaustible abundance of common salt, this substance constitutes the chief source from which all manufactured compounds of sodium are more or less directly derived; one other natural sodium-containing mineral, however, deserves mention as a source of sodium compounds—the mineral Cryolite, a double

fluoride of sodium and aluminum. Nitrate of sodium ($NaNO_3$), a somewhat deliquescent and very soluble salt, occurs abundantly on the surface of the soil in certain desert districts of Peru. When heated, this salt first fuses and then undergoes decomposition. It is employed in the manufacture of nitric and sulphuric acids and as a manure; but as a source of sodium-compounds it is comparatively insignificant.

479. *Chloride of Sodium* ($NaCl$).—There is but one chloride of sodium—common salt. This natural mineral is, when pure, a colorless, transparent, anhydrous stone, which crystallizes in cubes, dissolves readily in about three times its weight of cold water, and possesses a specific gravity of 2·15, and an agreeable taste, which, because familiar, is the representative or type of that peculiar savor called *saline*. A saline taste means a taste suggestive of that of common salt, just as the phrase "saline substance" characterizes a very large class of bodies which resemble more or less in appearance and properties the longest- and best-known of all such substances, common salt.

There are three sources of salt—the beds of the native mineral, saline springs, and sea-water. In all cases in which the salt is obtained from its solution in water, evaporation by fire, or by the heat of the sun in warm, sunny climates, is necessary. When pure enough, the rock-salt is mined like any other ore; but when it is mixed with earth or other impurities as it lies in its natural bed, the solubility of the chloride of sodium in water is availed of to free the salt from its insoluble impurities, and to facilitate the lifting of it to the surface of the earth. Water is let into the bed of salt, and allowed to remain there till it has become saturated; the brine is then pumped out and evaporated. Some natural brine-springs contain so small a proportion of salt that some cheaper mode of evaporation than by fire is essential to their profitable working. Such waters are concentrated by a process termed *graduation*. The brine is pumped up to a sufficient height, and then allowed to trickle slowly over large stacks of fagots, which are sheltered by a roof from rain, but are freely exposed to the prevailing wind. The brine, thus diffused over a very large surface, is rapidly concentrated by the draft of air. By repeating the process a moderate number of times, a weak brine may be brought to a degree of concentration at which evaporation by fire may be employed. Since almost all brines contain sulphate of calcium ($CaSO_4$) in solution, the surfaces of the fagots employed in the *graduation* process become covered with a stony coating of this comparatively insoluble salt. If the strong brine is boiled down rapidly, a fine-grained table-salt is obtained; if it is slowly evaporated, a hard, coarsely crystallized salt is the product. During the earlier stages of the evaporation a deposit is formed, consisting principally of sulphate of sodium and sulphate of calcium. Finally there remains a thick *mother-liquor,* from which no more chloride of sodium will crystallize, but which contains the more soluble salts of the original brine, such as chloride of calcium, and chloride and bromide of magnesium, besides a large proportion of common salt which cannot be separated from the liquor. Such mother-liquors are sometimes so rich in magnesium-salts as to be advantageously worked for these substances, and they are also sometimes profitable sources of bromine. Considerable quantities of magnesium-salts and of bromine have also been extracted from concentrated sea-water after all the available chloride of sodium has been withdrawn. The salt of commerce generally contains a small proportion of chloride of magnesium, which makes it slightly deliquescent and bitter.

Exp. 224.—Heat a few crystals of coarse salt on a piece of sheet-iron over the gas-lamp. The crystals will decrepitate forcibly, and the greater part of the salt will be thrown off the plate; what remains will melt as the temperature rises, and if the heat be strong enough, it will finally volatilize. The decrepitation is due to little particles of water, mechanically enclosed in the crystals, which, when expanded by heat, burst the crystals asunder.

Exp. 225.—Dissolve 9 grms. of fine salt in 25 c. c. of water at about 20°. Add to the solution another gramme of salt; it will not dissolve. Bring the solution to boiling; the added gramme of salt will barely dissolve. Chloride of sodium is scarcely more soluble in hot than in cold water, wherein it differs from the great majority of soluble salts. Evaporated brines deposit their salt with almost equal facility when hot and when cold; but the hot liquors will hold in solution a much greater proportion of the salts with which the chloride of sodium is associated, than the cold brines could retain. In the process of evaporation by fire, the associated magnesium, calcium, and sodium-salts therefore do not crystallize with the common salt, but remain in the hot mother-liquor.

Exp. 226.—Expose a saturated solution of salt in winter weather to a temperature of −10°. Large, transparent, six-sided tables, which contain a considerable proportion of water chemically combined, will crystallize from the solution. The warmth of the hand is sufficient to destroy this crystalline compound; the water separates, and the crystals are resolved into a mass of minute cubes.

480. The uses of common salt are manifold. Since it is a constituent of almost all kinds of food, and essential to the life of animals, it is not surprising that salt exists in small quantities in almost every spring, soil, plant, and animal. The antiseptic quality of salt is applied to the preservation of fish, meat, and wood. Salt is extensively employed in glazing earthenware, its volatility at furnace-heat (Exp. 224) combining with other qualities to fit it for this use. Immense quantities of salt are consumed in preparing sulphate of sodium, from which in turn common "soda" (carbonate of sodium) is made. From the carbonate of sodium thus obtained the greater number of other sodium compounds are prepared. Salt is also the source from which chlorhydric acid and chlorine are derived (§§ 101, 105).

481. *Bromide and Iodide of Sodium* (NaBr and NaI).—These salts bear a close resemblance to the chloride of sodium; they both crystallize in anhydrous cubes, and both occur native in sea-water, though in minute proportion. Many marine plants appropriate iodide of sodium from sea-water; sea-weeds are therefore the commercial source of iodine (§ 135).

482. *Sulphate of Sodium* (Na_2SO_4).—This compound is made in great quantities from common salt and sulphuric acid as a preliminary step in the manufacture of carbonate of sodium.

The process has two stages. The first operation is performed in large, covered, cast-iron pans, capable of holding 250 kilos. of salt, and an equal weight of sulphuric acid of the density of 1·7. A very gentle heat suffices to disengage from such a mixture enormous volumes of chlorhydric acid gas; this gas, which would be injurious to vegetation if suffered to escape into the air, is all absorbed by being passed through vertical stone towers, filled with lumps of coke, over which water is kept trickling. The reaction in the iron pan is by no means complete, much chloride of sodium remaining undecomposed. The reaction at this first stage may be represented as follows:—

$$2NaCl + H_2SO_4 = NaCl + NaHSO_4 + HCl.$$

The pasty mass is then pushed into an adjoining fire-brick chamber, which is strongly heated by flues from a furnace. The acid sulphate of sodium of the last reaction decomposes the remainder of the salt, and a further quantity of chlorhydric acid is disengaged to be condensed by the water in the coke-towers, while sulphate of sodium remains:—

$$NaCl + NaHSO_4 = Na_2SO_4 + HCl.$$

The sulphuric acid used in this process is not stronger than can readily be made by evaporation in leaden pans (compare § 230); it is always made on the spot. The crude, weak, chlorhydric acid, which is the incidental product of the manufacture of sulphate of sodium, has of course some value in the arts; a portion of the product is often immediately consumed in the same factory in the manufacture of "bleaching-powder" (§§ 105, 120). In some works the smoke and gases from the fires pass through the coke-towers in which the chlorhydric acid is absorbed; in others, the products of combustion are not suffered to mix with the liberated chlorhydric acid, but are conducted by separate flues around the pans and chambers to be heated, and thence into the main chimney. The latter mode of construction is the best.

The sulphate of sodium resulting from this process is a white anhydrous salt which dissolves easily in water at 30°. When a strong solution of the anhydrous salt, made at this temperature, is cooled, there separate large colorless crystals of a transparent salt, bitter and cooling to the taste. This salt, long known as Glauber's salt, contains, besides the elements of sulphate of sodium, ten molecules of water; it therefore answers to the formula $Na_2SO_4,10H_2O$. There is another hydrated sulphate of sodium, which contains only seven molecules of water. These hydrated salts *effloresce* in dry air, and crumble into an opaque powder of the anhydrous salt.

Exp. 227.—Place 10 grms. of crystallized Glauber's salt in a warm, dry place. When it is completely converted into a white powder, weigh the residue. Since Glauber's salt is more than half water, the dry residue will not weigh more than 4·5 grms.

Exp. 228.—In a flask holding about 250 c. c., heat 50 c. c. of water to a temperature of 33°, and keep the water at this temperature, as determined by a thermometer immersed in it. Add to the warm water 161 grms. of crystallized Glauber's salt. If this saturated solution be made hotter, the anhydrous sulphate of sodium crystallizes out in octahedrons of rhombic base; if, on the other hand, the solution be suffered to cool, crystals of the common hydrated Glauber's salt appear. This example forcibly illustrates the general fact that the relations of water to other bodies are greatly affected by temperature.

Exp. 229.—Dissolve 10 grms. of crystallized Glauber's salt in water of which the temperature has been previously observed; during solution the temperature falls; cold is produced, in consequence of the expenditure of some of the heat of the mixture in overcoming the cohesion of the crystallized salt. Dissolve a like quantity of effloresced

Glauber's salt (anhydrous sulphate of sodium) in a small bulk of water; heat will be developed. A part of the water is solidified by combining with the anhydrous sulphate to form the hydrated sulphate, and the heat which before kept that quantity of water fluid, being set free to do other work, raises by a certain amount the temperature of the mixture.

Exp. 230.—Dissolve 50 grms. of Glauber's salt in 25 c. c. of water in a small flask, by heating the contents of the flask until the liquid boils. Cover the mouth of the flask loosely with a card or piece of glass, and allow the liquid to cool, at perfect rest, to the ordinary temperature. No crystals will be deposited from the fluid, although the water holds in solution a much larger quantity of salt than it could dissolve at the atmospheric temperature. Such a solution is said to be *supersaturated.* The crystallization of such a solution may generally be brought about, almost instantaneously, by jarring the vessel which contains it, or by permitting some foreign body, like a glass rod, a wire, a crystal of the salt, or a grain of dust, to come in contact with the fluid. By touching with a stick or glass rod the clear solution prepared as above described, this sudden crystallization will be strikingly illustrated, the whole mass becoming solid.

Crystallized sulphate of sodium is rapidly soluble in chlorhydric acid, with great depression of temperature. A convenient refrigerating mixture may be prepared in climates where ice is dear, by pouring 5 parts of the commercial acid upon 8 of the crystallized sulphate. The effects of solution on temperature, and the phenomena of supersaturated solutions, though well exhibited by sulphate of sodium, are by no means peculiar to this substance; they are manifested to a greater or less degree by a large number of salts.

483. *The Double Sulphate of Sodium and Hydrogen* ($NaHSO_4$) is a very acid salt, to which the name of *bisulphate* is commonly applied. When heated it first gives up a molecule of water, and subsequently, at a higher temperature, a molecule of anhydrous sulphuric acid; the anhydrous salt is therefore employed as a convenient source of the teroxide of sulphur.

$$2(NaHSO_4) = H_2O + SO_3 + Na_2SO_4.$$

The formation of this double sulphate marks an intermediate stage in the making of sulphate of sodium and chlorhydric acid from common salt; and it is the residue of the nitric-acid manufacture whenever nitrate of sodium is employed.

484. *Carbonate of Sodium* (Na_2CO_3).—The manufacture of this

substance constitutes one of the most important branches of chemical industry. Immense quantities of it are consumed in the fabrication of glass and soap, in the preparation of the various compounds of sodium, and in washing, both by the manufacturer of cloth and in the household. The ashes of sea and sea-shore plants were formerly the source of the carbonate of sodium, but it is now chiefly made from common salt by a process called, from the name of its French inventor, the process of Leblanc.

The first stage of this process we have already studied; it consists in the preparation of sulphate of sodium from common salt. The second stage consists in the reduction of this sulphate to the condition of sulphide of sodium in presence of carbonate of calcium; by interchange of metals there results sulphide of calcium and carbonate of sodium. The sulphate of sodium is ground up with an equal weight of chalk and rather more than half its weight of coal, and the mixture is thoroughly melted by the flame of a reverberatory furnace. The black mass, which is called "black ball" or "black ash," is cast into blocks in iron wheelbarrows, cooled, broken up, and systematically washed with warm water until all the soluble portions are extracted. The black solution is evaporated in large iron pans by the waste heat of the reverberatory furnaces. The residue contains some caustic soda, mixed with the carbonate; the residue is therefore mixed with about one-seventh of its weight of sawdust, or like material, and roasted in a reverberatory furnace. The product of this heating is the *soda-ash* of commerce; it is almost white, and generally contains about 80 per cent. of pure anhydrous carbonate of sodium.

The so-called *crystals* of soda are obtained by dissolving the crude soda-ash in hot water, and suffering the hot solution to cool in large pans. In the course of five or six days, large transparent crystals are formed which contain 62·93 per cent. of water, and correspond to the formula $Na_2CO_3,10H_2O$. The impure mother-liquor, drained from the crystals, is used in the manufacture of caustic soda. These crystals effloresce in the air; they have a disagreeable taste, called alkaline, are soluble in very large proportion both in hot and cold water, and even melt at a moderate temperature in their own water of crystallization. The crystals readily part with all their water, and the dry residue melts at a bright-red heat; this residue is anhydrous carbonate of sodium, purified by the process of crystallization which it has undergone. In this case, as in all others, the process of crystallization consists essentially in the aggregation of *like* particles; the strong tendency is to exclude heterogeneous particles, or, in other words, impurities, from the crystallizing structure. There is no more universally applicable and valuable means of purification than the process of crystallization.

The purchaser of carbonate of sodium for the sake of the alkali which it contains will prefer soda-ash to soda-crystals, unless the purity of the material be an important consideration. The crystals are purer than the ash; but more than half their weight is water, which must be transported at the cost of the consumer. Effloresced crystals are more advantageous to buy by weight than crystals which have not lost their water by exposure to the air. There are several hydrates of carbonate of sodium, of different solubilities.

485. *Double Carbonate of Sodium and Hydrogen* ($NaHCO_3$).—When masses of crystals of hydrated carbonate of sodium (soda-crystals) are exposed to an atmosphere of carbonic acid gas, they absorb carbonic acid with an evolution of heat sufficient to expel the greater part of their water of crystallization. A white powder remains, whose dualistic formula is $Na_2O,H_2O,2CO_2$, whence its most familiar name—*bicarbonate of soda*. This substance is one of the ingredients in most of the artificial yeasts used for *raising* bread, cake, and puddings, and is known to grocers and cooks as "soda," although the constituent which is really utilized is its carbonic acid. This carbonate of sodium and hydrogen is much less soluble than the carbonate of sodium; it may be washed with cold water until it is freed from the sulphates and chlorides which generally contaminate the carbonate. The chemist resorts to this process in order to prepare from the purified bicarbonate pure sodium-salts. At a low red heat the double carbonate loses its water and half its carbonic acid, and is converted into the normal carbonate (Na_2CO_3). If the aqueous solution of the double carbonate be heated, it loses one quarter of its carbonic acid.

Weigh out two separate and equal portions of bicarbonate of sodium. Heat one of these portions for a few minutes in an iron spoon or porcelain capsule, to a low red heat. Then wrap each portion in a piece of paper and introduce each little roll into a cylindrical jar filled with mercury and standing inverted on the mercury trough. By means of a curved pipette (see Appendix, § 22) pass equal quantities of dilute sulphuric acid into each of the jars. The moment the sulphuric acid comes in contact with the carbonates of sodium, carbonic acid is evolved, and upon comparing the volumes of gas yielded by the two samples, both of which contain the same quantity of sodium, it will be found that the carbonic acid evolved from the unheated bicarbonate is twice as great as that yielded by the portion which was ignited.

Bicarbonate of sodium may be deprived of its carbonic acid by almost any acid or acid salt. In the experiment just described sulphuric acid displaced the gaseous carbonic acid; tartaric acid, or an acid tartrate like cream of tartar (tartrate of potassium), will effect the same displacement, as will also the acid sulphate of sodium ($NaHSO_4$), the acid phosphate of calcium ($CaO,2H_2O,P_2O_5$), or common alum. "Soda powders" should be made of bicarbonate of sodium and tartaric acid. "Rochelle powders" consist of bicarbonate of sodium in one paper and cream of tartar in another; when these two materials are mixed in water, carbonic acid is set free, and a double tartrate of sodium and potassium, called Rochelle salt and used as a purgative, remains in the liquid. When bread or cake is "raised" with "soda" and cream of tartar, the escaping carbonic acid is the agent in puffing up the dough, and the same Rochelle salt remains in the bread. Tartaric acid and cream of tartar having been dear in late years, a cheaper chemical yeast powder has been made from acid phosphate of calcium; when this substance reacts within the dough with bicarbonate of sodium, there remains in the bread a mixture of the phosphates of sodium and calcium. Alum is sometimes used for the same purpose. It is necessary to employ for such purposes, in connexion with the bicarbonate, acids or acid salts which are solid, and not so corrosive as to be obviously dangerous and harmful.

There exists a native sesquicarbonate of sodium called *trona* or natron; it is a saline efflorescence, always contaminated with the sulphate and chloride of sodium, and less soluble in water than the carbonate, but more soluble than the bicarbonate of sodium. All the carbonates of sodium have an alkaline reaction on vegetable colors; carbonic acid is too weak an acid to overcome the intensely alkaline reaction of caustic soda.

486. *Sulphides of Sodium.*—In the conversion of sulphate of sodium into carbonate of sodium (§ 484) one step is the reduction of the sulphate to the sulphide of sodium by coal. The oxygen of the sulphuric acid used in making the sulphate is thus combined with carbon and lost; the sulphur of the acid goes to the

waste-heap in a useless combination with calcium. Herein lies the wastefulness of the Leblanc process, which the experience of fifty years has failed to remedy.

Exp. 231.—Mix a little powdered anhydrous sulphate of sodium, obtained by drying Glauber's salt, with as much powdered charcoal, and make the mixture into a paste with a drop of water. Place a little ball of this paste, as large as a small pea, in a depression in a piece of charcoal, and heat it strongly with a blowpipe. The mixture effervesces and finally melts into a brownish mass. The glowing coal takes all the oxygen out of the sulphate, and the carbonic oxide, which results from the union of the carbon and oxygen, in escaping causes the effervescence which occurs; the sodium and sulphur alone remain united. The reaction may be thus formulated:—

$$Na_2SO_4 + 4C = 4CO + Na_2S.$$

If the brownish solid be removed from the coal, placed in a watch-glass, and moistened with dilute chlorhydric or sulphuric acid, sulphydric acid gas will be evolved, as may be recognized by the smell, or by the use of lead-paper:—

$$Na_2S + 2HCl = 2NaCl + H_2S.$$

Sulphur unites with sodium in more than one proportion; and it may indeed be doubted whether the actual reaction in the above experiment is so simple as the formula represents it to be. There probably exist five distinct sulphides of sodium—Na_2S, Na_2S_2, Na_2S_3, Na_2S_4, and Na_2S_5. All these sulphides have an alkaline reaction to test-paper, and evolve a more or less distinct odor of sulphuretted hydrogen. When they are brought into contact with an acid they are decomposed, sulphydric acid escapes, and a white precipitate of finely divided sulphur falls, in every case except that of the first sulphide, Na_2S. Besides these sulphides, a compound of sodium, hydrogen, and sulphur (NaHS) is known, which is perfectly analogous in composition to the combination of sodium, hydrogen, and oxygen with which we are already familiar under the name of caustic soda (NaHO).

487. *Sodium* (Na).—The element sodium is never found uncombined in nature, for the reason that in its elementary condition it cannot exist in contact with either air or water. It is, however, artificially prepared from the carbonate of sodium without serious difficulty, and it might be produced in considerable quantities, if there were any large use for the element.

A mixture of 20 parts of carbonate of sodium, 9 parts of coal, and 3 parts of chalk, placed in an iron bottle or cylinder, is heated to a very high temperature in a suitable furnace; a narrow iron tube connects the bottle, or cylinder, in the furnace, with a flat sheet-iron box outside the furnace; this box receives the sodium which distils from the hot mixture, and a hole in the front lower corner of the receptacle permits the melted metal to fall into a vessel containing petroleum-naphtha, beneath which the sodium can be preserved. The reaction is theoretically a conversion of all the oxygen in the carbonate of sodium into carbonic oxide, partly by a new combination with the carbon in the carbonate, and partly by union with the carbon which the coal supplies,

$$Na_2CO_3 + 2C = 3CO + 2Na;$$

but the amount of sodium practically obtained from a given weight of the materials is by no means as large as this simple formula would indicate. The chalk has no chemical effect, but is practically essential to the success of the operation. It prevents the mass from melting, and the gas which it gives off when heated assists in sweeping the sodium out of the bottle into the receiver. To purify the crude sodium thus obtained, it is melted under naphtha and cast into ingots in iron moulds. It must be kept under naphtha in tightly closed bottles.

The properties of the element, sodium, are very curious. The substance, when freshly cut, or when melted under naphtha or in an atmosphere artificially deprived of oxygen, has the brilliant, white metallic lustre of silver. Though possessing so eminently this characteristic property of the class of bodies called metals, and being like them a good conductor of heat and electricity, sodium is far from resembling the ordinary metals in other respects. Thus it is lighter than water, having a specific gravity of only 0·972, whereas the common metals are dense and heavy; again, it is as soft as wax at common temperatures, and melts at a temperature below that of boiling water, while it has none of the comparative permanence which characterizes lead, tin, copper, silver, gold, and other familiar metals. If exposed to the air, even for a few seconds only, it tarnishes, and soon becomes covered with a coating of oxide. Instead of being quenched by water, it takes fire when thrown into warm water. We have already seen that it decomposes cold water (Exp. 14), setting free its hydrogen, and combining with its oxygen.

Exp. 232.—Cover the bottom of a large bottle (at least a litre bottle) with hot water, drop in a piece of sodium as large as a small pea, and immediately cover the mouth of the bottle with a card or glass plate. The heat of the chemical combination between the sodium and the oxygen of the water is sufficient to inflame the hydrogen; the escaping hydrogen carries with it a small portion of the volatilized sodium, and therefore burns with an intensely yellow flame which is very characteristic of sodium-compounds. The metal swims rapidly about on the surface of the water, and is completely converted into caustic soda; at a little interval, after the flame has ceased to burn, a globule of caustic soda, which has escaped solution, bursts, and scatters in all directions; the mouth of the bottle should always be covered to avoid the possible projection of particles of hot soda out of the bottle. The water in the bottle, tested with litmus paper, will be found to possess a strong alkaline reaction. If the bit of sodium be previously wrapped up in muslin, it will take fire in cold water or even on ice. The muslin prevents the sodium from moving about, and the heat of combination is therefore concentrated upon one spot. The same effect may be produced without the muslin on cold water made viscid with gum.

At a high temperature sodium will remove oxygen from almost all bodies which contain it, whether solid, liquid, or gaseous. Hence the necessity of preserving the metal under some liquid which, like naphtha, contains no oxygen. Sodium enters directly into combination with all the elements of the chlorine and sulphur groups, and is capable of withdrawing these elements from nearly all the compounds into which they enter. In this substance, therefore, the chemist possesses a very potent agent for effecting chemical transformations. Its atomic weight is 23.

488. *Hydrate of Sodium* ($NaHO$).—When sodium is burnt upon water, a solution of hydrate of sodium, possessing an intensely alkaline reaction, remains behind; but in practice the hydrate is generally made from the carbonate.

Exp. 233.—Dissolve 100 grms. of crystallized carbonate of sodium (§ 484) in 400 c. c. of water. Slake 20 grms. of quicklime with water enough to make the slaked lime into a cream. Boil the solution of carbonate of sodium in an iron pan, and add to it, little by little, the cream of lime, until a small portion of the liquor, filtered off, produces no effervescence when poured into dilute chlorhydric or sulphuric acid. The calcium of the lime replaces the sodium in the carbonate of sodium; a white insoluble precipitate of carbonate of calcium is formed, and hydrate of sodium remains in the solution:—

$$Na_2CO_3 + CaH_2O_2 = 2NaHO + CaCO_3.$$

When this reaction is complete, no carbonic acid remains in the clear solution, which, therefore, causes no effervescence when mixed with an acid. When this test shows the reaction to be accomplished, extinguish the lamp; cover the pan, and let the carbonate of calcium settle to the bottom of the vessel. After several hours, draw off the clear supernatant liquid into a bottle by means of a siphon. A weaker lye may be obtained by boiling up the residue in the iron pan once more with water, and again decanting the clear liquid from the deposited precipitate.

The lye thus prepared contains a very little lime, inasmuch as the hydrate of calcium is not completely insoluble in a dilute solution of hydrate of sodium. If the lye is needed stronger, it is only necessary to concentrate it by evaporation in a clean iron pan to the requisite strength. Since soda-lye attacks ground glass with facility, the stoppers of bottles in which the lye is kept are apt to get stuck so fast that they cannot be removed. Such a bottle may be best closed with a caoutchouc-stopper, or a cork soaked in melted wax or paraffine.

In order to obtain the solid hydrate of sodium, its solution must be evaporated in a silver dish (since iron would color the concentrated lye) until the liquid in the dish flows smoothly like oil at a temperature near to a red heat. This thick, oily liquid solidifies when turned out upon a cold plate of metal or poured into metal moulds.

Fused caustic soda is a white, somewhat translucent mass, whose composition corresponds to the formula NaHO. Heat will separate from it no more water; if the attempt be made, it volatilizes in caustic vapors without change of constitution. The commercial caustic soda always contains much more water than the formula indicates. The solid hydrate is very soluble in water, and greedily absorbs both water and carbonic acid from the air, until the formation of a coating of the non-deliquescent carbonate of sodium arrests the process by protecting the enclosed hydrate. It is the prototype of the class of bodies called *bases*. It colors litmus blue and turmeric brown, and when mixed in due proportion with oxides of the opposite quality, called acid, a saline compound is formed which is neither acid nor alkaline, and which may bear no more resemblance to its proximate constituents than bread bears to flour and water, or rust to iron and oxygen.

From such reactions between acids and hydrate of sodium, water is always disengaged simultaneously with the saline product, and the reaction may almost always be as well considered an interchange of place between hydrogen and some other element as an act of combination between one oxide and another oxide, both of which, or one of which, contain also hydrogen. The following formulæ will illustrate the meaning of this statement:—

$$NaHO + NHO_3 = NaNO_3 + H_2O, \text{ or}$$

$$\left.\begin{matrix}Na\\H\end{matrix}\right\}O + \left.\begin{matrix}NO_2\\H\end{matrix}\right\}O = \left.\begin{matrix}NO_2\\Na\end{matrix}\right\}O + \left.\begin{matrix}H\\H\end{matrix}\right\}O;$$

$$2\,NaHO + H_2SO_4 = Na_2SO_4 + 2\,H_2O, \text{ or}$$

$$2\left.\begin{matrix}Na\\H\end{matrix}\right\}O + \left.\begin{matrix}SO_2\\H_2\end{matrix}\right\}O_2 = \left.\begin{matrix}SO_2\\Na_2\end{matrix}\right\}O_2 + \left.\begin{matrix}H_2\\H_2\end{matrix}\right\}O_2;$$

$$NaHO + C_2H_4O_2 = C_2H_3NaO_2 + H_2O.$$

Acetic Acid. Acetate of Sodium.

While recognizing the frequent occurrence of such reactions as are above represented between hydrated oxides, it must not be forgotten that many anhydrous saline compounds can be made by the direct combination, under appropriate conditions, of two oxides which contain no hydrogen. By heating one molecule of hydrate of sodium, or 40 parts by weight, with one molecule, or 23 parts by weight, of sodium, an oxide of sodium is obtained which contains no hydrogen; but this body has none of the properties described by the adjective alkaline, any more than the anhydrous teroxide of sulphur possesses the properties suggested to the mind by the term "acid":—

$$NaHO + Na = Na_2O + H.$$

Now the very same sulphate of sodium which results from the second of the above reactions, may be prepared by bringing together this anhydrous oxide of sodium and anhydrous sulphuric acid:—

$$Na_2O + SO_3 = Na_2SO_4.$$

There exists another anhydrous oxide of sodium, corresponding in composition to the formula Na_2O_2, and the same sulphate of sodium can be made by heating this oxide with sulphurous acid gas:—

$$Na_2O_2 + SO_2 = Na_2SO_4.$$

These facts show that a knowledge of the constituents from which a salt may be made is not sufficient to establish any presumption concerning the molecular constitution of the salt itself.

The preparation of solid caustic soda, for household and other uses, has lately become a considerable industry. Soap is made by boiling together grease or oil with caustic soda or potash; soda-lye yields a hard soap, potash-lye a soft soap. If the maker of soap starts from the carbonate of sodium or potassium, he must first make a solution of caustic lye by the method of Exp. 233, and then boil the lye, so obtained, with the grease which is the other essential ingredient of soap. The soap-maker is saved the trouble of converting the carbonate into the hydrate of sodium or potassium by the maker of the solid caustic alkalies, which need only to be dissolved in water to yield the requisite lye. The solid alkali is commonly put up for transportation in sheet-iron canisters, of all sizes. The manufacturer of caustic soda directly from the sulphate of soda has an advantage, in that he can avail himself of the caustic soda which the "black ash" always contains. He is not obliged, first to convert this caustic alkali into carbonate, and then to remove all the carbonic acid by lime; he can therefore dispense with the heating with sawdust which is necessary in the manufacture of soda-ash.

489. *Phosphates of Sodium.*—There are three sets of phosphates of sodium:—1. Common phosphates, which contain three atoms of hydrogen or an equivalent metal; 2. Pyrophosphates, which occupy an intermediate place between common phosphates and metaphosphates; 3. Metaphosphates, which contain only one equivalent of hydrogen or an equivalent metal. (See § 293.)

$$3H_2O,P_2O_5 = 2H_3PO_4; \quad 3Na_2O,P_2O_5 = 2Na_3PO_4$$
$$2H_2O,P_2O_5 = H_4P_2O_7; \quad 2Na_2O,P_2O_5 = Na_4P_2O_7$$
$$H_2O,P_2O_5 = 2HPO_3; \quad Na_2O,P_2O_5 = 2NaPO_3.$$

The most familiar of the ordinary phosphates is the rhombic phosphate of sodium, of the formula $Na_2HPO_4,12H_2O$, which is the salt commonly called phosphate of sodium.

Exp. 234.—Digest 8 grms. of powdered, white, burnt bones with 32 c.c. of water and 6 grms. of sulphuric acid, until a uniform paste is produced; strain the mass through a piece of muslin, stir up the residue with water, and squeeze the liquor through the cloth filter. Evaporate the filtered solution considerably, again filter off the separated sulphate of calcium, dilute the filtrate with water until it measures 48 c. c., and gradually neutralize the acid solution with solid carbonate of sodium. A slight excess of carbonate may be added, and the solution evaporated until it crystallizes; the crystals may be almost freed from adhering sulphate of sodium by washing them. By recrystallizing the salt it may be obtained in large, transparent rhombic prisms, which are decidedly efflorescent. They have a cooling, saline taste, are soluble in four parts of cold water, and readily melt in their water of crystallization; their solution has a faint alkaline reaction.

Exp. 235.—Add to a solution of the purified crystals of the last experiment a few drops of a solution of nitrate of silver; a yellow precipitate appears, and the liquid becomes distinctly acid in its reaction with litmus. The precipitate is phosphate of silver, and the acidity is due to the simultaneous liberation of nitric acid in the solution:—

$$Na_2HPO_4 + 3AgNO_3 = 2NaNO_3 + HNO_3 + Ag_3PO_4.$$

From the rhombic phosphate two other terbasic phosphates may be prepared:—by adding caustic soda, a terbasic, or termetallic, phosphate, $Na_3PO_4,12H_2O$; by adding phosphoric acid, a so-called "acid" phosphate, NaH_2PO_4,H_2O.

Exp. 236.—Heat 4 or 5 grms. of rhombic phosphate of sodium in a porcelain crucible to bright redness. The water of crystallization first escapes, then another portion of water is driven off at a higher heat, the residue melts, and on cooling solidifies again to an opaque, white, substance—the pyrophosphate of sodium, $Na_4P_2O_7$:—

$$2(Na_2HPO_4,12H_2O) = 24H_2O + H_2O + Na_4P_2O_7.$$

Exp. 237.—Dissolve the new salt of the last experiment in water and add a few drops of a solution of nitrate of silver; a chalky-white precipitate of pyrophosphate of silver will be produced, while the supernatant liquid is neutral:—

$$Na_4P_2O_7 + 4AgNO_3 = 4NaNO_3 + Ag_4P_2O_7.$$

The student will observe, in passing, that the pyrophosphate of sodium dissolves in water with more difficulty than the ordinary phosphate. From its aqueous solution the pyrophosphate crystallizes in prisms which contain ten molecules of water of crystallization, $Na_4P_2O_7$, $10H_2O$. These crystals are not efflorescent, but permanent in the air. Two of the atoms of sodium in neutral pyrophosphate of sodium can be replaced by hydrogen, and we thus obtain a salt of the formula $Na_2H_2P_2O_7$ which, like the neutral pyrophosphate, throws down the white pyrophosphate of silver from a solution of nitrate of silver, but, unlike the neutral salt, leaves behind a liquor acidified with free nitric acid.

Exp. 238.—Mix a hot solution of 18 grms. of common phosphate of sodium with a concentrated solution of 3 grms. of chloride of ammonium; common salt remains in solution, and transparent crystals of a complex phosphate of sodium, ammonium, and hydrogen separate from the liquid as it cools :—

$$Na_2HPO_4 + NH_4Cl = NaCl + NaNH_4HPO_4.$$

This complex triphosphate is known as *microcosmic salt*; it contains only one atom of sodium.

Exp. 239.—Heat the microcosmic salt obtained in the last experiment to redness. Ammonia and water are driven off with a great bubbling, and a glassy substance is left which corresponds to the formula $NaPO_3$, and is called the metaphosphate of sodium. This salt cannot be obtained by this process in crystals; it is deliquescent and very soluble.

Exp. 240.—Dissolve in water some of the metaphosphate just made, and add a few drops of the solution to a solution of nitrate of silver.

The precipitate produced is white, but gelatinous; its appearance is quite different from that of the pyrophosphate of silver. The metaphosphate of silver is soluble in an excess of metaphosphate of sodium, so that a sufficient quantity of the sodium-salt will redissolve the silver-salt precipitate which the first addition of the sodium-salt produced. It may be observed, that the solution of metaphosphate of sodium possesses a feebly acid reaction.

The metaphosphate of sodium exists in several distinct modifications; or, in other words, there are several sodium-salts which correspond to the formula $NaPO_3$. The most noteworthy modifications are the glassy salt just prepared, a crystallizable modification, and a variety which is almost insoluble in water though soluble in acids. Pyrophosphates and metaphosphates are alike converted into common ter-metallic phosphates by long boiling with water in an acid liquid, or by fusion with an excess of caustic alkali. The metaphosphate may also be converted into the pyrophosphate by mere heat in presence of water. The metaphosphate, when dried at 100°, retains a certain portion of water; but this water does not enter into the constitution of the salt, for, on again dissolving the salt, the solution gives the usual reactions of the metaphosphate. If, however, the water-containing salt be heated to 150°, the water enters into the constitution of the salt, which becomes the "acid" pyrophosphate of sodium, yielding, on solution, the reaction of the pyrophosphates:—

$$2\,NaPO_3 + H_2O = Na_2H_2P_2O_7.$$

We have here a striking illustration of the fact that two substances, which contain precisely the same elements in the same proportions, may have quite different properties in consequence of the different arrangement of their elements, or, in other words, of their different molecular constitution.

490. *Borax* ($Na_2B_4O_7,10H_2O = Na_2O,2B_2O_3 + 10H_2O$). — The dualistic formula of this useful salt accounts for its chemical name—*bi*borate of sodium. The greater part of the borax used in the arts is prepared by dissolving the native boracic acid of Tuscany (§ 443) in a hot solution of carbonate of sodium. After all the carbonic acid has escaped, the hot liquid is allowed to settle and cool. Either of two different crystalline forms of borax may be obtained, at will, by suitably regulating the density of the solution and the temperature at which crystallization takes place. One of these forms, prismatic or common borax, contains 10 molecules of water of crystallization; the other, called octahedral borax, only 5. Both varieties must be purified by recrystallization. The large crystals, in which borax is demanded for the market, can only be obtained by operating on very large masses of the salt,—a remark which applies to most of the crystalline salts which are used in the arts. The crystals of octahedral borax are harder and less fragile than those of ordinary borax. They are unalterable in dry air, but in a moist atmosphere they absorb water and are converted into common borax. When heated, they fuse to an anhydrous glass with less intumescence than common borax, and without cracking. For these reasons, octahedral borax is better than prismatic borax for many purposes, and its smaller proportion of water (30 per cent. against 47 per cent.) diminishes the cost of transport. Never-

theless the prismatic borax is generally preferred, probably because it is sold at a less price, weight for weight. The prismatic crystals are soluble in about half their weight of boiling water, but in twelve times their weight of cold water; as they are slightly efflorescent, they are generally covered with a white dust. Borax has a feebly alkaline taste and reaction. When heated it bubbles up, loses its water, and melts below redness into a transparent glass; this glass dissolves many oxides of the metals, acquiring thereby various colors characteristic of these oxides. Hence borax is much used as a blowpipe test for determining the presence of certain oxides of the metals.

Exp. 241.—Make a little loop, about as large as this O, on the end of a bit of fine platinum wire 6 or 8 c.m. long. Make the loop white-hot in the blowpipe-flame, and thrust it while hot into some powdered borax; a quantity of borax will adhere to the hot wire; reheat the loop in the oxidizing flame; the borax will puff up at first, and then fuse to a transparent glass. If enough borax to form a solid transparent bead within the loop does not adhere to the hot wire the first time, the hot loop may be dipped a second time into the powdered borax. When a transparent glass has been formed within the loop of the platinum wire, touch the bead of glass, while it is hot and soft, to a speck of black oxide of manganese no bigger than the period of this type; reheat the bead with the adhering particle of oxide in the oxidizing flame; the black speck will gradually dissolve, and on looking through the bead towards the light, or a white wall, when the black oxide has disappeared, the glass will be seen to have assumed a purplish-red color.

The same experiment may be performed with oxide of iron, which imparts to the glass a yellow color, or with oxide of copper, which imparts a bluish-green color. The oxidizing flame must be used in both these cases, as with the oxide of manganese.

The power which borax possesses of dissolving metallic oxides suggests an explanation of its use in brazing, and in soldering the precious metals. The solder will only adhere to a bright and clean metallic surface, and the borax which melts with the solder removes from the pieces of metal the film of oxide which would otherwise prevent the adhesion of the solder. Borax is also used by the assayer and refiner as a flux. In making enamels and glazes, it is frequently added for the purpose of rendering the compound more fusible, and it is largely employed in fixing colors on porcelain.

There is a normal, or neutral, borate of sodium, $NaBO_2$ or Na_2O,B_2O_3, which crystallizes with various quantities of water; and other borates of sodium are known; but the "biborate" is the only one of any practical importance.

491. *Silicates of Sodium* may be prepared by dissolving silicic acid in caustic soda, as in Exp. 220, or by fusing together silicic acid and carbonate of sodium, or a mixture of silicic acid, sulphate of sodium, and carbon. From alkaline solutions, the single crystallizable silicate of sodium, Na_2SiO_3, can readily be obtained in hydrated crystals. The silicate of sodium of commerce, called waterglass or soluble glass, is, however, a much more siliceous silicate, of varying composition. Some samples of it have very nearly the composition expressed by the formula $Na_2O, 2SiO_2$, while other specimens approximate to the formula $Na_2O, 4SiO_2$. Between these extremes, the acid and the alkali unite in all possible proportions to form compounds, all of which are soluble with more or less difficulty in boiling water.

The normal silicate of sodium (Na_2SiO_3) is readily soluble in cold water, and, like carbonate of sodium, which it closely resembles, may even be melted in its own water of crystallization; but the acid salt of commerce is as good as insoluble in cold water. Waterglass may, however, be completely dissolved by long-continued boiling in water, and the solution thus obtained is largely employed by calico-printers and by soapmakers. Waterglass is also used for hardening porous stones, or even for binding sand into artificial stone, for painting rough woodwork to protect it from the weather, and for diminishing the combustibility of wood, canvas, and other coarse stuffs, such as are used for the decorations of theatres. It has been suggested that the interior surfaces of wooden-roofed railway bridges might be protected from the sparks of the locomotive by washing them with a solution of waterglass. When employed as paint, the coating of waterglass may be washed over with a solution of chloride of ammonium, which decomposes the silicate, with deposition of free silicic acid; or the silicate may simply be left to the decomposing action of atmospheric carbonic acid. The chloride of sodium resulting from the decomposition in the one case, and the carbonate of sodium in the other, are subsequently washed away by the rain. Combustible matters, when covered with a coating of silica, or of silicate of sodium, as above, are prevented from burning freely, in the same manner that the carbon of the paper in Exp. 110 was kept from burning by the coating of phosphoric acid.

492. The chief use of silicate of sodium, however, is as a component of common glass. The various glasses of commerce are mixtures of a highly siliceous silicate of sodium, or of potassium, or of both these substances, with silicates of other metals, such as calcium, aluminum, and lead. In green bottle-glass the easily fusible silicate of iron replaces in part the silicate of sodium or of potassium.

It is a peculiarity of the alkaline silicates that, in changing from the liquid to the solid state, they pass through an intermediate pasty or viscous stage, and finally solidify in transparent amorphous masses, totally devoid of crystalline structure. While in this pasty, ductile state, these silicates may readily be moulded into almost any form; and transparent vessels might doubtless be made from the acid alkaline silicates without admixture of other materials. The alkaline silicates, however, are, by themselves, far too easily acted upon by air and moisture to admit of being used as substitutes for ordinary glass. But it has been found that, by combining the alkaline silicates with the silicates of certain other metals, such as calcium, there may be obtained compound glasses which, while they retain the plasticity of the alkaline silicates as well as their amorphous character and transparency, are capable of resisting the action not only of air and water, but even of acids and alkalies, to a very great extent. Though the ordinary glasses are so difficultly attacked by water that they may, for most practical purposes, be regarded as altogether insoluble, it is nevertheless true, as has been stated in § 464, that glass may be partially dissolved by long-continued contact with water, particularly if the water be hot and the glass in the condition of powder. After the smooth surface of glass, as it comes from the fire, has once been removed, the corrosion of the glass goes on more rapidly.

It is remarkable that mixtures composed of several different silicates melt at temperatures considerably lower than the mean of the melting-points of their several ingredients.

The different varieties of glass vary in composition, according to the purposes for which they are prepared; they cannot be regarded as chemical compounds, being really indefinite mixtures of various acid silicates. The composition of window-glass may be represented approximately by the formula $Na_2O, 2SiO_2; CaO, 2SiO_2$; and that of the hard, Bohemian glass, suitable for ignition-tubes, by the formula $2(K_2O, 3SiO_2); 3(CaO, 3SiO_2)$. The lustrous "flint" glass, employed for the nicer kinds of household ware, may, on the other hand, be represented by the formula $2(K_2O, 2SiO_2); 3(PbO, 2SiO_2)$; it is prepared from the purest materials attainable. Bottle-glass is composed of the silicates of calcium and aluminum, together with a small proportion of the silicates of iron, of potassium, and of sodium; in this glass, as in the other varieties above formulated, small quantities of various other silicates, such as the silicates of magnesium and of manganese, almost always occur.

In the preparation of the cheaper kinds of glass the materials are melted in large open crucibles of refractory clay; but the better sorts of glass, such as flint-glass, are made in pots so covered that no smoke or dust from the fire can come in contact with their contents. In both cases the thoroughly melted mixture is kept in the liquid state until it has become perfectly homogeneous and until all bubbles of air have escaped from it; it is then allowed to cool to the temperature at which it possesses the peculiar, pasty, ductile, condition in which it admits of being blown, pressed, and moulded.

493. After glass has been moulded into the shape desired, it must still be subjected to a process of annealing before it can be used. Glass which has been suddenly cooled after fusion is extremely brittle; and in general all glass which has been quickly cooled after heating is far more fragile than that which has been allowed to cool gradually. The operation of annealing is nothing but a process of slow baking, at a temperature which, though not so high as the melting-point of glass, is nevertheless high enough to allow the particles of softened glass to move among themselves, and to come into easy and natural positions as regards

one another; after the baking follows a process of slow cooling, during which the heated material contracts uniformly in all directions as it assumes the dimensions proper to it when cold.

Unlike the silicates of the alkali-metals, most metallic silicates have a tendency to assume crystalline form on cooling, and it is not difficult to bring about crystallization of silicate of calcium, or silicate of aluminum, from ordinary glass, particularly from the coarser kinds, such as bottle-glass. Glass, in which some of the constituents have thus crystallized, has the appearance of porcelain; it is said to be *devitrified*. The devitrification may readily be shown by imbedding bottle-glass in sand, heating the glass almost to the melting-point, and then allowing it to cool slowly. In annealing some kinds of glass care must be taken not to heat the ware too strongly, lest it be devitrified during the process.

494. Melted glass, like melted borax (Exp. 241), is capable of dissolving small quantities of many of the metallic oxides, a transparent and often colored silicate of the metal being formed, which imparts its hue to the entire mass of glass. In this way, glass may be obtained of almost any desired color. The green color of bottle-glass is due to silicate of protoxide of iron; but richer shades of green may be obtained by using protoxide of copper or oxide of chromium. Dioxide of copper gives a ruby-red color, and oxide of gold various shades of red, inclining to purple. The oxides of uranium, of antimony, and of silver yield yellow glasses; oxide of cobalt affords a beautiful blue, and the binoxide of manganese a violet glass; while mixtures of the oxides of cobalt and of manganese impart to the glass a black color.

495. *Hyposulphite of Sodium* ($Na_2S_2O_3, 5H_2O$).—This easily crystallized and tolerably permanent salt is of great use to the photographer, because its aqueous solution is capable of rendering soluble the chloride, bromide, and iodide of silver, compounds much employed by the photographer, and very insoluble in water. The photographic paper or glass, uniformly coated with some silver compound, is exposed to light in the camera or press, and then immersed in a strong solution of the hyposulphite, which forms with the silver compound, in those parts of the picture which have not been acted upon by the light, a double salt which

is soluble in water. This double salt and the superfluous hyposulphite must be washed away by soaking the picture several hours in water which is constantly renewed. Hyposulphite of sodium is also used as an "antichlore," or agent for removing the last traces of chlorine, or hypochlorous acid, from substances which have been bleached therewith. The salt may be best prepared by digesting sulphur with a solution of sulphite of sodium.

Exp. 242.—Dissolve 20 grms. of crystallized carbonate of sodium in 30–40 c. c. of water. Place the solution in a small Woulfe-bottle, and pass sulphurous acid gas (Exp. 96) through it until all the carbonic acid is expelled from the carbonate and effervescence ceases. The liquid then holds in solution sulphite of sodium (Na_2SO_3). Pour this solution into a bottle which can be tightly closed, and add to it 3 or 4 grms. of finely powdered sulphur; let this mixture stand corked up for several days in a warm place; the sulphur will gradually dissolve, and form a colorless solution, which on evaporation will yield crystals of hyposulphite of sodium. Time may be saved by keeping the solution of sulphite of sodium hot, but not boiling, during the digestion of the sulphur. The reaction has been already formulated (§ 243).

CHAPTER XXIV

POTASSIUM.

496. The proximate sources of sodium-compounds are the sea, and salt springs and deposits. Potassium-compounds, on the other hand, are derived indirectly from the soil. Arable soils are produced by the weathering and gradual decomposition of the common granitic rocks. Into the composition of these rocks there enter two minerals, called feldspar and mica, which are mixed silicates of potassium or sodium and aluminum or iron. The element potassium thus becomes a normal constituent of the earthy food of plants. The soil itself is not directly available as a source of potassium-salts, because no cheap and easy method has yet been devised for separating the potassium-compounds from the other ingredients of the soil. Plants, however, are able to pick out and assimilate the potassium-salts from these rocks and soils; so that by burning the plants and extracting the ashes with water a soluble potassium-salt is obtained. Plants thus concentrate the potassium from out great masses of earth, and make it accessible to us. The salt which is obtained from the ashes of plants by washing and evaporation, is called *potash*, or, if refined, *pearlash*, and it is from this substance that the bulk of potassium-compounds are obtained.

Exp. 243.—Place a handful of wood-ashes on a filter, and pour hot water over them, collecting the filtrate in a bottle and returning it upon the ashes two or three times, in order to obtain a stronger solution. To exhaust the ashes of their potash, they must, of course, be treated with successive portions of hot water. This solution has a strong alkaline reaction upon test-paper. A few drops of it, poured into a test-tube containing a little dilute acid, occasion a brisk effervescence, a reaction from which we readily surmise the truth, that the potassium-salt contained in the solution is the carbonate of potassium. Proof that the gas evolved is carbonic acid can readily be obtained by conducting the gas into lime-water, as in Exp. 170. By evaporating the rest of the solution to dryness in a porcelain dish, we obtain a small sample of crude potash. By treating this potash with a quantity of cold water, insufficient to dissolve any but the most soluble portions of the mass, letting the mixture stand some time, and evaporating the partial solution to dryness, a whiter, purer carbonate is obtained, the pearlash.

497. *Carbonate of Potassium* (K_2CO_3) is a hygroscopic and very soluble salt. When exposed to damp air it becomes moist, and finally deliquesces. In this respect it does not resemble soda-ash, which is not hygroscopic, and is, for this reason among others, better adapted than potash for transportation, storing, and most commercial uses. Carbonate of potassium fuses at a red heat, but cannot be decomposed by heat alone. At a red heat it is decomposed by silica, as is also the carbonate of sodium, carbonic acid being expelled with effervescence, whilst the silica unites with the alkali. Advantage is taken of this property in the analysis of minerals which contain a large quantity of silica, and are not easily decomposed by acids. The finely powdered mineral is fused with about three times its weight of carbonate of sodium or of potassium, or, better, with thrice its weight of a

mixture of $5\frac{1}{2}$ parts of carbonate of sodium with 7 parts of carbonate of potassium. The mixed carbonates produce a more fusible mixture than either alone (§ 492). The fused mass is then treated with dilute chlorhydric acid, which decomposes the alkaline silicates, and dissolves all the bases of the mineral which were before combined with the silica.

Carbonate of potassium was the most important source of alkali, until Leblanc's process made soda cheaper than potash. It is still largely consumed in the manufacture of soap, glass, caustic potash, and other compounds of potassium; but sodium-salts have, to a great extent, displaced potassium-salts in commerce and the arts.

498. *Carbonate of Potassium and Hydrogen* ($KHCO_3$).—This salt, commonly called the "bicarbonate" of potassium (K_2O, $H_2O, 2CO_2$), is prepared by passing a current of carbonic acid through a strong solution of carbonate of potassium; crystals of the bicarbonate will be deposited, which are permanent in the air, and require about 4 parts of cold water for solution. When the solution of this salt is long exposed to the air, or boiled, it loses one-fourth of its carbonic acid; when the dry salt is fused, it loses half its carbonic acid, and is converted into the carbonate. It is a valuable salt to the chemist and the apothecary, because it can be readily obtained in a state of purity; when itself made from a refined carbonate of potassium, it may be advantageously used as the material from which to prepare other pure compounds of this important element.

499. *Hydrate of Potassium* (KHO).—The manufacture of hydrate of potassium, from carbonate of potassium, resembles, in every detail, the preparation of caustic soda from carbonate of sodium (Exp. 233). The carbonate of potassium is dissolved in 10 or 12 times its weight of water, and decomposed by a cream of lime; carbonate of calcium is precipitated, and hydrate of potassium remains in solution. All that has been said of the concentration of the solution of hydrate of sodium (§ 488) is true, also, of hydrate of potassium.

Hydrate of potassium is a hard, whitish substance, possessing a peculiar odor and a very acrid taste. Like the hydrate of sodium, it rapidly absorbs moisture and carbonic acid from the air;

and since the carbonate of potassium thus formed is a deliquescent salt, this change will go on until the entire mass of hydrate is converted into a syrup of the carbonate; whereas, in the case of hydrate of sodium, the absorption of water and carbonic acid is soon arrested by the formation of a coating of non-deliquescent carbonate of sodium upon the surface of the lump of hydrate. In chemical industries and speculations, the question of success or failure often turns on such points as this; the advantage of a new material, for example, often depends upon just such differences as this between caustic soda and caustic potash.

500. The hydrate of potassium, cast into small sticks, is employed by physicians as a cautery,—a use which illustrates forcibly its destructive effect upon animal and vegetable matters. Like hydrate of sodium, its solution destroys ordinary paper, and cannot be filtered except through asbestos, or gun-cotton. A clear solution is best obtained by decantation from the subsided impurities. All vessels made of materials which contain silica are attacked by this caustic solution; and even platinum is slowly oxidized in its presence; vessels of gold and silver resist it best. This hydrate, like that of sodium, forms soaps with oils or fats; the sodium-soaps are hard, the potassium- soft. At a high temperature hydrate of potassium volatilizes without change; heat alone cannot decompose the caustic alkalies. In the chemical laboratory, solutions of caustic potash and caustic soda are in frequent use for absorbing acid gases, such as carbonic acid, and especially for separating the hydrates of other metals from solutions of their salts.

Exp. 244.—Dissolve a crystal of blue vitriol (sulphate of copper) in a few centimetres of cold water, and add to the solution several drops of a solution of caustic soda (or potash). The hydrate of copper is precipitated as a delicate, blue, insoluble powder, while colorless sulphate of sodium (or potassium) remains in solution.

$$\underset{\textit{Sulphate of Copper.}}{CuSO_4} + 2NaHO = \underset{\textit{Hydrate of Copper.}}{CuH_2O_2} + Na_2SO_4.$$

Exp. 245.—Place in a small flask 4 or 5 grms. of chalk or marble (carbonate of calcium), and 7 or 8 c. c. of water; then cautiously add chlorhydric acid, little by little, until effervescence ceases and the chalk is dissolved. When the effervescence is not violent, the flask may be warmed to facilitate the process of solution. A rather concentrated solution of chloride of calcium will thus be obtained.

$$CaCO_3 + 2HCl = CaCl_2 + H_2O + CO_2.$$

Carbonate of Calcium. *Chloride of Calcium.*

Add to this solution a few drops of a solution of caustic soda, which is free from carbonic acid. A white precipitate of hydrate of calcium will immediately appear, since this hydrate is insoluble in the menstruum, while chloride of sodium will be found in the clear solution.

$$CaCl_2 + 2NaHO = CaH_2O_2 + 2NaCl.$$

Hydrate of Calcium.

501. On account of this power of precipitating other hydrates from solutions of their salts, the caustic alkalies are often called *strong* bases, as he is the strongest wrestler who throws his adversary; but this term "strong" is applied to bases and acids so confusedly as to be frequently a hindrance rather than a help in classification. Thus, if the reaction which occurs in the preparation of caustic soda (Exp. 233), between carbonate of sodium and hydrate of calcium, be compared with the reaction last given between chloride of calcium and hydrate of sodium, it will be seen that in the first calcium displaces sodium, while in the second sodium displaces calcium; in the one case hydrate of sodium is eliminated from the reaction, and in the other hydrate of calcium. So of acids; we have before had occasion to remark that, of two acids, now the one and now the other will be stronger, according to the temperature at which the contest between them takes place, or other extrinsic conditions. Thus sulphuric acid is at certain temperatures capable of displacing phosphoric acid or boracic acid, but at high temperatures both these acids displace it (§§ 294, 448). It is obvious that the definition of the term "strength" must be very vague and unsatisfactory when applied to relations thus capable of actual reversal. Two general principles, however, have been arrived at through the comparative study of such reactions; these are:—1st. When from all or part of the elements of any mixture of liquefied materials a substance can be formed which is insoluble in the existing menstruum, that substance will separate in the solid state; 2nd. When from all or part of the elements of a solid or liquid mixture a substance can be compounded which is volatile at the existing, or induced,

temperature, that substance will be eliminated in the gaseous state. In either case such interchange of atoms as may be essential to the formation of the eliminated substance, takes place, and the remaining elements necessarily adjust themselves to new relations. Such insoluble precipitates often present peculiarities of color or texture by which they may be recognized; and such volatile gases may frequently be identified by their color, odor, or specific gravity, or by the chemical effects which they are capable of producing. If every chemical element were known to yield, under attainable conditions, a characteristic precipitate, or to evolve a peculiar and recognizable gas, the analytical chemist would possess the means of detecting every element with certainty. This is to a great extent the case, and chemical analysis is chiefly based upon a knowledge of the degrees of solubility and volatility which belong to a great variety of chemical substances with whose appearance and prominent properties the analyst has previously made himself acquainted; of these substances many are common, but not a few rare and useless, except to serve the purpose of the analyst.

There exists an anhydrous oxide of potassium, K_2O, and also a peroxide. The anhydrous oxide K_2O is a gray, inodorous, hard, brittle solid; it melts a little above a red heat, but volatilizes only at a very high temperature.

502. *Alkalimetry.*—Since the value of the carbonates and hydrates of sodium and potassium, as they are manufactured and consumed on the large scale in the chemical arts, is generally dependent upon the amount of alkali which they contain ready to enter into chemical combination, it is important to have some quick and easy method of determining how much available alkali any sample of these substances really contains. The impurities which most frequently contaminate the carbonate of potassium are the chlorides and sulphates of potassium and sodium, silicic acid, lime, alumina, and the oxides of iron; the commonest impurities of carbonate of sodium are the chloride and sulphate of sodium, as might readily be inferred from consideration of the process by which the carbonate is manufactured. Some sulphite of sodium also is not infrequently present in commercial soda-ash. Both carbonates are apt to contain small proportions of the hy-

drates; but as the hydrates are quite as valuable for most uses as the carbonates, weight for weight, this admixture is not inconvenient. In the common methods of testing the carbonates, the hydrates present are estimated as available alkali, but are made no separate account of.

It would be foreign to our purpose to enter upon the details of alkalimetry; the process consists essentially in ascertaining how much dilute sulphuric acid of a known strength is required to neutralize exactly a known weight of the sample examined. The requisites are a graduated burette (Appendix, § 21), a standard acid, pure carbonate of sodium wherewith to prepare this acid, and a colored solution, sensitive to both acid and alkali, to indicate the point of neutralization. The following experiment will give some idea of the manipulation required in this sort of analysis, which, on account of its rapidity, is of very general application in technical chemistry. The general method is called *volumetric*, because, when once the standard liquids are prepared, quantitative results are obtained, not by weighing, but by measuring the bulk of liquid consumed in the testing.

Exp. 246.—Weigh out accurately 5 grms. of pure anhydrous carbonate of sodium; transfer it to a flask or beaker having the capacity of about 400 c. c.; dissolve it in about 200 c. c. of water, and color the solution blue with about 2 c. c. of a violet tincture of litmus. To prepare this tincture, digest 1 part of litmus in 6 parts of water, on a water-bath, for several hours; filter; divide the blue liquid into two equal portions, and stir one half repeatedly with a glass rod dipped in very dilute nitric acid, until the color just appears red; then add the other blue half, together with one part of alcohol, and keep the tincture in a small *open* bottle. In a stoppered bottle the tincture fades.

Mix about 60 grms. of strong sulphuric acid with 500 c. c. of water and let the mixture cool (§ 233). Fill a 50 c. c. Mohr's burette (Appendix, § 21) up to the 0 mark with the cold dilute acid. Place the flask or beaker containing the soda solution beneath the burette, gently press the spring-clip, and allow the acid to flow gradually into the soda solution, stirring the while, until the color of the liquid changes to a wine-red; then place the flask or beaker over a lamp, and bring the liquid to ebullition; the dissolved carbonic acid will be driven out, and the liquid will again become blue; more acid is then added to the nearly boiling fluid, the vessel being occasionally placed over the lamp, until the color of the liquid becomes red, slightly inclining to yellow.

When the point of saturation is approaching, add the acid two drops at a time, and after each fresh addition dip a fine glass rod into the fluid, and make with it two spots on a slip of fine blue litmus-paper, reading the volume each time, and marking the number of centimetres between the two spots. Continue this operation until the spots on the paper appear distinctly red; then dry the paper and take for the correct number that figure which stands between those two spots which just remain red when dry.

For some eyes, tincture of cochineal possesses great advantages over tincture of litmus as a means of recognizing the point of neutralization. The tincture of cochineal is prepared by digesting 3 grms. of powdered cochineal in a mixture of 50 c. c. of alcohol and 200 c. c. of water, at the ordinary temperature, for several days. The tincture, which may be either decanted or filtered from the residue, has a ruby-red color. The caustic alkalies and the alkaline carbonates change the color to a violet-carmine; solutions of strong acids and acid-salts make it orange; to carbonic acid it is nearly indifferent. 10 or 15 drops of the tincture are sufficient to color 200 c. c. of liquid.

Dilute the acid which remains of the original 500 c. c. with enough water to give a fluid of which exactly 50 c. c. are required to saturate 5 grms. of pure carbonate of sodium. This dilution is effected as follows:—Suppose that 40 c. c. of the acid have proved sufficient to neutralize 5 grms. of the pure carbonate, then 10 c. c. of water must be added to every 40 c. c. of the acid. 460 c. c. of the original dilute acid remain; now, as 40 : 460 = 10 : number of c. c. water to be added = 115 c. c. Dilute, therefore, the acid with 115 c. c. of water, using some of this water to rinse the burette which contains the undiluted acid, the measuring-glass which may have been used to ascertain the extent of volume of the original acid, and any other vessel into which it may have been temporarily poured. Of this diluted acid, 50 c. c. should exactly neutralize 5 grms. of pure anhydrous carbonate of sodium. It may be tested by again weighing out 5 grms. of the pure carbonate, and repeating the volumetric determination precisely as above described. If the work has been well done, the standard acid will be found ready for use.

Weigh out 5 grms. of common soda-ash, and dissolve in about 200 c. c. of water whatever of the sample is soluble; repeat upon this liquid, without filtering, the volumetric operation of the last experiment. The number of *half* c. c. of standard acid used gives directly the percentage of pure anhydrous carbonate of sodium which the sample contains. If 50 c. c. or 100 half c. c. are used, the sample is pure carbonate of sodium; if 40 c. c. or 80 half c. c. are used, the sample is eighty per cent. pure carbonate, and twenty per cent. water and other impurities.

503. When an acid has been prepared, of which 50 c. c. exactly neutralize 5 grms. of pure carbonate of sodium, it is a matter of simple calculation only to determine how much hydrate of sodium, how much carbonate of potassium, and how much hydrate of potassium the same 50 c. c. acid will neutralize. The following are the proportions required, the atomic weight of potassium being 39·1:—

106 :	80 =	5 :	x = 3·773;
Combining weight of Na_2CO_3.	*Com. wt. of* $Na_2H_2O_2$.	*Grms.* Na_2CO_3.	*Grms.* $Na_2H_2O_2$.
106 :	138·2 =	5 :	x = 6·519
Combining weight of Na_2CO_3.	*Com. wt. of* K_2CO_3.	*Grms.* Na_2CO_3.	*Grms.* K_2CO_3.
106 :	112·2 =	5 :	x = 5·292.
Combining weight of Na_2CO_3.	*Com. wt. of* $K_2H_2O_2$	*Grms.* Na_2CO_3.	*Grms.* $K_2H_2O_2$.

The following table shows the quantities of these four substances which are equivalent in neutralizing- or combining-power:—

5·000 grms. of carbonate of sodium	are saturated by 50 c. c. of one and the same standard acid.
3·773 „ „ hydrate „ „	
6·519 „ „ carbonate of potassium	
5·292 „ „ hydrate „ „	

It is apparent from this table that a given weight of carbonate of sodium will saturate more silicic acid or more fat, or will give off more carbonic acid than the same weight of carbonate of potassium and that the hydrate of sodium is also more efficient, gramme for gramme, than the hydrate of potassium. These facts are to be inferred at once from the atomic weights of sodium and potassium, which are 23 and 39·1 respectively; they have had their weight in bringing about the rapid substitution of sodium-salts for potassium-salts in the chemical arts.

504. *Potassium* (K).—This element, like sodium, is made from its carbonate by heating intensely a mixture of the carbonate and charcoal, in accordance with the reaction,

$$K_2CO_3 + 2C = 2K + 3CO.$$

The apparatus employed is similar to that described in treating of sodium (§ 487); and as potassium closely resembles sodium, the same general method is followed, and the same precautions are observed. A second distillation of the crude potassium is absolutely essential, because, if it be neglected, a black, detonating compound of uncertain composition is formed, which explodes violently upon the slightest friction.

Potassium is a silver-white substance, of very brilliant lustre, which is brittle at 0°, soft as wax at ordinary temperatures, fuses at 62°·5, and is volatile at a red heat. In its soft state, two clean surfaces can be welded together like iron. It is lighter than water, having a specific gravity of only 0·865. It is almost instantaneously oxidized by air and water at the ordinary temperature, and, when heated, by nearly every gas or liquid which contains oxygen. Like sodium, it must therefore be collected and kept under naphtha, out of contact with the air. At moderate temperatures, potassium readily absorbs hydrogen, nitrogen, and carbonic oxide, and enters into direct combination with chlorine, bromine, iodine, sulphur, selenium, and tellurium. We have had occasion to avail ourselves of its intense chemical energy (§§ 85, 97, 411).

Exp. 247.—Throw a piece of potassium, as large as a small pea, upon some cold water in the bottom of a large bottle, and place a card or glass plate over the mouth of the bottle. The potassium decomposes the water, and evolves heat enough to kindle the hydrogen which is given off; this hydrogen burns with a purplish-red color, imparted to the flame by a minute quantity of vaporized solid. This color is characteristic of potassium compounds, as a yellow color is characteristic of sodium compounds.

505. It is not a matter of indifference from what kind of a mixture of carbonate of potassium and carbon potassium is prepared. The material which is best adapted to its preparation is the potassium-salt of a vegetable acid rich in carbon, which, when decomposed by heat in a vessel from which air is excluded, yields carbonate of potassium and a large quantity of free carbon. While grape-juice is being converted into wine by fermentation, a stony deposit, called "tartar," which is sometimes gray and sometimes reddish, fastens to the bottom and sides of the casks

which contain the fermented juice. When freed by recrystallization from adhering coloring-matters, this crystalline and difficultly soluble substance is a white salt, acid and cooling to the taste. It is an acid tartrate of potassium, and is commonly called "cream of tartar." When this substance, or crude tartar, is heated in a covered crucible until it ceases to emit combustible vapors, the cooled residue is found to consist of a porous mass of carbonate of potassium, intimately mixed with very finely divided carbon. This mixture is the best material from which to prepare potassium; it is also an excellent flux, useful in assaying the ores of the common metals. In wine-producing countries considerable quantities of excellent carbonate of potassium are prepared from the deposits of the wine-vats, by dissolving the carbonate out of the ignited tartar and purifying the salt, so extracted, by recrystallization. The carbonate so obtained is the purest source of hydrate of potassium for laboratory use.

506. *Chloride of Potassium* (KCl).—This salt is a subordinate source of potassium compounds. It is extracted in considerable quantity from the ashes of sea-weeds, and is largely used in the manufacture of common alum, which is a sulphate of aluminum and potassium. It occurs native, sometimes pure, but more frequently mixed or combined with other chlorides. The chloride of potassium is capable of uniting with most other metallic chlorides, forming crystallizable double salts. The native double chloride of potassium and magnesium ($KCl,MgCl_2,6H_2O$) has become of late years a productive source of potassium-salts. This mineral is dissolved in hot water; from the cooled solution the greater part of the chloride of potassium crystallizes out, while the chloride of magnesium remains in solution. Chloride of potassium occurs also with the chlorides of sodium, magnesium, calcium, and other salts in sea-water and brine-springs, and is obtained as a secondary product in the preparation of chlorate of potassium (§ 517), the purification of nitre (§ 514), and in several other manufacturing-operations. It may be prepared directly from potassium and chlorine, or from the carbonate or hydrate of potassium and chlorhydric acid.

Chloride of potassium crystallizes in anhydrous cubes, looks and tastes like common salt, is not acted upon by the air, decrepitates when heated, melts at a low red heat, and volatilizes unchanged at a higher temperature. It is somewhat more volatile than common salt, is more soluble in water, and produces a greater degree of cold in dissolving. This chloride enters into some highly crystalline compounds, of curious composition, of which the product of the following experiment may serve as an example:—

Exp. 248.—By the aid of a gentle heat, dissolve 6 grms. of powdered red chromate of potassium in 8 grms. of strong chlorhydric acid, avoiding evolution of chlorine. When the powder is dissolved, allow the solution to cool; flat, red prisms will crystallize from the liquid. This compound answers to the formula KCl,CrO_3. It is decomposed by solution in water.

$$K_2O,2CrO_3 + 2HCl = 2(KCl,CrO_3) + H_2O.$$

Red Chromate of Potassium.

507. *Bromide of Potassium* (KBr) is a very soluble salt, which crystallizes in cubes, and closely resembles in all its properties the chloride. Potassium and bromine unite directly, with inflammation and violent detonation, the bromide being the product of the reaction. When bromine is added to a solution of caustic potash until the liquid acquires a permanent yellowish tinge, a mixture of bromide and bromate of potassium is produced:—

$$6KHO + 6Br = 5KBr + KBrO_3 + 3H_2O.$$

The mixed salts are dissolved in water, and a current of sulphydric acid is passed through the solution to reduce the bromate:—

$$KBrO_3 + 3H_2S = KBr = 3H_2O + 3S.$$

The liquid is then gently warmed to expel the excess of gas, filtered from the deposited sulphur, and evaporated until the bromide crystallizes. The salt has lately come into use in medicine as a sedative.

508. *Iodide of Potassium* (KI).—This valuable medicine and photographic material may be procured by adding iodine to a solution of hydrate of potassium until the liquid turns brown, and gently igniting the residue obtained by evaporation. The process and the reaction are the same as in the preparation of the bromide, except that the iodate may be decomposed by heat alone

and does not require reduction by sulphuretted hydrogen. A better mode of preparing the iodide of potassium is to be found in the decomposition of iodide of iron by carbonate of potassium.

Digest 4 grms. of iodine and 2 grms. of iron filings in a stoppered bottle with 20 c. c. of water; iodide of iron (FeI_2) is formed under these conditions by direct union of the elements. The liquid is then transferred to a flask and boiled, and a solution of carbonate of potassium is added by small portions so long as a precipitate occurs. The solution, filtered from the insoluble carbonate of iron, yields on evaporation crystals of iodide of potassium.

Iodide of potassium ordinarily crystallizes in semiopaque cubes. It is permanent in the air, has a sharp taste, melts below a red heat, and volatilizes unchanged at a low red heat. It is very soluble in water, and in dissolving produces a considerable fall of temperature. Alcohol also freely dissolves this salt. The facility with which the salt is decomposed and its iodine liberated by chlorine, ozone, and nitric acid has been already amply illustrated (Exps. 70, 72, 76).

Iodide of potassium is much used in medicine; it is not poisonous even in doses of 5 to 20 grms. Its solution in water dissolves iodine to a large extent, acquiring thereby a dark-brown color. This brown solution is sometimes used in medicine as a vehicle for iodine; it is also useful to the photographer for removing from the skin the stains produced by nitrate of silver. Iodide of potassium is, further, employed in the photographic process upon glass, to produce in the substance of the film of collodion a deposit of iodide of silver by double decomposition with nitrate of silver.

509. *Cyanide of Potassium* (KCN).—The elements which make part of the whitish, soluble, fusible, and deliquescent solid which bears this name are potassium, carbon, and nitrogen. At a bright-red heat these three elements will come together to form this salt out of quite a variety of materials, and under quite various circumstances. When nitrogen gas is passed over a mixture of charcoal and hydrate or carbonate of potassium, at a bright-red heat, cyanide of potassium is formed. When nitrogenous organic matters are fused with hydrate or carbonate of potassium, the cyanide is formed. In presence of iron scraps or filings, this

mixture produces a common cyanide, containing potassium, iron, carbon, and nitrogen, and known in the arts as "yellow prussiate of potash," and to chemists as ferrocyanide of potassium. If this "prussiate" be ignited at a moderate red heat, it is decomposed into cyanide of potassium, nitrogen, and a compound of iron with carbon:—

$$4KCN,Fe(CN)_2 = 4KCN + 2N + FeC_2.$$

In the blast-furnaces in which iron ores are smelted with coal or coke, a considerable quantity of cyanide of potassium is often produced, the nitrogen being probably derived from the torrents of nitrogen which the blast of air carries into the furnace. Instead of igniting the yellow prussiate alone, a more economical process is to ignite a mixture of the prussiate with carbonate of potassium. This method saves all the cyanogen in the prussiate; but the salt thus obtained is always mixed with cyanate and carbonate of potassium:—

$$K_4Fe(CN)_6 + K_2CO_3 = 5K(CN) + K(CN)O + CO_2 + Fe.$$

The presence of these impurities does not injure the cyanide for many of its uses.

Cyanide of potassium is of great use in galvanic gilding and silvering, since the cyanides of gold and silver are both soluble in a solution of cyanide of potassium. Its solution dissolves the sulphide of silver, and has therefore been suggested for household use in cleaning silver-ware; photographers sometimes use it for removing stains of nitrate of silver from the hands; but both these applications of cyanide of potassium are dangerous and inexpedient. The cyanide is intensely poisonous, not only when taken internally, but also when brought into contact with an abrasion of the skin, a cut, or scratch. As a reducing agent, cyanide of potassium has great power; it is especially useful in blowpipe reactions (Exp. 132).

510. *Sulphides of Potassium.*—Potassium, heated in sulphur-vapor, takes fire readily, and burns brilliantly. There are supposed to be five different compounds of potassium and sulphur, corresponding to the formulæ K_2S, K_2S_2, K_2S_3, K_2S_4, and K_2S_5; and there is a sulphydrate KHS, analogous to the hydrate KHO.

Exp. 249.—Heat, gently, a thorough mixture of 10 grms. of dry powdered carbonate of potassium and 6 grms. of sulphur, in a covered earthen or iron crucible, until effervescence ceases and the mass flows quietly. The fused mass has a yellowish-brown color, and consists of tersulphide, quinquisulphide, and intermediate sulphides of potassium, mixed with sulphate, and often with carbonate of potassium; it is called "liver of sulphur." This substance, dissolved in water, gives a greenish solution, from which dilute acids liberate sulphuretted hydrogen, and precipitate milk of sulphur (§ 198). The carbonic acid of the air is strong enough to effect this decomposition; hence the solid substance and its solution, when exposed to the air, smell of sulphuretted hydrogen:—

$$K_2S_3 + H_2SO_4 = K_2SO_4 + H_2S + 2S.$$

The chief use of liver of sulphur is in the medical treatment of skin-diseases.

511. When sulphydric acid gas is passed to saturation into a solution of caustic potash, a colorless solution is obtained, which is supposed to contain the sulphydrate KHS. It has no permanency, quickly absorbing oxygen and turning yellow.

All the sulphides of potassium are brown solids, having an alkaline reaction to test-paper, and smelling of sulphydric acid. Their solutions are oxidized by exposure to the air, and sulphur is deposited from them.

512. *Sulphate of Potassium* (K_2SO_4).—This anhydrous salt crystallizes in transparent hexagonal prisms, terminated by hexagonal pyramids, and consequently bears some resemblance to common quartz crystals. The salt has a strong tendency to form double sulphates; a double sulphate of potassium and magnesium is of importance in the manufacture of potassium-salts from sea-water. It also enters into the composition of many of the double sulphates, which are called alums from the name of the commonest member of the class, the sulphate of aluminum and potassium.

513. *Sulphate of Potassium and Hydrogen* ($KHSO_4$).—This salt, commonly called the "bisulphate," is formed on a large scale as a residuary product whenever nitric acid is manufactured from nitrate of potassium. When ignited, its crystals lose half their acid:—

$$2(KHSO_4) = K_2SO_4 + H_2SO_4;$$

and the salt is therefore sometimes used as a flux, in cases where the action of a strong acid is required at a high temperature upon salts or oxides with which it may be fused. Platinum tools may be cleaned by fusing this salt in or upon them.

514. *Nitrate of Potassium* (KNO_3).—This valuable salt, commonly called saltpetre, is very widely diffused as a natural product, being indeed seldom wholly wanting in a productive soil, or in spring- or river-water. At many localities it is found in caverns or caves in calcareous formations; but the principal commercial source of the salt is the soil of certain tropical regions, especially of districts in Arabia, Persia, and India, where the nitrate is found disseminated through the upper portion of the soil, or forming an efflorescence upon the surface, but never extending lower than the depth to which the air can easily penetrate.

This natural production of nitrates appears to result mainly from the putrefaction of vegetable and animal matters, in presence of the air and of alkaline or earthy bases capable of fixing the nitric acid as soon as it is formed. The well-waters of towns, contaminated by neighboring sewers and cesspools, nearly always contain nitrates. The decay of the luxuriant vegetation of the tropics, promoted by a hot sun and a moist atmosphere, is a never-failing source of ammonia; but it is not certain that the production of ammonia is a necessary stage in the process of nitrification. In the artificial production of nitrates in temperate climates, the supposed natural conditions have been roughly imitated. In the old saltpetre "plantations" of European countries, nitrate of calcium was produced by mixing decomposing vegetable and animal matters with cinders, chalk, marl, and so forth, moistening the mass repeatedly with leachings of manure-heaps, exposing it freely to the air for two or three years, and then lixiviating. The nitrate of calcium, which was the main product, was then converted into saltpetre by double decomposition with carbonate of potassium.

The saltpetre is extracted from the earth which contains it by lixiviation, evaporation, and crystallization; but inasmuch as for most uses it is required in a very pure state, the crude salt must generally be refined. The common impurities of crude saltpetre are chloride of sodium and chloride of potassium. In order to separate these chlorides, advantage is taken of the fact that the nitrate of potassium is four times as soluble in boiling water as the chloride of potassium, and six times as soluble as the chloride of sodium. The crude saltpetre is treated with a quantity of water, sufficient to dissolve at boiling all the nitrate of potassium, but not all of the chloride of sodium beside. This residual salt is scooped out of the vessel in which the solution is effected, and the solution, after being somewhat diluted, is boiled with a little glue, to coagulate the coloring-matters and other soluble dirt, and sweep the liquid clean by means of the adhesive scum which rises to the surface. The strong, clear solution is then transferred to shallow crystallizing pans, and left at rest if large crystals are desired; if small crystals are preferable, the liquid is constantly stirred from the moment that crystallization begins; the saltpetre is then deposited in a crystalline powder, called saltpetre-flour. The chlorides of sodium and potassium are nearly as soluble in cold water as in hot; but nitrate of potassium is only one-eighth as soluble in water at the temperature of the atmosphere as in boiling water. Hence the chlorides remain in the mother-liquor, while the nitrate rapidly separates from the solution as it cools. The crystals of saltpetre are drained, and then washed with a solution of saltpetre saturated in the cold. This solution takes up the adhering chlorides, but leaves the pure nitrate of potassium undissolved.

Large quantities of saltpetre are now made by decomposing nitrate of sodium with carbonate of potassium. When, through governmental interference, the East-Indian supply of saltpetre is checked, this method is resorted to with advantage. Tartar, and the ashes of the beetroot-sugar manufacture, are good sources of potash to be applied to this purpose. Crude nitrate of sodium contains so much chloride of sodium, that it is desirable to purify it for this use by previous recrystallization; otherwise potash would be unprofitably consumed in converting chloride of sodium into chloride of potassium. One of the processes for converting the nitrate of sodium into nitrate of potassium consists simply in adding the nitrate of sodium to a hot concentrated solution of carbonate of potassium; a precipitation of carbonate of sodium takes place, and this precipitate is removed as fast as it forms, until no more appears; from the cooled liquid saltpetre-flour is deposited. The carbonate of potassium may be replaced in this process by chloride of potassium. $NaNO_3+KCl=KNO_3+NaCl$.

515. Nitrate of potassium is white, inodorous, and anhydrous, and has a cooling, bitter taste. When pure, it is permanent in the air,—a fact of great importance, inasmuch as the chief use of this salt is in the manufacture of gunpowder. Were it hygroscopic, like nitrate of sodium, it would not be applicable to this use. Saltpetre is one of those few potassium-salts which cannot be wholly replaced in the arts by the corresponding sodium-salt. It is very soluble in water, especially in hot water; it melts below

a red heat to a colorless liquid without loss of substance, but at a red heat it gives off oxygen and suffers decomposition. Its most marked chemical characteristic is its oxidizing-power. It deflagrates in the fire with charcoal, sulphur, phosphorus, and other combustible bodies; when ignited in contact with copper or iron (Exp. 46), it converts these metals into oxides; and it even oxidizes gold, silver, and platinum. It is on the oxidizing-power of saltpetre that its use in the manufacture of gunpowder and fireworks, and in the preparation of matches, depends.

Arrange 10 grms. of pure nitrate of potassium and 20 or 30 grms. of thin copper-turnings, or small bits of sheet copper, in alternate layers, in a covered copper crucible, and expose the mixture for half an hour to a moderate red heat. Dissolve out the cooled mass with water, and let the liquid stand in a tall, closed bottle until the oxide of copper has settled to the bottom. The supernatant liquid is a pure solution of caustic potash; indeed this is an excellent method of preparing pure hydrate of potassium for use in analysis:—

$$2KNO_3 + 5Cu + H_2O = 2KHO + 5CuO + 2N.$$

Exp. 250.—Heat 10 or 12 grms. of saltpetre, gently, in a small porcelain dish, until it melts; pour the melted salt out on a cold piece of iron or stone; break the fused mass into small fragments, and fill an ignition-tube, 12–15 c.m. long, one-third full with these bits. Heat the tube cautiously, taking pains to keep all the salt, when once melted, in a state of fusion. At a red heat, oxygen, pure at first, is slowly evolved, and may be collected at the water-pan; simultaneously nitrite of potassium (KNO_2) is formed; at a second stage this nitrite is itself decomposed, and the escaping oxygen is then contaminated with a certain proportion of nitrogen. A portion of the gas collected may be tested with a glowing splinter (Exp. 6); another portion may be mixed with coal-gas, and, with the mixture, bubbles may be blown, as directed in Exp. 30; the mixture will be found to be exceedingly explosive.

This experiment proves, in the first place, that saltpetre itself is not explosive, and, in the second place, affords an explanation of the fact that frightful explosions do often occur when storehouses containing saltpetre are burned. Carburetted hydrogen, such as was obtained in Exp. 151 (represented by the coal-gas in the last experiment), is evolved from the wood-work of the burning building, wherever the wood is heated out of contact with the air; meanwhile oxygen is given off from the ignited saltpetre, and whenever these two gases mix in the requisite proportions, and their mixture comes in contact with a flame, a violent explosion inevitably ensues.

Exp. 251.—Dissolve 5 grms. of saltpetre in 20 c. c. of water; dip strips of bibulous paper in the solution, and dry them; this paper, once kindled, will smoulder away till consumed. It is used in connexion with fireworks, and in the manufacture of pastilles and aromatic fumigating paper.

Exp. 252.—Mix 5 grms. of powdered saltpetre with 1 grm. of dry powdered charcoal; place the mixture on a piece of porcelain and ignite it with a hot wire. When the deflagration is over, a white solid will be found upon the porcelain. Dissolve this solid in a few drops of water; the solution will be alkaline to test-paper; add a few drops of a dilute acid; a brisk effervescence marks the escape of carbonic acid. The nitrate has oxidized the carbon to carbonic acid, part of which escaped with the nitrogen during the deflagration, while part entered into combination with the potassium :—

$$4KNO_3 + 5C = 2K_2CO_3 + 3CO_2 + 4N.$$

Exp. 253.—Place 30 grms. of saltpetre in a small beaker with 110 c. c. of water; insert a thermometer in the mixture, and observe the very considerable fall of temperature occasioned by the solution of the salt. In those countries where saltpetre is cheap and ice dear, this property of the salt is availed of for the refrigeration of drinks.

516. *Gunpowder* is an intimate mechanical mixture of soft-wood charcoal (§ 382), sulphur, and nitrate of potassium, in the proportions of 70 or 80 per cent. of the nitrate to 10 or 12 per cent. of each of the other ingredients. Though it is in no sense a chemical compound, we may, for convenience' sake, express the composition of gunpowder by the formula $K_2N_2O_6+S+3C$, and may roughly formulate the reactions which occur when it is burned, by the following equation :—

$$K_2N_2O_6 + S + 3C = 3CO_2 + 2N + K_2S.$$

Speaking in general terms, the oxygen of the nitrate combines with the carbon to form carbonic acid, or, at the least, carbonic oxide, while the sulphur is retained by the potassium, and nitrogen left free. Gunpowder burns at the expense of the oxygen contained in it; it has no need of air for its combustion, but can be burned in any closed space—as well, for example, in canisters under water, or tightly enclosed in the chamber of a gun, as in free air.

From the formula, it will be seen, at a glance, that a very large proportion of gas, as compared with the bulk of the solid powder, must be evolved when powder is burned. But gunpowder burns rapidly and with great evolution of heat, so that the volume of gas, large at any temperature, is enormously expanded at the moment of its formation; hence it happens that the gas set free in the barrel of a gun may be capable of occupying a thousand or fifteen hundred times as much space as the powder which generated it. An enormous pressure is thus engendered at the spot where the powder burns, and to this pressure some part of the matter which confines the powder must yield. In the case of the gun-barrel, it is the bullet which represents the weakest, or breaking side of the chamber in which the powder burns; but when rocks are blasted, then the packing, or "tamping," which represents the ball, is made so firm that it shall be stronger than the rocky sides of the drill-hole, which is equivalent to the barrel of the gun. In case the walls of the gun can be disrupted more readily than the firmly impacted bullet can be driven out, then, of course, the gun bursts; and, conversely, the tamping of a drill-hole is thrown out if it be less firm than the rock. In the case of the gun-barrel, a part of the effect of the explosion is felt in the kick or recoil of the gun.

Though the equation last given is useful in so far as it exhibits the gaseous products evolved during the combustion of gunpowder, it does not truly express the solid products of the reaction. The residue of the combustion really contains only a comparatively small proportion of sulphide of potassium; it consists mainly of sulphate of potassium and carbonate of potassium, together with some hyposulphite of potassium, and a trace of unburned carbon. Enough sulphide of potassium is always present, however, to impart the offensive odor which is perceived in washing a foul gun, and in powder-smoke.

Exp. 254.—Pulverize, *separately*, 23 grms. of nitrate of potassium, 4 grms. of sulphur, and 4 grms. of charcoal. Place a drop or two of water in a porcelain mortar, and grind into it, first, the charcoal, and then the other ingredients, taking care to add enough water to form a plastic dough. After the mass has been thoroughly kneaded, roll out small portions of it between two pieces of board, into long threads, of the thickness of a fine knitting-needle. With a knife, cut the threads into small fragments or granules, and leave the granules in a warm room to dry. The thoroughly dried product is gunpowder; and the manipulation in this experiment does not differ essentially from the mode of manufacture employed in the powder-mills, excepting that the granulation is there effected by passing the moist paste through cullenders.

The sulphur in gunpowder acts mainly as a kindling material (§ 200). In powder intended for use in guns, the proportion of sulphur is kept comparatively low, since any excess of it would corrode the metal of the gun.

Exp. 255.—Knead together, as in Exp. 254, 7 grms. of powdered nitrate of potassium and 1·5 grm. of moistened, finely powdered charcoal. Granulate and dry the product, as before, and compare its inflammability with that of the gunpowder prepared in Exp. 254, by touching small heaps of each with a red-hot wire. Mixtures of charcoal and nitrate of potassium, such as the foregoing, are much used in the manufacture of fireworks.

517. *Chlorate of Potassium* ($KClO_3$).—The basis of the large use now made of this beautiful salt in medicine, in calico-printing, in pyrotechny, in the match-manufacture, and in the chemical laboratory is its large oxygen-contents. It is an oxidizing agent of the most vigorous description.

It may be prepared by saturating a solution of 1 part of hydrate of potassium in 3 parts of water with chlorine, and heating the liquid some time to the boiling-point. The ultimate result may be expressed by the formula

$$6KHO + 6Cl = KClO_3 + 5KCl + 3H_2O;$$

but the process has two stages, which are sufficiently described in § 124. The hot solution, left to itself, deposits the greater part of the chlorate in anhydrous six-sided plates of a pearly lustre; the chloride of potassium remains in the mother-liquor. The chlorate is freed from adhering chloride by recrystallization. The success of the process depends upon the very different solubilities of the chlorate and the chloride of potassium. At the temperature of their saturated boiling solutions both salts are about equally soluble; 100 parts of water will dissolve between 60 and 67 parts of either salt; but at the ordinary temperature of the air 100 parts of water will dissolve 30–40 parts of chloride of potassium and only 6 or 7 parts of chlorate of potassium. We find here the explanation of the fact that chlorate of sodium has not replaced chlorate of potassium in the arts. The chlorate of sodium is more soluble in water at all temperatures than the chloride of sodium is, while both are exceedingly soluble, so that the two salts cannot be separated by crystallization. This process of crystallization is the chemical manufacturer's chief reliance in refining both his materials and his products; and the purchaser of chemicals finds his best guaranty of the purity of his commodities in the peculiar form, lustre, color, and degree of transparency which characterize the crystals of every crystallizable and permanent chemical compound. Hence an easily crystallized permanent salt, of characteristic appearance, like chlorate of potassium, will always have the preference over one which, like chlorate of sodium, can be crystallized and purified only with difficulty, and is not permanent when once obtained. The chlorate of sodium is deliquescent.

The waste product in the making of chlorate of potassium by the process just described is chloride of potassium, a comparatively dear salt. An economy is effected by substituting hydrate of calcium for hydrate of potassium, and thus making the secondary product chloride of calcium instead of chloride of potassium; one equivalent only of the chloride of potassium is then required instead of six of the hydrate of potassium. An excess of chlorine is passed into a mixture of 300 parts of quicklime, 154 parts of chloride of potassium, and 100 of water. The mass is heated by steam, stirred with agitators, filtered, and then evaporated nearly to dryness by steam-heat; the mass is then redissolved in hot water and set to crystallize:—

$$3CaO + KCl + 6Cl = KClO_3 + 3CaCl_2.$$

The mother-liquor, which contains all the chloride of calcium, may be decomposed with sulphate of potassium, in which event a very finely divided sulphate of calcium, available for "stuffing" in the manufacture of paper, is precipitated, and chloride of potassium is recovered, to be again applied to the production of the chlorate; or the chloride-of-calcium solution may be decomposed with carbonate of sodium, in order to precipitate a very finely divided carbonate of calcium, which is largely employed by the pharmaceutist and perfumer. In the latter case, chloride of sodium has to be thrown away. The whole manufacture is a good example of a technical chemical process.

518. Chlorate of potassium is easily decomposed by heat; at a moderate temperature it yields perchlorate and chloride of potassium (§ 124); but at a red heat it is resolved into chloride of potassium and oxygen (Exp. 7):—$KClO_3 = KCl + 3O$. Chlorate of potassium is so prompt an oxidizing agent that mixtures of it with combustible bodies often detonate violently when struck or heated (Exps. 113, 157). These combustions are dangerous unless very small quantities be used. It has been often proposed to replace gunpowder by such mixtures; a mixture of the chlorate with catechu, or some similar stable substance rich in tannin, is

the most promising of these suggestions. Strong acids, like sulphuric, nitric, and chlorhydric acids, decompose chlorate of potassium with evolution of oxides of chlorine, or of chlorine and oxygen. The decomposition is often attended with decrepitation, and sometimes with a flashing light; combustibles, like sulphur, phosphorus, sugar, and resin, are inflamed by the gases evolved. A mixture of chlorate of potassium and chlorhydric acid is used in toxicological investigations as an oxidizing agent for the destruction of organic matter (§ 329). The following formulæ will elucidate some of these reactions:—

$$3KClO_3 + 2H_2SO_4 = 2ClO_2 + KClO_4 + 2KHSO_4 + H_2O$$
$$8KClO_3 + 6HNO_3 = 6KNO_3 + 2KClO_4 + 6Cl + 13O + 3H_2O$$
$$4KClO_3 + 12HCl = 4KCl + 3ClO_2 + 9Cl + 6H_2O.$$

Exp. 256.—Pour into a conical test-glass 25–30 c. c. of water, and throw into the water some scraps of phosphorus, weighing together not more than 0·3 grm., and 3–4 grms. of crystals of chlorate of potassium. By means of a thistle-tube bring 5 or 6 c. c. of strong sulphuric acid into immediate contact with the chlorate at the bottom of the glass. Then withdraw the thistle-tube. In a moment the phosphorus is kindled, and burns with vivid flashes of light beneath the water. An evolution of chlorine accompanies the reaction.

Exp. 257.—Rub 4 or 5 grms. of clean chlorate of potassium, free from dust, to a fine powder in a porcelain mortar. In powdering chlorate of potassium, care must be taken that the mortar and pestle are perfectly clean, and the salt free from organic matter, and that violent percussion and heavy pressure upon the contents of the mortar be wholly avoided. Place the powdered chlorate on a piece of paper, add an equal bulk of dry powdered sugar to the pile, and, with the fingers and a piece of card, mix the two materials thoroughly together. Mixtures of chlorate of potassium and organic matter are liable to explode, if strongly rubbed or but lightly struck. Wrap the mixture in a paper cylinder, and place the cylinder on a brick in a strong draught of air; let fall upon the mixture a drop of sulphuric acid from the end of a glass rod; a very vivid combustion will ensue, with the violet-colored flame characteristic of potassium.

Exp. 258.—Mix together, on paper, with the precautions above described, 1 grm. of black oxide of copper, 1 grm. of sulphur, and 2·5 grms. of powdered chlorate of potassium. Place the mixture, inclosed in a paper cylinder, on the top of a brick, and touch it with a hot wire; it will burn vividly, and with a purple color which is prized in pyrotechny.

CHAPTER XXV.

AMMONIUM-SALTS.

519. The hypothetical metal ammonium (NH_4) is a device for explaining the constitution and properties of one well-defined class out of the several classes of compounds into which the gas ammonia enters. This class of compounds is that which results from neutralizing an aqueous solution of ammonia with acids, as in the following reactions:—

$$NH_3,H_2O + H_2SO_4 = (NH_4)HSO_4 + H_2O.$$

Sulphate of Ammonium and Hydrogen.

$$NH_3,H_2O + HNO_3 = (NH_4)NO_3 + H_2O.$$

Nitrate of Ammonium.

According to this hypothesis, the crystalline salts which result from such neutralizations contain a group of atoms (NH_4) which is analogous in its action to potassium and sodium, and which forms salts analogous in composition to the potassium-salts. Thus chloride of ammonium $(NH_4)Cl$ is analogous to chloride of potassium KCl; sulphate of ammonium $(NH_4)_2SO_4$ is analogous to sulphate of potassium K_2SO_4, and so forth (§ 91).

All the actual evidence we possess of the separate existence and metallic character of the group NH_4 is contained in the following curious but inconclusive experiment:—

Exp. 259.—Pour 8 or 10 c. c. of mercury into a small flask, and warm the mercury over a gas-lamp; drop upon the mercury six or eight bits of metallic sodium, no one of them larger than a hemp-seed. The sodium dissolves with some spattering in the warm mercury, and a sodium amalgam is thus obtained. Transfer the amalgam to a tall glass or bottle of at least 300 c. c. capacity, and pour over it a concentrated solution of chloride of ammonium. The amalgam immediately begins to swell up, and ultimately increases to 8 or 10 times its original bulk, in the cold, or to 20 or 30 times if the solution be hot, assuming a pasty consistency like that of soft butter, but preserving its metallic lustre. It begins to undergo spontaneous decomposition as soon as it is formed, and if it is placed in water, this decomposition is quite rapid;

hydrogen gas is given off in minute bubbles, and ammonia is found in the solution. This curious substance has been supposed to be a combination of ammonium (NH_4) with mercury; all attempts, however, to isolate the ammonium have been unsuccessful. The proportion of ammonium (?) present in the amalgam is extremely minute, notwithstanding the great change of bulk and properties experienced by the mercury. The amalgam is said to contain only 1 part of nitrogen and hydrogen to 1800 parts of mercury.

520. Ammonium-salts are generally isomorphous with potassium-salts. They have mostly a pungent, saline taste; they are colorless, like sodium- and potassium-salts, unless the acids are colored; the carbonates, and those salts which, like the chloride and iodide, contain no oxygen, are volatile at a moderate heat without decomposition; some salts lose their ammonia when heated; if the acid which neutralized this ammonia is a non-volatile substance, like phosphoric acid, it will remain behind undecomposed; others, like the nitrate (Exp. 34), yield simpler gases than ammonia, as, for example, nitrogen or nitrous oxide. An aqueous solution of an ammonium-salt, when exposed to the air or evaporated, generally loses ammonia and acquires an acid reaction; hence in crystallizing an ammonium-salt, ammonia-water must be occasionally added during evaporation. All ammonium-salts, whether solid or in solution, evolve ammonia when heated with the hydrates of sodium, potassium, calcium and a few other metals (Exp. 48).

Exp. 260.—Warm a few centimetres of a solution of chloride of ammonium in a test-tube, add a few drops of a solution of caustic soda, and boil the liquid. The gaseous ammonia can be detected by its odor. If in any case the ammonia evolved be in so small a quantity that its characteristic smell cannot be detected, it may be recognized by its property of restoring the blue color to reddened litmus-paper (§ 83), and of forming white fumes by contact with a rod moistened with somewhat dilute chlorhydric acid (Exp. 65). The reaction may be formulated as follows:—

$$NH_4Cl + NaHO = NaCl + NH_3 + H_2O.$$

521. The solution of ammonia gas in water (NH_3,H_2O) may be regarded as a solution of hydrate of ammonium $(NH_4)HO$, comparable with the solution of caustic soda, $NaHO$, or caustic potash, KHO. This solution produces, indeed, many of the effects which

the solutions of the caustic alkalies produce; it neutralizes acids, displaces the oxides of many metals from solutions of their salts, and combines with fats to form a soap; it is, in short, a powerful base.

Exp. 261.—Dissolve a small crystal of alum in 6–8 c. c. of water in a test-tube, and add ammonia-water until the solution, after being well shaken, smells strongly of ammonia. A gelatinous precipitate of the hydrate of aluminum will appear in the liquid.

Exp. 262.—Dissolve about 1 grm. of sulphate of zinc in 6–8 c. c. of water in a test-tube; add 4 or 5 drops of ammonia-water, and shake up the contents of the tube. A white, translucent precipitate of the hydrate of zinc will appear. Pour into the turbid liquid in the tube 3 or 4 c. c. more of ammonia-water; the precipitate will redissolve and the liquid again become clear. The zinc is at first displaced from its position in the sulphate by the group (NH_4); but the hydrate of zinc thus precipitated is soluble in an excess of ammonia-water. The hydrate of ammonium behaves in this way with not a few salts of metals.

Ammonium-salts are very numerous; but only the few which are of present importance in the useful arts will be here described.

522. *Chloride of Ammonium* (NH_4Cl).—This salt, commonly called sal-ammoniac, is found native in many volcanic regions. When nitrogenized animal matter and chloride of sodium are distilled together, this salt sublimes from the mixture; the commercial supply of the salt was formerly obtained from the soot resulting from the incomplete combustion of camel's dung.

The raw material whence ammonium-salts are now manufactured is derived from gasworks and boneblack-factories. Coal and bones contain a portion of nitrogen, which, during the process of distillation, is partially converted into ammonia (§ 92); this ammonia combines with the carbonic acid and sulphydric acid which are likewise products of the distillation, and these compounds are condensed into a somewhat watery liquor, contaminated with tarry and oily matters, from which the ammonium-salts are subsequently extracted. The impure carbonate is converted into chloride by the addition of chlorhydric acid, or of the mother-liquor from saltworks (a liquor containing the chlorides of magnesium and calcium); on evaporating the clarified solution, crystals of sal-ammoniac are obtained, but they are generally too dirty for use. They are partly freed from tarry matters by heating them to a temperature a little below their subliming-point, but high enough to drive off the tar, and are then redissolved in water; this solution, decolorized by being filtered through animal charcoal, is recrystallized; these crystals are sometimes further purified by sublimation.

Chloride of ammonium serves for the preparation of ammonia (Exp. 48), and of carbonate of ammonium. It is somewhat employed in dyeing, and also in certain processes with metals, such as tinning, soldering, and silvering copper and brass, and galvanizing (zincing) iron. The sublimed salt forms semitransparent, tough, fibrous masses; it is very soluble in water, and a great reduction of temperature occurs during its solution; hence it is employed as a refrigerant. Its taste is sharp and acrid. When heated, it sublimes much below redness, without undergoing fusion.

Exp. 263.—Heat a bit of sal-ammoniac on a piece of porcelain, and observe the low temperature at which the solid is completely converted into vapor.

Exp. 264.—Place a teaspoonful of powdered chloride of ammonium in the hollow of the hand, and pour upon it two or three teaspoonfuls of water. The cold produced by the solution of the salt will be very perceptible.

523. *Sulphate of Ammonium* ($(NH_4)_2SO_4$).—If the ammoniacal liquor from gasworks or animal-charcoal factories be neutralized with sulphuric acid, or if it be decomposed by gypsum (sulphate of calcium), the sulphate of ammonium will be obtained. In the latter case, the impure carbonate of ammonium in the liquor, on being filtered through powdered gypsum, yields carbonate of calcium and sulphate of ammonium. Another recent mode of utilizing the ammoniacal liquor of gasworks yields a crude sulphate of ammonium; the liquor is suffered to flow down the coke-towers which are now often connected with sulphuric-acid chambers (§ 228), and there absorbs all the acid-fumes which escape from the chambers. A crude chloride of ammonium may be prepared in a similar way, by substituting ammoniacal liquor for water in the coke-towers of sulphate-of-sodium furnaces (§ 482). The absorbent power of the ammoniacal liquor is, of course, much greater than that of water.

Sulphate of ammonium is colorless, and has a very bitter taste; it is soluble in twice its weight of cold, and in its own weight of boiling water. Its crystalline form is the same as that of sulphate of potassium, and the commercial article looks very much like sand, just as the crystals of sulphate of potassium have a superficial resemblance to quartz crystals. It forms a considerable number of double salts, which are isomorphous with the corresponding salts of potassium. Sulphate of ammonium is employed in the manufacture of ammonium-alum, as an ingredient of artificial manures, and as a source of other ammonium-salts.

524. *Nitrate of Ammonium* ($(NH_4)NO_3$).—The method of preparing this salt, and its complete decomposition by heat, have been already described (Exps. 33, 34, and § 91). The salt crystallizes in long needles; it has a pungent taste, is soluble in less than half its weight of boiling water, and in dissolving produces sharp cold. Between 230° and 250° it is decomposed into water and nitrous oxide; if it be heated hotter, or too rapidly, ammonia, nitric oxide, and nitrite of ammonium $(NH_4)NO_2$ are also formed. Nitrate of ammonium is formed by the action of dilute nitric acid on several metals, especially tin.

525. *Carbonates of Ammonium.*—Commercial carbonate of ammonium (sal-volatile) is a white, semitransparent, fibrous substance, with a pungent taste and a strong ammoniacal smell; it is prepared, on a large scale, by the dry distillation of bones, horn, and other animal matters. The product is purified from empyreumatic substances by repeated sublimation.

Exp. 265.—Mix thoroughly together 10 grms. of chloride of ammonium and 20 grms. of powdered chalk; heat the mixture in a small evaporating-dish placed upon a sand-bath. When white vapors begin to rise from the hot mass, place a wide-mouthed bottle over the fuming mixture. The white sublimate which collects in the bottle is a carbonate of ammonium; chloride of calcium remains in the dish.

This experiment illustrates a second, and very common, method of preparing the commercial carbonate, which simply consists in heating to redness a mixture of 1 part of chloride, or sulphate, of ammonium and 2 parts of carbonate of calcium. When this commercial carbonate is dissolved in strong ammonia-water at about 30°, a solution is obtained which yields large, transparent, prismatic crystals. These crystals, however, have no stability; they are rapidly decomposed in the air, giving off water and ammonia. "Sesquicarbonate of ammonia" is the name generally applied to this substance—a name deduced from the dualistic formula $2(NH_4)_2O, 3CO_2$. The commercial carbonate approximates to the composition represented by this formula; but it is an impure product, and, when exposed to the air, changes gradually into a more stable compound, the carbonate of ammonium and hydrogen, or "bicarbonate" $(NH_4)HCO_3$.

This bicarbonate may be obtained by saturating the solution of ammonia, or sesquicarbonate of ammonium, with carbonic acid; it forms large, transparent, prismatic crystals. When exposed to the air, it slowly volatilizes, giving off a slight ammoniacal odor. At the ordinary temperature, it is soluble in 8 parts of water; this solution, if heated above 36°, evolves carbonic acid. Even at the ordinary temperature, the solution gradually becomes ammoniacal on keeping. White crystalline masses of this bicarbonate have been found in guano deposits. It seems to be the most stable of the carbonates of ammonium; for the other carbonates change into it if left to themselves.

526. *Sulphides of Ammonium.*—At a temperature of −18°, two volumes of ammonia-gas combine with one volume of sulphydric acid gas to form a crystalline unstable substance, of strong alkaline reaction, which corresponds in composition with the sulphides Na_2S and K_2S.

$$2NH_3 + H_2S = (NH_4)_2S.$$

Exp. 266.—Pass a slow stream of washed sulphydric acid through 100 c. c. of strong ammonia-water, until the solution has a predominating and persistent odor of sulphuretted hydrogen. This solution is at first colorless, and contains a sulphide of ammonium and hydrogen $(NH_4)HS$; but when kept in contact with air it becomes yellow, owing to the formation of a higher sulphide of ammonium. The solution has the property of dissolving many of the sulphides of the metals, by forming with them double sulphides, and is a very useful reagent in the analytical laboratory.

The higher sulphides of ammonium are obscure bodies, to which the following formulæ have been assigned,—$(NH_4)_2S_2$, $(NH_4)_2S_3$, $(NH_4)_2S_4$, $(NH_4)_2S_5$, $(NH_4)_2S_7$. With the exception of the last, the septisulphide, all these sulphides are soluble in water. With the same exception, they correspond with the sulphides of sodium and potassium. They are in general unstable; the most permanent is the septisulphide, which forms ruby-red crystals, capable of resisting temperatures below 300°, and only slowly decomposable by water and chlorhydric acid.

CHAPTER XXVI.

LITHIUM—RUBIDIUM—CÆSIUM—THALLIUM.

SPECTRUM ANALYSIS.

527. *Lithium* (Li).—This rare metal occurs as a constituent of not a few minerals, especially micas and feldspars, but does not form a large percentage of any of them. The minerals lepidolite, triphylline, and petalite usually contain from 3·5 to 5 per cent. of lithium, and are the chief sources of the element. In much smaller proportion, it has been recognized in sea-water, mineral waters, and almost all spring-waters, in milk, tobacco, and human blood. It is therefore a widely diffused, but not abundant, substance. The metal, which is obtained by decomposing the fused chloride by the galvanic current, has the color and lustre of silver on a freshly cut surface, but quickly tarnishes on exposure to the air. It is harder and less fusible than sodium and potassium, but softer than lead; it may be welded, by pressure, at ordinary temperatures. It floats on naphtha, and is the lightest of all known solids which include no air, its specific gravity being only 0·59. The atomic weight of the element is also low, namely 7.

In its chemical relations, lithium closely resembles sodium and potassium, but is somewhat less energetic; it combines with the same elements to form analogous compounds to those of sodium and potassium; but the properties of these compounds, while presenting a striking general resemblance to those of the sodium and potassium compounds, nevertheless offer some special points of divergence from them. Thus, the hydrate of lithium (LiHO) has the same taste, causticity, and alkalinity as the hydrates of sodium and potassium, but is much less soluble in water. The fused hydrate attacks platinum more energetically than caustic soda and potash do. The carbonate and phosphate of lithium are rather sparingly soluble in water, while the corresponding sodium and potassium-salts are exceedingly soluble. The chloride of lithium (LiCl), produced when lithium burns in chlorine, or when the hydrate or carbonate of lithium is dissolved in chlorhydric acid, crystallizes in cubes, and has the taste of common salt; but it is more volatile than the chloride of potassium, and it deliquesces faster than any other known salt, whereas the chlorides of sodium and potassium are almost permanent when pure. All the volatile lithium-compounds color a gas-, alcohol-, or blowpipe-flame carmine-red. The most delicate reaction for the detection of lithium, the test which has revealed its existence in a great variety of substances which were never imagined to contain it, is the presence of one bright line, of a peculiar red, in the spectrum seen on looking through a glass prism at a flame colored with a lithium compound.

528. *Spectrum Analysis.*—We have had occasion to observe that certain chemical substances, like boracic acid and salts of sodium, potassium, and lithium, impart peculiar colors to the blowpipe-flame, or to any other hot and colorless flame. If these colored flames are looked at through a prism, a narrow pencil of the colored light being directed through a slit upon the prism, it will be seen that each different flame produces a peculiar *spectrum*, consisting of one or more distinct bright lines of colored light, and bearing no resemblance to the continuous band of rainbow-colors which constitutes the common spectrum produced by a pencil from any source of white light. Thus the spectrum of the yellow sodium-flame consists of a single, bright, yellow line; the purple potassium-flame gives a spectrum containing two bright lines, one lying at the extreme red and the other at the extreme violet end, and a second, fainter red line; while the lithium spectrum consists of a very characteristic red line and a fainter orange line.

These peculiar lines which characterize the spectrum of any element are invariably produced by that element, and never by any other substance, and not only the color and number of lines, but their position in the normal spectrum always remains unaltered. When the spectrum of a flame colored with a mixture of sodium and potassium salts is examined, the yellow line of sodium is seen in its place, and the red and purple lines of potassium are as visible in their respective positions as if no sodium had been present. This example illustrates one great advantage which the use of the prism gives,—the unaided eye cannot distinguish the potassium color in the presence of the intense sodium-yellow, the brighter color hiding the paler; but with the prism it is easy to detect each of several ingredients of a mixture by the appearance of its characteristic lines. Now, every elementary substance, whether metallic or non-metallic, solid, liquid, or gaseous, when heated to the point at which its vapor becomes luminous, emits a peculiar light produced by it alone, and the bright lines of the spectrum of this light are characteristic of this element in number, color, and position. Many metals require a much higher temperature than that of the common gas-flame to convert them into luminous vapors; but by the use of the electric lamp all the metals, even gold, silver, and platinum, may be made to yield peculiar spectra. The permanent gases also give characteristic spectra when they are heated by the passage of the electric spark; the spectrum of hydrogen, for example, consists of one red, one green, and one blue line.

A new method of analysis, of extreme delicacy, is based upon these facts. Spectrum analysis is competent to detect the $\frac{1}{2736000000}$ of a gramme of sodium, or the $\frac{1}{900000000}$ of a gramme of lithium, and many other elements in incredibly small proportions. So extreme is the delicacy of the method that it brings into plain sight minute quantities which altogether escape the coarser process of analysis, and reveals, as substances common in familiar things, elements which were long supposed to be of extreme rarity. Thus the presence of lithium, formerly considered a rare element peculiar to a few obscure minerals, has been demonstrated by spectrum analysis in many drinking-waters, in tea, tobacco, milk, and blood. A still more striking illustration of the value of spectrum analysis is to be found in the discovery of four new elementary bodies by its means.

529. Two new elements which closely resemble sodium and potassium, and are in nature associated with these alkali-metals, have been found in certain mineral waters, and in the mineral lepidolite. One of these elements gives a spectrum containing, among others of less mark, a superb, double, red line, and has thence been called *Rubidium*; the other produces a spectrum characterized by two beautiful blue lines, and has thence been called *Cæsium*. These two new metals resemble potassium so closely in all their chemical properties, that it would have been nearly impossible to detect them by the common analytical processes; yet their spectra are in the highest degree characteristic, exhibiting bright bands which exist neither in the potassium spectrum nor in any other known spectrum. The recently discovered metal *Thallium* was discovered and traced to its source in certain kinds of pyrites by observing a splendid green line which did not belong to any known substance. The new metal *Indium* was also detected, traced to its source in certain zinc ores, and successfully isolated, by the help of a dark-blue line which had not been previously observed.

The methods and processes of spectrum analysis are not applicable to colored artificial lights alone; they have been applied with encouraging success to the lights of various quality which emanate from the sun, the stars, and the nebulæ; but the details of these observations belong rather to physics than to chemistry.

530. *Rubidium and Cæsium* (Rb and Cs).—These two elements are always found together, and in association with potassium. Though extensively diffused, they generally occur in very minute quantities. Rubidium seems to be rather the most abundant. Ten kilogrammes of the mineral water in which these metals were first discovered yield not quite two milligrammes of chloride of cæsium, and about two and a half milligrammes of chloride of rubidium. Since the original discovery of the elements, they have been found in many other springs, in several kinds of mica and in other silicates, and in the ashes of beet-root, tobacco, coffee, and grapes. To separate the metals from potassium the analyst relies on the greater insolubility of the double chlorides which they form with platinum; if a mixture of the chlorides of potassium, rubidium, and cæsium be completely precipitated by chlo-

ride of platinum, and the yellow precipitate be repeatedly treated with boiling water, the insoluble residue will contain the new metals. The cæsium is separated from the rubidium by converting a mixture of their carbonates into their tartrates, the rubidium into the acid, or bitartrate, the cæsium into the neutral tartrate, and then exposing this mixture to very moist air. The neutral tartrate of cæsium deliquesces, the acid tartrate of rubidium remains solid, and the two salts are separated by filtration. Most of the salts of rubidium and cæsium are isomorphous with the corresponding potassium-salts. Their hydrates, RbHO and CsHO, are caustic alkalies, soluble in water and alcohol. Their carbonates are fusible, deliquescent, and strongly alkaline; the nitrates ($RbNO_3$ and $CsNO_3$) and sulphates (Rb_2SO_4 and Cs_2SO_4) are anhydrous crystalline salts, soluble in water; the sulphates form alums with sulphate of aluminum. The chloride of cæsium, CsCl, is a deliquescent salt, like chloride of lithium; the chloride of rubidium, RbCl, is permanent, like the chlorides of sodium and potassium. The fused chlorides are easily decomposed by the galvanic current.

The metal rubidium is white, and has the brilliant lustre of silver, but it rapidly oxidizes in the air; its specific gravity is 1·52, and its atomic weight 85·7. It may be prepared either by the electrolysis of its chloride, or, like potassium, by the reduction of its carbonate by ignition with carbon and chalk.

The properties of cæsium have only been studied in the amalgam with mercury resulting from the electrolysis of its chloride; the metal itself has not been isolated. Its atomic weight, deduced from the analysis of its chloride, is 133. There can be no question that the properties of both rubidium and cæsium differ from those of sodium and potassium not in kind, but only in degree. They are therefore classed with sodium and potassium as alkali-metals.

531. *Thallium* (Tl).—This metal was discovered, by means of spectrum analysis, in Lipari sulphur and in the deposit in the flue of a pyrites-burner—a furnace in which iron pyrites are roasted for the sake of the sulphurous acid they yield. The element is found to occur in not inconsiderable quantities in many specimens of iron pyrites, and appears to take the place of arsenic, which is

a common impurity of this mineral. Thallium presents the external characters of lead; it is heavier than lead, having a specific gravity of 11·85, and it is so soft that the thumb-nail can indent it; it is very malleable, and ductile enough to be drawn into wire; it melts at 200° and volatilizes at redness; its freshly cut surface has a bluish-white lustre; but it quickly tarnishes and is gradually oxidized in the air, so that it is best preserved under water. Water is not decomposed by it even at 100°. When strongly heated in oxygen, it takes fire and burns with a bright green flame.

Thallium dissolves in dilute acids, with evolution of hydrogen. There are several oxides of this metal, of which the most important is the oxide Tl_2O, corresponding in composition, and, to some extent, in properties, with the oxide of sodium Na_2O. This oxide is somewhat soluble in water, and yields a caustic alkaline solution, which absorbs carbonic acid from the air, and forms a well-defined series of salts. The sulphate, Tl_2SO_4, is a soluble salt, which forms an alum with sulphate of aluminum; the chloride, TlCl, is only slightly soluble in water, resembling, in this respect, the chloride of lead, and being quite unlike the soluble chlorides of sodium, potassium, rubidium, and cæsium. The carbonate of thallium is a soluble salt; but the sulphide of thallium, Tl_2S, is an insoluble black powder, which resembles the sulphide of lead, but is entirely unlike the sulphides of the alkali-metals. The soluble salts of thallium are very poisonous. In general, the properties of thallium are intermediate between those of lead and those of sodium and potassium. Like the alkali-metals, it replaces hydrogen atom for atom; its atomic weight is 204.

CHAPTER XXVII.

SILVER—THE ALKALI-METALS—QUANTIVALENCE.

532. Silver is a widely diffused and quite abundant element, but in its mode of occurrence it differs widely from the alkali-metals which we have just been studying. In the first place, it

frequently occurs native, both pure and alloyed with mercury, copper, and gold,—a mode of occurrence quite impossible for the alkali-metals, because of their readiness to combine with the elements of air and water. Native silver is found in various forms, sometimes crystallized in cubes or octahedrons, sometimes in filaments, both coarse and fine, and sometimes in shapeless masses. The metal more commonly occurs in combination with sulphur, mixed with sulphides of lead, antimony, copper, and iron. It is from argentiferous sulphides that the larger part of the silver of commerce is extracted; among ores of this kind the argentiferous sulphide of lead (galena) is the most abundant. Combinations of silver with selenium, tellurium, chlorine, bromine, and iodine are also to be enumerated among silver-containing minerals; of these the chloride (horn-silver) occurs in quantities large enough to make it valuable as an ore of the metal. It is noticeable that the only elements which are extracted in any quantity from their chlorides as ores are sodium, potassium, and silver. The chlorides of copper, mercury, and lead do, indeed, occur as natural minerals; but as sources of those metals they have no significance. A small proportion of silver exists in sea-water (about 1 milligramme in 100 litres), and its presence has been recognized in common salt, in chemical products in the making of which salt is used, in various sea-weeds, in the ashes of land-plants, in the ash of ox-blood, and probably also in coal. In sea-water it exists, as sodium and potassium do, in the form of chloride.

When silver is extracted from argentiferous sulphide of lead, the ore is first treated for lead, precisely as it would be if it contained no silver. The lead, so reduced, contains all the silver originally present in the quantity of ore treated. The subsequent separation of the metallic silver from the metallic lead depends upon the chemical properties of lead rather than of silver, for the silver remains unaltered during the whole process; this separation will therefore be described in the next chapter.

The mixed sulphides which contain silver have been heretofore generally reduced by a complicated process which depends ultimately on an amalgamation of the silver with mercury. The ore, after thorough washing and grinding, is mixed with a portion of common salt, and roasted for several hours; during this roasting, white fumes of arsenious and antimonious acids are expelled, the sulphides of copper and iron are partially converted into oxides, chlorides, and sulphates, and chloride of silver and sulphate of sodium are formed. The roasted product is then reduced to a very fine powder, and agitated in revolving casks with water and iron filings, or scraps, to which mercury is soon added. This operation lasts about 20 hours; during it, the iron decomposes the chloride of silver, and the mercury dissolves the silver to an amalgam; from this amalgam the excess of mercury is first squeezed out through leather or cloth filters, and the remainder is driven off by distillation. The residual spongy mass is silver, alloyed with a variable proportion of copper, derived from the ore and reduced to the metallic state by the same steps which have reduced the silver.

This process is European; the process of amalgamation as practised in Mexico and South America is quite different, and the reactions which it depends upon are somewhat obscure. The ore is not roasted, but, after being ground to powder, moistened with water, and mixed with from 1 to 5 per cent. of common salt, it is suffered to lie undisturbed for some days. From ¼ to 1 per cent. of roasted copper pyrites is then added, together with a considerable proportion of metallic mercury, and the mass is worked together and commingled by the trampling of mules or horses. After an interval of two or three weeks, a second dose of mercury is given, and after a still longer interval a third. By this last addition, a fluid amalgam is obtained, which is separated by washing and filtering, and distilled for the recovery of a portion of the mercury employed, and the isolation of the silver. In this process there is a great waste of mercury, because much of it is converted into a chloride of mercury (calomel) and lost. The recommendations of the process are mainly these—that it requires no fuel, except for the distillation of the amalgam, and that it leaves the silver in a condition of great purity. The whole process, though far from economical from the point of view of the theoretical chemist, was doubtless a legitimate outgrowth of the conditions under which it took birth.

Various processes have been patented for the extraction of silver without the use of the costly mercury, some of which have been successfully practised on a large scale. They depend, for the most part, either on the comparative stability, in the fire, of sulphate of silver when once formed, as compared with the sulphates of iron and copper, and the consequent possibility of dissolving sulphate of silver out of the roasted ore, or upon the fact that the chloride of silver, which results from the roasting of the ore with chloride of sodium, may be dissolved in solutions of the alkaline chlorides, and, indeed, in aqueous solutions of a great many other soluble salts, though it is by itself insoluble in water. Any aqueous solution containing, among other things, a silver-salt (whether in the condition of chloride, sulphate, or nitrate is indifferent) may be decomposed by digestion with metallic copper; the silver-salt will be decomposed, the corresponding copper-salt formed and dissolved, and the metallic silver will be precipitated.

533. *Silver* (Ag).—The element, silver, is much more familiarly known than any of its compounds; known from the earliest ages, this metal has always been prized as much for its beauty as for its rarity. White, brilliantly lustrous, susceptible of an admirable polish, wonderfully malleable and ductile, the best known conductor of heat and electricity, fusible only at a very elevated temperature and permanent in the air, whether hot or cold, wet or dry, it represents and embodies in the completest sense all that is commonly understood by the term *metal.*

This word metal cannot be strictly defined; it is a conventional term, vaguely used because expressing a vague idea. Thus metals would all be solid were not mercury, and perhaps cæsium, fluid; they are generally heavy; but lithium, sodium, and potassium float upon water; they have all a peculiar lustre, called metallic; but this lustre does not characterize metals alone, for coke and graphite, galena, molybdenite, and many other minerals often exhibit a similar lustre; they may all be said to be opaque; but gold may be beaten out so thin as to transmit a greenish light. While it is not possible to define the term metal with precision from chemical any more than from physical properties, one general chemical fact deserves attention in this connexion. We have seen that bodies which contain a large proportion of oxygen, such as SO_3, P_2O_5, N_2O_5, and CO_2, have a common tendency to unite with other bodies which are alike in that they contain a much smaller proportion of oxygen, such as K_2O, Na_2O, PbO, and CaO, to form more or less stable saline substances. The first class of bodies, which are usually rich in oxygen, have been called *acids*; the second class, which are usually poor in oxygen, have been designated collectively as *bases*. Now those elements which unite with oxygen to form acids alone are, as a rule, non-metallic, and those elements which unite with oxygen to form bases are, in the chemical sense of the term, the metals; but no sharp line of division between metallic and non-metallic elements can be established on this principle, inasmuch as some elements which possess the other characteristics of a metal form no basic oxide, while some metals, like antimony, form oxides which are at one time bases and at another time acids. The metal arsenic, for example, forms no basic oxide; and we shall hereafter meet with another illustration of the same difficulty of classification, in the element tungsten.

Melted silver possesses the curious property of absorbing a large volume of oxygen (twenty-two times its bulk), from the air, while it is liquid. This gas it gives out again on solidifying. When a globule of molten silver is cooled suddenly, the film of solid metal which forms upon its surface is burst open by the escaping gas, and the liquid silver within is apt to be projected outwards with the gas; this phenomenon is called *spitting*; it often occasions a loss in silver assays. When heated on lime before the oxyhydrogen blowpipe, silver gives off vapors which become oxidized if the blast of gas contain an excess of oxygen; a fine silver wire is dispersed in greenish vapors when a very powerful electric discharge is sent through it. Silver combines slowly with chlorine, bromine, and iodine, and promptly with sulphur. The tarnishing of silver is due to the formation of a thin film of the black sulphide over the metallic surface, by combination between the silver and the sulphur of the sulphuretted hydrogen which is often present in the air of towns and houses.

The best solvent for silver is nitric acid diluted with two or three parts of water; nitric oxide is evolved, and nitrate of silver remains in solution:—

$$3Ag + 4HNO_3 = 3AgNO_3 + NO + 2H_2O.$$

Chlorhydric acid acts upon it but slowly; for the chloride of silver is but slightly soluble in chlorhydric acid, whether strong or dilute. Boiling sulphuric acid dissolves silver, and forms the sulphate, sulphurous acid being evolved during the reaction:—

$$2Ag + 2H_2SO_4 = Ag_2SO_4 + 2H_2O + SO_2.$$

Neither the alkalies nor their nitrates have much effect on silver, whether they are in solution or are fused by heat; hence a silver dish is used in concentrating the caustic alkalies, and a silver crucible for fusing refractory minerals with the hydrate of sodium or potassium. The specific gravity of silver is 10·5, and its atomic weight 108.

534. The physical and chemical qualities of silver fit it to serve as a medium of exchange, and as the material of jewellery and plate. But as the pure metal would be rather too soft for ordinary use, it is hardened by combining with it a small proportion of copper. The proportion of copper in the "standard" silver employed for coinage varies in different countries. In the United States and in France it is 10 per cent.; in Great Britain it is 7·5 per cent.; in Germany it is 25 per cent.

Exp. 267.—Place one or two dimes in a small flask, and cover them with nitric acid diluted with two parts of water. Warm the flask gently in a place where there is a good draught of air; the coins will gradually dissolve, with evolution of nitric oxide, which, on contact with the air, produces the abundant red fumes which escape from the flask; add more nitric acid, from time to time, if necessary to complete the solution. The blue solution contains both the silver and the copper dissolved in nitric acid.

Place in the blue solution two or three copper cents, and leave the flask at rest for some days in a warm place. Then collect the little plates of pure silver, which have separated from the solution, upon a filter, and wash them, first with water, and then with ammonia-water, until the ammonia-water no longer shows any tinge of blue. This silver, washed finally with water and dried, is well nigh pure; two-thirds of it may be again dissolved in nitric acid; the solution will contain pure nitrate of silver.

535. *Nitrate of Silver* ($AgNO_3$).—This salt, as we have already seen, is obtained in solution by dissolving silver in nitric acid. When such a solution is evaporated to the point of crystallization, the nitrate is obtained in transparent, anhydrous, tabular crystals, which are soluble in their own weight of cold water, and in half their weight of hot water. The salt fuses easily, and when cast into cylindrical sticks is used in surgery as a caustic, under the name of *lunar caustic.*

Nitrate of silver, when pure, is not altered by exposure to sunlight; but if it be in contact with organic matter, light readily decomposes it, and a black, insoluble product is formed of no ordinary stability. Hence the solution of the nitrate stains the skin black, and the salt forms the basis of an indelible ink used for marking linen and other fabrics.

Exp. 268.—Dissolve 8 grms. of crystallized carbonate of sodium and

1 grm. of gum-arabic in 16 c. c. of hot water. Moisten a bit of linen or cotton-cloth with this preparatory solution, dry it, and press it smooth with a hot iron.

Dissolve 1 grm. of nitrate of silver and 0·1 grm. of gum-arabic in 1·75 c. c. of water, previously colored with India-ink.

Write with this silver solution upon the prepared surface of cloth, and expose the writing to the direct rays of the sun for a few hours. Then wash out the gum and carbonate of sodium with water; a very durable mark, which neither soap nor "soda" will obliterate, will remain on the cloth.

Nitrate of silver is even more completely decomposed by a red heat than nitrate of potassium, for nothing but metallic silver remains behind; in decomposing, it gives up a large quantity of oxygen; hence mixtures of combustibles, like sulphur, phosphorus, and charcoal, with nitrate of silver, detonate, explode, or deflagrate when struck sharply with a hammer or touched with a hot wire. (Compare §§ 515, 516.) Phosphorus, mercury, charcoal, grape-sugar, certain essential oils, and many other organic substances reduce metallic silver from solutions of nitrate of silver. Nitrate of silver is the material from which most other silver compounds are artificially prepared. It is largely consumed in photography.

The precipitation of metallic silver in a beautiful arborescent form is accomplished as follows:—Dissolve 2 grms. of nitrate of silver in 60 c. c. of water, and place the solution in a test-glass; pour 2 grms. of mercury into the liquid, and let the glass stand at rest for several hours. The nitrate of silver may be recovered by dissolving the precipitated silver in nitric acid, and evaporating the solution.

To illustrate the decomposition of a silver solution by an organic substance, dissolve 2 grms. of nitrate of silver in 60 c. c. of water, and immerse in the solution a horn or ivory paper-knife, which has been cleansed from grease with ammonia-water and rinsed in fresh water. Let the knife remain in the solution about an hour; it will turn yellow; take it out, rinse it in water, and expose it to the direct rays of the sun until it turns jet black; then burnish it with a piece of leather, and the silver will appear in the metallic state.

Exp. 269.—Wrap a piece of phosphorus, no bigger than a pin's head, with a small crystal of nitrate of silver, in a bit of paper; place the packet on an anvil and strike it with a hammer. The explosion will be sharp. The student will remember that nitrate of silver stains the fingers.

Exp. 270.—Mix 1 grm. of powdered nitrate of silver with 0·2 grm. of dry, powdered charcoal; place the mixture on a piece of porcelain, and touch it with a red-hot wire. The mixture deflagrates, and there remains behind metallic silver.

Exp. 271.—Add to a solution of nitrate of silver a solution of caustic soda, until no further precipitate is produced. The brownish precipitate is a hydrated oxide of silver.

536. *Oxides of Silver.*—Silver probably forms three oxides, Ag_4O, Ag_2O, and Ag_2O_2. The first is a very unstable black powder; the second forms with acids the ordinary silver salts; the third is a crystalline body obtained by electrolysis of nitrate of silver. The precipitate obtained in the last experiment is the hydrate of the oxide Ag_2O; this precipitate readily parts with its water, and at a temperature much below 100° becomes anhydrous. Unlike the oxides of sodium and potassium, this oxide of silver yields up its oxygen below a red heat, and metallic silver remains—as may be demonstrated by heating the product of the last experiment; light also reduces it, and hydrogen even at 100° has the same effect. Oxide of silver bears, however, certain striking resemblances to the oxides of the alkali-metals; thus it is a strong base, uniting with strong acids to form salts which are neutral to test-paper, and which are in some cases isomorphous with the corresponding salts of sodium. The oxide is slightly soluble in water, and the solution has a feeble alkaline reaction.

The oxide is freely soluble in ammonia-water, and the solution deposits, on exposure to the air, a black, micaceous powder which has received the name of *fulminating silver*, because of its explosive character. The same dangerous compound is formed when freshly precipitated oxide of silver is digested for some hours in ammonia-water, and it is also produced when an ammoniacal solution of chloride or nitrate of silver is precipitated with a solution of hydrate of sodium or potassium. It is necessary to be aware of these facts in order to avoid the risk of producing by accident this exceedingly dangerous substance. Its composition is not accurately known. Friction or slight pressure, even under water, may cause it to explode. The student should never venture to prepare this substance.

Exp. 272.—Fill three test-tubes one-third full of water, and pour into each a few drops of a moderately strong solution of nitrate of silver. Add to the first test-tube 2 or 3 c. c. of a solution of chloride of sodium, and shake the tube violently; a dense, white, curdy precipitate of the chloride of silver will be produced. Add to the second test-tube 2 or 3 c. c. of a solution of bromide of potassium, and shake the tube; a yellowish precipitate of bromide of silver will be thrown down. Add to the third test-tube 1 or 2 c. c. of a solution of iodide of potassium, and shake up the liquid; a pale-yellow flocculent deposit of iodide of silver will be formed.

Withdraw from each test-tube a portion of the precipitate it contains, and try to dissolve each precipitate in moderately strong nitric acid; the attempt will fail, for these silver salts are insoluble in nitric acid.

Withdraw from each test-tube another portion of the precipitate it contains, and treat each precipitate with ammonia-water; the chloride of silver will dissolve easily, the bromide less easily, the iodide with difficulty. Lastly, pour upon the remnants of the original precipitates in the three test-tubes a moderately strong solution of hyposulphite of sodium (§ 495); all three precipitates will immediately dissolve.

Exp. 273.—Precipitate some curdy chloride of silver by adding chloride-of-sodium solution, or chlorhydric acid, to a solution of nitrate of silver, so long as any precipitate is produced. Throw the precipitate upon a filter, and wash it with water; then open the filter, spread the chloride evenly over it, and place it in direct sunlight. The white precipitate rapidly changes to violet on exposure to the sun's rays, the depth of shade increasing as the action of the light continues. This coloration arises from a partial decomposition of the chloride of silver, the change of color being accompanied by a loss of chlorine. Upon the facts illustrated in this and the preceding experiments the main processes of photography depend.

537. *Photography.*—The chemical changes which the salts of silver undergo, when exposed to light, are the basis of the art of photography—not because these are the only salts which are affected by light, but because none are so advantageous on the whole. There are three different kinds of photographic process —that on silver, that on glass, and that on paper.

To produce a photograph on silver (a daguerreotype), a highly polished silver plate is exposed in a dark box to the diluted vapor of a mixture of bromine and iodine. A bronze-yellow film of brom-iodide of silver is thus produced upon the plate, which, at a certain stage, possesses a high degree of sensitiveness to light. The plate is then transferred to a camera, and exposed at the focus of the lens to the light radiated from the object to be copied. After remaining a few seconds in the camera, it is withdrawn, and immediately exposed in a warm box to the vapor of metallic mercury. When the plate is taken from the camera, the film looks as uniform as ever, and no image is visible upon it; but the exposure to mercury vapor immediately brings out an image. The mercury fixes itself strongly upon those parts which have received the light, while it takes no hold upon those parts of the film which the light has not decomposed. A strong solution of hyposulphite of sodium is then poured over the plate, in order to dissolve off the undecomposed brom-iodide. The highly polished silver, beneath, forms the shades, and the amalgam of mercury with silver forms the lights. The plate is washed, and a very dilute solution of chloride of gold in hyposulphite of sodium is poured over its surface and gently warmed. A thin film of gold, which, as it were, varnishes the picture, is thus deposited upon the plate; another washing completes the operation. The daguerreotype is the most perfect of photographs; but the polish of the surface prevents the image from being seen in all lights, and the plate is liable to be tarnished and ruined by sulphuretted gases.

In order to get a photograph upon glass, a transparent film capable of holding the necessary silver-salt must first be attached to the glass plate. Collodion (a solution of a variety of gun-cotton in a mixture of alcohol and ether) is the material of this film. To the collodion is added a solution of an iodide, either of potassium, cadmium, or ammonium, or a mixture of these; the bromides of ammonium and cadmium, or one of them, added in the proportion of one part of the bromides to three or four of the iodides, render the film more sensitive to yellow and red rays —a point of importance in cloudy weather or smoky towns. The collodion thus prepared is poured rapidly over a clean and dry surface of plate-glass; the volatile solvents evaporate rapidly, and as soon as the film is coherent the glass is plunged into a bath of nitrate of silver very slightly acidified with acetic or dilute nitric acid. This bath must be in a dark place; the plate remains in it for several minutes. A yellow layer of iodide or brom-iodide of silver is produced in the film, and nitrate of potassium, cadmium, or ammonium dissolves in the bath. The plate is then exposed in the camera for a few seconds. When removed no image is perceptible; but on pouring over the film a solution of gallic or pyrogallic acid in alcohol and acetic acid, or a solution of the green sulphate of iron, mixed with a few drops of a weak solution of nitrate of silver, the image will be developed, slowly or rapidly, according to the nature and strength of the developing-liquid, the degree of exposure, and the intensity of the light. The illuminated portions of the picture will appear, under the action of the *developer*, more or less black, while the shaded portions will retain the yellow colour of the iodide. As soon as the details of the shaded portions appear, the liquid is washed off and the development arrested. A saturated solution of hyposulphite of sodium is then poured over the film to dissolve off the yellow iodide of silver where it has not been affected by the light; only the reduced portions of silver remain, and they appear more or less opaque. The plate must finally be very thoroughly washed to remove all traces of the hyposulphite, and then dried and varnished on the collodion side to protect the film from injury.

Concerning the nature of the change which a film of iodide of silver undergoes when exposed to light, we cannot be said to have any exact knowledge. There is no perceptible alteration in the film; there is no loss of iodine; the iodide retains its solubility in hyposulphite of sodium; yet the impression is not of a temporary kind; for the invisible image produced on a plate may be developed many hours afterwards, if the plate is kept in the dark during the interval.

The photograph on collodion may be employed directly as a positive picture, if not too strongly developed, by placing it on a black background. Those portions which are opaque to light, or in other words those in which silver is deposited, will reflect light, and furnish the lights of the picture; while those on which the light did not act, and which are therefore transparent, will appear black from the nature of the background, and these will form the shades of the picture. In the daguerreotype the finished picture is inverted; in the collodion positive it is not inverted. If the development be pushed further, the image becomes so strongly defined that the deposited silver will more or less completely intercept the light. The collodion side of the plate is then placed in contact with the sensitive side of paper impregnated with chloride of silver by a process immediately to be described, and exposed to light in a pressure-frame. The light is arrested by the altered parts of the collodion, but is freely transmitted by the other portions; upon the paper, therefore, the lights of the real object are light and the shades are dark. Such a negative collodion picture may of course be copied on a second sensitive collodion film.

Two developing-solutions, used one after the other, produce a better effect than one. The green sulphate of iron may be used first, and pyrogallic acid with nitrate of silver subsequently; the iron solution must be completely washed off before the other is added. The picture may even be intensified by pyrogallic acid after the plate has been washed in hyposulphite of sodium.

Photographs were made on paper long before the film on glass came into use; but the paper process is now chiefly confined to the printing of positive impressions from collodion negatives on glass. The silver-salt which is preferred for photographic paper is the chloride, with or without albumen, but always accompanied with free nitrate of silver. The paper is floated for five minutes on a solution of chloride of sodium or ammonium; when dried, it is floated in a dark room, for five minutes, on its salted surface, in a solution of nitrate of silver; again dried, it is fit for use. When such paper is used to obtain a positive impression from a collodion negative, or from a paper negative made transparent with wax or a mixture of wax and paraffine, it is exposed to light under the negative to be copied, until the lights of the picture are of a pale lilac hue, and the shades of a deep bronze color. After being thoroughly washed, the paper is transferred to a "toning"-bath, which consists of a very dilute solution of bicarbonate of sodium, with a minute proportion of chloride of gold. The picture is kept in motion while in this bath; it remains there until its shades have acquired a deep purple-black color. It is only in those parts of the picture in which the silver has been well reduced that this toning effect is produced. The picture is again washed in water, and soaked for fifteen minutes in a solution of hyposulphite of sodium, in order to remove all the chloride of silver which is contained in the substance of the paper. Finally the picture must be soaked for twenty-four hours in water which is constantly renewed, in order to wash away every trace of the compound hyposulphite of sodium and silver. No photograph will keep long, unless the chloride of silver has been completely dissolved by the hyposulphite, and the compound hyposulphite washed away with a thoroughness that leaves no trace behind. If the first condition is not fulfilled, diffused daylight will alter the picture; if the second condition is not complied with, yellow or brown stains will ultimately destroy the picture.

As in every other art which embraces many details, and demands a trained eye and hand, eminent skill in photography can, as a rule, be acquired only by long practice.

538. *Chloride of Silver* (AgCl).—Native chloride of silver occurs, sometimes in cubical crystals, sometimes in compact semitransparent masses, which are sectile, and, from their general appearance, have given the mineral the name of horn-silver. Chloride of silver may be precipitated from any soluble silver-salt by adding to the silver solution chlorhydric acid, or the solution of any soluble chloride; or it may be obtained by passing over a dry silver-salt a stream of dry chlorine gas. This last reaction is the basis of a method of preparing anhydrous nitric acid. When a stream of dry chlorine is made to pass over perfectly dry nitrate of silver heated to 50° or 60°, the following reaction takes place:—

$$Ag_2N_2O_6 + 2Cl = 2AgCl + N_2O_5 + O.$$

The characteristics of precipitated chloride of silver have been already described (Exp. 272). The presence of an extraordinarily minute proportion of chloride of silver renders a clear liquid opalescent. It is easy to detect silver in a solution of which it forms only the $\frac{1}{200000}$ part, by adding to the solution a drop of chlorhydric acid or of a soluble chloride. An admirable method of determining the amount of silver present in any solution depends upon the insolubility of chloride of silver, the density and peculiar curdy quality of the precipitate, and the visibility of the smallest trace of it in a clear fluid. This method, now generally employed in mints and assay-offices, is applicable to the quantitative analysis of silver alloys; it is volumetric, and depends upon the measurement of the amount of a standard solution of chloride of sodium which is required to effect the complete precipitation, as chloride, of the silver contained in a given weight of the alloy. In a solution which is acidulated with nitric acid, and which contains no excess of the soluble chlorides, the chloride of silver is easily coagulated into dense flocks by agitation; so that the exact point at which the precipitate ceases to be formed is readily perceived.

Chloride of silver melts at about 260°. It is not decomposed when heated with carbon, but is easily reduced by hydrogen when heated in a current of the gas; zinc and iron reduce moist chloride of silver to metallic silver; when heated with carbonates or hydrates of sodium, potassium, or calcium, chloride of silver gives its chlorine to the other metal, and pure silver is set free.

These methods of reducing chloride of silver, except that by hydrogen, are turned to account in the refining of silver on a large scale. The coin, or bullion to be refined is dissolved in nitric acid, and to the solution chloride of sodium is added; the precipitated chloride of silver is washed until the washings are tasteless, and is then slightly acidulated with sulphuric acid; bars of zinc are placed in the moist mass, and the whole left at rest for two or three days. Chloride of zinc and metallic silver are the products. As soon as the reduced silver is entirely soluble in nitric acid, the reduction is complete. The reduced metal is digested for two or three days in dilute sulphuric acid, to remove adhering zinc-salts, and is then thoroughly washed, dried, and finally melted and cast into ingots. If an absolutely pure metal is desired, the first reduction should be made with pure zinc, and this refined silver may be again dissolved in nitric acid, thrown down as chloride, and reduced again from the washed chloride by fusion with carbonate of calcium.

539. The reduction of chloride of silver by hydrogen is the basis of one of the several determinations of the atomic weight of silver; and since silver forms a large number of anhydrous salts with acids, and has little or no tendency to form more than one salt with each acid, the silver-salt is often the best one to prepare and analyze whenever the combining-weight of an acid is to be determined. But it is clear that the accuracy of these determinations depends upon the accuracy with which the atomic weight of silver is known; hence extraordinary pains have been taken to arrive at the true atomic weight of silver. It has been found, by the most careful experiment, by heating chloride of silver in a current of hydrogen, that in 132·856 parts of that compound,

100 parts of silver are united with 32·856 of chlorine. If the atomic weight of chlorine be accepted as 35·5, a simple proportion leads to the atomic weight of silver.

32·856 :	35·5 =	100 :	x = 108·07.
Amount of Cl.	*At. Weight of* Cl.	*Amt. of* Ag.	*At. Weight of Silver.*

An entirely different experiment verifies this result; by burning finely divided silver in a current of perfectly dry chlorine, it is proved that 108 parts of silver combine with 35·505 of chlorine. The following round of experiment and simple calculation will illustrate the manner in which one atomic weight is derived from another, and all are verified by mutual comparison. Chlorate of potassium, when heated, gives off all its oxygen and chloride of potassium remains. Assuming that the formula of chlorate of potassium is $KClO_3$ and that the atomic weight of oxygen is 16, we derive the following proportion from the fact of experiment that 100 parts of $KClO_3$ yield 39·209 parts of oxygen.

39·209 :	48 =	60·791 :	x = 74·4208.
Amount of O.	3O.	*Amt. of* KCl.	*Molecular Weight of* KCl.

It is another experimental fact that 100 parts of chloride of potassium produce, when precipitated with nitrate of silver, 192·75 of chloride of silver.

100 :	192·75 =	74·4208 :	x = 143·446.
Amt. of KCl.	*Amt. of* AgCl.	*M. Weight of* KCl.	*Molecular Wt.* AgCl.

But it has been determined, as above stated, that 132·856 parts of chloride of silver contained 32·856 parts of chlorine, and accordingly

132·856 :	32·856 =	143·446 :	x = 35·476.
Amt. of AgCl.	*Amt. of* Cl.	*M. Weight of* AgCl.	*At. Weight of* Cl.

But if the molecular weight of chloride of silver is . .	143·446
we may deduct the atomic weight of chlorine . . .	35·476
and so obtain the atomic weight of silver;	107·970
and if the molecular weight of chloride of potassium is	74·4208
we may deduct the atomic weight of chlorine, . . .	35·476
and so obtain the atomic weight of potassium . . .	38·9448

These numbers will be found to be very nearly coincident with those previously given as the accepted atomic weights of these three very important elements.

540. *Bromide and Iodide of Silver* ($AgBr$ and AgI) are two rather rare minerals, usually associated with chloride of silver or with native silver. Their artificial preparation and such of their properties as have present importance have been already alluded to (Exp. 272). They are both easily fusible and insoluble in water, but soluble in concentrated solutions of the bromide and iodide of potassium.

541. *Cyanide of Silver* ($AgCN$) is a white powder, insoluble in water but soluble in ammonia-water, obtained by precipitating nitrate of silver with a soluble cyanide like the cyanide of potassium. Cyanide of silver is soluble in solutions of the cyanides of sodium, potassium, calcium, and other metals, forming double cyanides of the formula $MAgC_2N_2$. When such a solution is subjected to the action of a galvanic battery, metallic silver is deposited at the negative pole, in a compact, adherent layer, while at the positive pole, where a strip or plate of metallic silver is placed, a quantity of the metal equal to that which is deposited at the negative pole continually dissolves. A solution which contains $\frac{1}{50}$ of its weight of silver is found to be of convenient strength for the ordinary operations of electro-plating.

542. *Sulphide of Silver* (Ag_2S).—This compound is a principal ore of silver. The native mineral is sometimes crystallized, in cubes or octahedrons, and sometimes massive; it has a leaden lustre and color, and it is so soft that a knife will cut or a die impress it; it is fusible, and when roasted in the air yields silver (which remains in the metallic state) and sulphurous acid (the product of the combination of its sulphur with the oxygen of the air). The pure mineral is very easily recognized by these marked characteristics. Silver is readily tarnished by contact with moist gaseous sulphydric acid, or with a solution of a soluble sulphide; this tarnish is the sulphide of silver (§ 533). The sulphide may be artificially prepared by transmitting a stream of sulphuretted hydrogen through a solution of a salt of silver.

Exp. 274.—Place in a test-glass 8 or 10 c. c. of water to which 20 or 30 drops of a solution of nitrate of silver have been added, and pass through the dilute solution a slow stream of sulphuretted hydrogen. The black precipitate is the sulphide of silver.

Strong acids, especially when hot, dissolve or decompose this sulphide. It is not soluble in solutions of the sulphides of the alkali-metals; but by fusion it may be made to unite with many other sulphides of metals.

543. *Sulphate of Silver* (Ag_2SO_4).—When silver is boiled with strong sulphuric acid, the silver gradually dissolves, and there are formed the sulphate of silver, water, and sulphurous acid:—

$$2Ag + 2H_2SO_4 = Ag_2SO_4 + 2H_2O + SO_2.$$

The sulphate is dissolved by the excess of acid, but it is deposited in great part on the addition of water, for it is but slightly soluble in water. As gold is not soluble in sulphuric acid, small quantities of gold may be separated from large quantities of silver or silver alloys by boiling the metal, finely granulated, in cast-iron vessels with oil of vitriol; silver and copper dissolve, and the gold is left behind in a fine powder. The solution of silver is subsequently diluted, and the silver precipitated from the solution in the metallic state by means of metallic copper. (Exp. 267.) Old silver coin, containing not more than $\frac{1}{2000}$ of gold, has been profitably worked over by this process.

544. *The Alkali Group.*—The metals which must plainly be classed together under this head are sodium, potassium, (ammonium,) lithium, rubidium, and cæsium. Two other metals are better classed with this group than elsewhere; but their likeness to the alkali-metals is but partial, and in many respects their properties are quite unlike those of the six metals just enumerated; these two metals are silver and thallium. The common properties of the alkali-metals are mainly these:—they have the lustre of silver, are soft, easily fusible, and volatile at high temperatures; they unite greedily with oxygen, and decompose water with facility, forming basic hydrates which are very caustic and intensely alkaline bodies, not to be decomposed by heat; their carbonates, sulphates, sulphides, and chlorides, and, indeed, the vast majority of their salts, are soluble in water; and each metal forms but one chloride, one bromide, and one iodide; they all form basic oxides, and never an acid oxide; they occur in nature in modes analogous though not the same; their corresponding salts are often, though not always, isomorphous; lastly, there is a general, though not absolute, uniformity among the formulæ of the compounds into which these elements enter, so that, if a com-

pound of a given composition is proved to exist with one of these elements, the strong presumption is that analogous compounds with all the other elements of the group exist likewise, with properties similar though not identical.

Silver and thallium present, on the whole, so few points of resemblance to the alkali-metals that they would not be comprehended in the same group with them were it not for one consideration weighty enough to turn the balance when the discussion of other properties leaves the matter in doubt. Sodium, potassium, (ammonium,) lithium, cæsium, rubidium, silver, and thallium all replace hydrogen, atom for atom. All these elements are exchangeable for hydrogen and with each other, atom for atom, and in the present state of the science they must be regarded as the only metals thus equivalent to hydrogen. The atom of the elements of the chlorine group, including fluorine in that designation, and of the seven elements above enumerated, is exchangeable for one atom of hydrogen; it is *worth one* in exchange, and these elements are therefore said to be *univalent*, or, with less verbal precision, *monatomic.*

545. *Quantivalence.*—The chemical elements have not all the same *atom-fixing* power; thus, while an atom of chlorine combines with only one atom of hydrogen, an atom of oxygen has the power to drag two atoms of hydrogen into a molecule; an atom of nitrogen holds three atoms of hydrogen in firm chemical combination, and an atom of carbon four hydrogen-atoms. In all double decompositions an atom of sodium, potassium, or silver replaces one atom of hydrogen, but an atom of calcium or lead two atoms of hydrogen (§ 82). To indicate conveniently the atom-fixing power of each element a sign is needed and a name. The conventional sign is a Roman numeral, or an equivalent number of accents, placed above and at the right of the symbol of the element, in case its atom be worth more than one of hydrogen; and for the name to denote this atom-fixing power of the elements the word "quantivalence" may be used, or the less descriptive word "atomicity." The elements are called *univalent, bivalent, trivalent,* and *quadrivalent,* or *monatomic, diatomic, triatomic,* and *tetratomic,* according as their respective atoms are capable of saturating, or holding in firm chemical combination, 1, 2, 3, or 4

atoms of hydrogen. Thus, while the simple symbols Cl, Br, K, Ag indicate that chlorine, bromine, potassium, and silver are univalent, the symbols of nitrogen, antimony, and other trivalent elements may be written N‴, Sb‴, &c. In the same way the symbols O″ and Ca″ indicate that oxygen and calcium are bivalent, and the symbol C⁗ shows that carbon is quadrivalent.

The quantivalence of many of the elements is not yet determined with certainty; but the classification into groups of the elements we have thus far studied rests upon the quantivalence of the elements, as well as upon the other chemical resemblances, which have been dwelt upon in connexion with each group. The elements of the chlorine-group and the alkali-group are univalent; the elements of the sulphur-group and the majority of the metals, hereafter to be studied, are bivalent; the elements of the nitrogen-group are trivalent, and of the carbon-group quadrivalent.

It must not be supposed that the atom-fixing power of the elementary bodies is, under all circumstances, and in all compounds, invariably exerted to the fullest extent. Were the combination of the elements governed by any such law as this, it would evidently be impossible for any two elements to unite in more than one proportion. Thus trivalent nitrogen and bivalent oxygen could only combine in the proportions represented by the formula $N_2'''O_3''$, proportions which completely satisfy the atom-fixing power of both elements. But we know that these two elements actually form no less than five different compounds (§§ 75, 76), of which only one is marked by the complete equilibrium of the two elements; and this one is by no means the most stable member of the series; on the contrary, it is about the most unstable. The student must not imagine that a bivalent element has twice as strong affinities as a univalent element; the atom-fixing power of an element is no test or index of the avidity with which it seeks combination. Chlorine, which holds but one atom of hydrogen, is competent to decompose sulphuretted hydrogen, ammonia, and marsh-gas, although sulphur unites by preference with two, nitrogen with three, and carbon with four atoms of hydrogen.

CHAPTER XXVIII.

CALCIUM—STRONTIUM—BARIUM—LEAD.

CALCIUM.

546. This metal is a constituent of several of the commonest and most important minerals; it forms a very considerable portion (perhaps as much as one-sixteenth) of the solid crust of the earth. Before considering the properties of the metal itself, let us examine some of its familiar compounds.

547. *Carbonate of Calcium* ($CaCO_3$) occurs in nature in many different forms, called by a great variety of names, among which may be mentioned limestone, chalk, marble, calc-spar, and coral. There are whole ranges of mountains composed almost entirely of limestone, while in many extensive tracts of country the soil is calcareous and reposes upon limestone rocks. The shells of shell-fish are almost entirely composed of it, and it is an important constituent of dolomite, marl, and many other rocks and minerals. It is formed artificially, as has been seen (Exp. 168), when carbonic acid is brought into contact with lime-water; but it is noteworthy that carbonic acid will not unite with the anhydrous oxide of calcium (quicklime).

Carbonate of calcium, though tasteless, is slightly soluble in water, and the solution exhibits a faint alkaline reaction; it is, however, rather freely soluble in water charged with carbonic acid (§ 403).

Exp. 275.—Place in a test-tube 20 or 30 drops of lime-water, and as much pure water; immerse in the mixture the delivery-tube of a bottle from which carbonic acid gas is being evolved (Exp. 171). Carbonate of calcium will be thrown down at first; but after a while, as the water in the test-tube becomes saturated with carbonic acid, the precipitated carbonate will redissolve, and there will be obtained a perfectly clear solution, which, in spite of the large proportion of carbonic acid contained in it, has a decided alkaline reaction. By boiling the solution, so that a portion of its carbonic acid may be expelled, the carbonate of calcium can be again precipitated. So, too, if the liquid be left exposed to the air, it will gradually give off carbonic acid, and become turbid from deposition of carbonate of calcium.

The phenomena illustrated in this experiment often occur in nature. In many districts where limestone is abundant, the well- and river-waters are highly charged with carbonate of calcium held dissolved by carbonic acid; the water is thus made "hard" (see § 560), and is, comparatively speaking, unfit for washing and for many other purposes. When employed as a source of steam-power, such waters deposit carbonate of calcium as an incrustation upon the sides of the boilers as fast as the excess of carbonic acid is expelled by boiling. This scale, or incrustation, forms a more or less coherent coating upon the inner surface of the boiler, and, being a very poor conductor of heat, it greatly interferes with the heating of the water; the scale keeps the water away from the iron sides of the boiler, and the metal, being thus unduly heated, is rapidly oxidized, or "burnt out," as the fireman correctly states it.

The formation of calcareous petrifactions, of stalactites and stalagmites, of the stones called tufa and travertine, and of many deposits of crystallized carbonate of calcium, is directly referable to the escape of carbonic acid from calcareous waters. Whenever water, charged with carbonate of calcium, flows out from the earth into the open air, or trickles into hollows or caverns within the earth, carbonic acid is given off in the gaseous state, and carbonate of calcium is deposited. Stalactites are the pendent masses, like icicles, which hang from the roofs of caverns and the walls of cellars, bridges, and like covered ways; stalagmites are the opposite masses which grow up out of the drops of water which fall from the stalactites above them, before all the dissolved carbonate has been deposited. The waters of some mineral springs are so highly charged with carbonate of calcium, that, on being exposed to the air, they quickly deposit a considerable quantity of it upon any solid substance with which they come in contact. In case such waters flow over pieces of wood or other organic matter, the form of the wood will be preserved in the cast or "petrifaction," long after the wood itself has decayed and disappeared. Where such deposits are formed upon a scale so large as to be of geological importance, as is the case in some of the volcanic districts of Italy, the rock formed is called tufa when porous, and travertine if compact.

548. Carbonate of calcium dissolves also in aqueous solutions of several of the salts of ammonium, such as the chloride, nitrate, and sulphate, especially if it has only recently been precipitated and is still moist and incoherent.

Exp. 276.—Through 2 or 3 c. c. of lime-water, contained in a test-tube, blow, by means of a glass tube, a quantity of air from the lungs; to the milky liquid obtained, add, drop by drop, a cold, saturated aqueous solution of chloride of ammonium, until the cloudiness in the lime-water has disappeared—that is, until the carbonate of calcium has all been dissolved.

Exp. 277.—Place a drop or two of a solution of chloride of calcium in a test-tube, pour upon it several drops of a strong solution of chloride of ammonium; shake the mixture, and then add to it a few drops of a solution of carbonate of ammonium, and also a few drops of ammonia-water. If enough chloride of ammonium has been added to the liquid, no precipitate will be formed in it, though, in the absence of chloride of ammonium, a precipitate will at once be produced on mixing the other ingredients. A precipitate may, however, always be obtained by boiling the mixed solutions, unless a large excess of chloride of ammonium be present, or unless the chloride-of-calcium solution be very dilute.

By repeating this experiment under varied conditions, taking note, in each case, of the number of drops of the solutions of chloride of ammonium, chloride of calcium, and of water employed, and methodically increasing or diminishing each of these, the student will quickly perceive the real significance of the solvent power of the ammoniacal salt, and will appreciate the fact that, in testing for small quantities of either lime or carbonic acid, it is necessary for the analyst to exclude ammonium-salts from his solutions as far as may be practicable.

When boiled with solutions of the salts of ammonium (with chloride of ammonium for example), carbonate of calcium is rapidly decomposed and dissolved, carbonate of ammonium being given off, while the chloride (or some other salt) of calcium remains in solution.

549. Carbonate of calcium is remarkable not only for the very great diversity of external appearance which is presented by its several massive and amorphous varieties, but it is likewise found in a greater variety of regular crystalline forms than any other substance; more than 150 native varieties of it have been observed by mineralogists. As calc-spar, it occurs in rhombohedrons and other derivative forms of the sixth or hexagonal system (§ 191); but it is found also as the mineral arragonite, in forms of the trimetric system, and is consequently dimorphous.

The two forms of carbonate of calcium, calc-spar and arragonite, present many differences in their physical properties. Some specimens of calc-spar, called Iceland spar, are perfectly transparent and colorless, and exhibit to a remarkable degree the phenomena of double refraction. Transparent crystals of arragonite exhibit also the phenomena of double refraction; but arragonite has two axes of double refraction, calc-spar only one. Crystals of calc-spar are cleavable parallel to the faces of the rhombohedron which is the primary form of the mineral, and masses of it may often be broken up into more or less perfect rhombohedrons. Arragonite, on the contrary, presents two directions of distinct cleavage, parallel to the faces of a right rhombic prism. The fractures of the two minerals are therefore quite unlike. The specific gravity of calc-spar ranges from 2·7 to 2·75, while the specific gravity of arragonite is generally between 2·9 and 3·3. Arragonite is considerably harder than calc-spar, but its specific heat (0·1966) is less. When carbonate of calcium crystallizes from hot solutions it takes the form of arragonite, but from cold solutions it crystallizes as calc-spar. In like manner the precipitate formed by mixing boiling solutions of chloride of calcium and carbonate of ammonium is seen under the microscope to consist of acicular crystals of arragonite, while the precipitate obtained from cold solutions of the same salts is amorphous. In either case, however, if the moist precipitate be left to itself for some time in the cold, it will gradually assume the rhombohedral form of calc-spar, no matter whether it was at first acicular or amorphous.

In all its varieties carbonate of calcium is readily attacked by acids, even if they be dilute; the action is attended with effervescence, owing to the expulsion of carbonic acid and the escape of this gas through the liquid:—

$$CaO,CO_2 + 2HCl = CaCl_2 + CO_2 + H_2O.$$

Limestone is readily distinguished by this reaction from other rocks.

550. *Oxide of Calcium* (CaO).—On being heated, carbonate of calcium begins to give off carbonic acid at a low red heat, as has been seen in Exp. 170, and at full redness is completely resolved into oxide of calcium, commonly called quicklime, and carbonic acid.

Exp. 278.—Place a small fragment of marble upon a piece of charcoal and heat it strongly in the blowpipe-flame during several minutes. Or throw a lump of limestone upon an anthracite fire, and leave it there for half an hour or more. In either case, it will be found, upon examination, that the calcined product has lost the property of effervescing with acids; that it weighs less than the original limestone, and that it exhibits a distinct alkaline reaction when placed on wet test-paper.

For use in the arts, limestone is burned in special furnaces, of peculiar construction, called lime-kilns, some of which are so arranged that they may be kept in operation for years without intermission. When carbonate of calcium, instead of being heated merely in quiescent air, is heated in a current of air, or of any other gas, such as steam for example, it will give off all its carbonic acid very easily. It has been found in practice that limestone fresh from the quarry can be more readily burned than that which has been long dug out of the ground and has so lost its natural moisture; in damp weather, moreover, the burning is said to go on more satisfactorily than when the atmosphere is dry. If carbonate of calcium be ignited in a tube of iron, or other metal, closed hermetically, so that no carbonic acid can escape from the tube, the carbonate disengages carbonic acid until the pressure of the confined gas becomes so great as to arrest the further decomposition of the carbonate. Under these conditions, the temperature may be raised high enough to fuse the undecomposed carbonate; the cooled mass often presents the appearance of fine-grained marble. If the tube in which the experiment has been performed be very slowly cooled, the carbonic acid will be reabsorbed.

Of the anhydrous oxide of calcium little need here be said. It is infusible at the most intense heat at present at our command, and is therefore used for making crucibles in which the most refractory metals are melted by the aid of the compound blowpipe. It has no power to unite with dry carbonic acid at ordinary temperatures, but when exposed at very high temperatures to an atmosphere of carbonic acid possessing a certain tension, some of the gas is absorbed. It unites with water very energetically, and the product of this union combines readily with carbonic acid. When lumps of quicklime are exposed to the air they slowly absorb both water and carbonic acid, and after a while fall to powder. This powder is known as *air-slaked lime*; its composition may be represented by the formula $CaH_2O_2,CaCO_3$, or, dualistic, CaO,CO_2 ; CaO,H_2O.

551. *Hydrate of Calcium* (CaH_2O_2).—When water is brought into contact with oxide of calcium, the latter swells up and falls to powder; a large amount of heat is evolved, and there is obtained a compound of calcium, hydrogen, and oxygen, commonly called slaked lime, or in chemical language hydrate of calcium:—

$$CaO + H_2O = CaH_2O_2.$$

Exp. 279.—Place a lump of recently burned quicklime, weighing about 30 grms., upon a large earthen plate; pour upon the lime some 15 or 20 c. c. of water, and observe how much the lime increases in bulk as it is converted into hydrate of calcium. The heat of the mass may be shown by thrusting an ordinary friction-match into the middle of it; or, in case a considerable quantity of quicklime has been employed, by excavating a small hole in the dry powder and throwing in a few grains of gunpowder, inflammation will ensue in both cases. That much heat is evolved, may be shown also by covering the moistened quicklime with a not too tall inverted beaker glass or bottle, and observing that, after a considerable amount of aqueous vapor has been condensed upon the walls of the glass, the space within the latter will at last become filled with a hot, invisible atmosphere of steam; when the bottle is lifted, and the steam thus brought into contact with the cold external air, a dense cloud or fog is immediately formed.

So much heat is developed during the union of water with lime, that wood will quickly be brought to the kindling-temperature and inflamed, if it happen to be in contact with large masses of these substances reacting upon one another. Fires are very frequently occasioned by the access of water to ships or warehouses in which quicklime is stored. It has been noticed, when large quantities of quicklime are slaked in a dark place, that light as well as heat is evolved from the lime. Even when quicklime is brought into contact with ice, so much heat is evolved that the mixture sometimes becomes hot enough to boil water.

552. When hydrate of calcium is stirred into water, there is formed not only a true solution, lime-water, which may be obtained clear and colorless by filtration (See Exp. 168), but also a turbid liquor consisting of particles of solid hydrate of calcium diffused through the lime-water; this liquor is known as *milk* or *cream of lime*, according to its consistency. In slaking lime, only about half a part of water is really needed to convert one part of quicklime into hydrate of calcium; but in all cases where a fine,

smooth paste is desired, as in the preparation of milk of lime, or of mortar, and in general whenever hydrate of calcium is required in a very finely divided condition, it is best to pour two or three parts by weight of water upon one part of quicklime, so that the slaking may be quickly effected. By using hot water the process may be still further accelerated. The proportions of material given at the beginning of Exp. 279 are better adapted than these last for illustrating the evolution of heat; but if too little water be employed, the hydrate of calcium formed is liable to be granular and crystalline rather than powdery. Both milk of lime and dry powdery hydrate of calcium are largely employed for purifying the illuminating gas made from coal. They remove from the gas sulphydric and carbonic acids.

Exp. 280.—Provide two gas-bottles, one arranged for generating sulphydric acid (Exp. 86), the other for generating carbonic acid (Exp. 171). Connect with one of the gas-bottles a tube filled loosely with dry hydrate of calcium (Appendix, Fig. 15), and with the other a small bottle containing milk of lime. Pour chlorhydric acid into the gas-bottles, so that sulphydric and carbonic acids shall be freely evolved, and test, from time to time, with lead-paper (Exp. 90), and with lime-water (Exp. 170), as to whether these acids are completely absorbed by the dry hydrate of calcium and the milk of lime. After a while, change the places of the absorbing tube and bottle, so that the milk of lime shall now be where the dry hydrate was before, and again test the efficiency of the absorption, with lead-paper and lime-water. In actual practice it is found that, while the dry hydrate is a more efficient absorbent of carbonic acid than milk of lime, the latter is capable of taking up far more sulphydric acid than the former.

553. Hydrate of calcium may be obtained crystallized, in hexagonal prisms, by evaporating lime-water in the dry exhausted receiver of an air-pump. At a red heat it gives off its water, and is reconverted into quicklime. The residue in this case is left in an open, porous condition which well fits it for many chemical purposes (see § 120).

It is noteworthy that hydrate of calcium is somewhat less soluble in hot than in cold water. If a cold, saturated solution of lime-water be boiled, nearly half of its solid contents will be deposited; and in case none of the water has been driven off, the matter thus precipitated will slowly dissolve again after the liquid above it has become cold. In studying this point, the experimenter must take care that the solution is kept out of contact with the air, lest it absorb some of the carbonic acid which is always present in the atmosphere, and become turbid from deposition of carbonate of calcium. A familiar instance of this absorption is seen in cases where milk of lime is employed for whitewashing: the loosely adherent white coating, left after the liquid has become thoroughly dry, is no longer hydrate of calcium, but carbonate of calcium in a more or less pure condition.

554. Slaked lime is very largely employed for making mortar, as an ingredient of various cements, and for plastering. When mixed with enough water to form a thick paste, it is decidedly plastic, and admits of being spread and moulded like wax or clay. This paste sets, as it dries, to a firm, solid mass, which, when in thin layers, adheres firmly to any rough surfaces upon which it may have hardened. When, however, any considerable mass of the moist paste is allowed to solidify by itself, the dry product will gradually crack and fall to pieces. Lime-paste cannot, therefore, be employed as a mortar unless it be mixed with some substance like sand, which shall present numerous surfaces upon which the hardened product may adhere; by the addition of sand, moreover, the moist lime is prevented from shrinking too much as it becomes dry.

Mortar is commonly prepared by mixing 1 part of quicklime with water enough to form a thin paste, then adding 3 or 4 parts of coarse, sharp sand, and thoroughly incorporating these ingredients. The paste thus obtained is applied as a thin layer to the moistened surfaces of the bricks or stones to be united. The pasty mortar soon sets to the hard mass above described, and, on continued exposure to the air, it slowly absorbs carbonic acid at its surface, and is there converted into a compact compound of hydrate and carbonate of calcium. The stone-like mass thus obtained binds firmly together the bricks or stones between which it has been interposed. It has been asserted that the original mortar-paste sets more firmly if it contain a certain admixture of carbonate of calcium, than if it contain only the pure hydrate; this admixture is, of course, produced when mortar is left for some time in contact with the air before being used. In the course of time chemical combination occurs, to a limited extent, between the silicic acid of the sand and the oxide of calcium in the hardened mortar, though the process goes on but slowly; each grain of sand finally becomes covered with a thin layer of hydrated silicate of calcium, which contributes materially to the solidity of the mortar. The mortar taken from old buildings yields a certain proportion of gelatinous silica on being treated with chlorhydric acid (§ 466).

The conversion of the original mortar into hydrocarbonate and silicate of calcium is never completely accomplished; in the central portions of the mass, free hydrate of calcium will still be found after the lapse of many centuries. Samples of mortar, recently taken from the Great Pyramid, were found on analysis to contain a large proportion of the free hydrate.

555. The plastering used for finishing the walls and ceilings of rooms is mortar to which a quantity of hair has been added to increase its tenacity; in drying, it is, of course, subject to the same chemical changes as ordinary mortar. By absorbing carbonic acid from the air, it is gradually converted, in part, into carbonate of calcium, while water is set free:—

$$CaO,H_2O + CO_2 = CaO,CO_2 + H_2O.$$

Consequently the walls of recently plastered rooms cannot become permanently dry, until enough carbonic acid has been absorbed to expel the chemically combined water from their outside surfaces; hence the dampness so often perceived in new houses, when carbonic acid first comes to be freely generated in them by respiration and by burning lamps. In order to dry plastering, it would, doubtless, be better to employ open fires of charcoal, or of coke, and to deliver the products of the combustion directly into the room which is to be dried, instead of relying solely upon hot air, as is now usual.

556. Hydrate of calcium, like the hydrates of sodium and of potassium, exhibits a strong alkaline reaction when tested with moistened litmus-paper, and exerts a corrosive action upon most organic substances; hence it is often called *caustic lime.*

Exp. 281.—Add a few drops of water to a small quantity of dry hydrate of calcium, and rub it to a paste between the fingers. It will be felt that the alkali acts upon the skin; a little of the cuticle is really dissolved.

Exp. 282.—Wrap a handful of dry hydrate of calcium in a paper, or, better, in a piece of linen or cotton cloth, and set the packet aside for a week or two. After a while, the cloth or paper will become rotten and friable: the caustic lime, as the common phrase is, has eaten away their more corruptible portions, and has so destroyed the integrity of the whole. As a preliminary operation in tanning leather, hides are soaked in milk of lime to loosen the hair, so that it may be readily scraped off. The value of lime, as an ingredient of composts to be used as manure, appears to depend, in great measure, upon its power of hastening the decay and disintegration of organic matter.

Lime has been found to be specially valuable as manure when applied to soils rich in vegetable matter. The organic matters are decomposed or oxidized into carbonic and various other organic acids, which unite with the lime; sometimes, under special conditions, more or less nitrate of calcium is found among the products.

Lime is important, also, from being not only the cheapest alkali, but the cheapest of all the bases. Since its compounds with carbonic and sulphuric acids are nearly insoluble in water, it is largely employed for removing these acids from solutions in which their presence is not desired; it may itself be removed from any solution by means of the acids in question. It is used in the manufacture of the caustic alkalies (soda and potash), of ammonia-water and of bleaching-powders, as a flux in many metallurgical operations, in the refining of sugar, for preparing a lime soap in the manufacture of stearine candles, and for numberless other purposes. A noteworthy property of slaked lime is its power of dissolving freely in solutions of common sugar.

557. *Sulphate of Calcium* ($CaSO_4$) is found native in large quantities, as the minerals *gypsum* and *alabaster*. These minerals contain one-fifth their weight of water; their composition may be represented by the formula $CaSO_4 + 2H_2O$. The same hydrated salt may be obtained by adding sulphuric acid, or the solution of some sulphate, to a strong aqueous solution of almost any of the salts of calcium. This hydrated compound is the substance commonly meant when sulphate of calcium is spoken of. The anhydrous compound is also important: it is sometimes found in nature as the mineral *anhydrite*, and may be readily

prepared by heating the hydrated salt. There is still a third compound, the composition of which may be represented by the formula $2CaSO_4 + H_2O$, of which, however, but little is known.

Exp. 283.—Place in a porcelain evaporating-dish, or, better, in an iron pan, two tablespoonfuls of powdered gypsum; heat the gypsum moderately over the flame of the gas-lamp, and observe the movement of ebullition occasioned by the escaping water; stir the mixture as long as the vapor of water is seen to escape, and then set the residue aside to cool. The dry product is known as *calcined gypsum* or *plaster of Paris.*

As much as nine-tenths of the water which the gypsum contains may be readily expelled at temperatures between 100° and 120°; but, in order to drive off the last portions of the water, a temperature of nearly 300° is required. If the dry compound be heated to temperatures much higher than 300°, its particles appear to become agglutinated, and the chemical properties of the substance are somewhat changed; the gypsum is then said to be over-burned. At the temperature of redness, sulphate of calcium melts without decomposition, and, on cooling, assumes a crystalline structure similar to that of native anhydrite.

558. When powdered sulphate of calcium, which has been made anhydrous at a comparatively low temperature, is made into a paste with water, and then left to itself, it soon sets or hardens into a compact, coherent mass. This solidification is a consequence of the reassumption by the sulphate of calcium of the two molecules of water of crystallization which were driven off by heat when the substance was made anhydrous.

Exp. 284.—Place a small coin at the bottom of a cylindrical pasteboard pill-box a little wider than the coin; smear the coin and the interior of the box with a thin film of oil. Mix intimately two or three teaspoonfuls of the calcined gypsum of Exp. 283, with about half their volume of water, in a small porcelain dish, and quickly pour the mixture into the box, so that the coin shall be completely covered by it. The mixture, which is of the consistence of cream, should then be immediately stirred or puddled with a hair-pencil, or with a tuft of cotton tied upon a stick, or with the end of the finger, so that the bubbles of air which remain adhering to the surface of the coin may be pressed out, and the moist paste be made to come everywhere into contact with the metal. In the course of a few minutes the paste will solidify and become so hard that the pasteboard envelope may be torn away from it, and the coin removed. A perfect cast or copy of the stamp upon the coin will be found impressed upon the hardened gypsum. The impression in this first cast is, of course, reversed, but by smearing it with oil and then pouring over it a new portion of the gypsum-paste, precisely as was done with the coin, a fac-simile of the original coin may be obtained.

Plaster of Paris is largely used in this way for taking accurate copies of a great variety of objects. Thus, in the process known as stereotyping, a thin paste of plaster is poured upon the surface of the printers' types, after they have been set up and made ready for printing; the mould thus formed is dried and baked to expel the water from the gypsum, and is then plunged into a bath of a melted alloy of lead, antimony, and tin, known as stereotype-metal, in such manner that, on withdrawing the mould and allowing the metal within it to cool, there is obtained a fac-simile of the original types. From this durable metallic casting the page is finally printed.

As has been said above, the moist paste sets as soon as the water, which has been mechanically mixed with the anhydrous sulphate of calcium, enters into chemical combination with it. As in all other instances of chemical action, so here, heat is evolved as the water and plaster combine, as may readily be appreciated by operating upon considerable quantities of the materials. Since the plaster assumes crystalline form as it becomes hydrated, the paste increases in bulk as it hardens, and is thus pressed into the finest interstices of the moulds.

Gypsum sets the more quickly in proportion as the temperature at which it has been dehydrated was low. After it has been heated above 300°, it will no longer set on being mixed with water. Besides its use in taking casts, plaster of Paris, on account of this power of combining with water, is largely employed in the preparation of stucco and of various imitations of marble. The hydrated compound finds application also as a manure, in the manufacture of ammoniacal salts, and for various other purposes.

Exp. 285.—That the plaster paste expands considerably at the moment of solidification may be shown as follows:—Procure a cracked test-tube, or small flask, and fill it completely with a paste made of calcined gypsum and water, in the proportions of 12 pts., by weight, of the former, to 5 of the latter. In the course of 15 or 20 minutes it will be seen that the original crack in the glass vessel has extended in various directions, in consequence of the expansion of the mass within it. It will be noticed, also, that the vessel feels warm to the hand (compare Exp. 284). Finally, by breaking away the glass envelope, there may be obtained a cast of the glass vessel.

Exp. 286.—The power of sulphate of calcium to take up water, to solidify water as it unites with it to form the crystalline compound CaH_4SO_6, can be made manifest as follows :—Prepare two table-spoonfuls of a saturated aqueous solution of chloride of calcium, by dissolving 1 pt., by weight, of the dry chloride in 1·5 part of water; also prepare the same quantity of a saturated solution of sulphate of sodium (Exp. 228), and, finally, mix the two solutions. Sulphate of calcium will be formed, in accordance with the reaction

$$CaCl_2 + Na_2SO_4 = CaSO_4 + 2NaCl,$$

and will unite with the water in which the ingredients from which it has been formed were previously held in solution, so that an almost solid mass of $CaSO_4,2H_2O$ will take the place of the two liquids.

Ordinary hydrated sulphate of calcium is soluble in about 400 parts of water at the ordinary temperature of the air; but, like hydrate of calcium, sulphate of sodium, and a few other salts, it is less soluble in hot water than in cold. When an aqueous solution of sulphate of calcium is heated to 100° or more, a precipitate will soon be formed in it, even if the solution be very dilute and at temperatures as high as 140° or 150° the anhydrous compound is completely insoluble in water. In the same way as with sulphate of sodium (Exp. 228), it appears that the bihydrated sulphate of calcium cannot exist at temperatures much superior to 100°, and that above that temperature we have to deal with other compounds of different solubility. In other words, the water which is held in chemical combination in ordinary unburned gypsum may be expelled by heat even when the gypsum is dissolved in water. Whenever water containing sulphate of calcium in solution is strongly heated, as in steam-boilers, there is precipitated the half-hydrated compound, of composition $2CaSO_4+H_2O$, which has been mentioned above. Hence the formation of incrustations, or *scale*, of sulphate of calcium upon the walls of boilers fed with sea-water, or with other water containing the sulphate. It should be remarked that the incrustation in this case does not depend upon evaporation; the sulphate of calcium will be deposited the more rapidly in proportion as the water of the boiler is hot, and as more of the impure feed-water is pumped into the boiler.

559. Besides occurring in sea-water, sulphate of calcium is a very common impurity in spring-water. Water which contains

much of it is "hard," and is not well adapted either for washing or for cooking.

Exp. 287.—Dissolve a small bit of soap in hot water, and add to the solution an equal bulk of a solution of sulphate of calcium. The mixture immediately becomes turbid, and after a few moments there will be formed a greasy, flocculent, adhesive scum upon the surface of the liquor. This precipitate is a *lime soap*, formed by the union of the fatty ingredients of the soap and the base of the sulphate of calcium. Common soap is a compound of one or more organic acids, known as fatty acids, with caustic soda. This soda soap is soluble in water, but lime soap is insoluble; hence, when a soluble salt of calcium is added to a solution of soap, precipitation occurs. When soap is added to hard water, it will produce neither permanent froth nor cleansing effect, until the sulphate, or other lime-salt present, has all been decomposed; with such waters, much soap is consumed in removing the calcium compound, before the proper detergent action of the soap can be brought into play.

560. An excellent process for determining the relative hardness of several samples of water has been founded upon the behavior of water towards soap, as set forth in the foregoing experiment:—

Exp. 288.—Prepare a sample of water, of standard hardness, as follows:—Dissolve 0·5 grm. of white marble, or other pure carbonate of calcium, in dilute chlorhydric acid, evaporate to dryness, in order to expel the excess of acid, and dissolve the pure chloride of calcium obtained in 2 litres of water. Next prepare a solution of soap by digesting 7 grms. of Castile soap, or, better, white curd soap, in 1120 grms. of a mixture of 3 parts of alcohol, of 0·83 specific gravity, and 1 of pure water, until no more soap dissolves; filter the solution, and preserve it in a tight bottle. Measure off 100 c. c. of the water, of standard hardness, place it in a bottle of 200 or 250 c. c. capacity, and by means of a graduated burette (Appendix, § 21), or pipette, add to it the solution of soap by portions of 1 c. c. each. After the addition of each c. c. of the soap solution, replace the stopper in the bottle, and shake the latter violently, then place the bottle upon its side, and observe whether the bubbles, which form upon the surface of the liquid, quickly disappear. So long as the bubbles disappear immediately, new portions of the soap-liquor must be added; but as soon as a permanent froth is formed, the operation is finished. It is customary to consider the operation completed when the bubbles persist during three minutes. The number of c. c. of soap-liquor which has been employed in producing this result, is then carefully recorded.

Samples of well- and river-water may readily be compared with the water of known standard hardness. We have only to measure off 100 c. c. of the well-water, place it in a small bottle, as above, and add to it the soap-liquor, whose value has been determined, until a persistent froth is produced. If it be assumed that the standard chloride-of-calcium water represent 100° of hardness, the comparative hardness of any other sample of water will follow from the proportion:—As the quantity of soap-liquor required to produce persistent bubbles in the standard water is to 100, so is the quantity of soap-liquor which produces bubbles in any given sample of water to the relative hardness of the sample.

When the water under examination has a much higher degree of hardness than 100°, it is necessary to dilute it with from 1 to 5 times its volume of distilled water before adding the soap-liquor; for the curdy precipitate, which would form, if soap were added to the undiluted liquid, would interfere with the formation of froth, and so make it difficult to determine when a sufficient quantity of the soap-liquor had been used.

On being ignited in an atmosphere of hydrogen, or in contact with substances containing carbon, gypsum may readily be deoxidized and converted into sulphide of calcium:—

$$CaSO_4 + 4C = CaS + 4CO.$$

This reduction is readily effected, also, when aqueous solutions of gypsum are left in contact with decaying vegetable matter. Since, in this case, carbonic acid will necessarily come in contact with the sulphide of calcium as soon as it is formed, sulphuretted hydrogen gas will be set free, as may be perceived wherever the mud of docks and marshes is wet with sea-water:—

$$CaS + H_2O + CO_2 = CaCO_3 + H_2S.$$

561. *Phosphates of Calcium.*—There are several of these phosphates, comparable respectively with the various phosphates of sodium (§ 489); the most remarkable among them is the triphosphate ($3CaO, P_2O_5$), commonly called bone-phosphate, from being found in bones. It is the chief of the inorganic constituents of which the skeletons of animals are composed. Small portions of it are found in most rocks and soils (§ 262), it being a very widely diffused, though nowhere a very abundant, substance. Considerable masses of it have been found, however, in Spain, New Jersey, and Canada, and it is the principal ingredient of some kinds

of guano. No matter whence obtained, it is a valuable manure when reduced to a fine powder. Though as good as insoluble in water, it dissolves readily in acids and in solutions of various organic substances.

562. *Chloride of Calcium* ($CaCl_2$) may be prepared by dissolving chalk or marble in chlorhydric acid (as in Exp. 171), and evaporating the solution to dryness. It is produced in large quantities in the arts by heating chloride of ammonium with slaked lime in the preparation of ammonia-water (Exp. 48):—

$$2NH_4Cl + CaH_2O_2 = CaCl_2 + 2NH_3 + 2H_2O.$$

When dried at about 200°, chloride of calcium is left as a porous mass, which is largely employed in chemical laboratories for drying gases (Appendix, § 15). It absorbs water with great avidity, and is one of the most deliquescent substances known. When exposed to air at the ordinary temperature, it soon absorbs so much water that it dissolves completely. At a low red heat the anhydrous chloride melts to a clear liquid; if ignited for any length of time in contact with the air, it suffers decomposition to a slight extent, a little oxide and carbonate of calcium being formed. From highly concentrated aqueous solutions there may be obtained crystals of the hydrated compound $CaCl_2 + 6H_2O$.

Slaked lime may be dissolved in considerable quantity in a boiling aqueous solution of chloride of calcium, and the filtered solution deposits, on cooling, long, thin crystals of a compound known as oxychloride of calcium ($CaCl_2$, $3CaO + 16H_2O$), which is immediately decomposed when treated with pure water.

563. *Hypochlorite of Calcium* ($CaCl_2O_2$), as has been shown in § 120, is a component of the substance commonly called "chloride of lime." This important bleaching agent is prepared by passing chlorine gas into chambers filled with layers of finely powdered slaked lime, in accordance with the reaction already set forth. Chloride of lime, or bleaching-powder, is a dry, white powder, smelling feebly of hypochlorous acid; it always contains a certain excess of hydrate of calcium which has been unacted upon by chlorine; it is therefore only partially soluble in water. When exposed to the air, it slowly absorbs carbonic acid, and, at the same time, evolves chlorine; hence its employment as a disinfecting agent. If, instead of being left to be

slowly acted upon by the carbonic acid of the air, it be treated with a dilute acid (such as vinegar), a copious evolution of chlorine will immediately occur.

Exp. 289.—Place half a teaspoonful of bleaching-powder in a test-glass, cover the powder with water, and stir in enough of a solution of blue litmus to distinctly color the mixture. By means of a glass tube, blow into the mixture air expired from the lungs, and observe that the blue color of the litmus will soon be destroyed. The carbonic acid from the lungs decomposes the hypochlorite of calcium, and the chlorine set free destroys the color.

Exp. 290.—At the bottom of a large, tall beaker, or other wide-mouthed glass vessel, of the capacity of two or three litres, place a small bottle containing 15 or 20 grms. of bleaching-powder. Cover the beaker with a glass plate, or sheet of pasteboard, provided with a small hole at the centre: through this hole in the cover pass a thistle-tube down into the bottle of bleaching-powder, and pour upon it several small successive portions of sulphuric acid diluted with an equal volume of water. Chlorine gas will immediately be set free from the bleaching-powder, in accordance with the reaction

$$CaCl_2,CaCl_2O_2 + 2H_2SO_4 = 2CaSO_4 + 2H_2O + 4Cl,$$

and, falling over into the bottom of the large beaker, will gradually press out and displace the air therein contained, so that after a short time the beaker will be seen to be completely filled with the green gas. This is by far the easiest and most expeditious method of preparing chlorine. If desirable, the bleaching-powder may, of course, be placed in a flask, together with the acid, and the evolved gas collected at will, by means of suitable delivery-tubes; but many of the experiments of Chapter VIII. may be performed perfectly well in the jar of chlorine obtained as above. The heavy gas may be ladled out of the jar with a dipper made of any small bottle, and poured upon a solution of indigo to show its bleaching-power.

It will be noticed, in the above reaction, that by the addition of an acid all the chlorine of the bleaching-powder is expelled. The point is important as bearing upon the practical use of this agent.

Exp. 291.—Soak a bit of printed calico in a half-litre of water, into which 10 or 15 grms. of bleaching-powder have been stirred. Observe that the color of the calico slowly undergoes change; then transfer the cloth to another bottle filled with very dilute chlorhydric or sulphuric acid, and take note of the rapidity with which the color is discharged. If need be, again immerse the calico in the bleaching bath, and afterwards in the dilute acid. Finally, wash the whitened cloth thoroughly in water.

564. When heated, bleaching-powder gives off oxygen, while chloride of calcium is left as a residue. The reaction furnishes a cheap and convenient method of obtaining oxygen. Another method of procuring oxygen from the hypochlorite is to mix a solution of the latter with black oxide of manganese, red oxide of mercury, oxide of iron, or oxide of copper, or, better, with hydrated sesquioxide of iron, hydrate of copper, of nickel, or of cobalt, and to gently warm the mixture.

Exp. 292.—Fill an ignition-tube one-third full of bleaching-powder, and arrange the apparatus so that the gas may be collected over water. Heat the tube, and observe that the gas is expelled at a comparatively low temperature. 1 grm. of bleaching-powder yields 40 or 50 c. c. of oxygen gas.

Exp. 293.—Take as much bleaching-powder as was employed in Exp. 292, dissolve it in a small quantity of water, filter the solution, and place it in a small flask provided with a delivery-tube. Add to the contents of the flask two or three drops of the solution of a cobalt salt, connect the flask with an inverted bottle of water upon the water-pan, by means of the delivery-tube, then heat the flask to 70° or 80°, and observe that oxygen is freely evolved.

The cobalt solution employed amounts to the same thing as hydrated oxide of cobalt, since the latter is immediately precipitated from the cobalt salt by the caustic lime in the bleaching-powder. The action of the oxide of cobalt, or other metallic oxide, in this experiment, appears to be somewhat analogous to that of the higher oxides of nitrogen in the manufacture of sulphuric acid (§ 228). The oxide of cobalt probably takes oxygen from the solution of bleaching-powder, and combines with it to form a high, unstable oxide, which immediately decomposes again with evolution of oxygen. The oxidation and deoxidation of the cobalt compound thus goes on incessantly, and a very small quantity of the latter is sufficient to decompose any desired amount of bleaching-powder. It is important that the solution of the hypochlorite should be filtered as above directed, lest a quantity of it be lost by foaming over out of the flask.

565. The proportion of hypochlorite of calcium in bleaching-powder varies widely in different samples, according to the care with which the sample has been prepared, and to the length of time it has been exposed to the action of the air. The bleaching-power, or in other words, the money value of each special sample, should therefore be determined before it is sold. Of the several

methods of ascertaining the value of bleaching-powder, one of the simplest is to determine how much arsenious acid (As_2O_3) can be converted into arsenic acid (As_2O_5) by a given weight of the sample:—

$$As_2O_3 + 4Cl + 2H_2O = As_2O_5 + 4HCl.$$

To this end a weighed quantity of arsenious acid is dissolved in a certain definite quantity of a solution of carbonate of sodium in such manner that each c. c. of the liquor shall contain a known weight of arsenious acid. A considerable quantity of this standard solution may be prepared once for all, and kept for use in tightly closed bottles. A weighed sample of the bleaching-powder under examination is then dissolved in water, and into this solution of bleaching-powder the standard solution of arsenious acid is carefully poured, from a burette, so long as the arsenious acid continues to be oxidized and converted into arsenic acid.

In order to determine the precise moment when the oxidizing-power of the bleaching-powder solution ceases, a drop of this solution is taken out from time to time upon a glass rod and placed upon paper prepared with iodide of potassium and starch, as has been described in Exp. 71. When the paper no longer becomes blue on being touched with the solution, the operation is known to be completed. Towards the close of the experiment the solution of arsenious acid should only be added drop by drop to the solution of bleaching-powder, and some of the latter should be touched to the test-paper after each addition of the arsenious acid.

The number of c. c. of the solution of arsenious acid which have been employed is then carefully noted, and the amount of arsenious acid contained in them is computed. From these data the amount of chlorine in the sample of bleaching-powder is obtained by the following proportion:—

198	:	142	=	Weight of arsenious acid used.	:	Weight of chlorine in the sample.
Weight of one molecule of arsenious acid.		Weight of four atoms of chlorine.				

Of the other compounds of calcium may be mentioned the fluoride (§ 155), bromide, and iodide (analogous to the chloride), the peroxide CaO_2, several sulphides (§ 213), and the phosphide (§ 278). Nitrate of calcium (CaN_2O_6) is a very easily soluble, deliquescent salt, found in many soils, and in other localities where organic matters putrefy in contact with hydrate or carbonate of calcium. Sulphydrate of calcium CaS,H_2S, analogous to the hydrate, may be prepared by boiling monosulphide of calcium with water, or by passing sulphydric acid gas into a cold solution of the sulphide. The solution of this substance possesses a remarkable power of loosening hair from the skins of animals. After a skin has been soaked for a few minutes in a strong solution of this substance, the hair may readily be scraped off with any blunt instrument. A solution of sulphite of calcium ($CaSO_3$) in an aqueous solution of sulphurous acid, is sometimes employed, under the name bisulphite of lime, to check fermentation.

566. The metal itself may be obtained by decomposing fused chloride of calcium by means of the galvanic current, or by heating iodide of calcium with metallic sodium in a closed iron tube. It is a yellowish-white, lustrous, ductile metal, of 1·6 specific gravity, which suffers no change in dry air at the ordinary temperature. In moist air it oxidizes quickly, and it decomposes water with evolution of hydrogen. At a red heat it melts, and, if oxygen be present, takes fire and burns with a bright light. It is a bivalent element; the weight of its atom is 40.

STRONTIUM AND BARIUM.

567. The metals strontium and barium closely resemble calcium in appearance and properties, and may be prepared by methods similar to those used for calcium, as described in the last section. The specific gravity of strontium is 2·6, that of barium is 4. The atomic weight of strontium is 87·5, and that of barium 137. Like calcium, strontium and barium are both bivalent elements.

Most of the compounds of strontium and barium are closely analogous to the corresponding compounds of calcium. The oxides, peroxides, hydrates, carbonates, sulphates, nitrates, phosphates, chlorides, sulphides, &c. resemble in the main the corresponding calcium salts. The hydrates of strontium and barium, however, are more readily soluble in water than the hydrate of calcium, while their sulphates, nitrates, and chlorides are less soluble than those of calcium. Sulphate of barium is almost absolutely insoluble in water, and sulphate of strontium is only very slightly soluble. Sulphate of barium is found native, sometimes in considerable masses, as a very heavy white mineral called *barytes*, which, when powdered, is largely employed for adulterating white lead. The name barium comes from a Greek word meaning heavy.

From the carbonates of strontium and barium the carbonic acid cannot readily be driven off by heat alone, though when mixed with charcoal, and then ignited, these carbonates may be reduced to oxides. A better way of preparing the oxides is to heat the nitrates strongly in a porcelain crucible or retort. Unlike hydrate of calcium, hydrate of barium does not give off its water at the temperature of redness, but melts without undergoing decomposition. From hydrate of strontium the water may be expelled by heat, though with difficulty. Peroxide of barium (BaO_2) is of interest, since by means of it peroxide of hydrogen (§ 61) and antozone (§ 177) may be prepared.

568. In order to obtain peroxide of barium, a mixture of 1 part of oxide of barium, and 4 parts of chlorate of potassium may be thrown, little by little, into a crucible heated to low redness, and the fused mass subsequently washed with water to remove chloride of potassium; or a current of oxygen gas, or of air, may be made to flow over oxide of barium heated to low redness in a porcelain tube. As thus prepared the peroxide is never pure, being mixed with more or less protoxide. Peroxide of barium decomposes, with evolution of oxygen, at the temperature of bright redness, and in view of this fact it was at one time proposed to employ the substance as a means of obtaining pure oxygen from the air upon the large scale. A considerable number of tubes charged with protoxide of barium, having been suitably arranged in furnaces, half of the tubes were heated to dull redness, and a current of air was made to flow through them, until the protoxide had been converted into peroxide; the current of air was then transferred to the other tubes, while the first series was put in connexion with a gas-holder and heated to bright redness, until the second atom of oxygen had been driven out. The second series of tubes were next deprived of oxygen, while the tubes of the first series were put to their old work of absorbing oxygen from the air. The process thus

became a continuous one, and was really capable of furnishing large quantities of oxygen; it has, however, been superseded by cheaper methods (§ 242).

Strontium salts are commonly prepared from the native carbonate, a mineral called *strontianite*, while the various salts of barium are obtained either from the native carbonate (*witherite*), or more commonly from the sulphate. The finely powdered sulphate, after having been mixed with powdered charcoal and oil, is strongly heated in a covered crucible, and so reduced to the condition of sulphide of barium:—

$$BaSO_4 + 4C = BaS + 4CO.$$

Sulphide of barium is readily soluble in water, and on being treated with chlorhydric, nitric, or any other acid, it decomposes, sulphuretted hydrogen is given off, and there is formed chloride of barium, or some other salt, according to the acid employed:—

$$BaS + 2HCl = BaCl_2 + H_2S.$$

Several of the compounds of barium are useful reagents in the chemical laboratory. Sulphate of barium is employed as a pigment by artists in water-colors, under the name *permanent white*, also in the finishing of paper, pasteboard, &c., and for adulterating white lead. As a water-color it is valuable, since it is scarcely at all acted upon by any chemical agent; but when ground with oil it becomes translucent, and seriously impairs the opacity or covering power of the better pigments with which it is mixed.

Compounds of barium and of strontium are employed in the preparation of fireworks, for obtaining green and crimson flames respectively:—

The green barium-flame may be well shown by mixing with the fingers a gramme of powdered chlorate of barium with half a gramme of flowers of sulphur, and strewing the mixture upon a glowing coal. The green fire of the pyrotechnists may be prepared by mixing together 58 parts of nitrate of barium, 13 parts of sulphur, 6 parts of chlorate of potassium, and 2 parts of charcoal.

To exhibit the red strontium-flame, a mixture may be prepared by rubbing together in a mortar 30 parts of anhydrous nitrate of strontium, 10 parts of powdered sulphur, and 3 parts of sulphide of antimony; and to this mixture may be added, with the hand, taking care to avoid all violent friction, 7 parts of powdered, fused chlorate of potassium. The mixture may then be shaken loosely upon a piece of sheet iron and touched with a lighted stick or glowing coal.

Or the color may be shown upon a smaller scale by operating as follows:—

Exp. 294.—By means of iron wire, suspend three small bullets of well-burned coke from a ring of the iron stand. Heat the fragments in turn with the flame of the gas-lamp, and observe the slightly yellowish flame which will be produced in each case; then moisten one of the pieces of coke with a solution of chloride of calcium, the second with a solution of chloride of barium, and the third with a solution of nitrate of strontium, and again heat them in turn with the gas-flame. The calcium salt will impart a reddish-yellow color to the flame, the barium salt a green color, and the strontium salt a beautiful crimson. Instead of the bits of coke, platinum wire might of course be employed, as in Exp. 202.

569. As appears abundantly from the foregoing, the three elements calcium, strontium, and barium are intimately related one to the other, and are, as a family, clearly distinguished in several important particulars from the metals of the preceding group. Even before the metals of this family were discovered and isolated, it had long been customary among chemists to speak of the oxides of calcium, strontium, and barium as the *alkaline earths*, in contradistinction from the "alkalies," potash and soda, upon the one hand, and the "earths," such as the oxides of magnesium and aluminum, upon the other.

Each of the members of the alkaline-earthy group, now in question, decomposes water, even at the ordinary temperature, taking away its oxygen; and each of them forms two oxides—a neutral, insoluble binoxide, belonging to the class of antozonides (§ 182), and a more or less soluble protoxide, acting as a powerful base; their carbonates and sulphates are all difficultly soluble, and, like the other compounds of the three metals, are isomorphous with one another. In all their compounds, there may be seen the same progression of properties which has been met with in the groups previously studied. The barium compound will always be found at one end of the scale, the calcium compound at the other, and the strontium compound interposed between the two. The hydrate and the carbonate of calcium are both readily destroyed by heat, while the corresponding strontium compounds are decomposed with difficulty, and the barium compounds only at exceedingly high temperatures. The solubility of the oxides diminishes as we pass from baryta to lime, while that of the sulphates and carbonates follows the inverse order. The specific gravities of the metals are, Ca=1·6, Sr=2·6, Ba=4; and their atomic weights are 40, 87·5, and 137 respectively, that of strontium being nearly the mean of the other two. The specific gravities of their carbonates and sulphates are as follows:— $CaCO_3$ (arragonite) =2·95, $SrCO_3$=3·6, $BaCO_3$=4·33; $CaSO_4$= 2·33, $SrSO_4$=3·89, and $BaSO_4$=4·4. It should be observed that, in all these cases, the specific gravities of the strontium compounds approximate closely to the mean of the specific gravities of the corresponding barium and calcium compounds. The same remark applies also to the specific gravities of the three metals.

LEAD.

570. Almost all the lead which is employed in the arts is extracted from sulphide of lead, PbS, the mineral *galena*. This substance is tolerably abundant in many localities, and is often associated with sulphate of barium, fluor-spar, quartz, and other common minerals; it almost always contains a small proportion of sulphide of silver. In order to obtain metallic lead from galena, this mineral is mixed with a small quantity of lime, and then roasted at a dull-red heat in the flame of a reverberatory furnace. A portion of the sulphur burns off as sulphurous acid. Some oxide of lead, and more or less sulphate of lead is formed, while much of the ore remains undecomposed. After a time, the roasting-process is interrupted, air is excluded from the furnace, the oxide, sulphate, and sulphide of lead are thoroughly mixed together, and the heat of the furnace is suddenly raised. The undecomposed sulphide of lead then reacts upon the oxide and sulphate, sulphurous acid is given off, and metallic lead produced. The reactions may be thus formulated:—

$$2PbO + PbS = 3Pb + SO_2.$$
$$PbSO_4 + PbS = 2Pb + 2SO_2.$$

The lime is added for the purpose of forming a fusible slag with any siliceous matter which may be present in the ore.

Lead is a remarkably soft metal, of bluish-white color; it can be readily cut with a knife, and may even be indented with the finger-nail; it soils paper upon which it is rubbed. Its specific gravity is 11·4, and its atomic weight 207. It may be drawn into wire, and beaten into sheets, though, as contrasted with most of the other metals, it has but little tenacity. In comparison with other metals, it is a rather poor conductor of heat and electricity. It melts at about 325°, and contracts considerably in passing from the liquid to the solid condition. Its specific heat is 0·0314. Solid lead expands greatly when heated, though the heat be not carried near to the melting-point, and the expanded metal does not return again to its original dimensions when cooled. Melted lead begins to emit vapors at a red heat, and at very high temperatures the metal may even be distilled. Lead may be obtained crystallized in octahedrons, by slowly cooling the molten metal.

The ready crystallization of lead furnishes a very simple method of separating this metal from the silver with which crude lead is almost always contaminated as it comes from the smelting furnaces. When melted argentiferous lead is allowed to cool slowly, and is at the same time briskly stirred, a quantity of solid crystalline grains separate out after a while, and sink beneath the liquid metal, whence they may be dipped out in cullenders. These crystals are composed of lead nearly free from silver, while all but a trace of the silver contained in the original lead is left in that portion of the metal which has not yet solidified; in a word, the alloy of lead and silver melts at a lower temperature than pure lead. By methodically remelting and recrystallizing the lead crystals on the one hand, and the silver alloy on the other, it has been found profitable to extract the silver from lead so poor that it contained less than one thousandth part its weight of the precious metal.

When in thick masses, such as the common sheets and pipes of commerce, lead is scarcely at all acted upon by cold sulphuric acid, and is but slowly corroded by chlorhydric acid. Both these acids form, by uniting with lead, difficultly soluble salts; and so soon as a layer of the salt has once been deposited upon the surface of the metal, the latter is thereby protected from further corrosion. By hot, concentrated sulphuric acid, however, lead is dissolved rather easily. The best solvent of metallic lead is diluted nitric acid; strong nitric acid will not dissolve it readily, since nitrate of lead is well nigh insoluble in concentrated nitric acid.

571. *Oxides of Lead.*—When a compact piece of metallic lead is freshly cut, it exhibits considerable lustre, and this lustre may be preserved unimpaired by keeping the lead in perfectly dry air, or beneath the surface of pure water free from air (§ 47). But by exposure to ordinary air the brilliant surface soon tarnishes, in consequence of the formation of a thin coating of suboxide of lead; this incrustation protects the metal beneath from further oxidation. Finely divided lead, on the contrary, soon changes completely to suboxide when exposed to the air. If the metal be fine enough, it will oxidize instantaneously, with evolution of light and heat, and formation of yellow protoxide of lead.

Exp. 295.—Prepare a small quantity of tartrate of lead, as follows. Dissolve 0·25 grm. of common sugar of lead (acetate of lead) in 8 or 10 c. c. of water, also dissolve 0·1 grm. of tartaric acid in 4 or 5 c. c. of water, and mix the two solutions. Collect upon a filter the white precipitate of tartrate of lead which will be formed, wash it with water, then unfold the filter and spread it out with its contents to dry in the air, or, better, at a gentle heat upon a ring of the iron stand, high above a single gas-flame.

Fill an ignition-tube one-third full of the dry tartrate of lead, and heat it upon a sand-bath so long as any fumes escape, then cork the tube tightly and set it aside to cool. Holding the cooled tube high in the air, sprinkle its contents upon a plate, and observe that the black powder takes fire spontaneously, and burns with a red flash. The composition of tartrate of lead may be represented by the formula $C_4H_4Pb_2O_6$; on being heated this substance gives off water and carbonic oxide, as may be seen by lighting the fumes which escape from the tube, and there is left as a residue an intimate mixture of carbon and of metallic lead, so finely divided that it inflames in ordinary air. A spontaneously inflammable mixture such as this is called a *pyrophorus.*

572. Lead is far more readily oxidized by the continued action of air and water than by ordinary moist air. When exposed to the simultaneous or alternate action of these agents, a coating of the white hydrated protoxide of lead is rapidly formed; but as this compound is somewhat soluble in water, it is continually dissolved away and affords little or no protection to the lead beneath. The corrosive action of water upon lead is modified very mate-

rially by the presence of small quantities of various saline substances. Water containing traces of nitrates, nitrites, and chlorides corrodes lead more rapidly than pure water, while the corrosive action of pure water appears to be diminished by the presence of sulphates, phosphates, and carbonates, oxide of lead being scarcely at all soluble in water which contains these salts in solution. Water containing a solution of carbonate of calcium in carbonic acid, such as is frequently met with in nature, has been found to have remarkably little action upon lead; in such water a coating of insoluble, or nearly insoluble carbonate of lead is formed upon the metal, which protects it from further action. But, on the other hand, water which contains much free carbonic acid dissolves away the protective coating and exposes fresh surfaces of lead to corrosion. As a general rule, the acids, even when very dilute, greatly accelerate the oxidation of lead in the air; and the same remark is true of organic substances and of metals, the first by their decay, the second by the galvanic action which their presence excites.

Since solutions of lead are poisonous, and since the metal is employed to an enormous extent for cisterns and conduits, a knowledge of the action of water upon lead is very important in a sanitary point of view. The question has consequently engaged the attention of many chemists, and has been much discussed. It has been proved by numberless experiments that the action of natural waters upon lead is so general that it is rare to find any sample of water, which has been kept in a leaden cistern, wholly free from traces of that metal. The opinion of most chemists is at this time (1867) decidedly adverse to the use of leaden water-pipes in houses, in spite of the fact that the metal is nowadays employed for this purpose almost everywhere with apparent impunity.

When lead is melted in the air it oxidizes readily, with formation at first of gray suboxide, and afterwards of the yellow protoxide.

573. *Suboxide of Lead* (Pb_2O) may be prepared in a state of purity by cautiously heating oxalate of lead at a temperature not exceeding 300° in a retort from which air is excluded, so long as any gas is evolved:—

$$2PbC_2O_4 = Pb_2O + CO + 3CO_2.$$

After the retort has become cold, suboxide of lead will be found in it as a black velvety powder. It is decomposed by acids, with formation of salts of protoxide of lead, and separation of metallic lead.

574. *Protoxide of Lead* (PbO), commonly called *litharge*, may be obtained as a lemon-yellow powder by gently igniting nitrate, carbonate, or oxalate of lead upon an iron plate, or in an open porcelain crucible. The oxide fuses at a red heat, and when melted in vessels of porcelain or earthenware it rapidly destroys them by combining with their silica, an easily fusible slag or glass composed of double silicates of lead, aluminum, iron, &c. being formed. Silicate of lead is an actual constituent of the easily fusible variety of glass known as flint glass (see § 492, and Appendix, § 3).

In the arts, litharge is prepared upon the large scale by heating metallic lead in a current of air; the color and texture of the product varies considerably according to the temperature and the other conditions under which it has been prepared.

Exp. 296.—Heat a small fragment of lead upon charcoal in the oxidizing flame of the blowpipe, and observe the gray film of suboxide which forms at first, and the yellow incrustation of litharge which is obtained subsequently. The litharge may be melted if a strong, hot flame be thrown upon it.

This property of lead, of rapidly oxidizing when heated in the air, taken in connexion with the easy fusibility of the oxide, is the basis of the common method of separating lead and silver in the large way, known as *cupellation.* The iridescent film of litharge continually formed upon the surface of the molten metal as incessantly flows off, exposing new surfaces of the metal to the action of the air. The silver, on the other hand, undergoes little or no change, and when the lead has been completely burned the silver appears in all its brilliant whiteness.

The cupel is a shallow cup or basin, composed either of marl or of mixture of bone-ash and wood-ashes firmly compacted and beaten to a smooth surface, which may be placed either in the muffle of an assay furnace, upon the hearth of a reverberatory, or in any position where it can be strongly heated at the same time that a free current of air plays over its surface. A charge of argentiferous lead having been melted upon the cupel, new portions of the lead are added as fast as the melted litharge flows off from the convex surface of the metal and makes room for these additions, until an alloy very rich in silver has been obtained. This alloy is then cupelled until the last traces of lead have been removed, and the silver is left pure and glistening. In cupelling upon the small scale, for purposes of assaying, the cupel is made of bone-ash of such quality that the litharge may be absorbed into the substance of the cupel, and not flow off through gutters upon its edge, as is the case in large metallurgical operations.

575. Protoxide of lead unites readily with acids, and forms many important salts. When in the state of powder, it even absorbs a certain amount of carbonic acid from the air; hence the powdered litharge of commerce always contains more or less carbonate of lead, and therefore effervesces on being treated with acids, as has been seen in Exp. 42. As a general rule, it is far better to prepare the salts of lead by dissolving the protoxide in acids, than to treat the metal itself with acids. Protoxide of lead has a remarkable tendency to form basic salts; thus besides the normal nitrate (PbO,N_2O_5) there is a dinitrate ($2PbO,N_2O_5$), a trinitrate ($3PbO,N_2O_5$), a tetranitrate ($4PbO,N_2O_5$), and a hexanitrate ($6PbO,N_2O_5$).

Though a strong base as regards the acids, protoxide of lead behaves like an acid towards the alkalies and alkaline earths. For example, it dissolves readily in soda or potash lye, with formation of plumbite of sodium, or of potassium, as the case may be. The term plumbite, like the symbol of lead, Pb, is derived from *plumbum*, the Latin name of the metal.

576. *Peroxide of Lead* (PbO_2) may be prepared by oxidizing the protoxide—for example, by passing a current of chlorine gas through water in which protoxide of lead is kept suspended by agitation, or as follows:—

Exp. 297.—Place 8 or 10 grms. of very finely powdered sugar of lead in a capacious porcelain dish, cover the salt with a filtered solution of bleaching-powder (§ 563), heat the solution to boiling, and maintain it at this temperature, until the escaping vapors smell strongly of acetic acid; then pour the contents of the dish upon a filter, and wash with water the dark-brown powder of peroxide of lead which has been formed.

Peroxide of lead may be easily obtained also by digesting red lead with dilute nitric acid, as will appear in the following paragraph.

Peroxide of lead is a powerful oxidizing agent; it readily gives up oxygen to many organic substances, even at the ordinary temperature of the air. On being heated to redness, it loses half its oxygen, and is converted into the protoxide.

Exp. 298.—In a small porcelain mortar, rub together a mixture of 1 grm. of oxalic acid, and 1 grm. of peroxide of lead. Decomposition will occur, aqueous vapor and carbonic acid will be given off, and carbonate of lead, $PbCO_3$, will be left as a residue. When the peroxide is thus mixed with one-eighth its weight of sugar, or with one-sixth its weight of tartaric acid, so much heat is developed, that the mass in the mortar glows.

Peroxide of lead is decomposed by chlorhydric acid, with liberation of chlorine, and formation of normal chloride of lead—

$$PbO_2 + 4HCl = PbCl_2 + 2H_2O + 2Cl,$$

—and by hot sulphuric acid, with evolution of oxygen. It combines with sulphurous acid readily, and is often employed in the laboratory as an absorbent of this gas; and it is noteworthy that the product of this combination is sulphate of lead:—

$$PbO_2 + SO_2 = PbSO_4.$$

It is indifferent, therefore, whether we put together peroxide of lead and sulphurous acid, or protoxide of lead and sulphuric acid; the product will, in either case, be common sulphate of lead; for

$$PbO + SO_3 = PbSO_4.$$

As a rule, peroxide of lead does not readily enter into combination with acids, though compounds of it with acetic, phosphoric, arsenic, and some other acids have been obtained. With strong bases, however, it combines readily, forming salts known as plumbates.

577. *Red Lead or Minium.*—When protoxide of lead is kept at a low red heat for some time, in contact with air, it gradually absorbs one or two per cent. of oxygen, and acquires a brilliant red color. The product of this oxidation is extensively used as a pigment and in the manufacture of some kinds of glass ware; it may be regarded as a compound of PbO and PbO_2, in varying proportions. By digesting it for some time in dilute nitric acid, the protoxide of lead may all be dissolved out and converted into nitrate of lead, while the peroxide of lead is left as a residue:—

$$2PbO,PbO_2 + 2H_2N_2O_6 = 2PbN_2O_6 + 2H_2O + PbO_2.$$

Exp. 299.—Heat in an iron spoon 4 grms. of litharge and 1 grm. of chlorate of potassium, and observe that the color of the mixture soon changes from yellow to red. Throw the cooled product upon a filter, wash it with water, then dry it and compare its color with that of the original litharge. In this experiment, the red lead could be obtained as well by simply heating litharge without admixture in the air for many hours, at a temperature just below its melting-point; time alone is gained by employing chlorate of potassium as the source of oxygen. For commercial purposes, red lead is obtained by heating metallic lead in reverberatory furnaces, or, when a very pure article is needed, by heating carbonate of lead.

When heated strongly, red lead is resolved into protoxide of lead and free oxygen. Oxygen gas may be prepared from it in the same manner as from oxide of mercury (Exp. 5), though at a higher temperature.

578. *Sesquioxide of Lead* (Pb_2O_3) is recognized as a distinct oxide by some chemists, but is more generally regarded as a compound of the proto- and peroxides, PbO, PbO_2,—a plumbate of lead.

579. *Sulphides of Lead.*—There are several of these compounds, but the protosulphide, PbS, is the only one whose composition is accurately known. This sulphide is the native mineral galena (§ 570); it may be prepared artificially either by melting together lead and sulphur in atomic proportions, or by treating the solution of any lead salt with sulphydric acid (§ 209). The native mineral, like the compound obtained artificially by way of fusion, is of a leaden-gray color of 7·5 specific gravity, but the precipitate which forms when sulphydric acid is added to the solution of a lead salt, is black, or brown, or even red if the solution be dilute. On account of the deep color, as well as the insolubility of this precipitate, sulphydric acid is often made use of as a means of detecting lead; the test is, in fact, so delicate that solutions containing only a hundred thousandth of their weight of metallic lead will assume a brown color on being charged with sulphuretted hydrogen.

Exp. 300.—Dissolve quarter of a gramme of sugar of lead in 4 litres of water, add to the solution a few drops of nitric acid so that it shall exhibit a faint acid reaction with litmus-paper, pass into the solution a current of sulphydric acid gas until the solution smells strongly of it and observe the brown color imparted to the fluid after some time.

In testing for the presence of lead in excessively dilute solutions, such, for example, as water drawn from leaden pipes, it is well to evaporate the liquid to a small bulk in a porcelain dish, to acidulate the concentrated liquor very slightly with nitric acid, and then to transfer it to a beaker glass. The liquid should then be saturated with sulphuretted hydrogen gas, the beaker covered with a glass plate, and left to stand during several hours in a moderately warm room. If lead be present, it will be indicated after a while either by the brown coloration of the liquid, or by the actual separation of a black powder at the bottom or upon the sides of the glass.

580. Sulphide of lead is volatile at high temperatures, and is often found in the cracks and upon the walls of smelting furnaces, in the form of crystals, which have been deposited by sublimation. By virtue of the volatility of its sulphide, lead may be transported to very considerable distances from furnaces where ores containing galena are roasted or reduced. It has been found, moreover, that growing plants are capable of taking up the lead thus deposited, and of assimilating a certain portion of it in their tissues. Comparative experiments made at very high temperatures have shown that galena may lose as much as 3·7 per cent. of its weight by volatilization, while metallic lead, exposed to the same conditions, loses less than 0·1 per cent.

Sulphide of lead, like the sulphides of the alkali-metals and those of the alkaline-earthy metals, acts as a sulphur base; with sulphantimonic acid, for example, it unites to form a salt, $3PbS,SbS_5$, analogous to that formed by the union of antimonic acid and oxide of lead, $3PbO,SbO_5$.

581. *Chloride of Lead* ($PbCl_2$).—Metallic lead is but slowly acted upon by chlorine, or by chlorhydric acid, though hot chlorhydric acid dissolves a little of it with formation of chloride of lead and evolution of hydrogen, even when out of contact with the air; the chloride may, however, be readily prepared by digesting oxide or carbonate of lead in chlorhydric acid, or by mixing the solution of almost any lead salt with chlorhydric acid, or with a solution of some soluble chloride:—

$$PbN_2O_6 + 2NaCl = PbCl_2 + Na_2N_2O_6.$$

Chloride of lead is but sparingly soluble in cold water, and is still less soluble in water acidulated with chlorhydric acid; hence it may readily be precipitated as above described, and collected upon a filter. In hot water, however, it dissolves rather easily, and it is somewhat soluble in concentrated acid also.

Exp. 301.—Boil together, in a small flask, 1 grm. of litharge, 14 grms. of strong chlorhydric acid, and 14 grms. of water, during 15 or 20 minutes. Pour the mixture upon a filter supported in a funnel which has been gently warmed by holding it over the flame of the gas-lamp, and collect the clear filtrate in a warm bottle. As the solution cools, lustrous needle-shaped crystals of chloride of lead will form in it.

Exp. 302.—Pour off the cold supernatant liquor from the crystals of chloride of lead obtained in Exp. 301, place the crystals upon a fragment of porcelain, dry them at a gentle heat, and finally heat them more strongly. It will be found that the crystals melt very easily, and that on cooling they solidify to a soft translucent horny mass, whence the old name of this substance, *horn-lead.*

582. The compounds of lead with iodine, bromine, and fluorine are analogous to chloride of lead. The iodide is remarkable on account of its beautiful yellow color, which may readily be shown by adding a drop or two of a solution of iodide of potassium to a small quantity of a solution of nitrate of lead.

Of the numerous other salts of lead little need here be said. The nitrate and tartrate have already been prepared (Exps. 42, 295); we have obtained the sulphate also as a white, nearly insoluble powder by adding water to concentrated sulphuric acid, and it may be had in any quantity by mixing the solution of a lead salt with dilute sulphuric acid, or with the solution of a soluble sulphate. Acetate of lead, one of the most important of the lead salts, and the one most readily to be procured in commerce in a state of purity, is prepared by dissolving oxide of lead directly in acetic acid such as is obtained in the distillation of wood (§ 380), or indirectly by moistening plates of metallic lead with vinegar in vessels open to the air. It crystallizes readily, is easily soluble in water, and has a sweet, astringent taste, whence the name, *sugar of lead.* Like the other lead salts, it is highly poisonous. It is employed for many purposes in the arts, and is in particular much used in medicine. Carbonate of lead ($PbCO_3$), or rather compounds of carbonate of lead and of hydrate of lead in varying proportions, are used to an enormous extent as a white paint, under the general name of *white lead*. The composition of this substance may usually be expressed by a formula lying within the limits $PbCO_3,PbH_2O_2$, upon the one hand, and $3PbCO_3,PbH_2O_2$ on the other. As contrasted with the other white pigments it possesses remarkable covering-power and durability, and is consequently much esteemed in spite of its high cost, its injurious influence upon the health of workmen who have to do with it, and the fact that it is discolored by air containing sulphuretted hydrogen. White lead is often adulterated with sulphate of barium, with oxide of zinc, and with gypsum. It is usually prepared by bringing carbonic acid, obtained from decaying vegetable matter, or from the combustion of fuel, into contact with basic acetate of lead,—the latter being prepared in this case either by mixing litharge and vinegar to the consistence of a paste, or by exposing rolls of sheet lead to the simultaneous action of vapors of vinegar and air, or by actually dissolving an excess of litharge in vinegar. Sometimes the carbonic acid is made to act upon the subacetate at the very moment when it is being formed, while at other times the acetate is prepared by itself, and subsequently treated with carbonic acid.

583. Silicate of lead is of interest from being an important ingredient of flint glass; a certain proportion of it renders glass lustrous and very beautiful. Such glass is, however, soft, easily fusible, and incapable of bearing sudden changes of temperature; it is, moreover, rather easily acted upon by alkalies, acids, and other chemical agents, and is hence comparatively useless in the chemical laboratory.

584. In many points of chemical behavior the compounds of lead resemble more or less clearly the corresponding compounds of barium, strontium, and calcium. Its compounds are moreover isomorphous with those of the metals in question, and its atom, like the atoms of these metals, is bivalent. Lead is therefore classed as a member of the calcium group, though, as is the case with fluorine in the chlorine group, it differs in some respects from the other members of the family. The specific gravity of lead is 11·4, and its atomic weight 207. The specific gravity of carbonate of lead is 6·5, and that of sulphate of lead is 6·2.

CHAPTER XXIX.

MAGNESIUM—ZINC—CADMIUM.

MAGNESIUM.

585. This metal, or rather its oxide, was formerly classed with the group which comprises the alkaline earths, but it is now known to be more closely connected with zinc and cadmium than with any other of the elements. It is found widely diffused, and rather abundantly, in nature. The bitter taste of sea-water and of some mineral waters is due to the presence of magnesium salts, while silicate of magnesium and carbonate of magnesium are contained in a variety of minerals and in such common rocks as dolomite, serpentine, soapstone, and talc.

586. Metallic magnesium may be prepared by heating anhydrous chloride of magnesium with sodium in a crucible of porcelain or platinum, and subsequently dissolving out in cold water the chloride of sodium which results from the reaction. Magnesium is a lustrous metal, as white as tin; its specific gravity is 1·75, and its atomic weight 24. It does not tarnish in dry air, though in damp air it soon becomes covered with a film of hydrate of magnesium. It melts at a low red heat, and volatilizes at higher temperatures; it may be readily distilled at a bright red heat. When heated strongly in the air it takes fire and burns with a bluish-white light of great brilliancy and high actinic power. The metal is employed by photographers for illuminating caverns and other places into which sunlight cannot penetrate, and in cloudy weather it is even used by them as a substitute for daylight. The metal can be pressed into wire or into thin ribbons, and a considerable quantity of it is now used in both these forms for purposes of illumination, as above stated. Magnesium lanterns are much used in theatres for illuminating scenery and tableaux. The white light has the advantage of showing colors just as they look by daylight. For scenic effects the light may be modified by transmission through colored glass. Magnesium is only slowly acted upon by cold water, but is

rapidly oxidized by hot water and by water acidulated with almost any acid; oxide of magnesium is formed and hydrogen set free.

587. *Oxide of Magnesium* (MgO).—There is but one compound of magnesium and oxygen; it is obtained as a white amorphous powder when magnesium is burnt in the air, or when carbonate, chloride, or nitrate of magnesium is ignited.

Exp. 303.—Roll 10 or 12 c.m. of magnesium wire or thin ribbon into a coil around a small pencil; withdraw the pencil and place in its stead a piece of iron wire or a knitting-needle; holding this wire horizontally, apply a lighted match to the end of the magnesium coil; the magnesium will burn to the white oxide, which coheres in an imperfect coil, clinging to the iron wire. A portion of the oxide goes off as white smoke. The magnesium wire for this experiment may be procured at toy-shops as well as of dealers in fine chemicals.

The oxide is tasteless and odorless; it is soluble to a very slight extent in water, and the solution has an alkaline reaction. The specific gravity of the solid oxide, or magnesia, as it is often called, varies from 3·07 to 3·2 as ordinarily prepared; but on being very strongly ignited it becomes denser, and samples have been prepared in this way of specific gravity as high as 3·61.

The light powdery oxide of magnesium known as "calcined magnesia," which is prepared by gentle but prolonged ignition of the hydrated carbonate, differs materially in several particulars from the more compact oxide obtained by calcining nitrate or chloride of magnesium at high temperatures, or by intensely heating the powdery oxide. Common calcined magnesia is, for example, readily soluble in acids; but after the oxide has been exposed to very high temperatures it dissolves but slowly even in the strongest acids. Similar differences between the products obtained at high and at low temperatures are met with among the oxides of almost all the metals hereafter to be studied.

588. A compact variety of oxide of magnesium, obtained by heating the nitrate or chloride to bright redness, but no higher, exhibits remarkable hydraulic properties. On being wet it quickly combines with a portion of water, and is converted into a crystallized hydrate of compact texture, harder than marble, and of great durability.

A mixture of equal parts of the hydraulic magnesia and of chalk, or powdered marble, made into paste with water, yields a slightly plastic mass, which admits of being readily pressed into any desired shape; if the moulded material be then placed in water it will become, after some time, extremely hard and compact (§ 591).

Oxide of magnesium is altogether infusible at temperatures short of that of the oxyhydrogen flame. Very excellent crucibles for scientific purposes are prepared by compressing oxide of magnesium into suitable forms. These crucibles undergo far less change in the air than those made from lime; and like the lime crucibles they do not unite with oxide of iron and the other metallic oxides to form the fusible slags or glasses which are so annoying in the ordinary crucibles, of which silicic acid is an essential component.

589. *Chloride of Magnesium* ($MgCl_2$) is found in sea-water and in many saline springs. It is formed when magnesium is burnt in chlorine gas, when a current of chlorine is passed over a red-hot mixture of charcoal and oxide of magnesium, and, in combination with water, by dissolving oxide of magnesium in chlorhydric acid. It is remarkable that the hydrated chloride last mentioned cannot be made anhydrous by evaporation and ignition without some decomposition of the chloride; oxide of magnesium is formed and chlorine goes off in combination with hydrogen as chlorhydric acid.

590. *Sulphate of Magnesium* ($MgSO_4$), or rather the hydrated compound ($MgSO_4 + 7H_2O$), is largely employed as a medicament under the name of Epsom salts. It is obtained not only from the mineral spring at Epsom, in England, and from various other springs, but is also prepared from sea-water, and by dissolving the minerals serpentine (silicate of magnesium), magnesite (carbonate of magnesium), and dolomite (carbonate of magnesium and of calcium), in sulphuric acid. Hydrated sulphate of magnesium is a colorless crystalline salt, readily soluble in water, and possessing the peculiar bitter taste common to most of the soluble magnesium compounds. It is often employed in laboratories as the source from which to prepare other magnesium salts.

591. *Carbonate of Magnesium* ($MgCO_3$) is found as a mineral in nature, and with due care may be prepared artificially. As met with in commerce, however,—the *magnesia alba* of the shops, prepared by mixing hot solutions of sulphate of magnesium and carbonate of sodium,—it is mixed with varying proportions of hydrate of magnesium. This compound is employed as a medicament. A compound of carbonate of magnesium and carbonate of calcium occurs abundantly in nature as the mineral dolomite, constituting extensive beds in various regions. Dolomite is much more slowly soluble in acids than true limestone, but when heated with a dilute acid it effervesces readily. When burnt, at temperatures so low that the carbonic acid shall be expelled only from the magnesium salt, while the carbonate of calcium remains unaltered, dolomite affords an hydraulic cement, preferable in many respects to ordinary lime. The product of the calcination "sets" rapidly under water, and is converted into a hard compact stone (§ 588). Citrate of magnesium, a preparation made from carbonate of magnesium and citric acid, is also largely employed as a medicament.

Of the other salts of magnesium, none are of sufficient importance to be described in this manual. Most of them are easily soluble in water; hence the insoluble double phosphate of magnesium and of ammonium ($MgNH_4PO_4 + 6H_2O$), obtained by adding ordinary diphosphate of sodium to a mixture of ammonia-water and any magnesium salt, is of importance to the analyst, since by means of it magnesium may be separated from its solutions. It should be observed that the ready solubility of sulphate of magnesium is in marked contrast with the insolubility of the sulphates of the alkaline-earthy group of metals.

ZINC.

592. Ores of zinc occur in considerable abundance in several localities. The metal is extracted from the carbonate, oxide, silicate, and sulphide. The carbonate and sulphide are first roasted in order to convert them into oxides, and the oxide is then reduced by means of hot charcoal, in earthen retorts or in crucibles provided with iron delivery-tubes. Since metallic zinc

is volatile at high temperatures, it distils over from the retorts as fast as it is formed and is condensed in receivers.

Zinc is a bluish-white metal of crystalline texture, brittle at the ordinary temperature, and also when heated above 200°, but a temperature of about 130° or 140° it may easily be rolled out or hammered into sheets. The metal melts at 425° and boils at a bright-red heat; in presence of air the red-hot metal takes fire and burns with a brilliant bluish-white light and formation of a dense cloud of white oxide of zinc.

Exp. 304.—Melt 200 or 300 grms. of metallic zinc in a small Hessian crucible, or in an iron ladle, placed in an anthracite fire. Remove the crucible from the fire by means of appropriate tongs (Appendix, § 27), and pour its contents in a very fine stream into a pail full of cold water, taking care to hold the crucible at a distance of 5 or 6 feet above the pail. Replace the empty crucible in the fire, in order that it may be ready for Exp. 305.

The small thin pieces of zinc which will be found in the pail when the water is poured away, are known as *granulated* or *feathered* zinc. This process of granulation may be conveniently applied to any of the other easily fusible metals, such as bismuth, lead, or tin, when they are required in a finely divided condition.

Granulated zinc is much used in chemical laboratories, for a variety of purposes, but particularly for preparing hydrogen (§ 50). In order that it may be fit for this purpose, it is best to heat the melted metal nearly to redness before pouring it into the water; for it has been noticed that when zinc is melted at the lowest possible temperature and then immediately poured into water, the granules obtained are but slowly acted upon by dilute sulphuric acid, while another portion of the same metal, heated nearly to redness, and then granulated, is readily soluble in the acid. If the hot metal be poured upon a warm iron plate, it will be found to be still more readily soluble in acids than that which has been suddenly cooled by the water.

Exp. 305.—Dry 20 grms. of the granulated zinc of Exp. 304, and mix it intimately in a mortar with 40 grms. of crude saltpetre; remove the empty crucible of Exp. 304 from the fire, and place it in such position that any fumes which may subsequently be evolved from it shall be drawn into the chimney. By means of a spoon or ladle, project into the red-hot crucible the mixture of zinc and saltpetre, taking care to stand away as far as possible from the crucible. The metal will burn fiercely, at the expense of the oxygen in the saltpetre, for the most part, though a portion of it will be volatilized by the intense heat of combustion, and converted into oxide of zinc in the air. The residue in the crucible is a soluble compound of oxide of zinc and potash, known as zincate of potassium.

If a strip of thin sheet zinc be held in the flame of the gas-lamp, it can readily be burned to oxide. The experiment succeeds best with zinc leaf, which instantly burns with a vivid flame and formation of floating flocks of the white oxide. In oxygen gas, zinc burns with peculiar brilliancy.

Zinc is not much acted upon either by moist or dry air at the ordinary temperature; but a fresh, bright surface of it, when exposed in a moist atmosphere, soon tarnishes and becomes covered with a thin film of basic carbonate of zinc, which adheres closely to the metal, and protects it from further change. Owing to this durability, the metal is much used in the form of sheets. Sheet iron and iron wire are often covered with a protecting coating of zinc by simple immersion in melted zinc, and are then said (most improperly) to be *galvanized.* The specific gravity of zinc varies from 6·8 to 7·3; its atomic weight is 65.

593. Zinc is readily attacked and dissolved by acids, with evolution of hydrogen in most instances. The chemical action of dilute acids upon zinc is a very common source of that peculiar mode of force called a galvanic current. There are few, if any, chemical reactions which cannot be made to produce electricity, and, in general, the more powerful the chemical action, the more powerful is the electrical action which results.

Exp. 306.—Solder a piece of stout copper wire to one end of a strip of sheet zinc, 4 c.m. wide by 10 c.m. long. The soldering will be readily effected by rubbing the zinc and the wire, in the vicinity of the proposed place of contact, with a strong solution of chloride of zinc, before applying the melted solder. In the same way, solder a similar wire to a like strip of bright sheet copper. Place the strips of zinc and copper in a tumbler filled with water, acidulated with 1-12th to 1-10th its volume of sulphuric acid, in such a way that the two strips shall not touch each other either within or without the liquid. So long as the wires coming from the strips of metal do not touch each other, the copper remains quiescent, while the zinc is attacked, and bubbles of gas rise from its surface; but if the two copper wires are brought into close contact, by means of a binding-screw, or by the application of solder, the following phenomena occur:—1st. Minute bubbles of hydrogen gas will be evolved from the surface of the copper plate. 2nd. The zinc dissolves more rapidly than before; at the close of the experiment sulphate of zinc may be recovered from the liquid in the beaker. 3rd. This transfer of the hydrogen from the zinc to the copper instantly ceases if the contact between the wires is destroyed. 4th. If the two wires be connected with the two ends of the coil of wire which surrounds the magnetic needle of the common galvanometer, the deflection of the suspended needle will demonstrate the fact that an electric current is passing through the wires from one plate of metal to the other.

This conversion of chemical force into electrical force is a striking illustration of the doctrine that all physical forces are correlated. The preceding experiment well illustrates the principle on which a large class of batteries employed in telegraphing and in electro-metallurgy are constructed and worked, except that the corrosion of the zinc is generally hindered by coating it with mercury. Artificial products, like metals, acids, and saline solutions, are used to supply all the chemical force which is immediately converted into and utilized as electrical force in the useful arts. We have not yet succeeded in realizing as electricity any considerable proportion of the prodigious chemical force which is incessantly active in the common processes of combustion.

594. Zinc dissolves in hot solutions of the caustic alkalies as well as in acids; hydrogen is given off and a zincate of the alkali formed:—

$$Zn + 2NaHO = Na_2ZnO_2 + 2H.$$

When immersed in the solution of a lead salt, such as the nitrate or acetate, zinc dissolves and lead is deposited in the metallic state:—

$$PbN_2O_6 + Zn = ZnN_2O_6 + Pb.$$

Exp. 307.—Dissolve 10 grms. of acetate of lead in 250 c. c. of water, add a few drops of acetic acid in order to dissolve the cloudy precipitate of carbonate of lead, which is formed from the carbonic acid in the water, pour the solution into a wide-mouthed bottle and suspend in it from the cork a strip of sheet zinc. The zinc will soon be covered with a brilliant coating of crystalline spangles of metallic lead, and this crystalline vegetation, as it were, will shoot out or grow even as far as the sides of the bottle. In the course of 24 hours all the lead will have been deposited from the solution, and the latter will contain nothing but acetate of zinc. Under the conditions of this experiment, and as a general rule, zinc is, chemically speaking, a stronger or more basic element than lead; it is capable of displacing lead from its compounds. The growth of lead, witnessed in this experiment, is frequently spoken of as a *lead tree*; the experiment is often performed in chemical laboratories for the sake of the chemically pure lead which it furnishes.

Many other metals besides lead may be thus thrown down by zinc, and the zinc may itself be replaced by other metallic precipitants. The whole series of experiments of which the one here indicated may be taken as the type, is interesting as illustrating the general law of the replacement of metals one by another in atomic proportions, and from the fact that by means of these experiments the atomic weight of various metals may readily be determined. For example, if in the foregoing experiment the piece of zinc be weighed before and after its immersion in the acetate of lead, and if the precipitated lead be also weighed, it will be found that the weight of lead obtained is, to the weight of zinc dissolved, very nearly as 207 is to 65, the atomic weights of lead and zinc respectively. The atom of zinc dissolved has replaced in the solution the atom of lead which was precipitated. By the exercise of care in the manipulation, by employing boiled water free from carbonic acid so that the addition of acetic acid to the lead salt shall be unnecessary, and by finally drying the lead in an atmosphere of hydrogen, a close approximation to the numbers above given can be obtained.

595. *Oxide of Zinc* (ZnO).—Like magnesium, zinc forms but a single compound with oxygen. This compound may be readily obtained by burning the metal, or by igniting carbonate or hydrate of zinc. As thus prepared, oxide of zinc is an insoluble, white, amorphous powder, which, under the name of *zinc white*, has of late years been largely employed as a white paint. It lacks the opacity or covering-power of white lead (§ 582), but, on the other hand, has no injurious action upon the health of the workmen and does not blacken or become discolored when exposed to the fumes of sulphydric acid. When heated in a crucible, oxide of zinc exhibits a yellow color, but it becomes white again on cooling. The oxide dissolves easily in acids, with formation of salts of zinc.

596. *Chloride of Zinc* ($ZnCl_2$), obtained by dissolving metallic zinc in chlorhydric acid, is a compound readily soluble in water; it is somewhat extensively employed for preserving timber, and

as a disinfecting fluid. It is used also by tinmen as a wash to cleanse the surfaces of tin-plate before soldering.

597. *Sulphate of Zinc* ($ZnSO_4$) is one of the commonest of the zinc salts. The hydrated compound, $ZnSO_4+7H_2O$, known as *white vitriol*, is used to a certain extent in medicine, and for other purposes in the arts. The action of carbon upon sulphate of zinc differs somewhat from its action upon the sulphates previously studied. When a dry mixture of the sulphate of zinc and charcoal is heated to dull redness, carbonic and sulphurous acids are evolved in the proportion of two volumes of the former to one of the latter gas, and pure oxide of zinc remains:—

$$2ZnSO_4 + C = 2ZnO + 2SO_2 + CO_2.$$

It would be quite possible to obtain metallic zinc from the sulphate in one operation by employing an excess of carbon, heating the mixture gently at first until the sulphuric acid had all been decomposed, and then urging the fire in order to obtain the temperature requisite for the reduction of the oxide of zinc and volatilization of the metal. But if the mixture of sulphate of zinc and charcoal be quickly raised to a high temperature, then sulphide of the metal is formed and carbonic oxide set free:—

$$ZnSO_4 + 4C = ZnS + 4CO.$$

Zinc forms several valuable alloys; brass is an alloy of zinc and copper, and German silver is a brass whitened by the admixture of a small proportion of nickel.

CADMIUM.

598. Cadmium is a comparatively rare metal, found associated with zinc in nature; it is remarkably similar to zinc in its chemical relations. In the process of obtaining zinc from its ores, the small proportion of cadmium which these ores contain comes over with the first products of the distillation, since cadmium is more readily volatile than zinc.

Cadmium may be prepared either from this early distillate, or from the residues obtained when metallic zinc is dissolved in chlorhydric acid in the preparation of chloride of zinc for manufacturing purposes. These residues always contain a quantity of lead, which next to cadmium is the commonest impurity of commercial zinc; and if care has been taken to keep an excess of metallic zinc in the dissolving-vat, they will contain also all the cadmium with which the zinc was contaminated.

From either of these sources, cadmium salts may be prepared by dissolving the crude materials in dilute nitric acid, separating the lead by means of sulphuric acid, and throwing down the cadmium with sulphuretted hydrogen. Sulphide of cadmium is a bright-yellow powder, insoluble in dilute acids, while sulphide of zinc is readily soluble in acids. Once isolated, the sulphide of cadmium may be dissolved in boiling, concentrated chlorhydric acid; from the solution of chloride of cadmium thus obtained, carbonate of cadmium may be precipitated, and from the carbonate any of the other cadmium compounds can readily be prepared.

Metallic cadmium is of a white color tinged with blue; it is lustrous and takes a fine polish, but gradually tarnishes upon the surface when exposed to the air. Its specific gravity varies from 8·6 to 8·7. It melts and volatilizes at temperatures below redness. Heated in the air it takes fire and burns to a brown oxide.

When combined with other metals, such as lead or tin, cadmium forms alloys of remarkable fusibility; in this respect it far surpasses bismuth (§ 359). The most fusible alloy yet made contains cadmium, bismuth, tin, and lead; it melts at 63°–65°.

599. Cadmium is a volatile substance, and the specific gravity of its vapor has been experimentally determined to be 56·85; the weight of a unit-volume of the vapor is 56·85 times the weight of the same volume of hydrogen. Now we have seen that the specific gravity of the elementary gases and of the vapors of the elements included in the chlorine and sulphur groups are the same as the atomic weights of these elements. On the other hand, the specific gravities of the vapors of phosphorus and arsenic were twice the atomic weights of these elements. Cadmium presents still a new relation between the least combining weight and the unit-volume weight; for the specific gravity of its vapor, 56·85, is about one-half of 112, its accepted atomic weight. The significance of this fact may be illustrated from its chloride. Cadmium is bivalent, and forms the chloride $CdCl_2$, containing, as experiment has proved, 112·24 parts, by weight, of cadmium to 71 parts of chlorine; if the unit-volume weight of cadmium were the same as its atomic weight, two unit-volumes of chloride of

cadmium would contain one volume of cadmium and two volumes of chlorine; but were it possible, by experiment, to resolve the vapor of chloride of cadmium into its component vapors, it would be found that two volumes of cadmium were therein combined with two volumes of chlorine.

The atom of cadmium when converted into vapor occupies twice as much space as the atom of oxygen, or hydrogen, or chlorine does; and accordingly the product-volumes of its compounds are packed with one volume more than the product-volumes of the corresponding compounds of oxygen or any member of the sulphur group. Whether the bivalent metals in general resemble cadmium on the one hand or oxygen on the other, in regard to the relation between their vapor-densities and their atomic weights, is a point on which experiment has thus far thrown but little light. Mercury resembles cadmium; but it is certainly possible that these two elements constitute an exception to some general rule hereafter to be proved—a rule, for example, like that which many chemists are inclined to accept in advance of proof, namely that the combining weights and the unit-volume weights of the elements are normally identical.

Cadmium is so soft that paper may be marked with it; but it is flexible, malleable, and ductile. In dilute chlorhydric and sulphuric acids it dissolves with evolution of hydrogen, though less readily than zinc. Its best solvent is nitric acid. It does not dissolve in the caustic alkalies.

600. From the foregoing it is apparent that the members of the group of metals now under consideration resemble one another with respect to volatility and several other of their physical properties, besides being very closely related in most of their chemical characters. The order of progression is from magnesium to cadmium, zinc and its compounds occupying always an intermediate position. The specific gravities of the three metals are—Mg=1·75, Zn=7·1, Cd=8·6; and their atomic weights are—Mg=24, Zn=65, Cd=112. Magnesium volatilizes at a bright-red heat, cadmium at a low red heat, and zinc at temperatures between these extremes. Cadmium is very fusible, melting at about 360°, zinc melts at 425°, and magnesium at a moderate red heat. All of these metals are bivalent; each forms but one oxide, sulphide, and chloride.

CHAPTER XXX.

ALUMINUM—GLUCINUM—CHROMIUM—MANGANESE—IRON—COBALT—NICKEL—URANIUM.

ALUMINUM.

601. Next to oxygen and silicon, aluminum is perhaps the most abundant element upon the earth's surface. It is the most abundant of all the metals, as much as a twelfth of the solid crust of the globe being composed of it. It occurs in enormous quantities in combination with oxygen and silicon, in all the so-called primitive rocks, and indeed in most other rocks and soils. It is contained in clay, marl, and slate, as well as in feldspar, mica, and many other common minerals.

Oxide of aluminum, chloride of aluminum, and many salts of the metal may readily be prepared artificially from the native minerals; they have long been known to chemists, and made use of in the arts; but the metal itself is less readily obtainable. It is but a few years since metallic aluminum has been prepared upon a manufacturing scale. The metal is nowadays prepared by heating metallic sodium either with chloride or fluoride of aluminum, or with a double chloride or fluoride of aluminum and sodium. It is a bluish-white metal, of remarkable lightness. Its specific gravity, 2·56, is about the same as that of porcelain, and only about a quarter of that of silver. The metal is malleable, ductile, and tenacious, and may be beaten into thin sheets, like gold and silver, and drawn into fine wire. It melts at a temperature lying between the melting-points of zinc and silver, but is not volatile. It conducts electricity much better than iron, and heat even better than silver; after having been heated, it cools very slowly. It is remarkably sonorous, a bar of it suspended by a wire rings with a clear musical note on being struck.

Aluminum changes little in the air, even at a strong red heat; but when melted in open crucibles a film of oxide forms upon it that hinders the operation of casting. It is not acted upon by water at temperatures short of a white heat, so long as it is in the ordinary compact condition. Sulphydric acid has no action upon it. Nitric acid, whether dilute or concentrated, has no action upon aluminum at the ordinary temperature, but when boiling dissolves the metal slowly. Cold dilute sulphuric acid has scarcely any action upon it; but it is easily soluble in chlorhydric acid, either dilute or concentrated, at all temperatures. It is soluble also in aqueous solutions of caustic potash, soda, or ammonia. The vegetable acids, such as acetic and tartaric acids, exert no perceptible action upon it. Although soluble with evolution of hydrogen in aqueous solutions of the fixed caustic alkalies, aluminum is not acted upon by fused hydrate of sodium, or hydrate of potassium; nor is it even attacked by fused nitrate of potassium, except at temperatures high enough to decompose the nitre so completely that it gives off nitrogen; when this limit is reached the aluminum is immediately oxidized with incandescence.

602. Aluminum unites readily with many of the metals to form alloys, among which that of copper and aluminum, called *aluminum bronze,* promises to be of especial importance. Aluminum bronze, composed of 90 parts copper and 10 parts aluminum, is exceedingly hard, very malleable, as tenacious as steel, of a beautiful golden color, and susceptible of being highly polished.

603. By uniting with the non-metallic elements, aluminum forms only one class of compounds, of which the oxide Al_2O_3 may be taken as the type. The atom Al is trivalent, or, in other words, it is equivalent to three atoms of hydrogen, and of the same value as one and a half atom of oxygen. Since it would be inconvenient to employ fractional expressions in writing chemical formulæ, as well as illogical to speak of half atoms, it is customary to write the formula of oxide of aluminum Al_2O_3 as above, and not $AlO_{1\frac{1}{2}}$, as might perhaps at first sight seem best. For the sake of consistency, the formula of the chloride is in like manner written Al_2Cl_6 and not $AlCl_3$. Since it contains one and a half atom of oxygen for each atom of aluminum, the oxide is often called a *Sesqui* (one and a half) oxide.

If no other element analogous to aluminum were known, if

this metal were not intimately related to glucinum, iron, chromium, and the other metals to be considered in the present chapter, chemists might possibly have taken the atomic weight of aluminum at one-third of its present value, namely at 9·1 instead of 27·4. The formula of oxide of aluminum would then have been written Al_2O, and that of chloride of aluminum AlCl, corresponding respectively with the formulæ of the oxides and chlorides of the alkali-metals. But, as will appear directly from the study of the other members of the aluminum family of metals, and particularly from the isomorphism of their various compounds, the atomic weight 27·4, and the formulæ first given, must be regarded as the most probable. The atomic weight 9·1, and the formulæ derived from it, are inadmissible, since there is no analogy between the chemical properties of aluminum, whether simple or compounded, and those of the alkali-metals.

604. *Oxide of Aluminum* (Al_2O_3), commonly called *Alumina*, is found crystallized in nature as the mineral corundum. The sapphire and the ruby are also composed of this oxide together with a little oxide of iron. It may be prepared artificially by oxidizing the metal, or by igniting the hydrate, or almost any oxygen salt of aluminum. Though unalterable in oxygen so long as it is compact, powdered aluminum and aluminum-leaf burn brightly when heated to redness in the air, with formation of oxide of aluminum. It has been found by careful experiments that 53·3 parts of the metal unite with 46·69 parts of oxygen to form 100 parts of the oxide. Now, since oxide of aluminum is isomorphous with certain oxides of iron and of chromium, which are known to be sesquioxides, and is capable of replacing these oxides in any proportion in their compounds (§ 252), it is inferred that oxide of aluminum is likewise a sesquioxide. Upon this view the atomic weight of aluminum is directly derived from the foregoing experimental data by the equation:—

46·69	:	53·3	=	48	:	54·8
				wt. of 3 atoms of oxygen.		wt. of 2 atoms of aluminum.

605. *Hydrate of Aluminum* ($Al_2H_6O_6$) may be obtained as a gelatinous, flocculent precipitate, by adding ammonia-water to the solution of an aluminum salt, such as common alum.

When dried at a moderate heat it forms a soft, friable mass, which adheres strongly to the tongue like clay; when dried still more thoroughly, it forms a hard, yellowish, translucent, horn-like substance, and at a red heat gives off all its water. The volume of the original precipitate contracts to an enormous extent during the operation of drying; the bulk of the final anhydrous oxide is exceedingly small as compared with that of the moist hydrate from which it has been derived.

606. Anhydrous alumina may be melted in the flame of the oxyhydrogen blowpipe. It is neither decomposable by heat alone, nor can it be reduced by carbon or any of the more common deoxidizing agents. At a white heat, potassium decomposes it partially, and an alloy of aluminum and potassium is formed. Oxide of aluminum is insoluble in water, and after having been strongly heated it is scarcely at all acted upon by acids, excepting concentrated boiling chlorhydric and nitric acids. The crystallized native oxide is insoluble in all acids. The anhydrous oxide is insoluble in solutions of the caustic alkalies, but dissolves readily in water after having been fused at a red heat with either hydrate or carbonate of sodium or of potassium. Hydrate of aluminum, on the contrary, though insoluble in water, dissolves easily in acids and in solutions of the fixed caustic alkalies. Alumina is in fact capable of acting not only as a strong base, forming well-defined salts by uniting with acids, but it plays the part of an acid as well (compare § 350), and combines with the alkalies and with other metallic oxides to form salts known as aluminates. Aluminate of potassium (K_2O,Al_2O_3) and aluminate of sodium (Na_2O,Al_2O_3) are substances somewhat extensively used in the arts; the mineral spinel is an aluminate of magnesium (MgO,Al_2O_3); and a native aluminate of zinc (ZnO,Al_2O_3) is called gahnite by mineralogists.

Exp. 308.—Heat a small fragment of alum with water in a test-tube until it has completely dissolved, pour half the solution into another tube, and add to it, drop by drop, ammonia-water, until the odor of ammonia persists after the mixture has been thoroughly shaken. Hydrate of aluminum will be precipitated, in accordance with the reaction:—

$$Al_23SO_4 + 6(NH_4)HO = Al_2O_3,3H_2O + 3(NH_4)_2SO_4.$$

Pour two or three drops of the moist hydrate of aluminum into another test-tube and cover them with ammonia-water; no clear solution will be obtained, for hydrate of aluminum is but little soluble in ammonia-water.

Pour two or three drops of the moist hydrate of aluminum into still another test-tube, and cover them with a solution of hydrate of sodium; the precipitate will dissolve immediately; aluminate of sodium is formed, and this salt is easily soluble.

Exp. 309.—Take another portion of the clear solution of alum prepared in Exp. 308, and add to it, drop by drop, a dilute solution of caustic soda. A precipitate will soon fall, as in Exp. 308, and if no excess of hydrate of sodium were added over and above that necessary to form sulphate of sodium with the sulphuric acid of the alum, this precipitate would remain undissolved; but on adding more of the soda solution the precipitate dissolves at once, with formation of aluminate of sodium.

607. Hydrate of aluminum combines readily with many organic coloring-matters, forming compounds which are insoluble in water. The fibre of cotton, when impregnated with alumina, can be made to retain colors which the cotton itself has no power to hold; hence the use of aluminum salts as *mordants* in dyeing.

Exp. 310.—Boil a few crushed granules of cochineal in water until a considerable portion of their coloring-matter has been extracted; add to the filtered solution an equal bulk of a solution of alum, and to the mixture add ammonia-water. A colored precipitate, consisting of hydrate of aluminum and of the coloring-matter of the cochineal, will be thrown down; it is the substance called carmine-lake. Similar precipitates may be prepared by substituting almost any other organic coloring-matter for the cochineal of this experiment. Precipitates thus formed by the union of a metallic oxide and a coloring-matter are all classed as *lakes.*

608. *Chloride of Aluminum* (Al_2Cl_6) may be prepared, in the same way that the chlorides of boron and silicon are prepared (§§ 449, 470), by passing chlorine over a heated mixture of alumina and carbon. It is formed also when hot finely divided aluminum is brought into contact with chlorine gas. Hydrated chloride of aluminum ($Al_2Cl_6.12H_2O$) can be made very easily by dissolving hydrated oxide of aluminum in chlorhydric acid; but the anhydrous chloride cannot be prepared by heating this hydrate, since a great part of the chlorine is expelled

from it, together with the water, at a low heat. When obtained in the dry way, however, chloride of aluminum is readily volatile. The anhydrous chloride, prepared by the reaction,

$$Al_2O_3 + 3C + 6Cl = Al_2Cl_6 + 3CO,$$

previously described, is found condensed in the cold portions of the tube in which the materials have been heated, apart from the residue of undecomposed alumina and carbon. It occurs either as a flocculent powder, or as a transparent wax-like mass of crystalline texture. It is colorless when pure, very deliquescent, and soluble in water. When large masses of it are heated to dull redness, a portion of it liquefies, but at temperatures near the melting-point it volatilizes rapidly. Unlike oxide of aluminum, it may be readily decomposed by sodium and potassium at a heat below redness, metallic aluminum being set free.

Chloride of aluminum combines readily with several of the other metallic chlorides, forming compounds analogous to the chloride of aluminum and sodium ($2NaCl,Al_2Cl_6$), from which metallic aluminum is commonly manufactured.

609. *Sulphate of Aluminum* (Al_23SO_4) is a salt largely employed in the arts. It is commonly prepared nowadays by acting upon hot roasted clay with sulphuric acid. Clay is a silicate of aluminum not very easily attacked by acids so long as it remains in the native plastic condition, but after having been exposed for some time to a dull red heat it readily yields its alumina to acids. A solid mixture of sulphate of aluminum and free silicic acid obtained as the product of this reaction is known in commerce as alum-cake. By lixiviating alum-cake, sulphate of aluminum may readily be obtained in solution: and from this solution the salt crystallizes as a hydrate, the composition of which may be represented by the formula $Al_23SO_4+18H_2O$.

Until a comparatively recent period, sulphate of aluminum was sent into commerce neither in its free state nor mixed with silica, but in combination with sulphate of potassium in the form of alum. Common alum is a hydrated double sulphate of aluminum and potassium; its composition is represented by the formula $Al_2K_24SO_4 + 24H_2O$, or $K_2O,SO_3; Al_2O_3,3SO_3 + 24H_2O$.

It crystallizes very easily in large, compact, well-defined octahedrons, belonging to the first or regular system. The crystalline character of alum is important, since it is solely on account of this character that the salt has come into such general use. Neither the sulphate of potassium nor the water in alum plays any useful part in the reactions for which this salt is commonly employed. Since 100 parts of alum contain only about 36 parts of anhydrous sulphate of aluminum, it follows that 64 per cent. of the alum is, for all chemical purposes, simply inert matter, which has to be transported and manipulated for the sake of the 36 per cent. of real sulphate of aluminum. The reason why this waste of labor and loss of the potassium salt is tolerated is twofold:—Until a comparatively recent period, sulphate of aluminum could be more readily purified by crystallization in alum than in any other way. At the present time, when it is easy to obtain pure sulphate of aluminum from responsible manufacturers, alum is still prepared because its clean, sharply defined crystals afford a valuable criterion of purity. So long as it is left in the condition of crystals, alum cannot be adulterated with any foreign substance.

Potash-alum is still the common alum of the American market, but in Europe ammonia-alum ($Al_2(NH_4)_2 4SO_4, 24H_2O$) is at present largely employed; it is there prepared by adding sulphate of ammonium obtained from the ammoniacal liquor of the gas-works to sulphate of aluminum resulting from the action of sulphuric acid upon clay. Ammonia-alum crystallizes almost as easily as potash-alum; but it is remarkable that the corresponding double sulphate of aluminum and of sodium (soda-alum) is easily soluble in water, and crystallizes with comparative difficulty; hence it has never come into commerce.

Sulphate of aluminum is employed as the source of the various compounds of aluminum used in dyeing, calico-printing, and paper-making. Acetate of aluminum, for example, is largely employed by dyers, particularly for the red colors obtained from madder, under the name of red liquor.

Exp. 311.—Dissolve 3 grms. of sugar of lead in 4 c. c. of hot water; also dissolve 4 grms. of common alum in 6 c. c. of hot water; mix the hot solutions and filter off the insoluble sulphate of lead which is formed. The solution obtained consists of basic acetate of aluminum, together with some sulphate of aluminum and all the sulphate of potassium of the original alum. Such solutions are preferred in practice to those containing normal acetate of aluminum, to prepare which a much larger proportion of acetate of lead would be required than has been given above.

Exp. 312.—Soak a small piece of cotton cloth in the solution of acetate of aluminum prepared in Exp. 311, and another piece of similar cloth of equal size in pure water. Hang up both pieces to "age," best in a moist and warm atmosphere, for a day or two. During the process of ageing, a portion of the acetic acid escapes from the salt on the cloth, and there is left within and upon the fibres of the cloth a quantity of hydrate of aluminum, or at least a mixture of highly basic acetate and sulphate of aluminum. This deposit is the true mordant. When cloth impregnated with it is soaked in a solution of coloring-matter, the coloring-matter unites with the alumina precisely as in Exp. 310, and is thereby firmly attached to the cloth. It should be mentioned that several other oxides, besides the oxide of aluminum, are capable of acting as mordants; the sesquioxides of iron and of chromium for example, as well as the binoxide of tin, are largely used as mordants.

Exp. 313.—Place a quantity of a solution of extract of logwood in two small evaporating-dishes, heat the liquor to 40° or 50°, then place the mordanted cloth in one dish, the unmordanted cloth in the other and boil the liquor in both dishes. Continue to boil during 10 or 15 minutes, then take out the pieces of cloth and wash them thoroughly in water. It will appear that the coloring-matter remains firmly attached to the mordanted cloth, while the cloth which has received no mordant can readily be washed clean or nearly clean.

610. *Silicates of Aluminum.*—Of all the aluminum compounds the silicates are by far the most important. Clay in all its varieties is a hydrated silicate of aluminum, usually mixed with an excess of silica, besides other impurities derived from the rocks from whose decomposition the clay itself has been formed. The purer kinds of clay, such as kaolin or porcelain clay, are products of the decomposition of feldspar, a mineral composed of silicon, aluminum, potassium, and oxygen in the proportions $Al_2O_3,K_2O,6SiO_2$. When exposed to the atmosphere, many varieties of feldspar gradually decompose, an alkaline silicate is washed away, and silicate of aluminum remains. Clay is remarkable on account of its plasticity when moist, of the facility with which it

is converted into stone-like masses when strongly heated, and of its infusibility when pure.

Earthenware, bricks, and ordinary pottery are made from common clay, by mixing the clay with water enough to form a plastic paste, which is then moulded into any desired form, dried and intensely ignited. The porous ware resulting from this operation may be glazed, and so made impermeable to liquids, by coating it over with some fusible substance, such for example as a mixture of litharge and clay, and again heating it so intensely that the coating shall melt to a glass, which either fills up the pores of the clay, or at the least stops their openings. Porcelain proper, and the better kinds of stoneware, are made from the purest varieties of clay, and are glazed with feldspar. Common stoneware, such as is used for jugs, beer-bottles, and the like, is covered with the so-called salt-glaze:—Moist chloride of sodium is thrown into the kiln in which the ware is baking, and being volatilized by the intense heat, comes in contact with the hot stoneware; decomposition ensues; the water and the chloride of sodium are both decomposed, silicate of sodium is formed, and by mixing with the silicate of aluminum, forms a smooth hard glaze upon the surface of the ware.

For all vessels which are to be employed for chemical or culinary purposes the hard and durable salt-glaze is very much to be preferred to the lead-glaze prepared from litharge and clay; for the lead-glaze is readily acted upon by many chemical agents, and is liable to impart its poisonous properties to articles of food which have been left in contact with it.

Fire-bricks, crucibles, and similar refractory articles fitted to support very high temperatures without undergoing fusion, are prepared from pure varieties of clay, free from iron, lime, or magnesia, but containing an unusually large proportion of silica. Some varieties of fire-clay contain as much silica as is represented by the formula $Al_2O_3,6SiO_2$, while the composition of many of the common clays may be approximately represented by the formula $Al_2O_3,3SiO_2$, or better by the formula $Al_2O_3,2SiO_2$. In the manufacture of fire-bricks, and of many varieties of potters' ware, it is usual to incorporate with the original clay a certain proportion of foreign matter which prevents the moulded article from shrinking too much as it dries, and from cracking. In fire-bricks the coarse powder obtained by pulverizing old fire-bricks is employed for this purpose; in Hessian crucibles it is very easy to detect numerous grains of quartz sand; and, in general, finely powdered flint or quartz, as well as previously baked clay, is used for the same purpose in many varieties of pottery.

611. Silicate of aluminum is moreover a very important ingredient of the common hydraulic cement employed to replace lime mortar in constructions exposed to the action of water. It has been found that by carefully burning some varieties of impure limestone, containing from 10 or 12 to 30 or 35 per cent. of clay, and mixing the product with water, there is obtained, in place of ordinary mortar, a cement capable of "setting" or hardening, even under water, to a compact stone. Hydraulic cements may readily be prepared artificially by mixing with quicklime a suitable proportion of roasted clay, or by heating mixtures of clay and limestone; in fact, some of the best cements now in use are artificial. A porous volcanic stone called *pozzolana*, from the vicinity of Naples, consisting of silicates of aluminum, calcium, and sodium, was much used by the Romans to the same end. When powdered and mixed with ordinary lime the pozzolana yields an excellent hydraulic mortar. In many Roman ruins it may be seen to-day far more perfectly preserved than the bricks which it cements.

When treated with water hydraulic limes simply absorb the water and form a slightly plastic paste without greatly increasing in bulk; they do not slake or evolve much heat like ordinary quicklime. The moist paste soon begins to set, and is then ready for application. In order that the cement may harden properly under water it should not be submerged before it has begun to set; it should in any event be kept moist until it has become hard; otherwise it is liable to remain loose and porous.

The solidification of hydraulic limes appears to depend upon the formation of insoluble hydrated compounds of lime with silicic acid and alumina. Cements which contain from 25 to 35 per cent. of clay solidify in the course of a few hours; but those in which the proportion of clay is no more than 10 or 12 per cent. become hard only after the lapse of several weeks.

A mixture of hydraulic cement with coarse gravel constitutes the material known as *concrete*, employed for the foundations of buildings, and by the ancients for walls which have proved to be of great durability; this mixture soon concretes or hardens into a firm mass, well nigh impermeable to water. The influence of magnesia in the preparation of hydraulic mortars has already been indicated, in § 588.

GLUCINUM.

612. Glucinum is a rather rare metal, found, together with aluminum, in the emerald, in beryl, and a few other minerals. It closely resembles aluminum in its physical properties, and forms compounds analogous in composition to those of aluminum, and of similar chemical deportment. Like aluminum, metallic glucinum may be reduced from its chloride by means of sodium or potassium. There is but a single oxide of glucinum, Gl_2O_3, and a single chloride, Gl_2Cl_6. The salts of glucinum have a sweet taste, whence the name, from a Greek word meaning sweet. The atomic weight of glucinum is 14, and its specific gravity 2·1.

CHROMIUM.

613. Chromium is nowhere found in very large quantities, nor is it very widely disseminated in small portions like iodine and fluorine, but it is nevertheless found in sufficient abundance to admit of its compounds being rather extensively employed in the arts. The chief ore of chromium is a compound of oxide of chromium and oxide of iron ($FeCr_2O_4$) called chrome iron-ore. Metallic chromium may be reduced from its oxide by means of intensely heated charcoal, and from its chloride by means of sodium, potassium, magnesium, or zinc; but it has as yet been little studied. Its specific gravity is about 7, and its atomic weight 52·5.

614. *Oxides of Chromium.* — Chromium forms three well-defined oxides:—a protoxide, CrO; a sesquioxide, Cr_2O_3; and a teroxide, CrO_3, called chromic acid. Besides these there is a compound of the protoxide and sesquioxide ($Cr_3O_4 = CrO, Cr_2O_3$), another of the sesquioxide and teroxide ($Cr_2O_3, CrO_3 = 3CrO_2$), and an ill-defined compound containing more oxygen than chromic

acid, usually spoken of as perchromic acid (Cr_2O_7). Both the protoxide and the sesquioxide are bases, corresponding respectively to oxide of magnesium and oxide of aluminum; but the protoxide and all its compounds rapidly absorb oxygen from the air, with formation of the sesquioxide. On account of this instability, they are rarely prepared, and are mainly interesting from their analogy to the compounds of the protoxides of manganese, iron, cobalt, and nickel, hereafter to be studied. The sesquioxide, on the other hand, is a stable compound, closely resembling oxide of aluminum. Chromic acid is a strong, well-characterized acid, which combines with bases to form a great number of salts.

The most important of the chromium compounds is the bichromate of potassium; this salt is readily procurable in commerce, and is the source from which all the other compounds of chromium are commonly derived. Bichromate of potassium is itself prepared by heating finely powdered chrome iron-ore with carbonate and nitrate of potassium in a reverberatory furnace. The sesquioxide of chromium of the ore is oxidized and converted into the teroxide, chromic acid, which displaces the carbonic acid of the carbonate of potassium.

615. *Sesquioxide of Chromium* (Cr_2O_3).—Compounds of this oxide are more commonly met with than any other of the chromium salts, except the salts of chromic acid. As has been stated already, compounds of the protoxide must be regarded merely as chemical curiosities. By adding ammonia-water to the solution of a salt containing the sesquioxide, a bulky green precipitate of hydrated sesquioxide of chromium is thrown down, which, when collected and ignited, leaves the anhydrous oxide as a bright-green powder, unchangeable at the highest furnace-heat. It is employed in the decoration of porcelain, and is a valued pigment much used in painting and printing, under the name *chrome green.*

616. *Chlorides of Chromium.*—There are two of these compounds, the protochloride ($CrCl_2$) and the sesquichloride (Cr_2Cl_6). The latter compound is the more important, and is the substance usually meant when chloride of chromium is spoken of. Hydrated sesquichloride of chromium, obtained by dissolving the hydrated sesquioxide in chlorhydric acid, is the chloride most commonly met with.

617. *Sulphate of Chromium* ($Cr_2O_3,3SO_3$) is sometimes prepared in the pure state; but, like sulphate of aluminum, it ordinarily occurs in combination with sulphate of potassium or sulphate of ammonium, as a double salt, called chrome alum. Chrome alum is a compound of a beautiful violet color, crystallizing in well-defined octahedrons of the same form as the crystals of ordinary alum; its composition also corresponds to that of common alum, the formula of the chromium salt being $Cr_2K_24SO_4 + 24H_2O$, or $K_2O,SO_3; Cr_2O_3,3SO_3 + 24H_2O$.

Exp. 314.—Dissolve 15 grms. of powdered bichromate of potassium in 100 c. c. of warm water; cool the solution, and then add to it 25 grms. of concentrated sulphuric acid; cool the liquor again, and pour it into a porcelain dish, surrounded with cold water; slowly stir into the mixture 6 grms. of alcohol, and set the whole aside. In the course of 24 hours the bottom of the dish will become covered with well-defined octahedral crystals of chrome alum.

The alcohol in this experiment deprives the chromic acid, of the bichromate of potassium, of half its oxygen, and is itself converted for the most part into acetic acid and water.

618. It is remarkable that the salts of sesquioxide of chromium, as well as the oxide itself, occur in two isomeric conditions. One modification is known as the green, the other as the violet modification. As a rule, the violet compounds crystallize readily, while the green compounds do not. In the preparation of chrome alum it is important to guard against the formation of a green, soluble sulphate of chromium, which does not crystallize. In general, if the solution of a salt of the violet modification is heated nearly to boiling, the salt passes into the green modification and becomes uncrystallisable. Hydrated sesquioxide of chromium, as obtained by adding a caustic alkali to the cold solution of a chromium salt of either modification, is readily soluble both in acids and in cold solutions of caustic soda or potash; but on boiling the green alkaline solution, all of the chromium is precipitated as a hydrate of the green modification.

Exp. 315.—Place in a test-tube a few drops of a dilute solution of chrome alum, or of some other salt of sesquioxide of chromium; add, drop by drop, a solution of hydrate of sodium, until the precipitate which forms at first is completely redissolved. Boil the clear solution, and observe that the precipitate which again forms in the liquor is no longer soluble in alkalies.

By means of this reaction, oxide of chromium may readily be separated from oxide of aluminum; for, as has been seen in Exp. 309, alumina is readily soluble in alkaline solutions, and is not precipitated therefrom by boiling.

619. *Chromic Acid* (CrO_3) may be obtained by decomposing bichromate of potassium with sulphuric acid.

Exp. 316.—Mix 40 c. c. of a cold, saturated, aqueous solution of bichromate of potassium with 50 c. c. of oil of vitriol, in a small beaker standing in cold water, and observe that chromic acid is deposited in crystalline needles. It is remarkable that in sulphuric acid of 1·55 specific gravity, such as is obtained in the foregoing mixture, chromic acid is well nigh insoluble, though it is readily soluble both in water and in strong sulphuric acid. Cover the beaker, and set it aside for some hours; finally pour off the supernatant liquor with care, scrape out the chromic acid with a glass rod, and place it upon a dry, porous brick, under an inverted bottle, in order that the sulphuric acid which adheres to it may be absorbed. Preserve the dry crystals in a glass-stoppered bottle. Chromic acid deliquesces rapidly when exposed to the air. It is easily brought to the condition of sesquioxide of chromium, both by heat and by reducing agents, and is hence an oxidizing agent of considerable power.

Exp. 317.—Shake up in a small bottle enough strong alcohol to moisten its sides; then throw in half a gramme or less of chromic acid; a portion of the alcohol will be oxidized so quickly, and with evolution of so much heat, that the remainder will take fire and burn in the air.

Several of the salts of chromic acid, as well as the acid itself, are employed as oxidizing agents. A mixture of bichromate of potassium and of sulphuric acid, for example, is employed for bleaching certain fats. From the chromates, both oxygen and chlorine may be conveniently prepared.

Exp. 318.—Heat a mixture of 6 grms. of powdered bichromate of potassium and 9 grms. of concentrated sulphuric acid in a small flask, provided with a delivery-tube leading to the water-pan, and collect the oxygen which is freely evolved:—

$$K_2O,2CrO_3 + 4(H_2O,SO_3) = Cr_2O_3,3SO_3 + K_2O,SO_3 + 4H_2O + 3O.$$

Exp. 319.—Place a mixture of 1 grm. of powdered bichromate of potassium and 6 grms. of chlorhydric acid of 1·16 specific gravity, in a flask provided with a delivery-tube, as in Exp. 318. Heat the flask gently for a few seconds until its contents begin to react upon one another; then quickly remove the lamp and attend to the collection of the chlorine, which will continue to be evolved without further heating:—

$$K_2O,2CrO_3 + 14HCl = Cr_2Cl_6 + 2KCl + 7H_2O + 6Cl.$$

620. *Chromates.*—As has been already indicated, bichromate of potassium is the commonest and the most important salt of chromic acid. It is the material from which most of the other compounds of chromium are prepared, and is itself important in dyeing and calico-printing. It has of late years been used in the art of photolithography.

When a mixture of gelatine and bichromate of potassium is exposed to light, the chromic acid is reduced, and an insoluble compound of gelatine and sesquioxide of chromium is formed. In practice, albumenized paper is covered in a dark room with a mixed solution of bichromate of potassium and gelatine, then dried, pressed smooth, and kept always in the dark until wanted for use. If a sheet of this prepared paper be placed beneath a negative photographic picture (obtained in the usual way) and exposed to light for a short time, the chromic acid will be reduced in such wise that a positive picture will be obtained upon the gelatine paper. In this positive, as taken from the press, the parts acted upon by the light will be brown, while the other portions of the sheet retain their original yellow color. The positive is then washed with water in such manner that the unchanged portions of gelatine and of bichromate are dissolved away, and an insoluble, clearly defined impression of the original picture is left upon the paper. By means of pressure, the design is then transferred to the lithographic stone, and from the stone any desired number of copies may be printed upon paper with ink in the usual way.

Besides the bichromate of potassium, there are several other chromates important in the arts or useful to the analyst. The normal chromate of potassium (K_2O,CrO_3) is a yellow salt, readily obtainable by adding a molecule of carbonate of potassium to one of the bichromate:—

$$K_2O,2CrO_3 + K_2O,CO_2 = 2(K_2O,CrO_3) + CO_2.$$

It is isomorphous with normal sulphate of potassium (K_2SO_4), chromic acid, like sulphuric acid, being bibasic (§ 238). The salt is hence easily adulterated. Chromate of barium is insoluble in water and in acetic acid; chromate of strontium is soluble in acetic acid, though nearly insoluble in water; while chromate of calcium is soluble both in water and in acetic acid; hence an easy method of separating compounds of either of the three metals from mixtures which contain compounds of all three.

Chromate of lead is the pigment called chrome-yellow; it may easily be prepared by mixing solutions of bichromate of potassium and acetate of lead. An orange-colored dichromate ($2PbO,CrO_3$) may be obtained by boiling together yellow chromate of lead and slaked lime in the proportion of two molecules of the former to one of the latter. This process is used to fix a permanent orange upon calico. A still more brilliant color may be obtained by fusing one part of the yellow chromate of lead with five parts of nitre; chromate of potassium and dichromate of lead are formed, and the former may be washed away. Chromate of mercury, of a brick-red color, may be precipitated by adding bichromate of potassium to nitrate of protoxide of mercury, or, of an orange-yellow color, by adding the potassium salt to the nitrate of dioxide of mercury.

MANGANESE.

621. Black oxide of manganese, such as has been employed in the preparation of oxygen and of chlorine (§§ 14, 105), is a tolerably abundant mineral. Small quantities of manganese exist also in a great number of other minerals and rocks; so that the element is really very widely diffused in nature. It is often associated with ores of iron. By heating oxide of manganese very strongly with charcoal, it may be reduced to the metallic state, though not readily. The metal is of a grayish-white color, and is very hard and brittle. It oxidizes quickly when exposed to the atmosphere; it melts only at the strongest heat of a blast furnace. The specific gravity of manganese is 8, its atomic weight is 55. It slowly decomposes water at the ordinary temperature, and dissolves readily in dilute sulphuric acid with evolution of hydrogen. Like iron, it combines with carbon and silicon. Metallic manganese is not used in the arts; and the alloys which it forms with the other metals are of no commercial importance, except that a small proportion of manganese is present in a peculiar kind of iron largely used for making steel.

622. *Oxides of Manganese.*—Six well-defined compounds of

oxygen and manganese are known; two of them are bases, two are acids, and two may be regarded as salts, formed by the union of the oxides one with the other. Protoxide of manganese (MnO) is a powerful base, while sesquioxide of manganese (Mn_2O_3) is but a weak base. Manganic acid (MnO_3) and permanganic acid (Mn_2O_7) are well characterized as acids, though they are known only in combination; they have never been obtained in the free anhydrous state. On the other hand, binoxide of manganese ($Mn_2O_3,MnO_3=3MnO_2$) and the red oxide ($MnO,Mn_2O_3 =Mn_3O_4$) are both neutral or indifferent bodies; they exhibit neither acid nor basic properties.

623. *Protoxide of Manganese* (MnO) may be obtained by heating carbonate of manganese out of contact with the air, or by heating either of the higher oxides of manganese to redness in contact with charcoal or hydrogen. The protoxide is itself reduced to the metallic state by these agents only at a white heat. It unites freely with acids to form salts of considerable stability. The crystallized sulphate $MnSO_4+5H_2O$ and the chloride $MnCl_2 +4H_2O$ are commonly employed in the laboratory. Both of them may be prepared from the residues obtained in the preparation of chlorine and oxygen (§§ 105, 626). Hydrated protoxide of manganese may be precipitated from the chloride as follows:—

Exp. 320.—Dissolve a small crystal of chloride of manganese in water; add to the solution soda-lye until the liquor exhibits a distinct alkaline reaction when tested with litmus-paper. Collect the gelatinous white precipitate upon a filter, and observe that it soon becomes brown as it absorbs oxygen from the air; the brown product is sesquioxide of manganese.

Exp. 321.—Heat a portion of the precipitated hydrate of Exp. 320 to redness upon a fragment of porcelain; it will slowly absorb oxygen, and change to the deep-brown-colored sesquioxide.

Exp. 322.—To a solution of chloride of manganese, such as was prepared in Exp. 320, add a few drops of sulphydrate of ammonium (§ 526). A flesh-colored precipitate of sulphide of manganese (MnS) will fall down. Like the hydrate above described, this precipitate soon becomes brown by exposure to the air. It is often prepared by the analyst when testing for manganese.

624. *Sesquioxide of Manganese* (Mn_2O_3) occurs in nature in

the minerals braunite and manganite. It is prepared artificially by roasting the protoxide obtained from chlorine-residues, and is itself used to a considerable extent in the preparation of chlorine:—

$$Mn_2O_3 + 6HCl = 2MnCl_2 + 3H_2O + 2Cl.$$

It combines with acids to form a series of unstable salts analogous to the sesquisalts of iron, though far less permanent. A solution of the sesquisulphate, for example, $Mn_2O_3,3SO_3$, is reduced to the condition of protosulphate by mere boiling. In like manner the sesquichloride Mn_2Cl_6 is doubtless formed when the protochloride is treated with cold chlorine, or the sesquioxide is digested in cold chlorhydric acid; but the salt is decomposed with extreme readiness, and splits up into free chlorine and the protochloride even when but slightly heated. In the preparation of chlorine from the sesquioxide as above formulated, there is no doubt an intermediate reaction,

$$Mn_2O_3 + 6HCl = Mn_2Cl_6 + 3H_2O,$$

before the final breaking up of

$$Mn_2Cl_6 \text{ into } 2MnCl_2 + 2Cl.$$

625. Of the salts of the sesquioxide, the double compound of sulphate of manganese and of potassium, known as manganese alum, is one of the most interesting; it is of analogous composition to ordinary aluminum alum, and is isomorphous with this body, as it is with the corresponding alums of iron and chromium. The series of double salts known as alums, admirably illustrates the relationship of the several members of the group of metals now under discussion, and the law of isomorphism as well. It is interesting to observe, moreover, that the name alum, originally applied specifically to the compound of sulphate of aluminum and of potassium, has with the growth of chemical knowledge come to have a generic signification. Several salts are now classed as alums, into the composition of which neither aluminum nor potassium enters. The following list enumerates some of the best-known potassium alums:—

Common alum $= K_2SO_4,Al_23SO_4 + 24H_2O,$
Chrome alum $= K_2SO_4,Cr_23SO_4 + 24H_2O,$

$$\text{Manganese alum} = K_2SO_4,Mn_23SO_4 + 24H_2O,$$
$$\text{Iron alum} = K_2SO_4,Fe_23SO_4 + 24H_2O.$$

But as has been stated in § 609, the potassium in these compounds may be replaced by any metal isomorphous with potassium. There are ammonium alums and sodium alums corresponding to each of the potassium alums above enumerated, and there is evidence that potassium may be replaced in these alums by the rarer alkali-metals. Some alums, on the other hand, are composed of mixtures in various proportions of alkali-metals, and of the metals capable of forming sesquioxides. Besides these true alums, there are allied bodies which contain no alkali-metal whatsoever; such, for example, are the following:—

$$\text{Aluminum iron alum} = FeSO_4,Al_23SO_4 + 24H_2O,$$
$$\text{Aluminum magnesium alum} = MgSO_4,Al_23SO_4 + 24H_2O,$$
$$\text{Aluminum manganese alum} = MnSO_4,Al_23SO_4 + 24H_2O,$$

but these affiliated alums do not crystallize in the octahedral form which is characteristic of the alums proper.

626. *Binoxide of Manganese* (MnO_2) is a black compound found abundantly in nature, and largely employed in the arts for the purpose of evolving chlorine from chloride of sodium or chlorhydric acid (§ 105), as well as for decolorizing glass. It may readily be prepared artificially from the lower oxides by the action of oxidizing agents. By itself, at the ordinary temperature, binoxide of manganese is an inert chemical substance, though at higher temperatures it has considerable oxidizing-power. At a strong red heat it gives off one-third of its oxygen:—

$$3MnO_2 = Mn_3O_4 + 2O.$$

Formerly oxygen was often prepared in chemical laboratories by heating the black oxide of manganese in iron retorts; but the process has long been superseded by more convenient methods. The oxide, $Mn_3O_4 = MnO,Mn_2O_3$, which is left as a residue in this experiment, corresponds in composition with the magnetic oxide of iron, an important ore of iron. This oxide is the most easily obtained by artificial means of all the oxides of manganese; it is produced when the protoxide or its nitrate or carbonate is strongly heated in the air, or when either of the higher oxides is intensely ignited.

Black oxide of manganese is insoluble in nitric acid, but is decomposed by strong hot chlorhydric acid, with formation of protochloride of manganese and free chlorine, as has been already explained (§ 105), and by hot concentrated sulphuric acid with evolution of oxygen:—

$$MnO_2 + H_2SO_4 = MnSO_4 + H_2O + O.$$

Exp. 323.—In a small glass flask, provided with a suitable delivery-tube, heat a mixture of 15 grms. of powdered black oxide of manganese and 10 grms. of concentrated sulphuric acid, and collect the gas over water in the usual way.

After all the available oxygen has been obtained in this experiment, and the flask, together with its contents, has been allowed to cool, pour 15 or 20 c. c. of water into the flask, boil the mixture; pour it upon a filter, and evaporate the filtrate to dryness upon a water-bath, taking care to stir it constantly when nearly dry. Hydrated sulphate of manganese ($MnSO_4,4H_2O$) will be obtained as a reddish-white powder.

Exp. 324.—For the sake of comparing the old process of making oxygen with methods now in use, charge an ignition-tube, such as was used in Exp. 7, to one-third of its capacity, with black oxide of manganese, connect it with the water-pan in the usual way, heat it strongly over the gas-lamp, and observe the comparatively slow rate at which oxygen is evolved from it.

627. *Manganic Acid* (MnO_3) has not yet been obtained in the free state; it is known only as it occurs in combination with potash or some other base. Of the manganates, those of potassium, sodium, and barium are the best-known; they are isomorphous with the corresponding chromates, sulphates, and seleniates. The alkaline manganates are important compounds to the analyst.

Exp. 325.—Place upon a piece of platinum-foil as much dry carbonate of sodium as could be held upon half a pea; mix with it an equal quantity of powdered nitrate of potassium and a bit of binoxide of manganese as large as the head of a small pin. Fuse the mixture in the outer blowpipe-flame, and observe the bluish-green-colored manganate of sodium which is produced.

Exp. 326.—Melt together in an iron ladle over an anthracite or charcoal fire, 10 grms. of hydrate of potassium, and 7 grms. of chlorate of potassium; stir into the pasty liquid 8 grms. of very finely powdered black oxide of manganese, and maintain the mixture for a short time at a temperature just below visible redness, taking care to stir it frequently with an iron rod. When the crumbly mass has become cold, place some of it in a test-tube with a small quantity of cold water and shake the tube. As soon as the solid particles have settled, there will be seen a clear green liquid, which is a solution of manganate of potassium.

Exp. 327.—Pour off half of the green solution of manganate of potassium into another short test-tube, and leave it open to the air; the green color of the solution will gradually change to blue, then to violet and to purple, and finally to ruby red. The red color is that of a solution of permanganate of potassium, into which the manganate is converted by exposure to the air. The intermediate colors are merely mixtures of the manganate green and the permanganate crimson. On account of these remarkable changes of color, the name chameleon mineral has been applied to manganate of potassium, and by this term it is still commonly known.

Manganate of potassium is a very unstable salt, especially when in solution; it may be readily decomposed in a great variety of ways. It breaks up into permanganate of potassium and binoxide of manganese when the aqueous solution is mixed with a large quantity of water, and even strong solutions are rapidly decomposed in the same way by boiling:—

$$3K_2MnO_4 + 2H_2O = K_2Mn_2O_8 + MnO_2 + 4KHO.$$

By means of acids, the change from manganate to permanganate may be almost instantaneously effected; but by the presence of an excess of alkali the decomposition is always greatly retarded.

Exp. 328.—Add a few drops of sulphuric acid to the remaining portion of the solution of manganate obtained in Exp. 326, and observe that a quantity of the red permanganate of potassium is immediately produced.

628. *Permanganic Acid* (Mn_2O_7), or rather its hydrate $H_2Mn_2O_8$, may be obtained in aqueous solution by decomposing permanganate of barium with sulphuric acid. The solution bleaches powerfully, and the acid is rapidly destroyed by organic matter and other reducing agents. Of the compounds of this acid, that with potassium is by far the best-known.

Exp. 329.—Place 300 c. c. of water in a porcelain dish, heat it to boiling and add to it by portions the remainder of the powdered green manganate of Exp. 326; from time to time add small portions of hot water to replace that which evaporates, and continue to boil until the green color of the solution has changed to deep violet red, and the manganate of potassium has all been changed to permanganate.

In case the manganate contains a large excess of free alkali it cannot readily be converted into permanganate by boiling; it will therefore often be found necessary to neutralize with nitric acid a portion of the alkali which is in excess. As soon as the transformation has been completed, pour the mixture into a tall bottle, leave it at rest until the binoxide of manganese and other insoluble matters have settled; then decant the clear liquor into a glass-stoppered bottle, and preserve it for use in subsequent experiments. The insoluble deposit may be again boiled with water and allowed to settle; the clear liquor thus obtained may be added to that previously prepared.

In order to obtain crystals of the permanganate, a clear solution like that above described should be rapidly evaporated to a small bulk, then decanted from the binoxide of manganese which is precipitated during the process, and set aside to cool. Needle-shaped crystals of a dark purple-red color will soon be formed; they are soluble in 16 parts of water at 15°, and are permanent in the air. It is well to purify the first crop of crystals by washing them with a little cold water, then dissolving in the least possible quantity of boiling water, and again crystallizing in the cold. Neither the crystals nor the solution should ever be brought into contact with paper. Decantation will ordinarily be sufficient in order to separate the crystals from the mother-liquor; but if filtration be necessary in any case, an asbestos filter should be employed (Appendix, § 14).

629. The permanganates are isomorphous with the perchlorates (§ 125), and the potassium salts of the two acids are capable of crystallizing together in all proportions. These compound crystals are red-colored when they contain much perchlorate of potassium, but are black if they contain as much as half their weight of the permanganate.

In the same way that perchloric acid is a more stable acid than chloric acid, so permanganic acid is less readily decomposed than manganic acid. Both manganic acid and permanganic acid, however, give up oxygen to other substances with remarkable facility, and are much used as oxidizing agents. Even a piece of wood or paper thrown into the green or red solution of a manganate or permanganate, will quickly abstract oxygen from the solution and destroy its color. In filtering the colored solutions, paper is consequently inadmissible, as has been stated in Exp. 329; asbestos, sand, or some other inert filtering-material must be resorted to. Permanganate of potassium is largely employed for disinfecting putrid water and animal or vegetable matters in a condition of putrefaction. A solution of it, such as has been prepared in Exp. 329, is of great use in volumetric analysis, especially for testing the value of iron ores.

IRON.

630. Although iron is one of the most widely diffused and most abundant of the metals, it is rarely found native in the metallic state. Meteors, however, fall upon the earth from outer space, which consist mainly of metallic iron, contaminated with several other elements in small proportions. Minerals containing iron occur in great numbers; and there are indeed few natural substances, whether organic or inorganic, in which iron is not present. It is found in the ashes of most plants, and in the blood of animals. The natural compounds of iron which are available as ores of the metal, are chiefly oxides and carbonates. The most important varieties of these ores of iron are the following:—1. *Magnetic iron-ore,* the richest of the ores of iron, containing when pure, 72·41 per cent. of iron, and not infrequently approximating closely to this composition in large masses. 2. *Red Hæmatite,* consisting, when pure, of anhydrous sesquioxide of iron containing 70 per cent. of iron; this ore often yields from 60 to 69 per cent. of the metal. 3. *Specular iron-ore,* which is a crystalline form of the same anhydrous sesquioxide of iron. 4. *Limonite, or Brown iron-ore,* which consists essentially of hydrated sesquioxide of iron, containing 59·89 per cent. of iron; yellow ochre is a clayey variety of this very abundant ore; the numerous ores classed under this head yield from 25 to 55 per cent. of iron. 5. *Spathic iron-ore, or Carbonate of iron,* which contains in its purest state 48·27 per cent. of iron, but is so generally contaminated with manganese, calcium, and magnesium as to yield very various quantities of iron, ranging from 14 to 43 per cent. 6. *Clay iron-ore,* a name applied to a mixture of clay and carbonate of iron, which occurs very abundantly in the coal-measures; as this ore is a mixture in uncertain proportions, it yields various percentages of iron, ranging from 25 to 40 per cent.

From the richer iron-ores, like the magnetic and specular oxides, a very excellent iron can be obtained by simply heating the broken ore with charcoal in an open forge fire, urged by a blast. The ore is deoxidized by the carbon of the fuel, and the reduced iron is agglomerated into a pasty lump called a "bloom," while the earthy impurities contained in the ore combine with a portion of the oxide of iron to form a fusible glass or slag. The spongy bloom is freed from slag and rendered homogeneous and solid by hammering while still red-hot; by reheating and hammering, the iron is then converted into bars or shaped into any other desired form. This process is not economical in the chemical sense, for much iron is lost in the slag, and much fuel is burnt to waste in an open fire; but when well conducted, it yields an admirable quality of iron; and since the original outlay for the construction of a bloomary is small, and repairs upon it are always easy, the method has many advantages in regions where transportation is dear while rich ores, charcoal, and water-power abound. The bloomary process, in its crudest form, is easily practised by people possessing but little mechanical skill and no chemical knowledge; it is undoubtedly the oldest method of extracting iron from its ores.

631. In the extraction of iron from its common ores, the metal is usually obtained, not pure, but in a carburetted fusible state, known as cast iron or pig iron. The main features of the process are, first, a previous calcination or roasting to expel water, carbonic acid, sulphur, and other volatile ingredients of the ore; secondly, the reduction of the oxide of iron to the metallic state by ignition with carbon; thirdly, the separation of the earthy impurities of the ore by fusion with other matters into a crude glass or slag; and, lastly, the carbonizing and melting of the reduced iron. With the purer kinds of iron-ore, the preliminary calcination is not always essential; but with the majority of ores it is very desirable; not unfrequently all the drying necessary is effected in the upper part of the blast furnace itself, within which the three last steps of the process always take place.

The blast furnace for iron consists essentially of a huge cylindrical structure of masonry, 15 to 25 m. in height, and 5 to 6 m. in diameter at the central portion of the cylinder, but contracted to a less diameter both at the top or *throat* and at the bottom or *hearth*. Air is forced in at the bottom of the furnace to support the combustion, and it has been found advantageous in the majority of cases to heat this blast of air to about the melting-point of lead before it enters the furnace. The reduction of the oxides of iron being effected by the carbonic oxide resulting from the combination of carbonic acid with hot carbon (Exp. 181), it is not difficult to calculate the amount of carbon and the amount of air requisite to reduce to metallic state the iron contained in a given weight of an iron-ore of known composition. Thus the formula of specular iron-ore, or of red hæmatite, is Fe_2O_3, and, since the atomic weight of iron is 56, these ores are 70 per cent. iron; accordingly the following quantities are equivalent one to the other:—

1	=	1·429	=	0·3214	=	0·4285	=	1·865.
Iron.		Fe_2O_3.		Requisite weight of carbon in state of CO.		Weight of oxygen requisite to convert so much C to CO.		Air.

For every kilogramme of iron produced, nearly two kilogrammes of air must be supplied, and at least $\frac{1}{3}$ kilogramme of fuel, merely to accomplish the chemical reaction. The reduction of the oxide of iron, however, is not alone sufficient to secure the metal; iron-ores almost always contain earthy admixtures, consisting chiefly of silica, clay, and carbonate of calcium; and these substances are so intimately mixed with the reduced metal, that it is essential to melt them before the iron can separate by virtue of its greater specific gravity. Any one of these substances taken alone is infusible at the temperature of the furnace; they must be converted into fusible double silicates; and as it is rarely the case that the natural impurities of an ore are present in the proportions requisite for the formation of such double silicates, it is generally necessary to mix with the ore a substance intended to effect this result, and therefore called the *flux*. With ores in which the earthy ad ixture is chiefly calcareous, the flux must be clay or some siliceous material; but in the more frequent case of ores containing clay or silica, the flux will be limestone or quicklime. In either case, a fusible double silicate of aluminum and calcium is the essential constituent of the slag. With siliceous ores there is another reason for the addition of lime; the double silicate of aluminum and iron is very fusible, and a considerable quantity of iron might be lost in the slag, were not lime enough added to prevent the formation of this iron-containing silicate. Sometimes both calcareous and siliceous ores are within reach of an iron-furnace, and the smelter, by mixing the two varieties in due proportion, may avoid the necessity of adding a flux.

The blast furnace is charged at the top with alternate layers of the fuel (which may either be charcoal, anthracite, or coke), the ore, and the flux, which is generally lime; these materials are constantly supplied at the top, and air is constantly supplied in immense quantities at the bottom of the furnace, the actual weight of the air forced in being greater than the sum of the weights of the ore, the fuel, and the flux. Where the blast first touches the ignited fuel, carbonic acid is formed; this gas, rising with the unused nitrogen through the furnace, comes in contact with white-hot carbon, and is reduced to carbonic oxide (Exp. 181). The layers of solid material thrown in at the top of the furnace gradually sink down, and as soon as a stratum of ore has descended sufficiently to be heated by the hot mixture of nitrogen and carbonic oxide it becomes reduced to spongy metallic iron, which, mixed with the flux and the earthy impurities of the ore, settles down to hotter parts of the furnace, where it enters into a fusible combination with carbon, while the flux and earthy impurities melt together to a liquid slag. The liquid carburetted iron settles to the very bottom of the furnace, whence it is drawn out, at intervals, through a *tapping-hole*, which is stopped with sand when not in use. The viscous slag flows out over a dam, so placed as to retain the iron, but to permit the escape of the slag which floats on the iron, as fast as it accumulates in sufficient quantity. The fusion of the materials in the lower part of the furnace requires a great heat; and the amount of fuel consumed in getting this high temperature is much greater than the amount requisite for the reduction and carbonization of the metal. As charcoal is a much purer carbon than coal or coke, iron smelted with charcoal is generally purer than that smelted with coal; but as charcoal crumbles under great pressure, the furnaces in which charcoal is used are usually much smaller than those intended for anthracite or coke. The consumption of fuel in smelting 1000 k. of iron varies with the nature of the furnace, the blast, and the fuel, between 500 k. and 3000 k. The gases which issue from the mouth of the blast furnace are charged with an enormous heating-power; for besides being themselves intensely hot they contain, even after having effected the reduction, a large proportion of combustible gases, such as carbonic oxide, carburetted hydrogen, and hydrogen. This gaseous mixture takes fire whenever it comes in contact with the air; a part of its heat may be utilized in heating the air-blast and generating steam.

Two distinct varieties of cast iron exist, which differ in color, texture, and fusibility; these are white cast iron and gray cast iron. White iron is hard and brittle, of crystalline texture and shining fracture. Gray iron is slightly malleable, and has a granular texture; its fracture may be either coarse or fine-grained, and minute particles of black graphite are visible upon the broken surface. White cast iron melts at a lower temperature than gray, but does not become so liquid as the gray. Gray cast iron, when rapidly cooled, is converted into white iron; when a casting is made in an iron mould, the layer of metal in contact with the mould is chilled and converted into the hard white iron, while the interior of the casting will retain the condition of the stronger gray iron. Excellent shot and shell for rifled cannon have lately been cast on this plan. The chief chemical difference between white and gray cast iron consists in the different condition of the admixed carbon. In white iron the carbon seems to be dissolved in or combined with the iron, while in gray cast iron, on the other hand, the greater part of the carbon seems to be mechanically diffused through the solid iron, in the state of graphite. The two varieties, however, shade off into each other through a great variety of intermediate mixtures. White cast iron rusts much more slowly than gray cast iron. When white iron is heated with strong chlorhydric acid it entirely dissolves, but the combined carbon enters into combination with a portion of the nascent hydrogen, forming hydrocarbons which impart a peculiar smell to the gas evolved. Gray iron does not wholly dissolve in hot chlorhydric acid; a residue of graphite remains; but the gas evolved has the same smell as the gas evolved from white iron. When bars, plates, or implements of common cast iron are exposed to the slow action of dilute acids or of saline solutions such as sea-water, the iron is not infrequently wholly dissolved away, while the graphitic carbon remains untouched—a light, soft, sectile substance which often retains the color and form of the original article, but possesses neither hardness nor tenacity. The largest proportion of carbon found in cast iron is 5·75 per cent.; this large percentage occurs in a lustrous variety of white iron which contains manganese and is called specular iron. In gray iron the amount of carbon varies from 2 to nearly 5 per cent.

Silicon, sulphur, phosphorus, manganese, and copper are very common impurities in cast iron. The silicon comes from silica deoxidized in the furnace; its amount varies from 0·1 to 3·5 per cent. There is more of it in gray than in white iron, and more in hot-blast iron than in cold-blast. Sulphur is almost always present in cast iron, but only in very small quantity; its presence is supposed to conduce to the formation of white iron. The presence of phosphorus to the extent of 1 or 2 per cent. is not uncommon, and does not injure iron intended for castings, inasmuch as the phosphorus makes the iron more fusible, and more liquid when melted. Manganese is frequently present in cast iron, as is not unnatural considering the common association of manganese ores with iron-ores. Cast iron containing manganese appears to be especially suitable for the production of steel.

The production of malleable or "*wrought*" iron from cast iron, consists essentially in burning out the carbon, silicon, sulphur, and phosphorus which cast iron contains. This oxidation of the impurities of cast iron is effected either by blowing upon the melted metal with an air-blast in a small charcoal furnace called a "finery," or by stirring the melted iron in a reverberatory furnace in which the fuel does not come in contact with the metal, and into which air can be admitted at will; the latter process, now much the most important method of manufacturing wrought iron, is called "*puddling*." In puddling it is customary to add to the charge of pig iron a quantity of iron scale or other oxide of iron. The oxidation of the silicon, carbon, phosphorus, and other impurities is effected partly by the air and partly by the oxide added to the charge; the carbon burns to carbonic oxide, which heaves the seething mass as it escapes and burns in jets of blue flame; the other impurities form, with the oxides of iron and manganese, a cinder or slag which ordinarily contains sulphur, silicic acid, and phosphoric acid. When the cast iron is so far decarbonized as to be pasty in the fire, it is gathered into lumps on the end of an iron bar and carried from the furnace to a hammer or squeezer which expresses the liquid slag and welds into a coherent mass the tenacious iron. The hammered lump may be reheated and rolled or forged into any desired shape. The waste of iron in converting cast into malleable iron amounts to from 13 to 30 per cent.

Ordinary malleable iron has a gray color, and a specific gravity of about 7·6. Though less malleable than gold and silver, its malleability is very great, and the greater the purer the metal and the higher the temperature to which it is raised. At a red heat, separate pieces may be firmly united by hammering or rolling; the operation is called *welding*. Sulphur is said to render wrought iron brittle while hot or "red-short;" silicon and phosphorus render iron brittle at the ordinary temperature, or "cold-short," in technical phraseology. These common impurities of cast iron are therefore very prejudicial to wrought iron. Wrought iron is hammered or rolled while in a doughy condition; and the uniform, close, fibrous texture which is valued in malleable iron depends much upon the nature of this mechanical treatment, and the extent to which it is carried. Common malleable iron still contains from 0·25 to 0·5 per cent. of carbon; the smaller the amount of carbon the softer the iron. Wrought iron dissolves almost completely even in dilute acids; but the hydrogen evolved has the peculiar smell attributed to the presence of carbonaceous vapor.

632. *Steel.*—This invaluable substance is in composition intermediate between cast and wrought iron, containing less carbon than cast iron, but more than wrought. It may be made from wrought iron by heating bars of iron to redness for a week or more in contact with powdered charcoal in close boxes from which air is carefully excluded. Though the iron is not fused, nor the carbon vaporized, yet the carbon gradually penetrates the iron and alters its original properties; when the bars are withdrawn from the chests in which they were packed, the metal has become fine-grained

in fracture, more brittle, and more fusible. The bars, however, are far from uniform in composition, the outside being more highly carbonized than the interior; they are apt to show blisters of various sizes on the surfaces, and the steel thus prepared is called "blistered" steel. To obtain steel of a uniform quality, it must be cast into ingots. This process of preparing steel is called the "cementation" process; it is a curious instance of chemical action between solid materials which are apparently in a state of rest. Since the materials used in this process are the purest attainable —the best iron and the best charcoal—the steel obtained is of the best quality. Cheaper methods of preparing an inferior steel are, however, of great industrial importance. If the "puddling" process for preparing malleable iron should be arrested when the cast iron had lost from one-half to two-thirds of its carbon, the product would be an impure steel, impure because the silicon, phosphorus, sulphur, and other impurities of the cast iron would only be incompletely removed. Nevertheless there are uses in the arts for a steel of this quality which may be cheaply manufactured.

633. A new and very rapid method of preparing cast steel directly from cast iron is that known as the Bessemer process. From two to six tons of cast iron, previously melted in a suitable furnace, are poured into a large covered crucible, made of the most refractory materials, and swung on pivots in such a manner that it can be tipped up and emptied by means of an hydraulic press. Through numerous apertures in the bottom of the crucible a blast of air is forced up into the molten metal; an intense combustion ensues involving the carbon in the iron and a portion of the metal itself, and generating a most intense heat, which keeps the mass fluid in spite of its rapid approach to the condition of malleable iron. Such a quantity of specular iron or white cast iron is then added to the iron in the crucible as is necessary to give carbon enough to convert the whole mass into steel, and the melted steel is immediately cast into ingots. Six tons of cast iron can thus be converted into tolerable steel in twenty minutes. This steel is suitable for the manufacture of axles, cranks, rails, boiler-plates, and many other articles in which

great strength should be combined with hardness. A pure steel cannot at present be made by this process, inasmuch as the combustion in the crucible does not get rid of the sulphur and phosphorus in the cast iron nearly as perfectly as does the puddling process; for the same reason the manufacture of wrought iron by this method, though the original object of the invention, has been thus far found impracticable. It deserves mention that the nailer who keeps his nail hot, while hammering it, by a carefully regulated blast of cold air, applies the chemical fact involved in Bessemer's process. This ancient practice was indeed a prophecy of Bessemer's invention.

634. The two qualities of steel which are of greatest importance are its hardness and its elasticity. These qualities are developed by quickly cooling the heated metal; the delicate processes by which steel tools and springs are hardened, tempered, and annealed are exceedingly curious, but are rather physical than chemical phenomena. Many implements are sufficiently well made by converting their exterior surfaces into steel, leaving the interior of cast or wrought iron. Thus cast-iron tools may be heated with oxide of iron to remove a part of the carbon from their exterior and thus coat them, as it were, with steel. Tires for wheels are well made of wrought-iron bars which have been superficially converted into steel by the cementation process; such tires combine the toughness of malleable iron with the hardness of steel.

635. *Oxides of Iron.*—There are several definite compounds of iron and oxygen. The best-known of these oxides are the protoxide (FeO), or ferrous oxide, as it is often called, and the sesquioxide (Fe_2O_3), often called ferric oxide, and sometimes spoken of as peroxide of iron. There is another oxide, ferric acid FeO_3, which is an exceedingly unstable substance, known only as it exists in combination with potassium, as ferrate of potassium (K_2FeO_4), or with some other powerful base. Besides these oxides, there are several compounds of intermediate composition, which may be supposed to result from the union of ferrous and ferric oxides, in various proportions; they are called collectively ferroso-ferric oxides; the most important among them is the

magnetic oxide $Fe_3O_4 = FeO,Fe_2O_3$, which is the black oxide formed when iron is oxidized at high temperatures in oxygen gas, in air or steam (§§ 9, 18, 34).

636. *Ferrous Oxide* or *Protoxide of Iron* (FeO).—This compound is not easily obtained pure, since it absorbs oxygen from the air with great avidity and thus becomes contaminated with the sesquioxide. But by dissolving a ferrous salt (that is, a salt of protoxide of iron) in recently boiled water, and adding to the liquid a solution of caustic alkali, which has likewise been boiled to expel air, there will be precipitated a white ferrous hydrate, provided the operation be conducted out of contact with the air. If this hydrate be exposed to the air, as when the solution of a ferrous salt is mixed with the alkali without the precautions above enumerated, it will rapidly absorb oxygen and will exhibit various shades of light green, bluish green, and black, till, finally, it assumes the red color of hydrated ferric oxide (Exp. 341). The anhydrous oxide obtained by igniting ferrous oxalate in close vessels, absorbs oxygen so rapidly that it takes fire when brought into contact with the air.

Hydrated ferrous oxide is readily soluble in acids, forming salts known as the protosalts of iron or ferrous salts; many of these salts are of a pale green color; like the hydrate, they rapidly suffer decomposition by absorbing oxygen from moist air.

637. *Ferric Oxide* (Fe_2O_3).—This oxide, called also red oxide, sesquioxide, or peroxide of iron, occurs very abundantly and widely distributed in nature. Several of its varieties have been already mentioned as ores of iron (§ 630). It may be obtained also by igniting metallic iron or either of the lower oxides or hydrates in contact with the air. For use in the arts, it is prepared by igniting ferrous sulphate with or without addition of a small proportion of nitrate of potassium, or by roasting the native hydrate (yellow ochre). The better sort, known as rouge, is largely employed for polishing glass and jewelry, and all grades of it are extensively used as pigments. Red ochre is impure ferric oxide. As commonly met with, the oxide is amorphous and has a red, brown, or nearly black color, according to the method of its preparation. At a full white heat, it gives off a portion of its oxygen, and magnetic oxide of iron is formed. It

is easily reduced to the metallic state by hydrogen gas, even at temperatures below redness, and by carbon and carbonic oxide at a red heat, as has been stated in § 631. Ammonia gas reduces it also at a red heat.

Exp. 330.—In the middle of a tube of hard glass, No. 3, 10 c. m. long, and provided at both ends with corks carrying short, straight delivery-tubes, place a teaspoonful of red ochre or ignited iron-rust. Attach to one end of the tube a hydrogen-generator or gas-holder provided with a chloride-of-calcium drying-tube, and connect with the other end a U-tube. Support the tube containing oxide of iron upon a ring of the iron stand, cause a current of hydrogen to flow through it, immerse the U-tube in a bottle of cold water, and finally heat the oxide of iron. The hydrogen will combine with the oxygen of the red oxide of iron, water will be formed and will condense in the U-tube, while finely divided metallic iron will be left behind. After the reduction has been completed, allow the tube to become cold, and then scatter its contents through the air upon an earthen plate. They will take fire and burn again to the condition of red oxide.

Exp. 331.—Repeat Exp. 330, using carbonic oxide instead of hydrogen, the products will be iron and carbonic acid instead of iron and water.

638. The facility with which red oxide of iron gives up oxygen, taken in connexion with the readiness with which metallic iron and the protoxide take on oxygen, is a fact of great practical importance. It has been found that organic substances may be more rapidly incinerated by heating them in the air in contact with a small quantity of ferric oxide than in air alone; the oxide of iron appears to act as a carrier of oxygen, as it is alternately reduced by the combustible and again oxidized by the air. Even at ordinary temperatures, and with the hydrated oxide, the same reactions are witnessed, though in a less degree. The iron nails employed in the construction of ships, bridges, fences, or shoes, actually corrode, "eat up" or "burn out" the organic matter in contact with them, by absorbing oxygen from the air and transferring it to the carbon compound with which they are in contact. The rotting of canvas by iron-rust, or of a fishing-line by the rusty hook, are familiar instances of corruption by rust.

These reactions doubtless play an important part in the formation of soils by the oxidation of vegetable remains.

In the same way ferric oxide converts sulphide of calcium (CaS) into sulphate of calcium ($CaSO_4$) at the expense of the oxygen of the air. A useful cement has been prepared by mixing the residual oxysulphide of calcium of Leblanc's soda process with an equal weight of the ferric oxide left as a residue in burning iron pyrites for sulphuric acid. Hydrated sesquioxide of iron ($Fe_2O_3,3H_2O$) may readily be prepared by adding an excess of ammonia-water to the solution of almost any ferric salt.

Exp. 332.—Cover a teaspoonful of fine iron filings or small tacks with three or four times as much dilute sulphuric acid in a small bottle; wait until the evolution of hydrogen ceases, then decant the clear liquor into a small flask or beaker, add to it a few drops of strong nitric acid, and heat it to boiling. The liquor will soon be colored dark brown by the nitrous fumes resulting from the decomposition of the nitric acid, which are for a short time held dissolved by the liquid; but this deep coloration soon passes away, and there is left only the yellowish-red color of the ferric sulphate which has been formed. Add to the solution ammonia-water, until the odor of the latter persists after agitation, and collect upon a filter the flocculent red precipitate of ferric hydrate.

Besides this normal hydrate ($Fe_2O_3,3H_2O$) there are several other ferric hydrates, containing smaller proportions of water. They are found in nature, and may be obtained by heating the normal hydrate, or by suffering it to remain for a long while under water, or by boiling it for some time in water.

639. Generally speaking, hydrated sesquioxide of iron is easily soluble in acids, though some peculiar varieties of it dissolve only with difficulty. The anhydrous oxide also dissolves in acids, though less easily in proportion as it has been more strongly ignited; its best solvent is concentrated boiling chlorhydric acid.

By long-continued heating at 300° or 320°, ferric hydrate can be deprived of all its water, and still be readily soluble in acids; but when heated to dull redness, this powder suddenly glows brightly for a moment and contracts in bulk, without either losing or gaining weight, and is then attacked by acids only slowly and with difficulty. It has been observed, however, that the ignited oxide may still be dissolved rather easily by a hot mixture of chlorhydric acid and ferrous chloride, protochloride of tin, zinc, or some other reducing agent.

Ferric hydrate is somewhat used as a mordant in dyeing, and is largely employed for purifying coal-gas. As has been stated under arsenic, the recently precipitated hydrate acts as an antidote to arsenious acid, since when given in sufficient quantity it forms a basic arsenite of iron scarcely at all acted upon by water.

Exp. 333.—Dissolve half a gramme of arsenious acid in 40 or 50 c. c. of boiling water. Divide the solution into two portions, and stir into one of these portions a considerable quantity of moist ferric hydrate, such as was obtained in Exp. 332; filter the mixture, acidulate the filtrate with chlorhydric acid, and test it for arsenic by means of sulphydric acid (§ 340).

If a sufficient quantity of ferric hydrate has been employed, no precipitate of sulphide of arsenic will be obtained in the filtrate, though on adding a drop of chlorhydric acid, and afterwards sulphydric acid, to the original solution of arsenic, an abundant yellow precipitate will be at once thrown down.

Exp. 334.—Fill a tube, 30 c.m. long, with alternate tufts of cotton and loose layers of dry ferric hydrate, pass a slow current of sulphydric acid (Exp. 86) through the tube, and observe that the oxide gradually becomes black; it appears to be converted into ferrous sulphide, while water and sulphur are set free :—

$$Fe_2O_3,3H_2O + 3H_2S = 2FeS + S + 6H_2O.$$

After a good part of the ferric oxide has become black, remove the contents of the tube to a porcelain plate, and leave them exposed to the air in a place where no harm can be done in case they take fire. By the action of the air, sesquioxide of iron will be reproduced and sulphur set free within the mass; some sulphurous acid is given off at the same time, and heat is evolved, as will readily be perceived if the quantity of material be large. The following equation, though it does not fully express the complete reaction which really occurs, may still serve to give a general idea of these chemical changes :—

$$2FeS + 5O = Fe_2O_3 + S + SO_2.$$

In practice the impure illuminating gas is made to pass through layers of ferric oxide, often made porous by an admixture of sawdust; as soon as the oxide ceases to absorb sulphuretted hydrogen, it is "revivified" by forcing or drawing through it a current of fresh air or by spreading it in the air. The oxide is thus used over and over again, until so much sulphur has accumulated within it as to interfere, mechanically, with its absorbent power. The sulphur may readily be recovered from this mixture by distillation; or the spent oxide may be used instead of pyrites for making sulphuric acid, wherever enough of it can be obtained to repay the trouble of collecting.

The ferric salts are usually of a yellowish-brown or red color when hydrated, though some of them have a violet tinge; they are white when dry. The normal salts are usually soluble in water and deliquescent; and there are numerous soluble basic salts, besides other basic salts which are insoluble.

In the ferric salts iron plays the part of a trivalent element like aluminum, while in the ferrous salts it is bivalent like calcium or lead. The ferric salts closely resemble salts of aluminum, and are for the most part isomorphous with them. Besides acting as a base, ferric oxide, like oxide of aluminum, combines with several of the more powerful bases to form salts called ferrites; magnetic oxide of iron, for example, may be regarded as the ferrite of iron.

It is remarkable that ferric oxide may be displaced from many of its compounds by the protoxide of iron; thus, when hydrated ferrous oxide is added to a solution of ferric sulphate, hydrated sesquioxide is precipitated:—

$$Fe_2O_3,3SO_3 + 3(FeO,H_2O) = Fe_2O_3,3H_2O + 3(FeO,SO_3).$$

In like manner, carbonate of barium precipitates anhydrous ferric oxide from ferric salts, but upon ferrous salts it has no action:—

$$Fe_2Cl_6 + 3BaCO_3 = Fe_2O_3 + 3BaCl_2 + 3CO_2.$$

Sulphides of Iron.—There are several sulphides of iron, the most important of which are the protosulphide (FeS) and the bisulphide (FeS_2), found native as iron pyrites. There is a sulphide, Fe_3S_4, which is magnetic like the oxide to which it corresponds.

640. *Ferrous Sulphide* (FeS) is a substance of great value to the chemist as the cheapest source of the important reagent sulphydric acid (§§ 202, 210).

This sulphide may be prepared by igniting pyrites in a covered crucible, by rubbing roll brimstone against a white-hot iron bar, or by fusing together sulphur and iron turnings. The second method is to be recommended if the student have ready access to a blacksmith's forge. The sulphide is a waste product in those chemical works where sulphate of lead, obtained from dye-houses, is reduced to the metallic state by fusion with iron and coal. In the laboratory it may be prepared as follows:—

Exp. 335.—Heat a common Hessian crucible to redness in a fire of coke or anthracite, and project into it from an iron spoon successive small portions of a mixture of 7 parts of iron turnings and 4 parts of powdered sulphur, replacing the cover of the crucible after each addition of the mixture. The sulphur and iron combine with great energy, and the sulphide formed melts down to the liquid state. Since the molten sulphide is capable of dissolving both iron and sulphur, according as the one or the other may be present in excess, it is impossible to prepare a pure protosulphide by this method. But the product obtained as above described, though of variable composition, answers perfectly well for all ordinary purposes. When the crucible has become half-full of the molten sulphide, remove it from the fire, pour out its contents upon a brick floor, and, if more of the sulphide be desired, replace the crucible in the fire and proceed as before.

Where comparatively large quantities of the sulphide are required, it is well to bore a hole through the bottom of a plumbago crucible and set the latter upon the grate-bars of the furnace in such manner that the hole may remain open; fill the crucible with iron turnings; heat it to redness and throw in lumps of sulphur upon the hot iron. As fast as sulphide of iron forms it will melt and flow through the hole in the crucible into the ashpit below, which should be kept clean to receive it. In the preparation of sulphide of iron, wrought iron should always be employed. From the filings of cast iron, but little, if any, of the fusible sulphide can be prepared.

The foregoing experiment illustrates the practical methods of making ferrous sulphide; but several other reactions which produce it are of scientific interest (compare Exp. 85).

Exp. 336.—Arrange a bottle for generating sulphuretted hydrogen, as in Exp. 86, but in place of the delivery-tube in Fig. 41 attach to the bottle a jet for burning the gas. After the air has been completely expelled from the bottle, light the sulphydric acid gas at the jet, and hold in its flame a piece of fine iron wire; the iron will burn to ferrous sulphide, and if the wire be held in the axis of the flame, so that a considerable portion of it shall be kept red-hot, the globule of sulphide of iron formed will melt and flow backward upon the wire as fast as the end of the latter is consumed.

Exp. 337.—Mix 20 grms. of fine iron filings, 14 grms. of flowers of sulphur, and 7 grms. of water in a small bottle, and heat the mixture gently upon a sand bath, or set it aside in a warm place. Chemical action will soon set in, much heat will be evolved, and in the course of half an hour the mixture will become black from formation of sulphide of iron. If the porous black sulphide be left exposed to the air, it will absorb oxygen, and will be partially converted into ferrous sulphate.

A firm packing or lute for the joints of iron vessels is prepared by mixing together 60 parts of fine iron filings, 1 part of flowers of sulphur, and 2 parts of powdered chloride of ammonium. The mixture is made into a stiff paste with water, and immediately applied to the iron. It soon becomes hot and swells up, and sets to a hard compact mass, while ammonia and sulphuretted hydrogen are disengaged.

Exp. 338.—Dissolve a small crystal of ferrous sulphate (copperas) in water, and add to the liquid a drop or two of sulphydrate of ammonium (§ 526). Black sulphide of iron will be thrown down (compare Exp. 334).

The finely divided ferrous sulphide obtained in the wet way, as in the last two experiments, dissolves much more quickly in acids than the compact sulphide obtained by the way of fusion; in contact with acids it evolves gas so tumultuously that it would be inconvenient as a source of sulphydric acid.

The black earth between the stones of the pavements of cities, and at the bottoms of drains and cesspools, owes its color to sulphide of iron formed by the putrefaction of sulphuretted compounds in contact with ferric oxide contained in the earth.

641. *Bisulphide of Iron* (FeS_2) occurs abundantly in nature as the well-known mineral *iron pyrites*. Two distinct forms of it are met with:—the yellow cubical pyrites, crystallized in forms of the monometric system; and the white pyrites or marcasite, which crystallizes in trimetric forms. A third variety of sulphide of iron, called magnetic pyrites, is of different composition from the foregoing, and contains less sulphur than the bisulphide. Iron pyrites appears to have been sometimes formed in nature by the deoxidation of sulphates, such as the sulphate of calcium, by means of organic matter in presence of chalybeate waters. The formation of pyrites has often been noticed in solutions of sulphate of iron into which organic matters have fallen. But bisulphide of iron may be readily formed in the dry way also.

The compact forms of yellow pyrites, whether natural or artificial, are permanent in the air; but when finely divided the mineral oxidizes rather easily, with evolution of considerable heat.

White iron pyrites oxidizes rapidly in the air, no matter whether it be compact or friable. The spontaneous combustion of many kinds of coal is due to the oxidation of iron pyrites disseminated through the combustible. Alum and copperas are often prepared from pyritous shales, either by firing heaps of the shale artificially, or by allowing the heaps to take fire spontaneously through oxidation of the pyrites, and then regulating the combustion so that the largest practicable yield of sulphate of iron or of sulphate of aluminum shall be obtained. So long as the temperature of the burning pyrites remains comparatively low, ferrous sulphate and sulphuric acid are the principal products, the latter uniting with the alumina of the shale, if such be present; when the heap has become cold, the sulphates can be separated by lixiviating the mass with water. When pyrites is roasted at higher temperatures, as in the manufacture of sulphuric acid, sulphurous acid is given off, and ferric oxide left as the principal residue.

When distilled in close vessels, one atom of the sulphur in iron pyrites is expelled, and ferrous sulphide remains. Sulphur has sometimes been prepared in this way in a dearth of native sulphur.

642. *Ferrous Chloride* ($FeCl_2$) may be obtained by passing chlorine or dry chlorhydric acid gas over hot iron; in case chlorhydric acid be employed, hydrogen will be evolved. As commonly met with, however, the chloride is in the form of a hydrate ($FeCl_2 + 4H_2O$) obtained by dissolving metallic iron in dilute chlorhydric acid. It crystallizes easily, forms double salts by uniting with many other chlorides, and may be deprived of its water without decomposition when heated carefully out of contact with the air.

643. *Ferric Chloride* (Fe_2Cl_6).—As obtained by burning metallic iron in an excess of dry chlorine, this compound occurs in anhydrous, glistening scales, which volatilize easily when heated. It dissolves readily in water, with evolution of heat, and deliquesces rapidly in the air. It hisses when thrown into water. Once dissolved in water, it cannot be freed from the water by evaporation, since chlorhydric acid goes off with the water, and a basic compound of ferric oxide and ferric chloride remains. Hydrated ferric chloride may readily be obtained by boiling a solu-

tion of ferrous chloride with a small proportion of nitric acid, or by passing chlorine gas through a solution of ferrous chloride. From concentrated solutions, ferric chloride crystallizes with several different proportions of water. Ferric chloride combines with many of the metallic chlorides to form double compounds, among which the ammonium salts are perhaps the most stable.

644. *Ferrous Sulphate* ($FeSO_4$).—A hydrate of this compound, of composition $FeSO_4+7H_2O$, usually called *copperas* or *green vitriol*, is the most common of all the compounds of iron. It may readily be prepared by dissolving metallic iron or protosulphide of iron in dilute sulphuric acid. On the large scale it is commonly prepared by roasting iron pyrites at a gentle heat, or aluminous shales containing pyrites in the manner already indicated. Sometimes, however, it is manufactured directly from metallic iron and sulphuric acid; and it is obtained as a secondary product in certain metallurgical operations where copper is precipitated, by means of iron, from a solution of copper. The reaction is analogous to those employed for obtaining pure silver (Exp. 267) and pure lead (§ 594).

Exp. 339.—Dissolve 5 grms. of common blue vitriol (sulphate of copper) in 50 or 60 c. c. of water, acidulate the liquor with a few drops of sulphuric acid, pour it into a bottle, and place in it a rod of thick iron wire. Copper will immediately begin to be precipitated as a coating upon the iron, and in the course of an hour or two will be completely removed from the solution. The original blue color of the solution will disappear and be replaced by the faint green color of copperas, while a spongy mass of metallic copper will be obtained:—

$$CuSO_4 + Fe = FeSO_4 + Cu.$$

Decant the solution of ferrous sulphate from the precipitated copper, place in it a fragment of iron, and evaporate it to a small bulk; pour the concentrated solution into a wide-mouthed phial, cork the phial tightly, and set it aside in a cool place; the liquid will be converted into a mass of copperas crystals.

It has been proposed to prepare copperas from the "finery slag" of the puddling-furnaces (where cast iron is converted into wrought iron) by treating this slag with dilute sulphuric acid. The finery slag consists chiefly of basic silicate of protoxide of iron ($2FeO,SiO_2$).

645. When perfectly pure, the crystals of ferrous sulphate are

compact, transparent, and of a bluish-green color; but in dry air they effloresce and become covered with a white incrustation, the color of which subsequently changes to rusty brown through absorption of oxygen. The common commercial article is of a grass-green color, and is contaminated with more or less ferric sulphate. Besides the common hydrate containing 7 molecules of water, there are hydrates which contain 4, 3, and 2 molecules of water respectively. From all of these hydrates the water can easily be expelled by heat, and if the anhydrous salt thus obtained be still further heated it will decompose; two stages in this decomposition may be formulated as follows:—

$$\text{I. } 2FeSO_4 = SO_2 + Fe_2O_3,SO_3.$$
$$\text{II. } Fe_2O_3,SO_3 = Fe_2O_3 + SO_3.$$

Basic ferric sulphate is at first formed, while sulphurous acid is given off; and finally the ferric salt is itself decomposed into anhydrous sulphuric acid and ferric oxide. Upon this reaction the preparation of Nordhausen sulphuric acid depends (§ 239). Like ferrous hydrate, ferrous chloride, and all the ferrous salts, moist copperas, or an aqueous solution of copperas, rapidly absorbs oxygen from the air.

Exp. 340.—Pour a solution of copperas into an open capsule and leave it exposed to the air for a day or two; the solution will gradually become yellow as the oxidation proceeds, and after a while a rusty precipitate of ferric oxide, or of highly basic ferric sulphate, will fall. The oxide of iron which separates under these conditions is not readily soluble in dilute acids. It appears to be an isomeric modification of the easily soluble hydrate which is precipitated from cold ferric solutions by alkaline lyes. At all events the sulphuric acid of the copperas is insufficient to dissolve all of the ferric oxide formed during its oxidation. In most cases where a ferrous salt is to be converted into a ferric salt, it is best to add a certain proportion of free acid to the mixture, in order to prevent the separation of the oxide.

A difficultly soluble deposit, similar to the foregoing, may readily be obtained by boiling an exceedingly dilute solution of almost any of the soluble ferric salts. It is possible that these sediments should be regarded rather as highly basic salts than as mere hydrates. Their inertness may perhaps be due to the presence of small proportions of the acids of the salts from which they have been derived, still held in chemical combination; but there is at present less evidence in favor of this view than of the one previously stated.

Exp. 341.—To a teaspoonful of a solution of copperas add a few drops of soda lye, and observe that the hydrate rapidly absorbs oxygen, and changes color as has been set forth in § 636.

Exp. 342.—Mix a few drops of a solution of copperas with a drop or two of a solution of tannic acid, such for example as tincture of nutgalls, or of oak- or hemlock-bark; a light violet-colored precipitate will be formed and will remain suspended in the liquid; by exposure to the air this color soon changes to black. The violet precipitate is ferrous tannate, and the black precipitate ferric tannate; if these finely divided precipitates were produced in liquids made slightly viscous by the addition of gum or sugar, they would remain suspended in the liquor, which could then be used as writing-ink.

Ink may be prepared as follows:—Powder separately 12 grms. of nutgalls, 5 grms. of copperas, and 5 grms. of gum-arabic. Boil the nutgalls two or three hours in a flask with 75 c. c. of water, taking care to add hot water, by small portions, to supply that lost by evaporation. Allow the mixture to settle, and decant the clear liquor into a clean bottle. Dissolve the gum-arabic in a small quantity of water, and mix the mucilage thoroughly with the solution of nutgalls. Dissolve the copperas in another portion of water, and incorporate this solution with the mixture of nutgalls and gum. Add enough water to make the volume of the mixture equal to 100 c. c. Preserve the ink in a tight bottle. If the color of the product be lighter than is desired, the liquid may be left exposed to the air until it has acquired a deeper tint. When first applied to paper, the color of fresh ink is comparatively pale, but the writing darkens gradually in proportion as it absorbs oxygen.

In the course of the foregoing experiments, dip a small piece of cotton cloth in the solution of nutgalls, and allow it to become dry; then dip it in the solution of copperas and hang it up in damp air. Black, insoluble tannate of iron will be so firmly precipitated in and upon the fibres of the cloth, that it cannot be washed away.

The experiment illustrates one general method of dyeing, by means of which blacks and grays of various shades may be applied to cloth or leather, though in practice other astringent dye-stuffs, such as catechu, cutch, or gambier, are commonly employed in place of nutgalls.

Ferrous sulphate is largely employed in dyeing, sometimes directly, as in the foregoing experiment, but often as the source of other compounds of iron which are employed as mordants; ferrous acetate, for example, obtained by decomposing ferrous sulphate with acetate of calcium, is a compound much used by dyers. It should be remarked, however, that acetate of iron is sometimes made directly by dissolving scraps of iron in vinegar or pyroligneous acid (§ 380).

646. *Ferric Sulphate* ($Fe_2 3SO_4$) is interesting chiefly from its analogy with sulphate of aluminum. Like sulphate of aluminum, it combines with the sulphates of the alkali-metals to form well-defined alums. (Compare § 625). Ferric sulphate occurs as a waste product in the mother-liquors from which copperas and alum have crystallized. By drying these liquors and igniting them, red ochre of excellent quality can be obtained, in accordance with the second reaction of § 645. Fuming sulphuric acid is commonly manufactured nowadays by distilling pure ferric sulphate, instead of copperas as formerly at Nordhausen. The ferric salt is obtained by dissolving ferric oxide in weak sulphuric acid, and evaporating the solution to dryness; the residue of ferric oxide left after the ignition of the sulphate is thus reconverted into ferric sulphate, and the sulphate is used over and over again as often as it is decomposed.

647. *Ferrous Nitrate* ($FeN_2O_6 + 6H_2O$) is a compound of considerable scientific interest, which may readily be procured by dissolving ferrous sulphide in cold dilute nitric acid, or by decomposing a solution of copperas with an equivalent quantity of nitrate of barium. It may also be obtained, mixed with nitrate of ammonium, by dissolving iron in cold dilute nitric acid. The metal dissolves without evolution of gas, in a manner which may be thus formulated:—

$$4Fe + 10HNO_3 = 4FeN_2O_6 + (NH_4)NO_3 + 3H_2O.$$

The aqueous solution of ferrous nitrate decomposes readily when heated, and in warm weather changes spontaneously to a ferric compound.

648. *Ferric Nitrate* ($Fe_2 3N_2O_6$) may be obtained in hydrated crystals containing 18 molecules of water, by dissolving metallic iron in nitric acid, of 1·29 specific gravity, till the liquor has taken up about 10 per cent. of the metal, and then adding an equal volume of nitric acid of specific gravity 1·43. The solution will deposit, on cooling, rhombic prisms of ferric nitrate, which are sometimes colorless, but often of a faint lavender-blue color. They are slightly deliquescent, and very soluble in water, but are only slightly soluble in cold nitric acid. By adding nitric acid to a syrupy solution of ferric nitrate, there may be obtained another

hydrate, containing only 12 molecules of water, crystallized in cubes or square prisms. By mixing a solution of ferric nitrate, or, for that matter, almost any other of the normal ferric salts, with recently precipitated ferric hydrate, or by partially abstracting the acid of the salt by means of an alkali, deep-red solutions of various basic compounds may readily be obtained. A basic ferric nitrate is employed in dyeing, under the name *iron mordant.*

649. *Silicates of Iron.*—Several native silicates of iron are known; but none of them are of special interest. The green tinge of ordinary glass is due to the presence of a ferrous silicate, and by increasing the proportion of the ferrous salt, a deep bottle-green colour may be imparted to glass. Ferric silicate, on the other hand, has comparatively little coloring-power, though when a considerable quantity of it is present it imparts a yellow color to glass. It is sometimes used for coloring porcelain. To destroy the green color of the ferrous silicate, binoxide of manganese, or some other oxidizing agent, is often added to glass in the process of manufacture; the ferrous silicate is thus converted, for the most part, into ferric silicate, and a nearly colorless glass produced.

650. *Cyanides of Iron.*—There is a ferrous cyanide ($Fe(CN)_2$), known as a yellowish-red precipitate, which takes up oxygen and becomes blue when exposed to the air; and a ferric cyanide ($Fe_2(CN)_6$) has been obtained in solution. But by far the best-known of the cyanides of iron are certain double compounds, which constitute the peculiar pigments known collectively as Prussian blue. Common Prussian blue, for example ($Fe_7C_{18}N_{18}+18H_2O$), may be regarded as a double compound of ferrous and ferric cyanides, $3Fe(CN)_2, 2(Fe_2(CN)_6)+18H_2O$; it may be prepared as follows:—

Exp. 343.—Add to an exceedingly dilute solution of almost any ferric salt, such, for example, as the ferric sulphate of Exp. 332, a drop of ferrocyanide of potassium (§ 509). A beautiful blue precipitate will form, and will remain suspended in the liquor for a long while. Another variety of Prussian blue, known as Turnbull's blue, may be obtained by mixing a solution of red prussiate of potash, known to chemists as ferricyanide of potassium, with a solution of copperas or other ferrous salt.

Since the yellow prussiate of potash will give no blue coloration with ferrous salts, and since the red prussiate yields no blue with ferric salts, it is evident that the two solutions may be used as tests by which to detect the presence of ferrous and ferric salts, respectively, in any solution.

Exp. 344.—Soak a piece of cotton cloth in a solution of ferric sulphate (Exp. 332), and then immerse it in an acidulated solution of yellow prussiate of potash. Prussian blue will be precipitated upon the cloth and will remain firmly attached to it. Prussian blue is largely employed in dyeing and calico-printing in a variety of ways.

Now that we have discovered a ready means of detecting ferrous and ferric salts, it will be well to determine experimentally how easily the members of either of these classes may be changed to salts of the other class.

Exp. 345.—Dissolve 4 or 5 grms. of iron tacks or wire in dilute chlorhydric acid in a test-tube, pouring off the liquid from time to time as it becomes nearly saturated. Test a few drops of the solution first with ferro- and then with ferricyanide of potassium, in order to prove that it is pure ferrous chloride. Boil the rest of the liquid with a few drops of nitric acid to convert it to ferric chloride, and determine when the conversion has been completed by testing as before. Finally, divide the ferric solution into three portions. Through the first portion pass sulphydric acid gas; sulphur will be deposited and ferrous chloride formed,

$$Fe_2Cl_6 + H_2S = 2FeCl_2 + 2HCl + S;$$

to the second portion add small fragments of protochloride of tin, until a drop of the mixture, tested with the ferrocyanide, will no longer give a blue coloration,

$$Fe_2Cl_6 + SnCl_2 = 2FeCl_2 + SnCl_4;$$

boil the third portion with a fragment of metallic zinc, and determine the fact of reduction as before,

$$Fe_2Cl_6 + Zn = 2FeCl_2 + ZnCl_2.$$

By leaving either of these reduced solutions in the air, or by heating them with a little chlorate or nitrate of potassium, nitric acid, or other oxidizing agent, they may be readily converted again to the condition of ferric salts.

COBALT AND NICKEL.

651. Cobalt and nickel are two metals remarkably similar to one another both in physical and chemical properties. They are found together in nature in the same ores, in combination with sulphur and arsenic, and are both ingredients of meteoric iron. They can be reduced from their oxides by charcoal and by hydrogen at high temperatures, and the metals thus obtained can be melted about as readily as pure iron. Both cobalt and nickel resemble iron more closely than any other common metal; they are very tenacious, hard, and refractory; like iron they are magnetic, and when hot they may be forged; they rust less readily than iron, but resemble it closely in most of their chemical properties. The atomic weights of cobalt and of nickel are identical; the same number (58·8) applies to both. The specific gravities of the two metals also are equal or nearly so, varying in different samples from 8·2 to 8·9. Cobalt is not used in the metallic state; but several of its compounds are remarkable for the beauty of their color, and find important applications in the arts as pigments, especially for coloring glass and porcelain. A blue glass containing silicate of cobalt, obtained by fusing oxide of cobalt with ordinary glass, is largely employed, under the name of *smalt*, as a vitrifiable pigment. This coloration may readily be exhibited by adding a minute particle of any cobalt compound to a borax bead (§ 490) upon a loop of platinum wire, and again placing the bead either in the oxidizing or in the reducing flame of the blowpipe. Nickel, on the other hand, is used in the metallic state as an ingredient of various alloys, of which the alloy known as German silver, composed of copper, zinc, and nickel, is one of the most important. A whitish alloy, obtained by adding nickel to copper, is sometimes employed for coin of low denominations.

652. Both cobalt and nickel form protoxides (CoO and NiO), protochlorides, and protoxide salts, like those of iron, except that the protosalts of cobalt and nickel are far more stable than the salts of protoxide of iron; so that the protoxides of cobalt and nickel must be regarded as the principal oxides of these metals. Like iron, chromium, and the other metals of the family now under discussion, cobalt and nickel also unite with oxygen to form sesquioxides (Co_2O_3 and Ni_2O_3), and these sesquioxides, or at least the sesquioxide of cobalt, combine with bases to form salts; but these salts and the sesquioxides themselves are comparatively unstable bodies; they are far more easily decomposed

than compounds of the protoxides of cobalt and nickel, or than compounds of the sesquioxides of the other metals of the group. Hence, in the matter of nomenclature, the salts of the protoxides of cobalt or nickel take precedence of the salts of the sesquioxides. When, for example, nitrate of cobalt is spoken of, nitrate of protoxide of cobalt is the substance referred to; whereas when nitrate of iron or of chromium is mentioned, without further specification, we must infer that the nitrate of the sesquioxide is the substance meant. The use of terms in the loose manner referred to in the foregoing examples is of course always to be deprecated; but, in order to avoid the chance of being misunderstood, some chemists have extended to all metals having two salifiable oxides the use of the terminations *ous* and *ic*, which has been exemplified under iron by the terms ferrous and ferric oxides. Thus the terms cobaltous and cobaltic oxides, and nickelous and nickelic oxides have been applied by some writers to the oxides of cobalt and nickel; and there is at present a tendency to adopt and amplify this system of names; but they are as yet too little employed in the literature of science to find appropriate place in an elementary manual.

URANIUM.

653. Uranium is a rare metal, found in but few localities. It can be reduced from its chloride by means of hot potassium, but not from its oxide by means of hydrogen. Metallic uranium is of a steel-white color, and is somewhat malleable; it does not oxidize in air or in water at ordinary temperatures, but burns brilliantly when strongly heated in air. It dissolves in chlorhydric or sulphuric acid, with evolution of hydrogen, and, in general, is closely analogous to iron and manganese in its chemical behavior. The atomic weight of uranium is 120; its specific gravity is 18·4.

There are two principal oxides of uranium, capable of uniting with acids to form salts (a protoxide UrO, and a sesquioxide Ur_2O_3), and two other intermediate oxides, formed by the union of the proto- and sesquioxides in different proportions. The sesquioxide also plays the part of a weak acid towards strong bases. Uranium is never used as a metal; but compounds of it are somewhat extensively employed for coloring glass, and to a certain extent in photography also. Sesquioxide of uranium imparts a beautiful greenish-yellow color to glass, and the glass thus colored is to a high degree fluorescent; the protoxide, on the other hand, gives a fine black, highly esteemed for painting porcelain.

654. The salts of sesquioxide of uranium are remarkable in that they constitute an exception to the general rule, that, to form a normal salt, as many molecules of the acid are required as there are atoms of oxygen in the base employed. The normal sulphate of calcium, for example (§ 241), may be formed by the union of CaO and SO_3, and the normal sulphate of sesquioxide of iron is composed of Fe_2O_3 and $3SO_3$; but in sulphate of sesquioxide of uranium we find only Ur_2O_3,SO_3, and analogous formulæ express the composition of the nitrate and other salts of this oxide. But in spite of this peculiarity, uranium has many properties in common with the other members of the sesquioxide group of metals. One characteristic, for example, of this aluminum-iron group, which is shared by uranium, is that the sesquioxides are capable of uniting with acids, not only in the fixed and definite proportions requisite for the normal, crystallized salts already described, but also in very numerous indefinite proportions to form soluble basic compounds, incapable of crystallization for the most part, and solidifying in tough shining masses like gum when their solutions are allowed to evaporate spontaneously in the air. Nitrate of iron, for example, may be made as basic as the compound $8Fe_2O_3,N_2O_5$, and still be soluble in water; and between this limit, on the one hand, and that of the crystallized normal salt ($Fe_2O_3,3N_2O_5+18H_2O$) upon the other, sesquioxide of iron and nitric acid can combine chemically in every conceivable proportion. The compounds of sesquioxide of iron with other acids, and the nitrates and other salts of the sesquioxides of the other metals of the group, all behave in a similar way, the compounds of uranium being no exception to the rule. This tendency to form soluble, gummy, polybasic sesquisalts, so strikingly exhibited by the members of the group of elements now under discussion, is evidently one of those obscure manifestations of the chemical force which we have already met with when dis-

cussing the phenomena of solution (§ 49), and the law of multiple proportions (§ 76, end).

655. The most important point of difference between uranium and the other members of the sesquioxide group of elements is the fact, already alluded to, that one molecule of its sesquioxide unites with but one molecule of base to form crystallized salts, whereas the sesquioxides of the other members of the group all unite with acids in the proportion of one molecule of base to three molecules of acid to form their normal crystallized salts. Since the alums are formed by the union of a normal sulphate of some metal of the alkali group with a teracid sulphate of some metal of the sesquioxide group, there is no such thing as a uranium-alum, because the teracid uranium-sulphate is wanting. Sesquioxide of uranium, in fact, behaves among bases somewhat as metaphosphoric acid does among acids; it stands in much the same relation to the other teracid bases of its class, as metaphosphoric acid to the ordinary terbasic phosphoric acid.

656. *The Sesquioxide Group.*—The bond of union between the metals included in this class is the fact that they all form salifiable sesquioxides. Most of them form also salifiable protoxides; and if we arrange the metals in the order of their atomic weights,

$$Gl = 14, Al = 27{\cdot}4, Cr = 52{\cdot}5, Mn = 55, Fe = 56,$$
$$Ni = 58{\cdot}8, Co = 58{\cdot}8, Ur = 120,$$

it will be apparent that the sesquioxides of the metals at the head of the list are the most stable of the sesquioxides, and that the protoxides of nickel and cobalt are the most stable of the protoxides, while with manganese and iron both forms of oxide are well represented; uranium does not conform to this arrangement. Glucinum and aluminum have no protoxides at all; and the protoxide of chromium is very unstable. Some of the metals of the group are usually bivalent, others trivalent, while others are both bi- and trivalent.

The class of salts called alums affords strong evidence of the existence of a natural relation between the members of the alkali group, on the one hand, and the members of the sesquioxide group, on the other. These highly crystallized isomorphous salts are all moulded upon one pattern, and their atomic volumes (§ 252) are very nearly equal, as the following table will illustrate for some of the alums:—

Alums.	Atomic Weight.	Specific Gravity.	Atomic Volume.
$KAlS_2O_8,12H_2O$	474·5	1·722	275·6
$NaAlS_2O_8,12H_2O$	458·4	1·641	279·2
$(NH_4)AlS_2O_8,12H_2O$	453·4	1·621	279·6
$KCrS_2O_8,12H_2O$	499·6	1·845	270·7
$(NH_4)CrS_2O_8,12H_2O$	478·5	1·736	275·5
$(NH_4)FeS_2O_8,12H_2O$	482·0	1·712	281·4

It is a fact not unworthy of notice, that the compounds of this group of metals, with the exception of those of glucinum and aluminum, are for the most part colored, independently of the colors of the substances with which they are united. The metals of the sodium, calcium, and magnesium groups produce colorless compounds, unless when joined with an acid possessing a color of its own. Glucinum and aluminum produce, in like manner, colorless compounds; but the oxides, hydrates, chlorides, bromides, iodides, sulphides, and oxygen-salts of chromium, manganese, iron, nickel, cobalt, and uranium are all more or less colored in themselves, and every color of the spectrum, from the violet at one extremity to the red at the other, can be matched from among the innumerable tints exhibited by the various compounds of the last six members of the sesquioxide group.

657. With the members of the group now under discussion are commonly classed a number of rare metals, more or less nearly related to aluminum and iron. They are all, however, of subordinate interest, and need only be named in this manual. The following is a list of these elements, together with their symbols and their atomic weights, so far as the latter have been determined:—Yttrium, Yt, 61·6; Erbium, Er, 112·6; Terbium, Tb=(?); Zirconium, Zr=67·2; Norium, No=(?); Cerium, Ce=92; Lanthanum, La=92·8; Didymium, Di=95; Thorium, Th=231·5 (?).

CHAPTER XXXI

COPPER AND MERCURY.

COPPER.

658. Though by no means one of the most abundant metals, copper is nevertheless very widely diffused in nature, and is largely employed by man. Traces of it exist in almost every soil, whence it is taken up by plants, in which it may almost always be detected by refined testing. Traces of it have repeatedly been found also in the various animal organs and secretions. Many natural waters contain minute quantities of copper; its presence may often be recognized in the deposit of oxide of iron which separates from chalybeate waters. Since the metal occurs native in many localities, several of its valuable properties were early recognized and made use of. Long before the discovery of methods of reducing iron from its ores, tools and weapons made of native copper were employed by many barbarous nations.

Besides occurring in the native state, copper is found in a great variety of combinations; the most common of its ores, however, is the sulphide, or rather a compound of sulphide of copper and sulphide of iron in varying proportions, known as copper pyrites. The processes of obtaining copper from its ores vary greatly, according to the quality of the ore. The oxides and carbonates may be readily reduced by heating the ore in contact with some carbonaceous material and a flux suitable to remove the impurities of the ore. The treatment of ores containing sulphur is far more complicated. Such ores are roasted in the first place, in order to convert a considerable portion of the sulphides of copper and iron into oxides; a proper flux is then added to the roasted ore, and the whole is melted down in either a reverberatory or blast furnace. The oxide of copper formed by roasting is reconverted into sulphide, while much of the sulphide of iron which had escaped oxidation before is now changed to oxide and passes off in the slag. Sulphide of copper comparatively free from iron is thus obtained; in other words, the copper ore is very much concentrated by the operation. If need be, the concentrated product is subjected to a series of roastings and meltings, until it has been almost completely freed from sulphide of iron and other impurities. The pure or nearly pure sulphide of copper is then roasted in a current of air, until a certain proportion of the sulphide has been converted into oxide. Finally the mixture of sulphide and oxide is strongly heated to a temperature at which its ingredients react upon one another in such manner as to yield sulphurous acid and metallic copper:—

$$Cu_2S + 2CuO = SO_2 + 4Cu.$$

Sometimes copper is obtained by precipitating it with iron from solutions of its salts, as has been shown in Exp. 339. The copper thus thrown down by iron is known as cement-copper, and is frequently obtained from the drainage-water of certain mines, in which a small proportion of sulphide of copper is oxidized by the air to sulphate of copper, and so carried into solution. In some localities, low-grade copper-ores are lixiviated with chlorhydric acid, obtained as a waste product from the manufacturers of soda-ash, and the copper solution subsequently made to flow over fragments of scrap iron.

The common method of assaying copper-ores is another application of the precipitation of copper by means of iron.

659. Copper is a rather hard metal, of a well-known red color; it is very tenacious, ductile, and malleable. The specific gravity of the metal when free from air-bubbles varies between 8·92 and 8·95. Copper melts less readily than silver, but more readily than gold; its melting-point has been estimated to be about 1170°. At an intense heat it volatilizes, though for all ordinary purposes it may be regarded as non-volatile. It is one of the best conductors of heat and electricity known. Its specific heat is 0·09515. Copper combines with oxygen far less readily than iron. Even at a bright-red heat it is not capable of decomposing water, excepting to a very slight extent. Finely divided copper, however, soon becomes oxidized on being exposed to the air, though, as is well known, solid masses of the metal suffer little or no change, at the ordinary temperature, in air free from sulphydric and carbonic acids. When strongly heated in the air, copper quickly becomes covered with a coating of black oxide of copper (see Exp. 12). Metallic copper is not very readily acted upon by acids, excepting those rich in oxygen. The weaker acids, such as acetic acid, have no action upon it, unless air be present, in which event the metal is soon corroded;

and the same remark applies to dilute chlorhydric and sulphuric acids. Finely divided copper, however, slowly dissolves with evolution of hydrogen in hot concentrated chlorhydric acid; and in hot oil of vitriol the metal dissolves readily, as has been seen in the preparation of sulphurous acid. (See Exp. 96.)

Copper is readily soluble in somewhat diluted nitric acid, such as is commonly found in commerce (see Exp. 37); but the strongest nitric acid, of specific gravity 1·52, does not act upon it. When immersed in such acid the metal remains bright, and no bubbles of gas arise from its surface. The phenomenon is explained by the fact that nitrate of copper is insoluble in monohydrated nitric acid, though readily soluble in water and in dilute acid. Ammonia-water and many salts, such as chloride of sodium and the various salts of ammonium, corrode copper rather rapidly when in contact with air. Finely divided copper takes fire in chlorine gas; and at a red heat the metal unites directly with bromine, iodine, sulphur, silicon, and the various metals. It does not appear to unite directly with carbon or with nitrogen at any temperature.

Several of its compounds with other metals are of great importance in the arts. Brass and the yellow metal used for sheathing ships are alloys of zinc and copper; bronze, gun-metal, and bell-metal are alloys of tin and copper; and various *compositions* are produced by mixing these alloys with brass. German silver in its various forms is an alloy of nickel, zinc, and copper; and copper is an essential ingredient of all the common coins, implements, and ornaments of gold and silver.

660. *Dioxide of Copper* (Cu_2O) is sometimes found in nature as *Ruby copper*; it may readily be obtained by heating protoxide of copper with finely divided metallic copper, or other reducing agents; so, too, when masses of metallic copper are gently heated in the air, they become covered with a thin film of the dioxide.

Exp. 346.—Dissolve in a test-tube a few drops of honey or a bit of grape-sugar in a little water. Add to the solution two or three drops of a rather dilute solution of sulphate of copper, and then pour in enough soda-lye to redissolve the precipitate which is at first produced by the lye.

Slowly heat the clear blue solution, and observe that a yellow precipitate of hydrated dioxide of copper soon separates, first at the uppermost part of the column of liquid, but soon in all parts of the tube, as its contents become sufficiently hot. When the liquor is heated to boiling, the hydrated yellow precipitate changes after a time to anhydrous red dioxide.

Most of the dilute acids decompose dioxide of copper with formation of salts of the protoxide and separation of metallic copper. But it dissolves in concentrated chlorhydric acid and in ammonia-water, forming colorless solutions. The ammoniacal solution may be employed as a test for the presence of oxygen in any mixture of gases; oxygen is immediately absorbed by the solution and a compound of protoxide of copper and ammonia (Exp. 352) of characteristic deep blue color is formed. Dioxide of copper is employed to a certain extent for coloring glass ruby-red.

661. *Protoxide of Copper* (CuO) may be prepared by heating the metal or the dioxide in a current of air, or by igniting carbonate, hydrate, or nitrate of copper.

Exp. 347.—Bind a bright copper coin with wire, in such manner that a strip of wire 8 or 10 c.m. long shall be left projecting from the coin; thrust the free end of the wire into a long cork or bit of wood, and by means of this handle hold the coin obliquely in a small flame of the gas-lamp. A beautiful play of iridescent colors will appear upon the surface of the copper, particularly if it be moved to and fro. Thrust the hot coin into water, and observe that it is at this stage covered with a coating of red suboxide of copper. Replace the coin in the lamp and hold it in the hot oxidizing portion of the flame (Exp. 195); it will soon become black from the formation of protoxide of copper. After a rather thick coating of oxide has been formed, again quench the coin in water; the black coating or scale of oxide will fall off, and beneath it will be seen a thin film of the dioxide firmly adhering to the metal. This film of dioxide is intentionally produced upon the surfaces of many copper implements by the manufacturers. If the coin were heated long enough it would all be converted, first into the red dioxide, and then into black protoxide of copper. The scales which fall off when hot metallic copper is beaten or rolled, like those obtained from the coin in this experiment, always consist of a mixture of the two oxides.

Exp. 348.—Evaporate to dryness in a porcelain dish upon a sand-bath some of a solution of nitrate of copper prepared from copper as in Exp. 37. There will be left as a residue a green basic nitrate of copper. Place a small quantity of this residue upon a fragment of porcelain, and ignite it until red nitrous fumes are no longer given off. Pure protoxide of copper will be left upon the porcelain.

Though no oxygen can be expelled from protoxide of copper by mere exposure to heat, all its oxygen may, nevertheless, be removed with great facility by means of reducing agents. Oxide of copper is, in fact, one of the most convenient oxidizing agents in the chemist's possession, and is largely employed to this end in the analysis of organic compounds. When heated with carbonaceous substances, it converts all their carbon into carbonic acid; and, in like manner, hydrogen is immediately oxidized by it and converted into water. Since carbonic acid and water can readily be collected and weighed, and since their composition is accurately known, the determination of the amounts of carbon and hydrogen in any substance, through the agency of oxide of copper, is merely a matter of mechanical detail.

Exp. 349.—Repeat Exp. 330, with the exception that a teaspoonful of black oxide of copper (Exp. 348) is placed in the tube instead of the iron rust. Water may be collected in the U-tube as before, and metallic copper will be left in the reduction-tube. But, unlike the easily oxidizable iron, the reduced copper will not take fire in the air.

662. Protoxide of copper is soluble in most acids, with formation of salts which are blue or green when hydrated, but white when thoroughly dried. From the solutions of most of these salts hydrated oxide of copper may be precipitated by means of any of the strong soluble bases.

Exp. 350.—Place in a test-tube, or small bottle, 8 or 10 c. c. of a cold dilute solution of sulphate of copper, and add to it enough of a solution of caustic soda to render the mixture alkaline to test-paper. A light-blue precipitate will fall; hydrate of copper is insoluble in water and in soda-lye.

Exp. 351.—Repeat Exp. 350, with the difference that the solutions of caustic soda and sulphate of copper are both heated to boiling and are mixed while hot. Instead of the blue hydrate, black protoxide of copper will now be thrown down; for hydrate of copper readily parts with its water when heated, even if it be all the while immersed in water; it does not again combine with water after it has become cold. Instead of mixing boiling solutions of the alkali and copper salt, the moist precipitated hydrate of copper of Exp. 350 might be changed to black oxide by simple boiling; but the transformation would be, comparatively speaking, slow, and the experiment less striking than the one here described.

Exp. 352.—Again repeat Exp. 350, but instead of soda-lye add to the copper salt ammonia-water, drop by drop, and shake the tube after each addition of the ammonia. Hydrate of copper will be precipitated as before, in accordance with the reaction

$$CuSO_4 + 2(NH_4)HO = (NH_4)_2SO_4 + CuH_2O_2;$$

for, as has been said, this hydrate is insoluble in water; but, since hydrate of copper is readily soluble in ammonia-water, the precipitate will redissolve as soon as more of this agent than is needed to decompose the copper salt is added. The ammoniacal solution of copper has a magnificent azure-blue color.

663. *The Sulphides of Copper* (Cu_2S and CuS) are interesting from their occurrence as ores, and from the reactions already briefly explained, which are so important in the industry of copper-smelting (§ 658). The protosulphide, as obtained by adding sulphydric acid to acidulated solutions of the salts of copper, is an important substance to the analyst; it is a black powder, insoluble in water, in dilute acids, and in alkaline lyes.

664. *Dichloride of Copper* (Cu_2Cl_2) may be obtained by treating a mixture of protoxide of copper and finely divided metallic copper with concentrated chlorhydric acid, or by boiling a solution of the protochloride with sugar or some other reducing agent. It is a white compound, insoluble in water, but soluble in strong chlorhydric acid, and in aqueous solutions of chloride of sodium, chloride of potassium, and many other chlorides.

665. *Protochloride of Copper* ($CuCl_2$) is formed when copper is burned in an excess of chlorine. It may readily be prepared in the hydrated condition ($CuCl_2 + 2H_2O$) by dissolving oxide, hydrate, or carbonate of copper in chlorhydric acid, and evaporating the solution upon a water-bath. Anhydrous chloride of copper is brown; but the hydrated salt forms green, needle-shaped crystals. The concentrated aqueous solution is green; when diluted with water it becomes blue, but turns green again on being boiled. The dry salt fuses when heated, and at a red heat gives off half its chlorine and is changed to the dichloride. Chloride of copper is soluble in alcohol and ether; if some of the alcoholic solution

is poured upon a tuft of cotton and then ignited, it will burn with a green flame, which is characteristic of copper.

666. *Sulphate of Copper* ($CuSO_4$) has been already obtained in solution as a secondary product in Exp. 96. It may also be readily prepared by dissolving oxide of copper, copper-scale for example, in moderately dilute sulphuric acid. Much of it was formerly prepared in this way for use in the arts; it is an incidental product also in the process of refining gold and silver, and in certain metallurgical operations. As it crystallizes from aqueous solutions, sulphate of copper holds in chemical combination 5 molecules of water, and may then be represented by the formula $CuSO_4+5H_2O$. This hydrated salt, known as blue vitriol, is the commonest salt of copper; most of the various pigments and other preparations of copper, medicinal or chemical, are made from it; and it is itself used to a considerable extent by dyers and calico-printers, and largely by the electrotypers.

Exp. 353.—Tie a piece of bladder over one end of a lamp-chimney, or over the mouth of a wide-mouthed bottle or beaker, off which the bottom has been somewhat evenly broken. Solder a piece of thick copper wire to a strip of stout sheet zinc, just wide enough to enter the chimney or bottle, and a little longer than the bottle is deep. Place the zinc in the bottle or chimney, and sink the bottle or chimney, with the closed end down, in a beaker or large tumbler containing a strong solution of sulphate of copper. Fill the bottle or chimney with di ute nitric acid, attach to the copper wire a clean medal or coin, of which one side has been varnished, and the other rubbed over with plumbago, and bend the wire so that the medal or coin may be immersed in the sulphate-of-copper solution contained in the beaker or tumbler. Thorough contact must be secured between the copper wire and the unvarnished side of the coin or medal. In a few hours a coherent film of solid, malleable copper will be firmly deposited on that face of the coin or medal which was not protected by the non-conducting varnish. The shell of copper may be readily detached from the coin or medal, the plumbago ensuring the ready separation of the two metallic surfaces; it is a perfect reverse of the object copied.

This experiment illustrates on a small scale the important art of electro-metallurgy. Plating in gold and silver, as well as in copper, is extensively performed by a process perfectly analogous to that of this experiment. Woodcuts, type, medals, maps, and engravings are accurately copied by means of this deposition of metals from their solutions under the action of the galvanic current.

It is remarkable that the blue color of sulphate of copper depends upon the presence of water.

Exp. 354.—Heat 1 c. c. of powdered blue sulphate of copper upon a piece of porcelain; as it loses its water, the light-blue powder will turn white. A drop of water upon the anhydrous powder will restore its blue color.

If concentrated sulphuric acid be poured over the blue crystals, it will abstract water from them, and a quantity of the white anhydrous salt will be formed.

Since the protoxides of iron and of copper are isomorphous, or rather since the metals iron and copper are capable of replacing one another in many of their compounds, it is not surprising that the sulphates of iron and copper should crystallize together to form a compound which may contain almost any proportion of either salt. This isomorphous mixture of the two salts was formerly largely employed in the arts, and is still somewhat used to meet the requirements of old receipts for dyeing. In a similar way, sulphate of copper crystallizes together with the sulphates of nickel, cobalt, zinc, and magnesium.

667. *Nitrate of Copper* (CuN_2O_6) may be obtained crystallized, by allowing the blue solution, obtained in Exp. 37, to evaporate spontaneously in dry air. It is interesting as an example of a salt ready to give up oxygen on slight provocation. If a piece of tin-foil about 20 c.m. square be twisted firmly around a rather large crystal of nitrate of copper, then pierced in several places with a needle, and moistened with water, or with a few drops of common spirits of wine, a powerful reaction will soon ensue. The tin will be oxidized, at the expense of the oxygen of the nitrate of copper, much heat will be evolved, and smoke, or even flame, produced.

668. *Acetates of Copper* are formed by the action of acetic acid upon metallic copper, exposed to the air. They are commonly called *verdigris*. Purified verdigris is the normal acetate of copper; and common verdigris is a hydrated basic acetate. Verdigris is usually prepared by packing plates of copper between woollen cloths steeped in vinegar; but sometimes, in wine-growing countries, the refuse of the wine-press is suffered to ferment

in contact with the copper plates. From time to time the verdigris is removed from the surface of the copper, the plate of metal being again and again subjected to the action of acetic acid so long as any of it remains.

In ordinary language, the term verdigris is often incorrectly applied to the green coating of carbonate of copper, which forms upon copper long exposed to damp air, or to the rust formed upon copper by the combined action of air and almost any acid. A compound of acetate of copper and arsenite of copper constitutes the beautiful and vivid green color known as Schweinfurt green.

MERCURY.

669. Small globules of metallic mercury are sometimes found in nature; but the principal ore of this metal is the sulphide (HgS), called cinnabar. From this sulphide the metal is readily extracted, by distilling a mixture of it and quicklime, or iron-turnings, in cast-iron retorts. The sulphur is retained by the lime or iron, as the case may be, while metallic mercury passes off in the state of vapor into receivers containing water, beneath which it condenses to the liquid state.

A rougher method of manufacture is to heat the coarsely broken sulphide on a perforated brick arch, by a quick fire of brush-wood; the sulphur in the ore is kindled, and, by its combustion, maintains the heat necessary to continue the distillation. The liberated mercury is condensed in wide and long earthen pipes, which slope first down, and then up.

670. At the ordinary temperature of the air, mercury is a brilliant, mobile liquid, of 13·6 specific gravity, which vaporizes slowly, even at ordinary temperatures, and boils at about 360°. The vapor-density of mercury does not coincide with its atomic weight. As is the case with the metal cadmium (§ 599), the atomic weight of mercury is the weight of two unit-volumes of its vapor, and is therefore double the vapor-density, instead of identical with it. Into the product-volume of any compound of mercury, one more volume is condensed than would be contained in the product-volume of a corresponding compound of oxygen or sulphur. The atomic weight of mercury is 200, and *two* unit-

volumes of its vapor weigh 200 times as much as one unit-volume of hydrogen; its vapor-density should, theoretically, be 100; and the results of experiment approximate closely to this number. (Compare § 259.)

When cooled to —39°·4 mercury freezes. In solidifying, liquid mercury contracts considerably, and there results a ductile metal of tin-white color and granular fracture, which may be cut with a knife. Perfectly pure mercury undergoes no change in air or in oxygen gas at the ordinary temperature, even when shaken about in the gas for a long while; but if mercury containing traces of foreign metals, such, for example, as that ordinarily met with in commerce, be exposed to the air, a gray pulverulent coating will, after a while, appear upon its surface. This coating is composed of oxides of the contaminating metals mixed with finely divided metallic mercury. A similar gray powder of finely divided mercury may be obtained by triturating mercury with sulphur, tallow, and a variety of other substances, or simply by shaking it with water or oil of turpentine. When heated in the air to temperatures near its boiling-point, even pure mercury absorbs oxygen, and is converted into the red oxide (§ 672). Metallic mercury combines directly with chlorine, bromine, iodine, and sulphur.

Chlorhydric acid has no action upon mercury, not even when it is hot and concentrated. Dilute sulphuric acid has scarcely any action upon it; but hot concentrated sulphuric acid converts it into solid sulphate of mercury, while sulphurous acid is evolved (Exp. 96). Nitric acid, even when dilute, dissolves it easily.

Large quantities of mercury are used in extracting gold and silver from their ores, for silvering mirrors, and in the process of fire-gilding. Preparations of mercury are employed also as medicaments, and for various purposes in the useful arts. The fluidity of the metal makes it valuable in the construction of certain philosophical instruments, of which the thermometer and barometer are familiar examples.

There are two oxides of mercury—a black dioxide, and the red protoxide; each of these oxides unites with acids to form a peculiar class of salts.

671. *Dioxide of Mercury* or *Mercurous Oxide* (Hg_2O) is best

prepared by decomposing one of its salts, calomel for example (§ 674), by means of caustic soda. Though a rather powerful base when in combination, it decomposes readily when in the free state, mere exposure to light or to gentle heat being sufficient to decompose it into metallic mercury and the red oxide. With acids it combines readily, with formation of salts of the dioxide of mercury, or mercurous salts as they are often called.

672. *Protoxide of Mercury* or *Mercuric Oxide* (HgO) may be prepared by heating metallic mercury in the air as above described, or, better, by heating nitrate of mercury at a temperature high enough to drive off oxides of nitrogen, but at the same time too low to decompose the oxide of mercury, or, again, by precipitating the solution of a salt of protoxide of mercury by means of a caustic alkali. As obtained by the first two methods, it is a compact, granular, almost crystalline, glistening powder, of bright brick-red color; but when prepared by the last method it is, when dry, a soft, light-orange-colored powder. Very considerable differences in the chemical deportment of these red and yellow varieties of the oxide have been noticed, though the differences are hardly so great as are usually found between the isomeric modifications of other substances. The precipitated yellow oxide is, for example, more readily decomposed by heat and by chlorine than the red oxide. The red oxide, however, is the substance known as oxide of mercury in commerce and the laboratory.

Considered as a source of oxygen, red oxide of mercury is peculiarly interesting, since by means of it oxygen may be derived directly from the air (§ 12); but it neither affords the gas cheaply, nor yields an abundant supply. Since red oxide of mercury contains only a single atom of oxygen for each atom of mercury, and since the atomic weight of mercury (200) is comparatively high, any given weight of the oxide can, of course, contain but a small proportion of oxygen:—for

216	:	16	=	1	:	0·074
Weight of a molecule of HgO.		Weight of an atom of O.		Grm.		Grm.

Although it contains so small a proportion of oxygen, compared with the nitrates and chlorates commonly employed for effecting oxidation, yet, from the facility with which it gives up its oxygen, mercuric oxide is still an oxidizing agent of considerable power. If a small portion of it be mixed with a little sulphur, and then heated, the mixture will explode; so, too, if it be mixed with a small fragment of phosphorus, and struck with a hammer upon an anvil, a similar explosion will ensue; the violent action depends upon rapid oxidation in both cases.

Most acids unite readily with oxide of mercury to form salts, often spoken of as mercuric salts. Both the oxides of mercury are, like the protoxide of lead (§ 575), remarkable for the facility with which they form basic salts. In general, the compounds of mercury unite with one another readily to form a great variety of double compounds and abnormal basic salts, such as the oxychlorides xHgO,$HgCl_2$, and the chlorosulphide 2HgS,$HgCl_2$. The properties of several of these complex substances are interesting; but none of them fairly fall within the scope of this manual.

673. *Disulphide of Mercury* (Hg_2S) is a compound nearly as unstable as the dioxide; but the *protosulphide* (HgS) is a permanent substance of considerable importance in the arts. Artificially prepared for use as a pigment, it is known under the name of vermilion. It is the most important ore of mercury, as has been already stated (§ 669).

674. *Mercurous Chloride* (HgCl), commonly called *calomel*, is extensively used as a medicament.

It may be prepared either by heating together a mixture of metallic mercury and corrosive sublimate (§ 675) until the dichloride sublimes,

$$Hg + HgCl_2 = 2HgCl,$$

or by subliming an intimate mixture of equal parts of sulphate of dioxide of mercury and common salt,

$$Hg_2SO_4 + 2NaCl = 2HgCl + Na_2SO_4.$$

By the way of precipitation it may be made by mixing together solutions of common salt and nitrate of dioxide of mercury:—

$$Hg_2N_2O_6 + 2NaCl = 2HgCl + Na_2N_2O_6.$$

In place of the corrosive sublimate used in the first method, intimate mixtures of sulphate of protoxide of mercury and common salt, or of common salt and black oxide of manganese or sulphate of sesquioxide of iron may be substituted.

Calomel is a heavy white powder, which volatilizes at temperatures below redness without previous fusion.

The vapor-density of calomel is, by calculation, 117·75.

Weight of one atom, or two unit-volumes, of mercury .	200·00
Weight of one atom, or one unit-volume, of chlorine .	35·5
Weight of two unit-volumes of mercurous chloride . .	235·5
Weight of one unit-volume of mercurous chloride . .	117·75

The vapor-density has been determined by experiment at 120·49. By slowly cooling its vapor, prismatic crystals of calomel may readily be obtained. Unlike metallic mercury, calomel does not volatilize at ordinary temperatures. It is tasteless, odorless, and as good as insoluble in water. Alkaline lyes decompose it readily, and it is slightly soluble in many saline solutions.

675. *Mercuric Chloride* ($HgCl_2$), better known by the name *corrosive sublimate,* may be prepared by burning mercury in an excess of chlorine gas, by dissolving protoxide of mercury in chlorhydric acid, or by dissolving metallic mercury in aqua regia, and evaporating to dryness.

In practice, it is commonly prepared by sublimation, by carefully heating an intimate mixture of sulphate of mercury and common salt:—

$$HgSO_4 + 2NaCl = HgCl_2 + Na_2SO_4.$$

Another method is, to add concentrated chlorhydric acid to a strong boiling solution of nitrate of dioxide of mercury, as long as a precipitate is formed, and to subsequently boil this precipitate with a quantity of chlorhydric acid equal to that used in preparing it:—

$$HgNO_3 + 2HCl = HgCl_2 + H_2O + NO_2.$$

Beautiful crystals of mercuric chloride will be deposited as the hot solution cools.

676. Mercuric chloride commonly occurs in commerce, in translucent crystalline masses; but crystals of it may readily be obtained, by careful sublimation, as well as by slowly cooling hot solutions. It melts at about 265°, forming a colorless liquid which boils at 293°; the fumes are acrid, and, like the salt itself, exceedingly poisonous.

The vapor-density of corrosive sublimate is, by calculation, 135.

Weight of one atom, or two unit-volumes, of mercury . .	200
Weight of two atoms, or two unit-volumes of chlorine . .	70
Weight of two unit-volumes of mercuric chloride	270
Weight of one unit-volume of mercuric chloride	135

The best experiments assign to the vapor of the salt the density of 141. Mercuric chloride is rather easily soluble in water and alcohol; with the alkaline chlorides it unites to form salts which are easily soluble in water. These double salts are so numerous and well defined, that they are regarded as chlorine salts comparable with the sulphur salts (§ 340) and the oxygen salts. In this view, protochloride of mercury would be called chloromercuric acid, and its compounds with the alkalies chloromercurates. The compound $NaCl,HgCl_2$, for example, would be called chloromercurate of sodium, and the compound $NaCl,2HgCl_2$ the bichloromercurate. Corresponding double chlorine salts are formed by the union of the chlorides of gold and of platinum with the chlorides of other metals and compound radicals, as will appear in the sequel.

677. Mercuric chloride unites with many organic substances to form compounds insoluble in water and imputrescible. It coagulates albumen, for example, and the more perishable portions of wood; hence the employment of raw whites of eggs as an antidote in cases of poisoning by corrosive sublimate, and the use of the mercury salt for preserving wood,—a purpose for which it would, no doubt, be largely employed were it not for its high cost. Collections of dried plants, and of other objects of natural history, are preserved both from decay and from the attacks of insects by brushing over them a solution of the chloride in alcohol. It is worthy of mention that the compound of albumen and chloride of mercury, though insoluble in water, is soluble in an excess of albumen.

Mercurous and mercuric bromides, iodides, fluorides, and cyanides are, in general, analogous to the corresponding chlorides. Mercuric iodide undergoes remarkable changes of color when heated or subjected to friction.

Exp. 355.—Dissolve half a gramme of iodide of potassium in a small quantity of water; also dissolve 0·4 of a gramme of corrosive sublimate in a little water, and mix the two solutions. Collect upon a filter the beautiful red precipitate which is formed, wash it carefully with water and dry it in the air. Place a portion of the dry red powder in a porcelain capsule; invert over the capsule a small glass funnel, and heat the capsule moderately upon a sand-bath; the iodide will melt, sublime, and finally be deposited upon the cold walls of the funnel in yellow crystals. On rubbing these crystals with a glass rod, their color will change again to red. Indeed the change of color often occurs of itself as the crystals cool, without friction. The composition of the iodide is neither changed by the sublimation nor by the friction; the change of color is due to a change of crystalline form—mercuric iodide being dimorphous, and exhibiting a red color in its octahedral form, and a yellow color when crystallized in rhombic prisms.

The change of coloration may be shown in another way, by dissolving some of the precipitated iodide in alcohol. The alcoholic solution is colorless and appears to contain the yellow modification of the iodide; on pouring it into water, iodide of mercury is precipitated as a yellow powder, which soon changes to red.

678. *Sulphates of Mercury.*—There is a sparingly soluble sulphate of dioxide of mercury (Hg_2SO_4), a normal sulphate of the protoxide ($HgSO_4$), and a basic sulphate of the protoxide (of composition $3HgO,SO_3$). Normal mercuric sulphate may be prepared by dissolving metallic mercury in an excess of boiling concentrated sulphuric acid, and evaporating the solution to dryness. It is the material from which many other compounds of mercury are derived. It is decomposed by water; an insoluble trisulphate is thrown down, while but a small proportion of mercury remains dissolved in the dilute sulphuric acid which is formed.

679. *Nitrates of Mercury* are numerous. There are at least four nitrates of the dioxide, and as many of the protoxide namely the normal salts and three basic salts in either case. Both of the normal salts are soluble in water, and are commonly kept in the laboratory as examples of the mercury salts. The nitrate of the dioxide is prepared by digesting an excess of metallic mercury in cold moderately strong nitric acid. The solution should be kept in closed bottles containing a few globules of metallic mercury. The nitrate of the protoxide may be readily obtained by dissolving red oxide of mercury in an excess of nitric acid.

680. *Amalgams.*—Mercury unites with most of the other metals, forming alloys, many of which are pasty, or liquid when the proportion of mercury contained in them is large. These

alloys are commonly called amalgams, in contradistinction to the ordinary solid alloys of the other metals, in which mercury has no place. The liquid amalgams are true solutions of other metals, or of solid amalgams, in the fluid mercury. The so-called silvering of mirrors is an amalgam of tin.

Mercury may be detected in almost any soluble salt of the element by introducing into a solution of the salt a piece of clean copper.

Exp. 356.—Place a drop of a solution of either of the nitrates or chlorides of mercury upon a clean copper coin and rub the liquid over its surface. A white coating of metallic mercury will be deposited upon the metal.

681. Copper and mercury are classed together partly because of certain resemblances between the two metals, but also because neither of them falls naturally into either of the other groups of elements. They are alike in that they are not readily acted upon by air, excepting at high temperatures, that they do not decompose water at any temperature, and that they both form two salifiable oxides, and two chlorides of analogous composition. They are both acted upon in the same way by nitric and by sulphuric acids, the acid being reduced to a lower degree of oxidation, while the metal is dissolved, as has been seen in Exp. 96. As the formulæ of their compounds have doubtless already suggested, mercury and copper are univalent, like the alkali-metals, in the mercurous and cuprous compounds, but bivalent in the mercuric and cupric compounds.

CHAPTER XXXII.

TITANIUM—TIN.

TITANIUM.

682. This comparatively rare metal is found in several minerals, such as rutile and titaniferous iron, in the condition of titanic acid, TiO_2. None of its compounds are employed in the arts, and

the element itself is here mentioned mainly on account of the analogies which it bears to tin. Titanic acid is isomorphous with stannic acid (§ 685), and resembles it closely in its chemical deportment. Sesquioxide of titanium, Ti_2O_3, corresponds to sesquioxide of tin, Sn_2O_3 (§ 685); and in the same way that the latter may be regarded as a stannate of tin, SnO,SnO_2, the titanium compound may be considered a titanate of titanium, TiO,TiO_2.

The bisulphide of titanium, and the bichloride, bibromide, and bifluoride, correspond in like manner to the tin compounds.

TIN.

683. Though by no means widely diffused in nature, and though ores of it occur in but few localities, tin is one of the metals which have longest been known to man. The fact admits, however, of ready explanation; for the specific gravity of the ores is high, and the metal is easily reduced from them by simple heating with charcoal. From the manner of the occurrence of many of these ores, in the beds of torrents, it is evident that their great weight would be likely to attract attention, and that their behavior towards fire would soon be noticed. The simplest possible metallurgical operation, and the one most likely to suggest itself to savage men, is the heating of a heavy stone in a wood fire.

The principal ore of tin is the binoxide, called tin-stone. In order to extract the metal from it, the tin-stone is mixed with powdered coal and heated upon the hearth of a reverberatory furnace in a reducing flame. The reduced metal melts readily, and is then run out of the furnace into iron moulds. Tin is a lustrous white metal, soft, malleable and ductile, though not very tenacious. Its ductility varies greatly with the temperature; at 100° the metal may be drawn into thin wire, but at 200° it is very brittle. When a bar of tin is bent, it emits a peculiar crackling sound, and if the bending be repeated, the metal becomes decidedly warm. These phenomena appear to depend on the disturbance of interlaced crystals contained in the bar, and upon the friction of these crystals one against the other. Tin always exhibits a great tendency to assume crystalline form in passing from the liquid to the solid condition. Upon this peculiarity is founded a method of ornamenting tinned iron.

Exp. 357.—Heat a piece of common tinned iron over the gas-lamp until the tin has melted, thrust the plate into cold water in order that the tin may harden quickly, then remove the smooth surface of the metal by rubbing it first with a bit of paper moistened with dilute aqua regia, and then with paper wet with soda-lye. By this treatment there will soon be laid bare a new surface covered with beautiful crystalline figures, like frost upon a window-pane. The plate should then be washed thoroughly with water, dried quickly, and covered with some transparent varnish.

The same crystalline structure can be brought out, though less conspicuously, by removing the outside polished surface of almost any piece of tin plate by means of warm dilute aqua regia, without first heating the plate as in this experiment.

Tin does not tarnish in the air at ordinary temperatures, no matter whether the air be moist or dry; but when strongly heated it oxidizes rapidly, and even burns with a brilliant white light. The specific gravity of tin is about 7·3; its atomic weight is 118. It melts at about 230°—at a lower temperature than any of the other common metals. At very high temperatures it is slightly volatile.

On account of its brilliant lustre, and its power of resisting atmospheric action, tin is largely employed for coating other metals,—copper, for example, as in ordinary pins, cooking-vessels, and bath-tubs—and iron, as in common sheet-tin, of which the so-called tin ware is manufactured.

Exp. 358.—Thoroughly clean the surface of a copper coin, or of a small piece of sheet-copper, by means of dilute sulphuric acid; place the copper over the gas-lamp, and melt upon it a bit of tin as large as a pea. Rub the melted tin over the copper with a rag. It will not adhere to the copper; for although the latter was once carefully cleaned, it afterwards became coated with oxide of copper in such manner that the tin could not come in contact with the metal.

Repeat the experiment as before; but when the tin has melted, strew over the copper some finely powdered chloride of ammonium. On now rubbing the tin against the copper, the two metals will adhere firmly. The chlorine of the chloride of ammonium unites with the copper of the oxide of copper to form fusible, volatile chloride of copper, while ammonia and water are set free, as may be perceived by the odor. The excess of tin should be wiped off with a rag, so that a smooth surface may be left upon the coin.

When in contact with dilute acids, or with alkalies, tin slowly absorbs oxygen from the air and goes into solution. Of the strong acids, nitric acid acts upon it violently, with formation of insoluble hydrated binoxide of tin; a certain amount of water is decomposed as well as the nitric acid, in this reaction, and some nitrate of ammonium is formed; the ammonium comes from the union of the nascent hydrogen and nitrogen of the deoxidized water and nitric acid (see § 92). Hot concentrated chlorhydric acid gradually dissolves tin, and gives off hydrogen. Boiling concentrated sulphuric acid converts it into sulphate of tin, with evolution of sulphurous acid; but dilute sulphuric acid has no action upon it out of contact with the air. When heated with concentrated soda or potash lye, tin slowly dissolves, with formation of soluble stannate of sodium or of potassium, and evolution of hydrogen.

There are two prominent oxides of tin, a protoxide and a binoxide, besides intermediate oxides compounded of these two. The binoxide occurs, moreover, in combination, in different isomeric modifications.

684. *Protoxide of Tin* (SnO) may be obtained as a black powder by heating its hydrate in an atmosphere of carbonic acid or other inert gas. The hydrated protoxide is prepared by adding the solution of an alkaline carbonate to a solution of tin in chlorhydric acid; the hydrate is thrown down as an insoluble precipitate while carbonic acid escapes:—

$$SnCl_2 + Na_2CO_3 + H_2O = SnH_2O_2 + 2NaCl + CO_2.$$

The hydrate rapidly absorbs oxygen from the air when moist, but is tolerably permanent when dry. The anhydrous oxide undergoes no change in air at the ordinary temperature; when touched with a glowing coal, it takes fire and burns vividly, being converted into the binoxide. The hydrate also burns in the same way when lighted. It is remarkable that the anhydrous oxide is more readily soluble in acids than the hydrate; but in alkaline lyes only the hydrate is soluble. Most of the salts of protoxide of tin greedily absorb oxygen from the air and from many oxygenated substances. They are much employed as reducing agents.

Exp. 359.—To 5 or 6 c. c. of a solution of corrosive sublimate add a few drops of protochloride of tin, and heat the mixture; a gray powder will separate; it is metallic mercury, very finely divided, which has been reduced from the mercuric chloride. The powder boiled with chlorhydric acid agglomerates into visible globules:—

$$HgCl_2 + SnCl_2 = SnCl_4 + Hg.$$

The protochloride of tin has abstracted the chlorine from the chloride of mercury.

685. *Binoxide of Tin*, or *Stannic Acid* (SnO_2).—This oxide occurs in nature as the principal ore of tin, as has been stated in § 683, and may readily be prepared artificially by roasting the metal in a free current of air, or by igniting hydrated binoxide of tin. It is insoluble in water, acids, and alkalies, and in general is not readily acted upon by chemical agents. When fused with caustic soda, however, it combines with it to form a soluble compound. Like the hydrated oxides of phosphorus and antimony, hydrated binoxide of tin (SnH_2O_3) is remarkable for the different chemical properties it exhibits when prepared in different ways. As obtained by heating metallic tin with concentrated nitric acid, it is almost absolutely insoluble in some acids, and dissolves only with difficulty in others. The hydrate obtained by precipitation from solutions of bichloride of tin and of an alkaline carbonate, is very readily soluble in acids. By ignition, the soluble hydrate is converted into insoluble anhydrous oxide of tin.

The difference in chemical behavior above mentioned is not confined to the hydrates alone, it exists as well in the compounds formed by the union of these hydrates with other substances, both acids and bases; hence the names stannic acid, applied to the soluble hydrate, and metastannic acid, applied to the insoluble modification. The generic terms stannate and metastannate applied to the compounds of these two varieties of the oxide are employed precisely as in the case of phosphoric and antimonic acids (§§ 293, 352). Both modifications are soluble in alkaline lyes, but the one representing metastannic acid is far less readily soluble than the other.

The stannates of the alkali-metals, sodium and potassium, crystallize readily from their aqueous solutions; but the corresponding metastannates do not crystallize, they are insoluble in saline solutions, and may be precipitated as gelatinous masses by adding almost any neutral salt of an alkali-metal to their aqueous solutions. Among the stannates, that of tin ($SnO,SnO_2=Sn_2O_3$) is worthy of mention, since it is often described as the sesquioxide of tin. Stannate of sodium is somewhat extensively employed in the printing of mousselines-de-laine.

A good method of preparing it is to boil granulated tin, or scraps of tinned iron, in a solution of litharge in an excess of caustic soda:—

$$Sn + Na_2Pb_2O_3 = Na_2SnO_3 + 2Pb.$$

Or a mixture of caustic soda, nitrate of sodium, and metallic tin may be melted in an iron kettle, a certain portion of chloride of sodium, or of stannate of sodium, from a previous fusion, being added to the mixture in order to mitigate the force of the reaction.

686. The *Sulphides of Tin* (SnS, Sn_2S_3, and SnS_2) correspond to the oxides. The protosulphide (SnS) is precipitated as a dark-brown powder when sulphydric acid is added to the solution of a salt of protoxide of tin. The bisulphide (SnS_2) when prepared in the dry way is a beautiful yellow compound, known as *mosaic gold*, or bronze powder, and is somewhat employed in decorative painting.

Exp. 360.—Prepare a quantity of tin amalgam by heating together in a glass flask 12 grms. of granulated tin and 6 grms. of mercury. Rub the amalgam in a porcelain mortar together with 7 grms. of sulphur and 6 grms. of chloride of ammonium until the different ingredients have been thoroughly incorporated one with the other. Place the mixture in a small, long-necked, glass flask, and slowly heat it to low redness upon a sand-bath. After an hour or two there will be found at the bottom of the flask a quantity of bisulphide of tin, in the condition of soft, beautiful, golden-yellow powder, of flaky texture, while in the neck of the flask there will be found a deposit of chloride of ammonium contaminated with sulphur, sulphide of mercury, and protochloride of tin.

Instead of the amalgam and the proportions of the other ingredients as given above, there may be heated in the flask an intimate mixture of 2 grms. of dry protosulphide of tin, half a grm. of sulphur, and 1 grm. of chloride of ammonium.

The part played by the chloride of ammonium in these experiments is not well understood; it is known only that the presence of this salt promotes the formation of a brilliant golden-colored product, though there is no evidence that the chloride either undergoes or produces any chemical change. It is by no means improbable, however, that by volatilizing at the right moment the chloride of ammonium may so moderate the heat engendered by the combination of the sulphur and the tin that the temperature of the mixture is prevented from reaching a point at which the bisulphide would be decomposed.

The Chlorides of Tin are perhaps more important than any other compounds of the metal.

687. *Protochloride of Tin* ($SnCl_2$) is obtained by dissolving granulated tin in boiling concentrated chlorhydric acid. On evaporating the solution, and allowing it to crystallize, prismatic hydrated needles are obtained of the composition $SnCl_2+2H_2O$. These crystals are largely used by dyers and calico-printers, under the name of *tin-salt*. Protochloride of tin, whether in the condition of crystals or in solution, rapidly absorbs oxygen from the air, and is converted into a mixture of bichloride of tin, and an insoluble oxychloride. It must therefore be kept in tight packages. The pure salt is completely soluble in a small quantity of water; but when this solution is mixed with a large quantity of water, it decomposes; a highly acid solution of protochloride of tin in chlorhydric acid remains in solution, while a precipitate of oxychloride of tin ($SnO,SnCl_2+2H_2O$) subsides.

Protochloride of tin is a powerful reducing agent, as has been shown in Exp. 359; it combines readily either with oxygen or with chlorine, and is frequently employed to remove these elements from their compounds. By means of it, the oxides or chlorides of arsenic, antimony, gold, silver, or mercury may be reduced to the metallic state; the salts of sesquioxide of iron, and of protoxide of copper, may be reduced to the degree of protoxide and of dioxide respectively, while acids such as chromic and manganic are reduced to the condition of basic oxides. Sulphurous acid is reduced by it in such wise that a precipitate of sulphide of tin is formed, when solutions of protochloride of tin and of sulphurous acid are mixed; it converts blue indigo to white indigo, and is capable of abstracting oxygen from a host of other substances.

At the temperature of 100°, all the water may be expelled from the crystallized salt; but some of the chlorhydric acid is liable to go off at the same time, so that it is not easy in this way to obtain the anhydrous salt in a state of purity. A better way of obtaining the anhydrous salt is to heat together equal weights of finely divided tin and corrosive sublimate:—

$$HgCl_2 + Sn = SnCl_2 + Hg.$$

The dry chloride of tin remains as a residue, while metallic mercury goes off. The anhydrous salt may itself be distilled at a full red heat.

Protochloride of tin unites with many of the metallic chlorides to form double compounds, which may be called chlorostannites.

688. *Bichloride of Tin* ($SnCl_4$).—When anhydrous, this compound is a fuming, volatile, colorless liquid, of 2·7 specific gravity. It does not solidify at —20°, but boils at 115°. When exposed to the air it gradually absorbs water, and, after a while, hydrated crystals are formed. When mixed with about one-third its weight of water, it solidifies to a mass of hydrated crystals, much heat being at the same time evolved. These crystals are readily soluble in a small quantity of water; but when treated with much water they decompose, hydrated binoxide of tin is precipitated, and free chlorhydric acid passes into solution.

In order to prepare anhydrous bichloride of tin, chlorine gas may be passed over hot chloride of tin, or melted metallic tin; or an intimate mixture of 1 part of tin filings and 4 or 5 parts of corrosive sublimate may be distilled in a retort. The hydrated bichloride and in general all solutions of the bichloride are obtained, either by dissolving tin in dilute aqua regia, by passing chlorine into solutions of protochloride of tin, or by heating the protochloride with chlorhydric acid to which a little nitric acid has been added. The anhydrous salt may be prepared from the hydrate, by distilling the latter with concentrated sulphuric acid, which retains the water.

Like the protochloride, bichloride of tin is largely employed in dyeing. It combines also with the alkaline and other chlorides to form salts, known as chlorostannates. The substance called pink salt, in commerce, employed in the preparation of pink colors upon calicoes, is a chlorostannate of ammonium, $2NH_4Cl,SnCl_4$. There are, of course, many other salts of tin, but none of them are of sufficient interest to be mentioned here.

689. The alloys of tin are important. The composition of bronze, bell-metal, &c. has been already mentioned under copper (§ 659), that of stereotype-metal under antimony (§ 348), and that of tin amalgam under mercury (§ 680). Of the other alloys of tin, those formed by its union with lead are most remarkable. *Plumbers'* solder consists commonly of equal parts of lead and tin, though some kinds of it contain only one-third their weight of lead, and others only one-third their weight of tin. *Pewter* is composed of tin, together with a small proportion of lead.

690. With tin and titanium may be classed the two exceedingly rare metals Columbium (niobium) and Tantalum. The principal source of columbium is the rare American mineral columbite. Tantalum is procured from the Scandinavian mineral tantalite. The four metals form binoxides and volatile chlorides containing four atoms of chlorine, and are therefore sometimes quadrivalent. In this respect the tin group differs from all the groups heretofore studied, excepting the group composed of carbon, boron, and silicon. Tin and titanium have a like mode of occurrence in nature.

CHAPTER XXXIII.

MOLYBDENUM—VANADIUM—TUNGSTEN.

MOLYBDENUM.

691. This rare element is generally found in nature in combination with sulphur, as bisulphide of molybdenum. This bisulphide is a mineral closely resembling graphite and galena in appearance. The name molybdenum is derived from a Greek word sometimes applied to galena. Molybdenum is a white metal almost as lustrous as silver, of specific gravity 8·6. Its atomic weight is 96.

There are three oxides of molybdenum—a protoxide (MoO) and a binoxide (MoO_2), both acting as bases, and a teroxide (MoO_3), which is a strong acid known as molybdic acid. Molybdate of ammonium is a salt much valued by the analyst, since by means of its solution very small quantities of phosphoric acid may be detected; a double compound of molybdate and phosphate of ammonium is deposited as a yellow crystalline precipitate.

VANADIUM.

692. Vanadium is a metal somewhat resembling molybdenum on the one hand, and having certain analogies with chromium upon the other. It appears to be related to the members of the nitrogen group also. Though seldom found in large masses, it appears to be rather widely diffused in nature, traces of it often accompanying the ores of iron, for example. It has four oxides, viz., V_2O_2 — V_2O_3 — V_2O_4 and V_2O_5. The last acts as an acid and forms salts by combining with bases. The atomic weight of vanadium is 51.3.

TUNGSTEN.

693. The element tungsten is far less rare than the other members of the group now under discussion. It is found in considerable quantities in combination with oxygen, iron, and manganese, in the mineral wolfram, whence the Latin name of the element wolframium, and the symbol W. The mineral scheelite also contains tungsten, in combination with oxygen and calcium. Metallic tungsten may be reduced from its oxides by means of hydrogen gas at a bright-red heat, or by charcoal at a white heat. It is a hard iron-gray metal, of specific gravity 17·6, and very refractory. Its atomic weight is 184. The metal has been employed to a certain extent in the preparation of steel; a small quantity of it added to steel has been found to greatly increase the hardness of the steel, and to impart to it other valuable properties.

694. There are two oxides of tungsten:—a binoxide (WO_2), which does not unite with acids to form salts, but acts rather as an acid; and a teroxide (WO_3) called tungstic acid. Tungstic acid by uniting with bases forms a large number of salts, many of which are of very complex composition. The most important ore of tungsten, wolfram, is a mixture in varying proportions of the tungstates of the protoxides of iron and of manganese. The general formula of the mineral may be written RO, WO_3, in which

R stands for either iron or manganese; but there are nevertheless two special varieties of the mineral, one tending to correspond with the formula $2(FeO,WO_3), 3(MnO,WO_3)$, in which the proportions of iron to manganese are as 2 to 3, and the other to the formula $4(FeO,WO_3); MnO,WO_3$, in which the relation of the iron to the manganese is as 4 to 1. The first variety, richer in manganese, is of lower specific gravity than the variety rich in iron. But, since the atomic weights of iron and of manganese are nearly equal, the proportion of tungstic acid is almost absolutely the same in both varieties of the mineral. Wolfram is a very heavy mineral, its specific gravity being as high as 7·3. Indeed the name tungsten is derived from Swedish words meaning heavy stone.

Tungstate of sodium has been employed to a small extent for the purpose of rendering cotton and linen uninflammable. If a weak solution of the tungstate be added to the starch employed to stiffen light fabrics, the cloth therewith impregnated may be exposed to fire without inflaming; it will simply be slowly charred.

The compounds of tungsten are remarkably similar to those of molybdenum. The metal resembles both molybdenum and vanadium in forming an acid teroxide, a binoxide, and a volatile terchloride. Like molybdenum and vanadium, it decomposes water at high temperatures.

CHAPTER XXXIV.

GOLD AND PLATINUM.

GOLD.

695. Though generally found only in small quantities, gold is very widely diffused upon the surface of the globe. Traces of it may be found beneath the sandy beds of most rivers, and it occurs in many of the crystalline rocks and in the soils resulting from their decomposition. Many varieties of iron pyrites, in particular, contain appreciable quantities of gold, and silver is never

found in nature altogether free from it. It occurs in the lead and copper of commerce, as well as in the ores from which these metals are derived and in many of the salts obtained from them, and has been detected in various other metals; it is, in short, almost everywhere. The chief source of the metal as an article of commerce is native gold; this is sometimes found in a condition of purity, but is usually alloyed with more or less silver. It is collected, either directly by mechanically washing away the lighter substances with which it is associated, or, in the case of poorer ores, the gold is dissolved out chemically by means of quicksilver, and is subsequently recovered from the amalgam by filtration and distillation.

The separation of gold from the rocks and sands in which it occurs is a process attended with much labor; hence gold is one of the costliest of metals. The price of a gramme of gold is about sixteen times that of a gramme of silver, and twice as great as that of a gramme of platinum.

696. Pure gold is remarkable as being the most malleable of the metals, and as being the only metal of a decided yellow color; also for its softness, which is nearly as great as that of lead. It has, however, much tenacity, and may be drawn into extremely fine wire; 1 grm. of gold can be made to yield as much as 3 kilometres of wire. The metal can be beaten into leaves which are not more than one ten-thousandth of a millimetre thick. Very thin leaves of gold are transparent, transmitting a green polarized light.

Next to platinum, gold is the heaviest of the ordinary metals; its specific gravity varies from 19·26 to 19·37, according as it has been more or less compressed. Its atomic weight is 196. It melts somewhat less readily than copper or silver, at a temperature estimated to lie between 1200° and 1250°. Its power of conducting heat and electricity is greatly inferior to that of silver. It is not volatile to any great extent at the melting temperature; but at higher temperatures, such as it is subjected to in the ordinary processes of melting and refining, the metal wastes considerably; and at the temperature obtained by the oxyhydrogen blowpipe the metal goes off as a thick vapor.

697. In the air, gold undergoes no change at temperatures

lower than its melting-point; and upon this fact, taken in connexion with the beautiful color and lustre of the metal, and its comparative rarity, its principal uses depend.

On account of this indestructibility, gold was regarded by the earlier chemists as the king of metals; together with platinum and silver, it is still spoken of as a noble metal. Few chemical agents, excepting melted metals, have any action upon gold. None of the common acids, when taken singly, can dissolve it, though the metal is completely soluble in a mixture of chlorhydric and nitric acids (§ 104), and is not completely insoluble in nitric acid contaminated with nitrous or hyponitric acid. The elements chlorine and bromine, however, unite with it in the cold; and when hot it is attacked by phosphorus and arsenic.

As commonly met with in coins or jewelry, gold is far from being pure; coin, for example, usually contains at least 10 per cent. of copper.

In order to prepare pure gold, a piece of coin may be dissolved in aqua regia (Exp. 50), the solution evaporated to dryness upon a water-bath, in order to expel the excess of acid, the residue taken up with water and filtered, to remove any chloride of silver which may be present, and the gold finally precipitated as a brown powder, by adding to the solution some sulphate of protoxide of iron dissolved in water:—

$$2AuCl_3 + 6FeSO_4 = 2Au + Fe_2Cl_6 + 2(Fe_2O_3,3SO_3).$$

The powder may then be collected and dried, and, if desirable, melted and cast into solid masses.

Upon the large scale, fine gold is obtained from its alloys by removing the baser metal by means of either sulphuric or nitric acid.

When an alloy of gold and silver, or copper is boiled with concentrated sulphuric acid in iron kettles, the silver and copper dissolve with evolution of sulphurous acid, while the gold remains undissolved, and the iron vessel is not acted upon. In order to recover the silver from the solution of mixed sulphates, sheets of copper are placed in this solution, and the silver is precipitated upon them, as has been shown in Exp. 267.

The solution of sulphate of copper is then evaporated, and the salt obtained in the crystallized condition fit for sale.

The treatment of the alloy of gold and silver with nitric acid is based upon the fact that silver is soluble, while gold is insoluble, in this acid. But it has been found necessary, in order to obtain a complete separation of the two metals, that the proportion of silver to that of the gold in the alloy should be as much as 2 or 3 to 1; otherwise portions of the silver would, after a while, become so covered with gold as to be protected from the action of the acid, and the two metals could not be completely separated from one another. In practice, whenever the alloy to be treated is found to contain more than a quarter of its weight of gold, enough silver is added to reduce it to this proportion; hence the term *quartation*, by which this method of parting gold and silver is commonly known.

Finely divided gold obtained by precipitation, as above indicated, is employed to a considerable extent for gilding porcelain. The surface to be gilt is first painted with an adhesive varnish, then covered with a mixture of the gold powder and a fusible enamel, and exposed to intense heat; on being subsequently burnished, the gold takes a high polish.

There are two series of gold salts, corresponding to the two oxides—a protoxide AuO, and the teroxide AuO_3. These oxides are rather acids than bases; the teroxide in particular unites with many metallic oxides to form compounds known as aurates. The chlorides, bromides, and iodides of gold also readily combine with other metallic chlorides to form chloraurates, chloraurites, and the analogous bromine and iodine compounds.

698. *Terchloride of Gold* ($AuCl_3$) is the compound of gold most commonly employed in the laboratory. The manner of preparing it has been already indicated in § 697. It serves as a valuable test for tin.

699. *Gilding.*—There are several methods of attaching a film of metallic gold to surfaces of the baser metals.

In the old process of fire gilding, the object to be gilt was first heated to redness, then washed with dilute acid to cleanse its surface, and with a solution of nitrate of mercury in order to amalgamate it slightly; it was then rubbed with a pasty amalgam composed of two parts of gold and one part of mercury. After a portion of the gold amalgam had thus been attached to the surface of the article to be gilt, the latter was heated to drive off the mercury, and the gold left upon it was polished with a burnishing-tool.

In the more modern method of electro-gilding, the object to be gilt is attached to the negative pole of a galvanic battery, a bar of gold is fastened to the positive pole of the battery, and both the object to be gilt and the bar of gold are placed in a mixed solution of cyanide of gold and cyanide of potassium. Under the action of the current, the solution is decomposed; gold is deposited from it upon the object at the negative pole of the battery, while the other ingredients of the solution go to the positive pole, there to dissolve gold from the bar, and thus make good to the solution the metal it has lost. Compare Exp. 353.) Articles of silver, copper, bronze, brass, or platinum, may thus be gilt directly; but with iron, steel, or tin, it is necessary first to immerse the article attached to the battery in a solution of cyanide of copper and of potassium, in order to cover it with a film of copper to which the gold may adhere.

Even without a battery, gold can be deposited upon silver or copper by placing either of these metals in a hot solution of the double cyanide of gold and potassium. Copper trinkets are also sometimes gilt by boiling them in a liquor prepared by mixing a solution of chloride of gold with a solution of an alkaline carbonate.

In general, the compounds of gold have but few properties which are of chemical interest. What has been said of the permanence of the metal implies, of course, that it is a weak chemical agent, having but little affinity for other substances.

700. *Alloys of Gold.*—Gold unites with most of the other metals; but its most important alloys are those of copper, silver, and mercury. Pure gold is so soft that articles of jewelry made of it would quickly wear out if used; such articles, as well as coins and watches, are therefore always made of gold which has been alloyed with copper, in order to increase its hardness. The standard alloy for coin in this country and in France is nine parts by weight of gold to one part of copper; in England it is eleven parts of gold to one of copper. These alloys, as well as the alloys of silver and gold, are more fusible than pure gold, but less ductile. Native gold is an alloy of gold and silver, the proportion of the latter metal varying from 0·2 to 62 per cent. Amalgams of gold play an important part in the metallurgy of gold (§ 695), and in the process of fire gilding above described.

PLATINUM.

701. Platinum is a metal which, like gold, has little affinity for the other chemical elements. It is commonly found in the native state, alloyed with gold and with other metals. Like gold, it is obtained by washing away the earth and sand with which it is found mixed. It is a very heavy metal, the specific gravity of cast platinum being 21·15. Its atomic weight is 197·4. The color of platinum is intermediate between the white of silver and

the gray of steel; its lustre is far less brilliant than that of silver. It is as soft as copper, very malleable and very tenacious; it may be drawn into wire so fine that its diameter is only $\frac{1}{1200}$ of a millimetre. It is not fusible in ordinary furnaces, but may be fused in the blowpipe-flame (Exps. 26, 203), and is nowadays melted in considerable quantities in lime crucibles by means of a blowpipe-flame obtained from common coal-gas and oxygen. At very high temperatures it may be volatilized. Like wrought iron, platinum admits of being forged and welded at temperatures far below its melting-point. When heated, it expands less than any other metal, and is hence well adapted for the construction of apparatus in which metal and glass must be fused together. It conducts heat and electricity much less readily than gold, silver, or copper, standing in this respect not far from iron.

702. Platinum does not oxidize in the air at any temperature, nor is it attacked by any of the common acids taken separately; in aqua regia (§ 104) it dissolves slowly—much less readily than gold. Chlorine-water dissolves it, but neither bromine nor iodine has any action upon it. When heated to redness in the air, in contact with the fixed caustic alkalies or alkaline earths, it is slowly corroded, in consequence of the formation of an oxide which unites with the alkali. Phosphorus and arsenic unite readily with hot finely divided platinum, forming very fusible compounds; sulphur also combines with it, though far less readily. At high temperatures, platinum is easily acted upon by silicon (compare § 463). A platinum crucible should consequently never be placed in direct contact with a hot mixture of a carbon compound and silicic acid. If the crucible is to be heated in a coal fire, it should first be placed in an earthen crucible lined with some infusible earth, such as magnesia.

With most of the other metals platinum unites readily, forming alloys which in many instances are more fusible than platinum itself; hence, in using platinum vessels in chemical experiments, care must be taken not to touch the platinum while hot with easily fusible metals, or to place in the hot vessels any reducible compound of a metal. Most of the alloys of platinum are not only fusible, but they are also soluble in acids. Platinum which has been alloyed with 10 or 12 times its weight of silver, for example, is as completely soluble in nitric acid as the silver itself.

From its comparative inertness as a chemical agent, taken in connexion with its infusibility, platinum is an extremely useful metal to the chemist. It is employed in the scientific laboratory for crucibles, evaporating-dishes, stills, tubes, spatulæ, forceps, wire, blowpipe-tips, and the like; and, in the manufacture of oil of vitriol, large platinum stills, together with cooling-siphons of the same metal, are employed in the process of concentrating the acid.

703. A remarkable property of platinum, of inducing various gases to combine chemically one with the other, has already been repeatedly alluded to and illustrated (§§ 224, 240, 387). This power of causing combination is possessed even by clean surfaces of the ordinary solid metal, though to a much greater degree by spongy platinum (Exp. 364), and still more by the very finely divided powder known as platinum black (§ 706).

Platinum forms two series of compounds, corresponding respectively to the protoxide PtO and to the binoxide PtO_2. Its chlorides are well-defined compounds; but with the oxygen acids it forms comparatively few salts, and none of these are at present of much importance.

704. *Protochloride of Platinum* ($PtCl_2$) is a compound insoluble in water, obtained by carefully heating the bichloride to 230° upon an oil-bath. It dissolves in alkaline lyes, and the solution thus obtained may be used for making platinum black (§ 706). At a red heat chloride of platinum is completely decomposed to metallic platinum and chlorine. With the other metallic chlorides protochloride of platinum unites, to form compounds known as chloroplatinites; the general formula of these compounds is $2MCl,PtCl_2$.

705. *Bichloride of Platinum* ($PtCl_4$) is the platinum compound most commonly employed in the laboratory. It is a deliquescent substance, readily soluble in water, alcohol, and ether; the aqueous solution is of a reddish-brown color. When heated to 230°, or thereabouts, the salt loses half its chlorine, as has been already stated. The aqueous solution of bichloride of platinum is much used as a test for potassium and ammonium, and for preparing certain organic compounds suitable for analysis.

Exp. 361.—Cut half a gramme, or more, of worn-out platinum foil or wire into small fragments, and boil them with a teaspoonful of aqua regia so long as the metal appears to be acted upon, then decant the liquid into a porcelain dish, add to the fragments of platinum another teaspoonful of aqua regia, and proceed as before, repeating the treatments until all the metal has dissolved. By the repeated action of successive small portions of the solvent, platinum and other comparatively speaking insoluble substances can be dissolved much more readily than if all the liquid necessary for its solution were added at once. Evaporate the solution to dryness upon a water-bath, take up the residue with water, and preserve the solution in a bottle provided with a glass stopper.

Exp. 362.—Pour a teaspoonful of a solution of chloride of potassium, or of almost any other salt of potassium, into a test-tube, acidulate the liquid with chlorhydric acid, and add to it a drop of the solution of bichloride of platinum obtained in the preceding experiment. A yellow, insoluble powder will soon be precipitated. It is a double chloride of potassium and platinum, and its formula may be written $2KCl,PtCl_4$. This test serves to distinguish potassium from sodium, and, if need be, to separate potassium from solutions in which it is mixed with sodium; for the double chloride produced with chloride of sodium and bichloride of platinum is easily soluble in water.

Exp. 363.—Repeat Exp. 362, but substitute chloride of ammonium for the chloride of potassium. A yellow precipitate, similar to that obtained in Exp. 362, will separate immediately, or, if the solutions employed are dilute, after a short time. The composition of this precipitate may be represented by the formula $2NH_4Cl,PtCl_4$. Again repeat the experiment, and this time take enough of the platinum solution and of the chloride of ammonium to make half a teaspoonful of the yellow precipitate, taking care that at last there shall be a slight excess of free chloride of ammonium rather than of chloride of platinum in the supernatant liquid. Allow the precipitate to settle, separate it from the clear liquid by decantation, and dry it partially at a gentle heat. When the precipitate has acquired the consistence of slightly moistened earth, transfer it to a cup-shaped piece of platinum foil, and heat it to redness in the gas-flame, as long as fumes of chloride of ammonium continue to escape. All the chlorine, hydrogen, and nitrogen will be driven off, and there will remain upon the foil a gray, loosely coherent, sponge-like mass of metallic platinum; it is called *platinum sponge*.

Exp. 364.—Hold the dry platinum sponge of Exp. 363 in a stream of hydrogen or of common illuminating gas issuing from a fine jet.

The metal will soon begin to glow, and in a moment will become hot enough to inflame the mixture of air and gas in contact with it. Before friction-matches were employed, this property of spongy platinum, of inflaming hydrogen, was sometimes made use of for striking a light. The mode of action of the platinum in this experiment is obscure; it has already been alluded to in § 387.

From platinum sponge, solid articles of platinum may be manufactured by compression. If the spongy platinum be first rubbed to powder under water, the particles of metal of which it is composed can be readily compacted into solid bars by subjecting the powder to powerful pressure in appropriate moulds. The pressed bar is then heated intensely in a coke-fire with strong draught, and forged by striking it with the hammer upon its ends—the process of heating and forging being several times repeated, until the bar has become sufficiently condensed. The metal may then be wrought into any desired shape by heating and hammering, in the same way as any other malleable metal. This process of working platinum was for a long time the common method, and is still employed to a certain extent.

706. *Platinum Black* is a term applied to metallic platinum even more finely divided than the sponge above described.

By dissolving protochloride of platinum in hot concentrated potash-lye, and pouring into the hot liquor alcohol, by small successive portions, platinum will be thrown down as a black powder looking like soot. The powder should be freed from the supernatant liquor by decantation, and then boiled successively with alcohol, chlorhydric acid, potash-lye, and water, in order to free it from all impurities.

A capacious vessel must be chosen for the reaction of the alcohol upon the alkaline solution of chloride of platinum; for much carbonic acid is generated while the components of the alcohol are reducing the solution of platinum, so that lively effervescence occurs. Platinum black is capable not only of absorbing and storing up many times its own bulk of oxygen gas; it is also capable of giving away this oxygen to many other substances. If easily oxidizable liquids, such as alcohol or ether, are dropped upon platinum black which has previously been exposed to the air, the liquids will be oxidized and converted into new substances, while the powder becomes red-hot from the heat evolved during the act of oxidation.

707. Besides forming with the chlorides of potassium and ammonium the insoluble compounds above described, bichloride of platinum unites with many other chlorides, both of metals and of organic radicals, to form analogous salts of the general formula

$2MCl,PtCl_4$, or $MCl_2,PtCl_4$. These compounds are commonly called chloroplatinates; by means of them the composition and combining weights of many organic compounds have been determined. It is only necessary to ignite a weighed portion of the chloroplatinate, and to weigh the residue of pure platinum which is left after the organic matter has all been driven off, in order to ascertain how much platinum is contained in the compound. This fact having been determined, the quantity of the organic radical, or rather of the chloride of the radical, which was combined with the chloride of platinum in the chloroplatinate, may be readily calculated.

708. With gold and platinum are classed several rare metals which are never found except in association with platinum, and which closely resemble that metal. They are commonly called platinum metals, and the group may be appropriately termed the platinum group. The whole group consists of Rhodium (atomic weight =104), Ruthenium (104), Palladium (106·5), Gold (196·7), Platinum (197·4), Iridium (198), and Osmium (199). Palladium is used to impart to brass gas-fixtures a peculiar reddish tint, sometimes called salmon-bronze. Iridium is used for the very hard tips of gold pens. Osmium forms, among other oxides, a volatile compound OsO_4, whose vapors are intensely poisonous. The metals of this group are noble metals; they withstand the action of the atmosphere; none of them are acted upon by nitric acid, though they dissolve in chlorine and in aqua regia. Their oxides part with all their oxygen when simply heated, leaving the metal behind.

CHAPTER XXXV.

ATOMIC WEIGHTS OF THE ELEMENTS—CLASSIFICATION.

709. An alphabetical list of the sixty-five recognized elements, with their symbols and atomic weights, is here given for convenience of reference. The names of those elements which are so rare as to be at present of little importance are printed in italics:—

Element	Symbol	Atomic weight
Aluminum	Al	27·4
Antimony	Sb	122
Arsenic	As	75
Barium	Ba	137
Bismuth	Bi	210
Boron	Bo	11
Bromine	Br	80
Cadmium	Cd	112
Cæsium	Cs	133
Calcium	Ca	40
Carbon	C	12
Cerium	Ce	92
Chlorine	Cl	35·5
Chromium	Cr	52·5
Cobalt	Co	58·8
Columbium (Niobium)	Nb	94
Copper	Cu	63·4
Didymium	D	95
Erbium	E	112·6
Fluorine	Fl	19
Glucinum	Gl	14
Gold	Au	196
Hydrogen	H	1
Indium	In	72
Iodine	I	127
Iridium	Ir	198
Iron	Fe	56
Lanthanum	La	92·8
Lead	Pb	207
Lithium	Li	7
Magnesium	Mg	24
Manganese	Mn	55
Mercury	Hg	200
Molybdenum	Mo	96
Nickel	Ni	58·8
Nitrogen	N	14
Norium	No	?
Osmium	Os	199
Oxygen	O	16
Palladium	Pd	106·5
Phosphorus	P	31
Platinum	Pt	197·4
Potassium	K	39·1
Rhodium	Rh	104
Rubidium	Rb	85·7
Ruthenium	Ru	104
Selenium	Se	79·5
Silicon	Si	28
Silver	Ag	108
Sodium	Na	23
Strontium	Sr	87·5
Sulphur	S	32
Tantalum	Ta	182
Tellurium	Te	128
Terbium	Tb	?
Thallium	Tl	204
Thorium	Th	231·5 (?)
Tin	Sn	118
Titanium	Ti	50
Tungsten	W	184
Uranium	Ur	120
Vanadium	V	51·3
Yttrium	Yt	61·6
Zinc	Zn	65
Zirconium	Zr	67·2

710. In the following table the elements are arranged in what are believed to be natural groups. Without accepting any one infallible criterion of classification, or insisting upon any systematic arrangement of the elements in groups with that strenuousness which is apt to make classification rather a hindrance than a help, the student may provisionally use this subdivision of the elements into groups as a help in remembering facts, as a guide to the prompt recognition of general properties and general laws, and as a suggestive compend of his whole chemical knowledge:—

Element	Weight
Fluorine	19
Chlorine	35·5
Bromine	80
Iodine	127
Oxygen	16
Sulphur	32
Selenium	79·5
Tellurium	128
Nitrogen	14
Phosphorus	31
Arsenic	75
Antimony	122
Bismuth	210
Carbon	12
Boron	11
Silicon	28
Hydrogen	1
Lithium	7
Sodium	23
Potassium	39·1
Rubidium	85·7
Silver	108
Cæsium	133
Thallium	204
Calcium	40
Strontium	87·5
Barium	137
Lead	207
Magnesium	24
Zinc	65
Cadmium	112

Element	Weight
Glucinum	14
Aluminum	27·5
Chromium	52·5
Manganese	55
Iron	56
Cobalt	58·8
Nickel	58·8
Yttrium	61·6
Erbium	112·6
Terbium	?
Zirconium	67·2
Norium	?
Cerium	92
Lanthanum	92·8
Didymium	95
Uranium	120
Thorium	231·5 (?)
Copper	63·4
Mercury	200
Titanium	50
Columbium	94
Tin	118
Tantalum	182
Molybdenum	96
Vanadium	51·3
Tungsten	184
Rhodium	104
Ruthenium	104
Palladium	106·5
Gold	196
Platinum	197·4
Iridium	198
Osmium	199

711. *Atomic Heats of the Elements.*—The power of heat to cause changes of temperature is not the same for any two substances, but varies with the nature of the substance submitted to its action. Each chemical element is peculiarly affected in this respect by heat. The quantity of heat needed to raise the temperature of a certain weight of water from 0° to 1° being called unity, the quantity of heat required to raise the temperature of the same weight of any element by the same amount is the specific heat of that element (§ 31). In the second column of the following table will be found the specific heats of a number of representative elements, selected from each group of elements except the carbon group, and arranged in the order of their atomic weights. (Compare § 710.)

Name.	*Specific Heat.*	*Atomic Weight.*	*Atomic Heat.*
Lithium ..	0·94080	7	6·59
Aluminum	·21430	27·5	5·89
Sulphur ..	·20259	32	6·48
Iron	·11380	56	6·37
Copper....	·09515	63·4	6·04
Arsenic ..	·08140	75	6·11
Silver	·05701	108	6·16
Cadmium	·05669	112	6·35
Tin	·05623	118	6·63
Iodine	·05412	127	6·87
Tungsten..	·03342	184	6·15
Gold......	·03244	196	6·36
Lead	·03140	207	6·50
Bismuth ..	·03084	210	6·48

The preceding table contains only solid substances. It has been found that the specific heat of the same body is commonly greater in the liquid than in the solid state, and always less in the gaseous than in the liquid state. Accordingly in instituting any comparison between different bodies, based on their specific heats, it is essential to compare them in the same physical condition, solids with solids, liquids with liquids, gases with gases. The second column of the following table contains the specific heats of the four elements which are gaseous at the ordinary atmospheric temperature and pressure.

Name.	*Specific Heat.*	*Atomic Weight.*	*Atomic Heat.*
Hydrogen..	3·4090	1	3·4090
Nitrogen ..	0·2438	14	3·4132
Oxygen....	·2175	16	3·4800
Chlorine ..	·1210	35·5	4·2955

On comparing together the numbers in the second and third columns of the preceding tables, it will be noticed that the lower the atomic weight of an element is, the higher is its specific heat, and *vice versâ*. In the fourth column of the tables, under the name of *atomic heat*, will be found the product of the specific heat by the atomic weight of each element. The *atomic heats* of the elements represent the quantities of heat which are required to cause equal alterations of temperature in atomic proportions of the several elements. But the tables show that these quantities of heat are nearly the same for each and all of the solid elements compared in the first table, and are again approximately equal for the gaseous elements which are grouped in the second table.

This striking principle is deducible from the foregoing considerations—namely, that while the capacities for heat of the same weights of the various elements are very different, the capacities for heat of the atoms, or atomic proportions, of the elements are nearly identical, provided that the elements compared be in the same physical condition. In other words, those weights of the elements which are assumed to represent the relative weights of their atoms require approximately the same amount of heat to raise them through an equal number of degrees of temperature; while the amounts of heat required to raise *equal* weights of the elements through an equal number of degrees are expressed by very different numbers (the specific heats). It is essential, however, that the elements compared should be in the same physical condition.

It is true that the numbers representing the atomic heats show considerable discrepancies; but when it is remembered that there are unavoidable errors attaching to the determinations both of the specific heats and of the atomic weights, that many of the elements cannot yet be obtained in a condition of purity, and that the two factors of the product (specific heat and atomic weight) vary in the proportion of 1 to 30, it will be seen that the accordance is distinct enough to indicate the existence of a general law. The atomic heats of carbon, boron, and silicon, however, do not conform to this law.

712. *Atomic Heats of Compounds.*—The same kind of relation between the specific heat and the molecular weight obtains within certain classes of compound bodies of like constitution. Only those compounds which have an analogous atomic constitution can possess approximately the same molecular heat; and there are many admitted exceptions even to this rule. The following table contains the specific and atomic heats of a number of inorganic compounds, arranged in groups according to their chemical composition.

Formula.	*Specific Heat.*	*Molecular Weight.*	*Atomic Heat.*
1. OXIDES MO.			
PbO	0·05089	223	11·35
HgO	·05179	216	11·19
CuO	·14201	79·5	11·19
ZnO............	·12480	81	10·11
2. M_2O_3.			
Fe_2O_3	·16695	160	26·71
Cr_2O_3	·17960	153	27·47
Bi_2O_3	·06053	468	28·33
Sb_2O_3	·09009	292	26·31
3. CHLORIDES MCl.			
NaCl	0·21401	58·5	12·52
KCl	·17295	74·5	12·88
HgCl	·05205	235·5	12·26
CuCl	·13827	99	13·69
4. MCl_2.			
$MgCl_2$..........	·19460	95	18·49
$ZnCl_2$	·13618	136	18·52
$PbCl_2$	·06641	278	18·46
$MnCl_2$..........	·14255	126	17·96
$SnCl_2$	·10161	189	19·20
5. MCl_4.			
$SnCl_4$	·14759	260	38·37
$TiCl_4$	·19145	192	36·76

Formula.	Specific Heat.	Molecular Weight.	Atomic Heat.
6. SULPHIDES MS.			
FeS	0·13570	88	11·94
NiS	·12813	90·7	11·62
CoS	·12512	90·7	11·36
SnS	·08375	150	12·56
7. MS_2.			
FeS_2	·13009	120	15·61
SnS_2	·11932	182	21·72
MoS_2	·12334	160	19·73
8. M_2S_3.			
Sb_2S_3	·08403	340	28·57
Bi_2S_3	·06002	516	30·97
9. CARBONATES M_2CO_3.			
K_2CO_3..........	0·21623	138	29·84
Na_2CO_3	·27275	106	28·91
10. MCO_3.			
$BaCO_3$	·11038	197	21·74
$SrCO_3$..........	·14483	147·6	21·38
$FeCO_3$	·19345	116	22·44
11. SULPHATES M_2SO_4.			
K_2SO_4..........	0·19010	174	33·08
Na_2SO_4	·23115	142	32·82
12. MSO_4.			
$CaSO_4$..........	·19656	136	26·73
$MgSO_4$	·22159	120	26·59
$PbSO_4$..........	·08723	303	26·43

There are considerable departures from equality in the atomic heats of the bodies of similar constitution grouped in the above table; but when it is observed that the two factors of the product are very wide apart, the close coincidence in most cases will seem much more noteworthy than the occasional discrepancies.

713. *Value of Specific Heats.*—The determination of the specific heat of a substance is a difficult experiment in Physics. It is obvious that, if the laws concerning atomic heat which have been above illustrated were invariable and certain, the physicist

would determine for the chemist the atomic weights of the elements in experimentally fixing their specific heats; he would also indicate the true classification of chemical compounds by determining the specific heats of these compounds. To obtain the atomic weight of any element, it would suffice to divide the atomic heat common to all the elements in the same physical state by the specific heat of that element. To determine the class of compounds to which a given compound belonged, it would be sufficient to know its atomic heat. A given carbonate, for example, would belong to the class M_2CO_3 or to the class MCO_3, according as its atomic heat were about 29 or about 22.

Although the principle of the equality of the atomic heats of the elements and of compounds of like constitution is not established with that certainty which would give to the experimental determination of the specific heat the conclusive weight just indicated, yet this equality certainly affords a very strong presumption in favor of the correctness of the atomic weights of the elements contained in the above tables, and of the molecular formulæ on which the classification of compounds in the third table is based.

714. *Two Sets of Atomic Weights in use.*—The student will find in many books on chemistry weights, other than those used in this manual, assigned to the majority of the elements, and variously called combining, equivalent, or atomic weights. To all the elements, except those of the chlorine, nitrogen, and alkali groups, and the elements boron and gold, are assigned weights which are respectively the halves of the numbers given in §§ 709, 710. The weight assigned to sulphur, for example, is 16, to lead 103·5, to cadmium 56, to iron 28, to copper 31·7, to tin 59, to oxygen 8, and so forth. If these smaller numbers were accepted as the true atomic weights of the elements to which they are respectively assigned, it is obvious that the relation of equality between the atomic heats of all elements in the same physical state, presented in the tables of § 711, would disappear; furthermore, many of the relations indicated in the table of § 712 would be no longer visible. The atomic heats of the solid elements would be divisible into two classes, in one of which the atomic heat would be about double what it was in the

other. If the atomic weight of lead were 103·5, the formula of the white chloride of lead (§ 581) would be PbCl, analogous to NaCl, and there would be no clear reason why its atomic heat should not be that of the chlorides of the formula MCl, namely, about 12·5; but its atomic heat would be only 9·23. If the atomic weight of iron were 28, that of potassium being 39·1, there would be no assignable reason for the marked difference between the atomic heats of the two carbonates; the formulæ of the carbonates of iron and of potassium would be alike—either both M_2CO_3 or both MCO_3.

Another consequence of using the smaller atomic weights must not pass unnoticed. If the atomic weight of oxygen is 8 and of sulphur 16, the coincidence of the atomic weight and the unit-volume weight of those eight elements for which this equality has been affirmed (§ 259) ceases to be true; and the simple rule that the molecule of every compound gas or vapor occupies a volume twice as large as the combining volume of hydrogen, oxygen, chlorine, and so forth, will lose the universality which constitutes its chief value.

The student will remember that the determination of the *least* combining proportion by weight of any element which cannot be converted into vapor is not a matter of direct experiment simply (§§ 395, 603). The natural analogies of the element and its compounds, the greater or less simplicity of the formulæ which result from one assumption or another, the indications of isomorphism, and of atomic heat, have all to be consulted. That the best guides have thus far failed to lead to an unquestionable determination of the real *least* combining weights (or atomic weights) of the majority of the elements, may be inferred from the diversities of usage on this subject in chemical literature. In order to indicate that they mean the atomic weights which have been given in this manual, many chemists write the symbols of those elements for which two different weights are in use with a line drawn through them, thus O̶, S̶, F̶e̶, P̶b̶. This practice is almost essential in periodical publications to which writers of different theoretical views contribute.

715. In the midst of the doubts and discussions which to-day envelope chemical theories, the student will do well to remember

that all these questions lie without the sphere of fact. They do not affect the actual composition or properties of a single element or compound; they are questions of interpretation, classification, and definition. The existence of atoms is itself an hypothesis, and not a probable one; all speculations based on this hypothesis, all names which have grown up with it, all ideas which would be dead without it, should be accepted by the student provisionally and cautiously, as being matter for belief but not for knowledge. All dogmatic assertion upon such points is to be regarded with distrust. The great majority of chemists, devoted to the applications of chemistry in mineralogy, metallurgy, dyeing, printing, and the manufacture of chemicals, remain completely indifferent to discussions of chemical theories. Hence the student will find that in technical chemical literature the older notation and the corresponding smaller atomic weights are almost invariably employed.

Theories, however, are of great importance to the progress of the science and to the clear ordering of the ground already won. It is, on this account, very much to be wished that the great attention now devoted to the discussion of the best methods of representing symbolically the constitution of chemical substances and the changes to which they are subject, may result in the elaboration of a system good enough to command general acceptance.

APPENDIX.

CHEMICAL MANIPULATION.

1. *Glass tubing.*—Two qualities of glass tubing are used in chemical experiments—that which softens readily in the flame of a gas or spirit-lamp, and that which fuses with extreme difficulty in the flame of the blast-lamp. These two qualities are distinguished by the terms *soft* and *hard* glass. Soft glass is to be preferred for all uses except the intense heating, or ignition, of dry substances. Fig. I. represents the most convenient sizes of glass tubing, both hard and soft, and shows also the proper thickness of the glass walls for each size.

Fig. I.

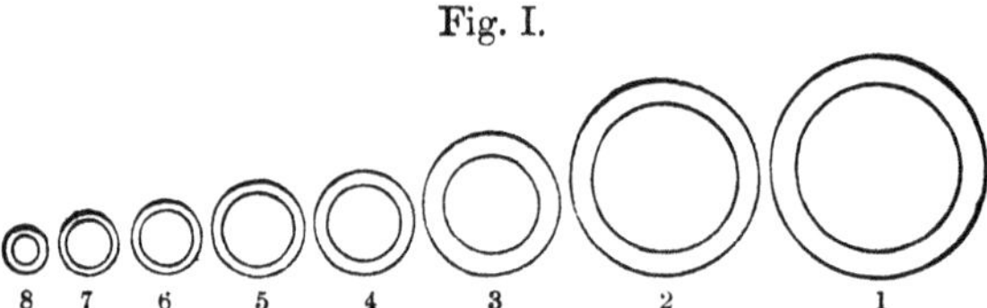

2. *Cutting and cracking glass.*—Glass tubing and glass rod must generally be cut to the length required for any particular apparatus. A sharp triangular file is used for this purpose. The stick of tubing, or rod, to be cut is laid upon a table, and a deep scratch is made with the file at the place where the fracture is to be made. The stick is then grasped with the two hands, one on each side of the mark, while the thumbs are brought together just at the scratch. By pushing with the thumbs and pulling in the opposite direction with the fingers, the stick is broken squarely at the scratch, just as a stick of candy or dry twig may be broken. The sharp edges of the fracture should invariably be made smooth, either with a wet file, or by softening the end of the tube or rod in the lamp. (See Appendix, § 3.) Tubes or rods of sizes four to eight inclusive may readily be cut in this manner; the larger sizes are divided with more difficulty, and it is often necessary to make the file-mark both long and deep. An even fracture is not always to be obtained with large tubes. The lower ends of glass funnels, and those ends of gas delivery-tubes which enter the bottle or flask in which the gas is generated, should be filed off, or ground off on a grindstone, obliquely (Fig. II.), to facilitate the dropping of liquids from such extremities.

Fig. II.

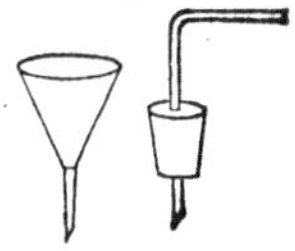

In order to cut glass plates, the glazier's diamond must be resorted to. For the cutting of extremely thin glass tubes and of other glass ware, like flasks, retorts, and bottles, still other means are resorted to, based upon the sudden and unequal application of heat. The process divides itself into two parts—the producing of a crack in the required place, and the subsequent guiding of this crack in the desired direction. To produce a crack, a scratch must be made with the file, and to this scratch a pointed bit of red-hot charcoal, or the jet of flame produced by the mouth blowpipe, or a very fine gas-flame, or a red-hot glass rod may be applied. If the heat does not produce a crack, a wet stick or file may be touched upon the hot spot. Upon any part of a glass surface except the edge, it is not possible to control perfectly the direction and extent of this first crack; at an edge a small crack may be started with tolerable certainty by carrying the file-mark entirely *over* the edge. To guide the crack thus started, a pointed bit of charcoal or slow match may be used. The hot point must be kept on the glass from 1 c.m. to 0·5 c.m. in advance of the point of the crack. The crack will follow the hot point, and may therefore be carried in any desired direction. By turning and blowing upon the coal or slow match the point may be kept sufficiently hot. Whenever the place of experiment is supplied with common illuminating gas, a very small jet of burning gas may be advantageously substituted for the hot coal or slow match. To obtain such a sharp jet, a piece of hard glass tube, No. 5, 10 c.m. long, and drawn to a very fine point (see Appendix, § 3), should be placed in the caoutchouc tube which ordinarily delivers the gas to the gas-lamp, and the gas should be lighted at the fine extremity. The burning jet should have a fine point, and should not exceed 1·5 c.m. in length. By a judicious use of these simple tools, broken tubes, beakers, flasks, retorts, and bottles may often be made to yield very useful articles of apparatus. No sharp edges should be allowed to remain upon glass apparatus. The durability of the apparatus itself, and of the corks and caoutchouc stoppers and tubing used with it, will be much greater, if all sharp edges are removed with the file or, still better, rounded in the lamp.

3. *Bending and closing glass tubes.*—Tubing of sizes five to eight inclusive can generally be worked in the common gas- or spirit-lamp; for larger tubes the blast-lamp is necessary (see Appendix, § 6). Glass tubing must not be introduced suddenly into the hottest part of the flame, lest it crack. Neither should a hot tube be taken from the flame and laid at once upon a cold surface. Gradual heating and gradual cooling are alike necessary, and are the more essential the thicker the glass; very thin glass will sometimes bear the most sudden changes of temperature, but thick glass and glass of uneven thickness absolutely require slow heating and annealing. When the end of a tube is to be heated, as in rounding sharp edges, more care is required, in consequence of the great facility with which cracks start at an edge. A tube should therefore always be brought first into the current of hot air beyond the actual flame of the gas or spirit-lamp, and there thoroughly warmed before it is introduced into the flame itself. If a blast-lamp is employed, the tube may be warmed in the smoky flame, before the blast is turned on, and may subsequently be annealed in the same manner; the deposited soot will be burnt off in the first instance, and in the last may be wiped off when the tube is cold. In heating a tube, whether for bending, drawing, or closing, the tube must be *constantly* turned between the fingers, and also moved a little to the right and left, in order that it may be uniformly heated all round, and that the temperature of the neighboring parts may be duly raised. If a tube, or rod, is to be heated at any part but an end, it should be held between the thumb and first two fingers of each hand in such a manner that the hands shall be below the tube or rod, with the palms upward, while the lamp-flame is between the hands. When the end of a tube or rod is to be heated, it is best to begin by warming the tube or rod about 2 c.m. from the end, and thence to proceed slowly to the end.

The best glass will not be blackened or discolored during heating. The blackening occurs in glass which, like ordinary flint glass, contains oxide of lead as an ingredient. Glass containing much of this oxide is not well adapted for chemical uses. The blackening may sometimes be removed by putting the glass in the upper or outer part of the flame, where the reducing gases are consumed, and the air has the best access to the glass. The blackening may be altogether avoided by always keeping the glass in the oxidizing part of the flame.

Glass begins to soften and bend below a visible red heat. The condition of the glass is judged of as much by the fingers as the eye; the hands feel the yielding of the glass, either to bending, pushing, or pulling, better than the eye can see the change of color or form. It may be bent as soon as it yields in the hands, but can be drawn out only when much hotter than this. Glass tubing, however, should not be bent at too low a temperature; the curves made at too low a heat are apt to be flattened, of unequal thickness on the convex and concave sides, and brittle.

In bending tubing to make gas-delivery-tubes and the like, attention should be paid to the following points: 1st, the glass should be equally hot on all sides; 2nd, it should not be twisted, pulled out, or pushed together during the heating; 3rd, the bore of the tube at the bend should be kept round, and not altered in size; 4th, if two or more bends be made in the same piece of tubing (Fig. III., *a*), they should all be in the same plane, so that the finished tube will lie flat upon the level table.

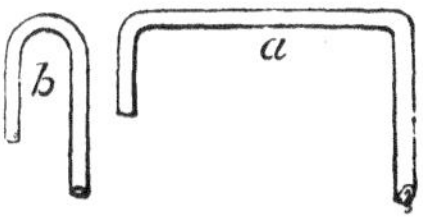

When a tube or rod is to be bent or drawn close to its extremity, a temporary handle may be attached to it by softening the end of the tube or rod, and pressing against the soft glass a fragment of glass tube, which will adhere strongly to the softened end. The handle may subsequently be removed by a slight blow, or by the aid of a file. If a considerable bend is to be made, so that the angle between the arms will be very small or nothing, as in a siphon, for example, the curvature cannot be well produced at one place in the tube, but should be made by heating, progressively, several centimetres of the tube, and bending continuously from one end of the heated portion to the other (Fig. III., *b*). Small and thick tube may be bent more sharply than large or thin tube.

Fig. III.

In order to draw a glass tube down to a finer bore, it is simply necessary to thoroughly soften on all sides one or two centimetres length of the tube, and then, taking the glass from the flame, pull the parts asunder by a cautious movement of the hands. The larger the heated portion of glass, the longer will be the tube thus formed. Its length and fineness also increase with the rapidity of motion of the hands. If it is desirable that the finer tube should have thicker walls in proportion to its bore than the original tube, it is only necessary to keep the heated portion soft for two or three minutes before drawing out the tube, pressing the parts slightly together the while. By this process the glass will be thickened at the hot ring.

To obtain a tube closed at one end, it is best to take a piece of tubing, open at both ends, and long enough to make two closed tubes. In the middle of the tube a ring of glass, as narrow as possible, must be made thoroughly soft. The hands are then separated a little, to cause a contraction in diameter at the hot and soft part. The point of the flame must now be directed, not upon the narrowest part of the tube, but upon what is to be the bottom of the closed tube. This point is indicated by the line *a* in Fig. IV. By withdrawing the right hand, the narrow part of the tube is attenuated, and finally melted off, leaving both halves of the original tube closed at one end, but not of the same form; the right-hand half is drawn out into a long point, the other is more roundly closed. It is not possible to close handsomely the two pieces at once. The tube is seldom perfectly finished by the operation; a superfluous knob of glass generally remains upon the end. If small, it may be got rid of by heating the whole end of the tube, and blowing moderately with the mouth into the open end. The knob, being hotter and therefore softer than any other part, yields to the pressure from within, spreads out and disappears. If the knob is large, it may be cut off with scissors while red-hot, or drawn off by sticking to it a fragment of tube, and then softening the glass above the junction. The same process may be applied to the too pointed end of the right-hand half of the original tube, or to any misshapen result of an unsuccessful attempt to close a tube, or to any bit of tube which is too short to make two closed tubes. When the closed end of a tube is too thin, it may be strengthened by keeping the whole end at a red heat for two or three minutes, turning the tube constantly between the fingers. It may be said in general of all the preceding operations before the lamp, that success depends on keeping the tube to be heated in constant rotation, in order to secure a uniform temperature on all sides of the tube.

Fig. IV.

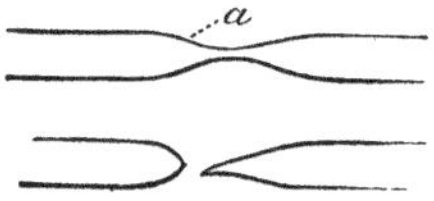

a

4. *Blowing bulbs and piercing holes in tubing.*—If the bulb desired is large in proportion to the size of the tube on which it is to be made, the walls of the tube must be thickened by rotation in the flame before the bulb can be blown. If the bulb is to be blown in the middle of a piece of tubing, this thickening is effected by gently pressing the ends of the tube together while the glass is red-hot in the place where the bulb is to be; if the bulb is to be placed at the end of a tube, this end is first closed, and then suitably thickened by pressing the hot glass up with a piece of metal until enough has been accumulated at the end. The thickened portion of glass is then to be heated to a cherry-red, suddenly withdrawn from the flame, and expanded while hot by steadily blowing, or rather pressing air, into the tube with the mouth; the tube must be constantly turned on its axis, not only while in the flame, but also while the bulb is being blown. If too strong or too

sudden a pressure be exerted with the mouth, the bulb will be extremely thin and quite useless. By watching the expanding glass, the proper moment for arresting the pressure may usually be determined. If the bulb obtained be not large enough, it may be reheated and enlarged by blowing into it again, provided that a sufficient thickness of glass remain.

It is sometimes necessary to make a hole in the side of a tube or other thin glass apparatus. This may be done by directing a pointed flame from the blast-lamp upon the place where the hole is to be, until a small spot is red-hot, and then blowing forcibly into one end of the tube while the other end is closed by the finger; at the hot spot the glass is blown out into a thin bubble, which bursts or may be easily broken off, leaving an aperture in the side of the tube.

It is hoped that these few directions will enable the attentive student to perform, sufficiently well, all the manipulations with glass tubes which ordinary chemical experiments require. Much practice will alone give a perfect mastery of the details of glass-blowing.

Fig. V.

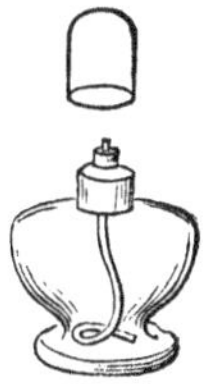

5. *Lamps.*—The common glass spirit-lamp will be understood without description from the figure (Fig. V.). This lamp does not give heat enough for most ignitions; for such purposes a lamp with circular wick, of some one of the numerous forms sold under the name of Berzelius's Argand Spirit Lamp (Fig. VI.), is necessary. These argand lamps are usually mounted on a lamp-stand provided with three brass rings; but the fittings of these lamps are all made slender, in order not to carry off too much heat. When it is necessary to heat heavy vessels, other supports must be used.

Fig. VI.

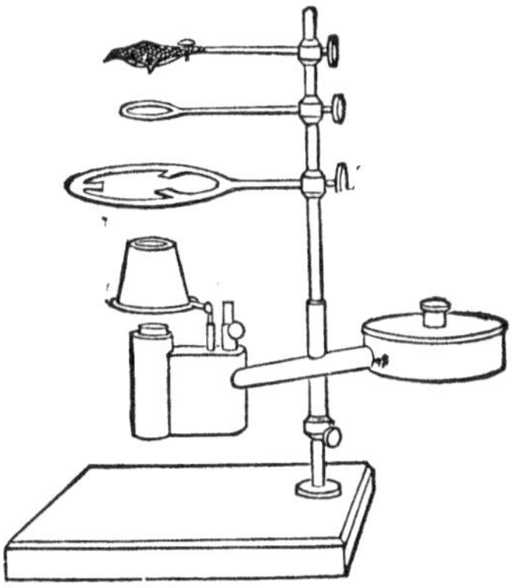

Whenever gas can be obtained, gas-lamps are greatly to be preferred to the best spirit-lamps. For all common chemical experiments, except a few for which ignition-tubes must be prepared or in which considerable lengths of tubing must be heated, the gas-lamp known as Bunsen's burner is the best lamp. The cheapest and best construction of this lamp may be learned from the following description with the accompanying figures. (Fig. VII.) The single casting of brass *a b* comprises the tube *b* through which the gas enters, and the block *a* from which the gas escapes by two or three fine vertical holes passing through the screw *d*, and issuing from the upper face of *d*, as shown at *e*. The length of the tube *b* is 4·5 c.m. and its outside diameter varies from 0·5 c.m. at the outer end to 1 c.m. at the junction with the block *a*. The outside diameter of the block *a* is 1·6 c.m., and its outside height without the screws is 1·8 c.m. By the screw *c*, the piece *a b* is attached to the iron foot *g*, which may be 6 c.m. in diameter. By the screw *d*, the brass tube *f* is attached to the casting *a b*. The diameter of the face *e*, and therefore the internal diameter of the tube *f*, should be 8 m.m. The length of the tube *f* is 9 c.m. Through the wall of this tube, four holes 5 m.m. in diameter are to be cut at such a height that the bottom of each hole will come 1 m.m. above the face *e* when the tube is screwed upon *a b*. These holes are of course opposite each other in pairs. The finished lamp is also shown in Fig. VII. To the tube *b* a caoutchouc tube of 5 to 7 m.m. internal diameter is attached; this flexible tube should be about 1 m. long, and its other extremity should be connected with the gas-cock through the intervention of a short piece of brass gas-pipe screwed into the cock. In cases where a very small flame is required, as, for example, in evaporating small quantities of liquid, a piece of wire gauze somewhat larger than the opening of the tube *f* should be laid across the top of the tube, and its projecting edges pressed down tightly against the sides of the tube before the lamp is lighted. In default of this precaution the flame of a Bunsen's burner, when small and exposed to currents of air, is liable to pass down the tube and ignite the gas at *d*.

Fig. VII.

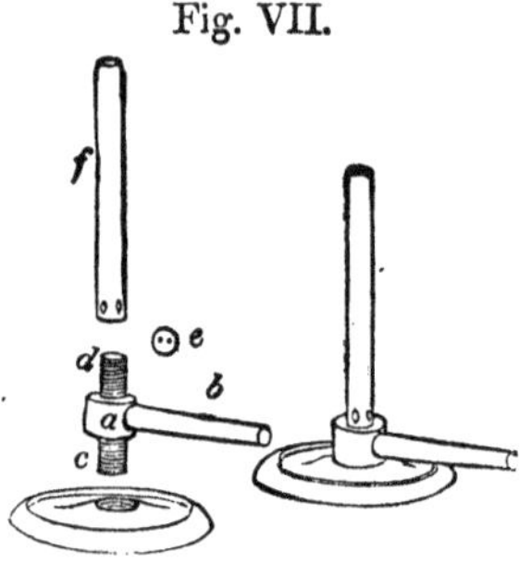

A lamp to give a powerful flame 8 or 10 c.m. long, suitable for heating tubes, may be very simply constructed by boring two holes, entering the side and issuing at the upper face, through a block of compact hard wood, 10 c.m. by 6·5 c.m. by 5·5 c.m., and fitting short pieces of brass tubing into the holes so formed. To the tubes at the side are attached the caoutchouc tubes which deliver the gas, and from the tubes at the top

Fig. VIII.

the gas issues under a sheet-iron funnel closed at the top with wire gauze. Above this gauze, the mixture of gas and air is to be lighted. The iron funnel will be readily understood from the accompanying figure, and the following dimensions:—Length of the wire gauze 10 c.m.; width of the gauze 5 c.m.; width at *a b* 9 c.m.; height of the line *a b* from the table 8·5 c.m.; whole height of the funnel 21 c.m. A partition parallel to *a b* divides the funnel into two equal parts from the gauze to the level of *a b*.

6. *Blast-lamps and Blowers.*—For drawing, bending, and closing large glass tubes, a blast-lamp is necessary. The best form is that sold under the name of Bunsen's Gas Blowpipe. Its construction and the method of using it may be learned from Fig. IX.; *a b* is the pipe through

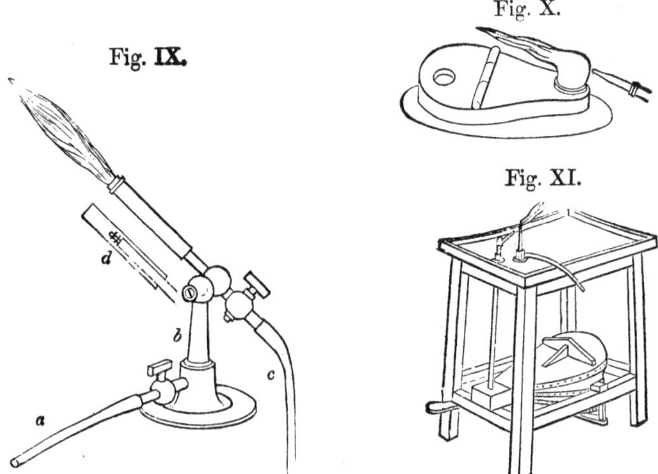

Fig. IX. Fig. X. Fig. XI.

which the gas enters, *c* is the tube for the blast of air; the relation of the air-tube to the external gas-tube is shown at *d*; there is an outer sliding tube by which the form and volume of the flame can be regulated.

If gas is not to be had, a lamp burning oil or naphtha must be employed. Fig. X. represents a common tin glassblower's lamp, suitable for burning oil. A large wick is essential, whether oil or naphtha be the combustible.

For every blast-lamp a blowing-machine of some sort is necessary. To supply a constant blast it is essential that the bellows be of that onstruction called double. Figs. XI. and XII. represent two forms of blowpipe-table; their height is that of an ordinary table, from which dimension the other proportions may be inferred. A small double-acting bellows is now made for the use of dentists, which is available at any table by the help of a caoutchouc tube to conduct the blast to the jet. These bellows are too small to give a steady flame of large size, but will nevertheless answer for most of the glass-blowing necessary in the execution of the experiments described in this manual.

Where an abundant supply of water is at command, the following blowing apparatus is very convenient. A tin pipe, *a b* (Fig. XIII.),

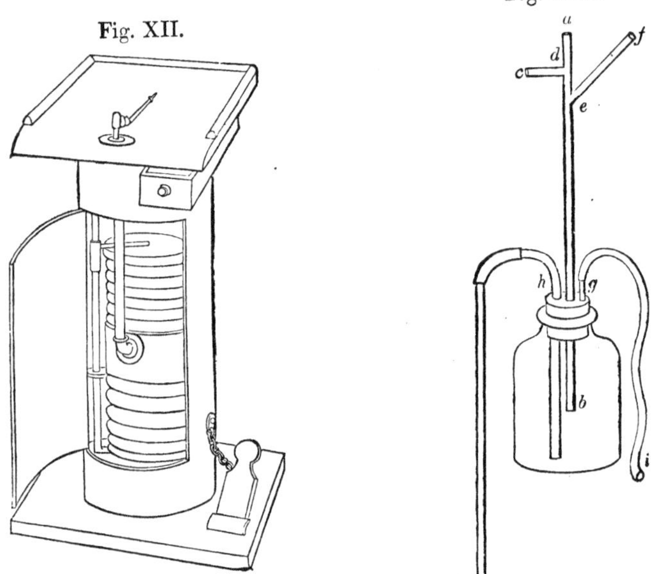

Fig. XII. Fig. XIII.

about one metre long and about 13 m.m. in diameter, has two smaller pipes, 12 to 16 c.m. long, soldered into it; these small pipes are 8 m.m. in diameter; one, *c d*, is inserted at right angles 12 c.m. from the end, the other, *e f*, 2·5 c.m. lower, at an angle of 45°. The lower end of the tube passes through the cork of a wide-mouthed five-litre bottle, extending rather more than halfway down. Two glass tubes also pass through the cork of the bottle—a short small tube *g*, No. 4, which should reach some 16 c.m. above the cork, but should not project into the bottle, and a larger tube *h*, No. 2, extending to the bottom of the bottle. The outer end of the tube *h* bends over and is connected by caoutchouc tubing with a straight tube of equal diameter. This last arrangement forms the siphon. To the tube *g* a caoutchouc tube, *g i*, is attached to convey the blast to any desired point. To produce a blast, the water-cock is connected with the tube *c d* by means of a caoutchouc tube. When the water is turned on, the caoutchouc tube *g i* is closed for a moment with the thumb and finger. This starts the water through the siphon, and immediately a continuous and powerful blast of air rushes out through the tube *g i*, and may be carried directly to the blowpipe. The siphon must be capable of carrying off a larger stream of water than that which is allowed to enter, so that there shall never be more than 3 or 4 c.m. of water in the bottle. By regulating the water-cock, the proper supply of water may be determined.

The same apparatus may be used as an aspirator. When the instrument is to be used to draw air through any apparatus, the tube *g i* is closed by inserting a glass rod; the upper end of the tube *a b* is closed with a cork, and the tube *e f* is connected with the apparatus through which the current of air is to be drawn. The force of the current of air is to a certain degree affected by the size of the tube *a b*; to diminish the effective calibre of this tube, in case a gentle current of air is required, a glass rod as long as the tube may be passed down the tube through a cork inserted at *a*. The same apparatus may thus be made to produce a gentle or a powerful current of air.

7. *Caoutchouc.*—Vulcanized caoutchouc is a most useful substance in the laboratory, on account of its elasticity and because it resists so well most of the corrosive substances with which the chemist deals. It is used in three forms:—(1) in tubing of various diameters comparable with the sizes of glass tubing; (2) in stoppers of various sizes to replace corks; (3) in sheets. Caoutchouc tubing may be used to conduct all gases and liquids which do not corrode its substance, provided that the pressure under which the gas or liquid flows be not greater or its temperature higher than the texture of the tubing can endure. The flexibility of the tubing is a very obvious advantage in a great variety of cases. Short pieces of such tubing, a few centimetres in length, are much used, under the name of connectors, to make flexible joints in apparatus of which glass tubing forms part; flexible joints add greatly to the durability of such apparatus, because long glass tubes bent at several angles and connected with heavy objects, like globes, bottles, or flasks full of liquid, are almost certain to break even with the most careful usage; gas-delivery-tubes, and all considerable lengths of glass tubing, should invariably be divided at one or

more places, and the pieces joined again with caoutchouc connectors. The ends of glass tubing to be thus connected should be squarely cut, and then rounded in the lamp, in order that no sharp edges may cut the caoutchouc; the internal diameter of the caoutchouc tube must be a little smaller than the external diameter of the glass tubes; the slipping on of the connector is facilitated by wetting the glass. In some cases of delicate quantitative manipulations, in which the tightest possible joints must be secured, the caoutchouc connector is bound on to the glass tube with a silk or smooth linen string; the string is passed as tightly as possible twice round the connector and tied with a square knot; the string should be moistened, in order to prevent it from slipping while the knot is tying.

Caoutchouc stoppers of good quality are much more durable than corks, and are in every respect to be preferred. The German stoppers are of excellent shape and quality; the American, being chiefly intended for wine-bottles, are apt to be too conical. Caoutchouc stoppers can be bored, like corks (see the next section), by means of suitable cutters, and glass tubes can be fitted into the holes thus made with a tightness unattainable with corks. German stoppers may be bought already provided with one, two, and three holes. It is not well to lay in a large stock of caoutchouc stoppers; for, though they last a long time when in constant use, they not infrequently deteriorate when kept in store, becoming hard and somewhat brittle with age. These stoppers must not be confounded with the very inferior caps which were in use a few years ago.

Pieces of thin sheet caoutchouc are very conveniently used for making tight joints between large tubes of two different sizes, or between the neck of a flask, or bottle, and a large tube which enters it, or between the neck of a retort and the receiver into which it enters. A sufficiently broad and long piece of sheet caoutchouc is considerably stretched, wrapped tightly over the glass parts adjoining the aperture to be closed, and secured in place by a string wound closely about it and tied with a square knot.

8. *Corks.*—It is often very difficult to obtain sound, elastic corks of fine grain and of size suitable for large flasks and wide-mouthed bottles. On this account, bottles with mouths not too large to be closed with a cork cut across the grain should be chosen for chemical uses, in preference to bottles which require large corks or bungs cut with the grain, and therefore offering continuous channels for the passage of gases, or even liquids. The kinds sold as champagne-corks and as satin corks for phials are suitable for chemical use. The best corks generally need to be softened before using; this softening may be

effected by rolling the cork under a board upon the table, or under the foot upon the clean floor, or by gently squeezing it on all sides with the well-known tool expressly adapted for this purpose, and thence called a cork-squeezer. Steaming also softens the hardest corks.

Corks must often be cut with cleanness and precision; a sharp, thin knife, such as shoemakers use, is desirable for this purpose. When a cork has been pared down to reduce its diameter, a flat file may be employed in finishing; the file must be fine enough to leave a smooth surface upon the cork; in filing a cork, a cylindrical, not a conical form should be aimed at.

In boring holes through corks to receive glass tubes, a hollow cylinder of sheet brass sharpened at one end is a very convenient tool. Fig. XIV. represents a set of such little cylinders of graduated sizes, slipping one within the other into a very compact form; a stout wire, of the same length as the cylinders, accompanies the set, and serves a double purpose. Passed transversely through two holes in the cap which terminates each cylinder, it gives the hand a better grasp of the tool while penetrating the cork; and when the hole is made, the wire thrust through an opening in the top of the cap expels the little cylinder of cork which else would remain in the cutting cylinder of brass. That cutter whose diameter is next below that of the glass tube to be inserted in the cork is always to be selected; and if the hole it makes is too small, a round file must be used to enlarge the aperture; the round file, also, often comes into play to smooth the rough sides of a hole made by a dull cork-borer. Cutters which have been dulled by use may be sharpened by filing or grinding down their outer bevelled edges and then paring off any protuberance or roughness which may remain upon the inside of the edge with a sharp penknife. A pair of small callipers is a very convenient, though by no means essential tool in determining which size of cutter to employ. A flask which presents sharp or rough edges at the mouth can seldom be tightly corked, for the cork cannot be introduced into the neck without being cut or roughened; such sharp edges must be rounded in the lamp. In thrusting glass tubes through bored corks, the following directions are to be observed:—(1) The end of the tube must not present a sharp edge capable of cutting the cork. (2) The tube should be grasped very close to the cork, in order to escape cutting the hand which holds

Fig. XIV.

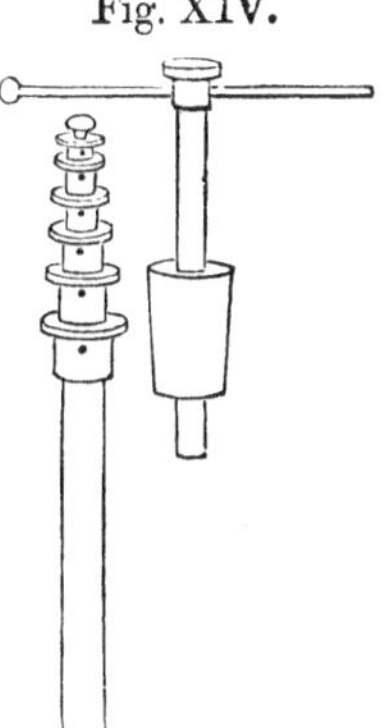

the cork, should the tube break; by observing this precaution the chief cause of breakage, viz. irregular lateral pressure, will be at the same time avoided. (3) A funnel-tube must never be held by the funnel in driving it through a cork, nor a bent tube grasped at the bend, unless the bend comes immediately above the cork. (4) If the tube goes very hard through the cork, the application of a little soap and water will facilitate its passage; but if soap is used the tube can seldom be withdrawn from the cork after the latter has become dry. (5) The tube must not be pushed straight into the cork, but screwed in, as it were, with a slow rotary as well as onward motion. Joints made with corks should always be tested before the apparatus is used, by blowing into the apparatus and at the same time stopping up all legitimate outlets. Any leakage is revealed by the disappearance of the pressure created. To the same end, air may be sucked out of an apparatus and its tightness proved by the permanence of the partial vacuum. To attempt to use a leaky cork is generally to waste time and labor and to insure the failure of the experiment. When, however, a leak is only discovered during the actual progress of the experiment, it is sometimes possible to save the experiment by using a lute; for this purpose wax applied with a warm knife, or a paste made of rye-meal and water may be used; common sealing-wax also is sometimes a useful makeshift.

9. *Iron Stand, Sand-bath, and Wire Gauze.*—To support vessels over the gas-lamp, an iron stand is used consisting of a stout vertical rod fastened into a heavy cast-iron foot, and three iron rings of graduated sizes secured to the vertical rod with binding-screws; all the rings may be slipped off the rod, or any ring may be adjusted at any convenient elevation. The general arrangement is not unlike that of the stand which makes part of the Berzelius lamp (Fig. VI.), although simpler and cheaper. As a general rule, it is not best to apply the direct flame of the lamp to glass and porcelain vessels; hence a piece of wire gauze is stretched loosely over the largest ring, and bent downwards a little for the reception of round-bottomed vessels; on this gauze, flasks, retorts, and porcelain dishes are usually supported. In a few cases, in which a very gradual and equable heat is required, the wire-gauze is replaced by a small shallow pan, beaten out of sheet-iron, and filled with dry sand. This arrangement is called a sand-bath. With the aid of annealed iron wire, the iron stand may be made available for supporting tubes over the lamp. Crucibles, or dishes, too small for the smallest ring belonging to the stand, are conveniently supported on an equilateral triangle made of three pieces of soft iron wire twisted together at the apices; this triangle is laid on one of the rings of the

stand. An iron tripod (that is, a stout ring supported on three legs) may often be used instead of the stand above described; but it is not so generally useful, because of the difficulty of adjusting it at various heights; with a sufficiency of wooden blocks wherewith to raise the lamp or the tripod as occasion may require, it may be made available.

10. *Pneumatic Trough.*—The pneumatic trough is a contrivance which enables us to collect and confine gases in suitable vessels, and to decant them from one vessel to another. Its efficiency depends on the pressure of the atmosphere, which, as we know, is capable of supporting a column of water 10·33 metres long or a column of mercury 76 c.m. long (see § 7), provided that the liquid column be so arranged that the atmospheric pressure shall be fully felt upon the foot of the column, but not at all upon its head. If a tube, closed at one end and open at the other, and of any length less than 10·33 m., be completely filled with water, and then inverted so that its open end shall dip beneath some water held in a basin or saucer, the tube will remain full of water when the thumb or cork, which closed the open end while the immersion was effected, is withdrawn. What is true of a tube is equally true of a bell, or other vessel closed at one end, of any diameter or shape, provided its height be not greater than 10·33 m.; and the principle which applies to water is equally applicable to mercury, except that the height of the mercury column which the average atmospheric pressure can hold up is only 76 c.m., because mercury is 13·596 times as heavy as water. If a few bubbles of any gas insoluble in water should be delivered beneath the open end of a tube, or bell, thus standing full of water in apparent defiance of gravitation, the gas would rise to the top of the tube, by virtue of being lighter than the water, and the exact volume of water displaced by the gas, small or large, would drop into the basin or saucer beneath. If the gas were thus delivered continuously beneath the tube or bell, we should finally get the tube or bell full of gas, without admixture of air, and sealed at the bottom by the water in the basin or saucer. If mercury were the liquid, the operation would be precisely the same, except as regards the height of the tube or bell. Even this difference of possible height is not noticeable in practice, because bells and bottles more than 50 c.m. high are very seldom used with either liquid. On account of its costliness, mercury is rarely used, unless the gas to be collected, or experimented upon, be soluble in water. A trough for mercury is made as small as possible for the same reason. It is obvious that the object of a pneumatic trough may be accomplished under a great variety of forms. Any bucket or tub with a hanging shelf in it may

be made to serve. It will be sufficient to describe two convenient forms of the apparatus.

A cheap pneumatic trough is represented in Fig. XV. Its materials are durable and its capacity sufficient. It consists of two pieces, 1st, a stoneware pan, about 30 c.m. in diameter on the bottom, with sides sloping slightly outwards and rising to the height of about 10 c.m.; 2nd, a deep flower-pot saucer about 15 c.m. in diameter, with one hole bored through the middle of the bottom, and a second arched hole nipped out of its rim; this saucer is inverted in the pan. If this second piece be made expressly for this purpose, it should be made about 5 c.m. high, and its interior should be rounded to the hole in the centre, while the outside is left flat like the flower-pot saucer. For the saucer may be conveniently substituted two slabs or blocks of any stone, like soapstone, sandstone, or marble, made of even thickness and laid side by side in the water-pan with an interval between them which permits the gas-delivery-tube to come beneath the mouth of an inverted bottle or cylinder supported on the two blocks over the intervening crack. To use this apparatus, the pan is filled with water to a level about 2 c.m. above the top of the inverted saucer; the bottle, cylinder, or bell which is to receive the gas is completely filled with water from a pitcher or water-cock, then closed with the hand of the operator or with a flat piece of glass or wood, inverted into the pan, and placed on the saucer over the hole in its centre; the end of the gas-delivery-tube is thrust through the side hole in the saucer; and the gas, rising through the centre hole, bubbles up into the bottle or cylinder placed to receive it. While one bottle is filling with gas, another is made ready to replace it; and when the first is full, it is pushed off the centre hole of the saucer, and the second bottle is brought over the hole. A bottle full of gas may be removed from the trough by slipping beneath the mouth of the bottle a shallow plate or dish, and then lifting plate and bottle out of the pan together in such a manner that water enough to seal the mouth of the bottle shall remain in the plate. The gas in one bottle may be decanted upwards into another, by filling the second bottle with water, and then carefully inclining the bottle containing the gas so as to bring its mouth under the mouth of the bottle which is full of water, keeping the mouths of both bottles all the time beneath the surface of the water in the pan. If the gas which has been collected is heavier than air, a bottle of it may be withdrawn from the

Fig. XV.

water-pan and got at for use, by simply slipping a flat piece of glass or wood beneath its mouth so as to close it rather tightly, and then standing the bottle, mouth upward, upon the table. If the cover be then removed from the bottle, the gas will not flow out, though it will slowly diffuse into the air. As the water with which the bottles or cylinders are filled falls into the pan when displaced by gas, it is possible that the pan may become inconveniently full if many large bottles are used; this difficulty must be remedied by dipping water out of the pan, and so restoring the true level.

Where considerable quantities of gas are frequently to be handled, and large vessels are therefore necessary, a large apparatus, shown in Fig. XVI., is much more convenient than the small pan, which suffices for all common experiments. The form of this larger pneumatic trough and the mode of using it will readily be understood from the figure; the depth and width of the tank or well must be determined by the size of the bells and cylinders which are to be sunk in it, and the length and breadth of the shallow part or shelf by the number of bells or jars of gas which are likely to be in use at any one time. The deep groove in the shelf permits a glass or caoutchouc tube to pass without compression under a bell whose rim projects over the groove. Such a trough is best made of zinc or lead. It is very convenient to have it sunk in a table, and permanently provided with a water-cock and drain-pipe. A chief merit of this instrument is that the glass vessels used can be filled with water by sinking them in the well much more conveniently than from a pitcher or water-cock.

Fig. XVI.

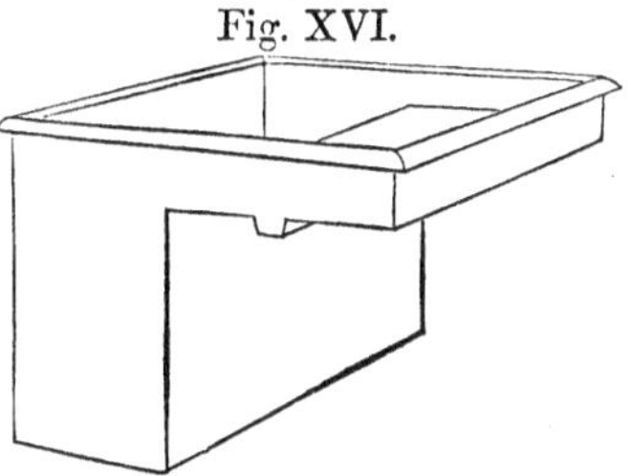

A pneumatic trough for mercury may be made either of wood, iron, or stone. For all common uses, it is very well cut out of a solid block of compact hard wood, which will not split. Small cylinders or bells only can be used, and the well of the trough should be scooped out but a little larger than the bell or cylinder selected, with its principal dimension horizontal, and its bottom curved to fit the cylindrical bell which is to be laid in it; the shelf, too, should have but a small area, sufficient only for four or five bells of 3 or 4 c.m. diameter.

In using a pneumatic trough, of any construction or dimensions, the student should be on his guard against two difficulties of possible occurrence—against the *sucking back* of the liquid in the trough into the gas-generating apparatus, and against the leakage sometimes induced by the pressure created by thrusting the gas-delivery-tube deep under water or mercury. The first of these difficulties is the most serious. When the flow of gas from a heated flask or tube is suddenly arrested, in consequence of some reduction of temperature, or from any other cause, it often happens that the volume of gas in the generating apparatus contracts, and the cold water or mercury from the trough rises in the delivery-tube to fill the void; if the contraction is so considerable as to suffer the cold liquid to penetrate into the hot flask or tube, an explosion almost inevitably ensues, which fractures the apparatus, if it does no worse damage. In collecting over water a gas somewhat soluble in that liquid, this danger is especially imminent. The occurrence of such accidents may be effectually guarded against by paying attention to the following directions:—(1) Whenever it is proposed to stop an evolution of gas which has been going on from a hot flask or tube, withdraw the delivery-tube from the water *before* extinguishing the lamp, and shake off from the bent end of the tube the drops of water which are apt to adhere to it; the lamp may then be safely put out, for air can enter the apparatus through the open tube. (2) When the flow of gas from a hot apparatus is observed to slacken, watch closely the escape of the gas from the delivery-tube, and as soon as any tendency to reflux of water is detected, lift the delivery-tube quickly out of the water, or, better, slip off the caoutchouc connector, which should always be found between the flask and the water-pan on every such piece of apparatus; if there be no connector, the cork must be loosened in the neck of the flask. Air will thus be admitted to the hot flask or tube.

These precautions apply more particularly to the cases where gas is evolved from dry materials, as in making oxygen or nitrous oxide; when a liquid is contained in the generating flask, a safety-tube is a sure protection against the danger of sucking back. The atmospheric pressure can force air into a flask, in which a partial vacuum has been created, through the safety-tube, by lifting and displacing a column of the liquid whose height is the length of that portion of the safety-tube which dips beneath the liquid. Unless the liquid in the flask be extraordinarily dense, the force required to do this will be very much less than that required to lift a column of water whose height is determined by the elevation of the highest point of the delivery-tube above the level of the water in the pan.

When the gas coming from the generating flask has to force out and keep out of the delivery-tube a column of water measured from the lowest point of the tube to the surface of the water in the pan, a pressure determined by the height of this column is established upon the interior of the flask and upon every joint of the apparatus. Hence an apparatus will sometimes leak, and refuse to deliver gas at the desired point, when its delivery-tube is deeply immersed, while it does not leak if the tube merely dip beneath the surface of the water. With mercury the pressure of a few centimetres is very considerable, on account of the high specific gravity of the fluid, so that this difficulty is more likely to occur with this metal than with water. Tight joints prevent the occurrence of this difficulty. A partial remedy is to dip the delivery-tube as little as possible below the surface of the fluid in the trough.

11. *Gas-holders.*—A small gas-holder, very convenient for many uses, is made from a common glass bottle in the following manner:—*A* (Fig. XVII.) is a bottle of 4–6 litres capacity; through the cork in its neck pass two glass tubes (No. 6), of which one reaches the bottom of the bottle, while the other merely penetrates the cork; with the outer end of the first tube a caoutchouc tube *c* is connected, with the outer end of the second a common gas-cock *a*. The bottle being first completely filled with water, the apparatus which generates, or contains, the gas to be introduced into the holder is connected with the tube carrying the cock *a*; this cock is open. As the gas presses in, the water mounts in the long tube, and flows out by the siphon *c*. In order to relieve the gas from this pressure at the beginning, it is only necessary to suck a little at *c*. The tube *c* should of course be thrust into a sink or drain-pipe.

Fig. XVII.

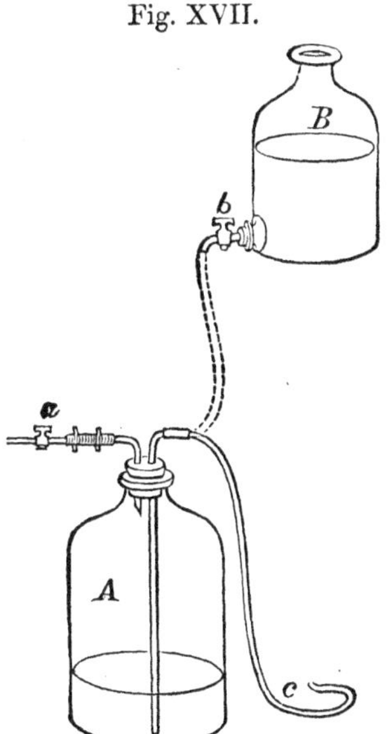

To get gas out of the bottle, thus charged, the cock *a* is closed, and the flexible tube *c* is lifted up and connected, as shown in the figure, with a bottle of water *B* placed on a shelf, or stand, somewhat above the bottle *A*. When the cock *b* is opened, the gas in *A* is pressed upon by the weight of the superincumbent column of water, and may therefore be made to issue at will from the cock *a*. The higher *B* is placed above *A*, the greater will be the force with which the gas will issue.

If a moderate or easily regulated water-pressure is at hand, supplied by city water-works or a reservoir in the upper part of the building, the bottle *B* is unnecessary, and the flexible tube *c* may be connected with such a water-supply whenever gas is to be pressed out of the gas-holder *A*.

When larger quantities of gas are to be stored for use, a metallic gas-holder, whose construction and proportions are shown in Fig. XVIII., is advantageously employed. The open cistern *B* is supported over the vessel *A* on two columns *c c*, and two tubes *a* and *b*; of these tubes the first, *a*, reaches from the bottom of *B* nearly to the bottom of *A*, while the second, *b*, starts from the bottom of *B* and *just* enters the arched top of *A* without projecting into it; *d* is a short large tube, sloping upwards and outwards, and capable of being tightly closed with a cork or caoutchouc stopper; *g* is a glass gauge to show the height of the water in the vessel *A*; *e* is the discharge-pipe. To fill the gas-holder with water, close *d*, open the stopcocks *a*, *b*, and *e*, and pour water into the cistern *B*; the water entering *A* will expel the air through *b* and *e*; when the water begins to flow through *e*, close that stopcock and expel the rest of the air through *b*. The gas-holder may now be filled with gas by displacing the water in the following manner:—Close all the stopcocks, withdraw the cork or stopper from *d*, and introduce the tube which delivers the gas through that opening. A short piece of caoutchouc tubing makes the best end for the gas-delivery-tube; but glass tubing will answer the purpose if the end be slightly bent upwards. The water flows out at *d* as fast as the gas enters, and the gas-holder should therefore stand in a shallow metal tray provided with a drain-pipe. When the desired quantity of gas has been introduced, close *d*. To draw gas out of a gas-holder of this construction, the cistern *B* is filled with water and the cock *a* is opened; under the pressure thus established the gas may be drawn off through *e*, or allowed to rise through *b* into bottles or bells filled with water and held over the mouth of the tube *b* in the cistern *B*; in this last case *B* answers the purpose of a pneumatic trough.

Fig. XVIII.

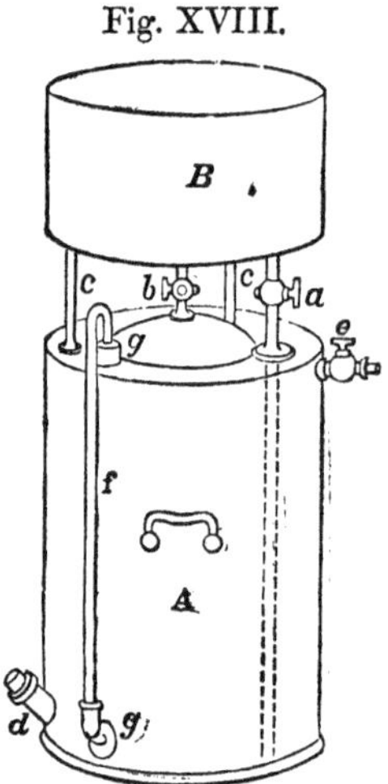

This gas-holder may be cheaply made of zinc; any gas-fitter can supply the necessary stopcocks; care must be taken that the glass tube which constitutes the gauge is fitted air-tight to the gas-holder. The stopcock *e* need not end in a screw; tubes may be as well connected with it by caoutchouc. The available pressure under which the gas in the holder streams out at *e* is of course limited by the elevation of *B* above *A*, which must always be moderate. When a stronger pressure is desirable, as in getting the oxyhydrogen blowpipe-flame, for example, a heavier water-column may be obtained by screwing a tall tube with a capacious funnel on the top of it into the tube *a*, where it opens into the bottom of the cistern *B*. A piece of common iron or copper gas-pipe about a metre long, answers this purpose very well; the funnel at the top should hold two or three litres, and must be kept full of water from a cask or tub provided with a cock and placed just above the funnel. Where a water-supply, with moderate pressure, is obtainable, it may be used to keep the funnel full, or to replace the funnel altogether, if directly connected with the tube *a*. A gas-holder, measuring not more than 50 c.m. in total height, is not too heavy to be portable, and during the process of filling may be placed over a tub; but a gas-holder of much larger proportions is better made a fixture, and provided in a permanent manner with drain-pipe and water-supply. The gas-holder thus described is that which is the most generally useful; it may be charged from any glass flask, retort, or bottle, without any pressure being exerted upon the glass vessel; and unused gas contained in any sort of bell, bottle, or flask can be very readily transferred to such a gas-holder without waste and with very little trouble.

Fig. XIX.

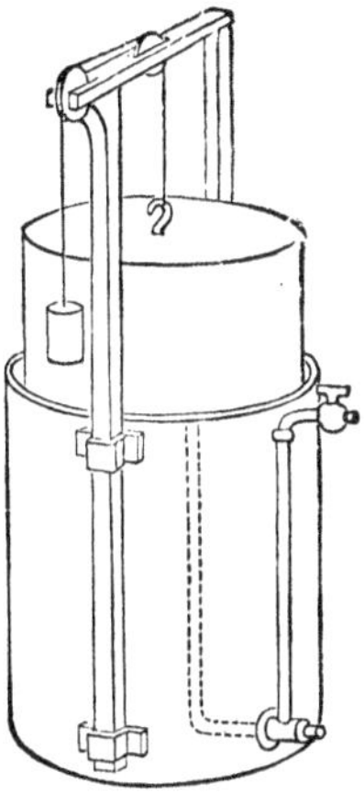

A cheaper gas-holder may be made on the plan of the large gas-holders, improperly called gasometers, used in gas-works. Fig. XIX. represents a gas-holder of this sort. Over a tank of water, which may be a cylinder of zinc as shown in the figure, or a headless pork- or oil-barrel, or any other water-tight tub, is balanced by pulleys and weights a tight bell of zinc, not too large for complete immersion in the tank. The U-tube, shown in the figure, which may be either of lead or brass, serves both to introduce and deliver the gas. To fill such a gasometer, open the cock, lift the counterbalancing weight, and let the bell sink into the water; then connect the vessel from which the gas is delivered with the tube of the holder, counterpoise the bell, and the gas coming from the generator will gradually lift the bell out of the water. To force the gas out of the holder it is only necessary to remove the counterbalancing weight; the weight of the bell forces out the gas; and if this pressure be not sufficient, additional weights may be placed on the top of the bell. Gas-holders of this construction, unless very small, are too heavy, when filled with water, to be carried about; but this difficulty may be obviated, when economy is not specially to be regarded, by placing within the lower cylinder, or tank, a second air-tight cylinder as a core, so as to leave only a narrow space between the inner and outer cylinders for the water into which the upper bell dips. Elegant, but not cheap, gas-holders are thus made, which are convenient for some uses, but are not so generally to be recommended as those of the construction first described. The vessel from which a gas-holder with counterpoised bell is charged is always subjected to some pressure, slight if the pulleys, cords, and weights are in perfect order, but more frequently considerable on account of the difficulty of maintaining such an apparatus in perfect condition.

12. *Deflagrating-Spoon.*—The little cup which holds combustible material, to be burnt in a bottle or jar of gas, is called a deflagrating-spoon; it may be cheaply made by hollowing a hemispherical cup out of a cube of chalk about 2 c.m. on a side, and attaching a stout iron or brass wire to the chalk, in such a manner that the cup will be right side up when hung by the wire in a jar of gas; the upper end of this wire should be straight, that it may be thrust through the cork or piece of wood which covers the mouth of the bottle or jar. The piece of chalk may be replaced by a bit of the cylindrical chalk crayon commonly used with blackboards. A piece of crayon 1·5 c.m. in length will make a sufficient spoon. A small cupel is a convenient ready-made substitute for the chalk cup. Brass deflagrating-spoons are also to be had of philosophical-instrument-makers.

13. *Platinum Foil and Wire.*—A piece of platinum foil about 3 c.m. square is useful in experiments involving the evaporation of a drop or two of liquid and the ignition of the residue; it is an essential tool in qualitative analysis. The foil should be at least so thick that it does not crinkle when wiped; and it is more economical to get foil which is too thick than too thin, for it requires frequent cleaning. A bit of platinum wire, not thicker than a No. 10 needle, and 20 c.m. long, will last a long time with careful usage. No other metal, and no mixture of substances from which a metal can be reduced, must ever be heated on platinum foil or wire, for platinum forms alloys with other metals which are much more fusible than itself. If once alloyed with a baser metal, the

platinum ceases to be applicable to its peculiar uses. Platinum may be cleaned by boiling it in either nitric or chlorhydric acid, by fusing the acid sulphate of sodium or potassium upon it, or by scouring it with fine sand. Aqua regia and chlorine-water dissolve platinum; the sulphides, cyanides, and oxides of sodium and potassium, when fused in platinum vessels, slowly attack the metal.

14. *Filtering.*—Filtration is resorted to in order to separate a finely divided solid from a liquid. The filter may be made of paper, cloth, tow, cotton, asbestos, and other substances. Paper is the substance oftenest used. A good filtering-paper must be porous enough to filter rapidly, and yet sufficiently close in texture to retain the finest powders; and it must also be strong enough to bear, when wet, the pressure of the liquid which must be poured upon it. For delicate experiments it is also necessary that filtering-paper should contain no soluble salts, and but a very small proportion of incombustible material, which would remain as ash were the filter to be burned. Filtering-paper (which is generally sold in sheets) is first cut into circles of various diameters, adapted to the various scales of operation and quantities of liquids to be filtered. To prepare a filter for use, one of these circles is folded over on its own diameter, and the semicircle thus obtained is folded once upon itself into the form of a quadrant; the paper thus folded is opened so that three thicknesses shall come upon one side, and one thickness upon the other, as shown in Fig. XX.; the filter is then placed in a glass funnel, the angle of which should be precisely that of the opened paper, viz. 60°; after being wetted, it is ready to receive the liquid to be filtered. The paper may be so folded as to fit a funnel whose angle is more or less than 60°; but this is the most advantageous angle, and glass funnels should be selected with reference to their correctness in this respect.

Fig. XX.

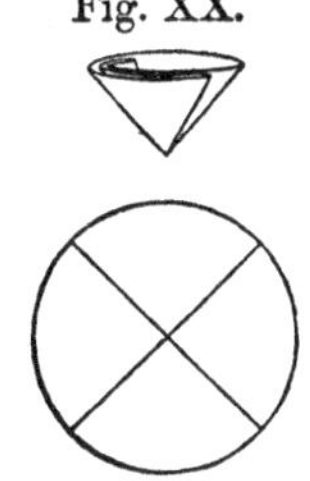

Coarse and rapid filtering can be effected with cloth bags, also by plugging the neck of a funnel loosely with tow or cotton. If a very acid or very caustic liquid, which would destroy paper, cotton, tow, or wool, is to be filtered, the best substances wherewith to plug the neck of the funnel are asbestos and gun-cotton, neither of which is attacked by such corrosive liquids.

Fig. XXI.

The glass funnel which holds the filter generally requires an independent support; for it is seldom judicious, or possible, to support the funnel directly upon the vessel which receives the *filtrate*, as the clear liquid which runs through the filter is called. The iron stand may be used for this purpose; and wooden stands, of a similar construction, adapted expressly for holding funnels, are very convenient and not expensive. In general, care should be taken that the lower end of the funnel touch the side or edge of the vessel into which the filtrate descends, in order that the liquid may not fall in drops, but run quietly down without splashing. Sometimes there is no objection to thrusting a funnel directly into the neck of a bottle or flask; but in this case an ample exit for the air in the bottle must be provided (Fig. XXI.).

15. *Drying Gases.*—It is often desirable to remove the aqueous vapor which is mixed with gases, collected over water or prepared from materials containing water. It very seldom happens that a gas can be prepared at one operation in so dry a state as to contain no vapor of water; this vapor must ordinarily be removed by a subsequent or additional process. Experience has shown that some gases are more easily dried than others; thus air, hydrogen, and common oxygen are thoroughly dried with great ease, but gaseous mixtures which contain antozone only with great difficulty; chlorine is three times as hard to dry as carbonic acid. These and similar facts must be borne in mind in constructing drying-apparatus. The common drying-process depends upon bringing the moist gas into contact with some liquid or solid which greedily and rapidly absorbs aqueous vapor. The three substances most used for this purpose are concentrated sulphuric acid, chloride of calcium, and dry quicklime. Sulphuric acid may be used in two ways: the gas may be made to bubble through a few centimetres depth of the liquid acid, or it may be forced to pass through the interstices of a column of broken pumice-stone which has been previously soaked in the acid. The latter method is the most effectual, because it secures a more thorough contact of the gas with the hygroscopic acid than is possible during the rapid bubbling of the light gas through a shallow layer of the dense liquid. The column of fragments of pumice-stone may be held in a U-tube, arranged like that shown in Fig. XXII.; but the vertical cylinder shown in the same figure is better adapted for this use, because the acid as it becomes dilute from absorption of moisture, gradually trickles from the pumice-stone, and is apt to collect in such quantity at

Fig. XXII.

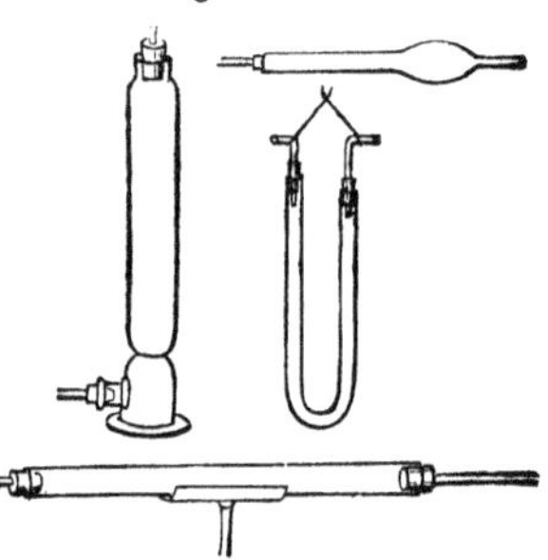

the bottom of the U-tube as to completely close the tube. In preparing the upright cylinder for use, the portion below the contraction is not filled with pumice-stone; it receives the drippings from the pumice-stone column. The gas to be dried enters by the lower lateral opening, and goes out at the top of the cylinder. Though especially adapted to the column of acid-soaked pumice-stone, this cylinder may very well be used with either of the other drying-agents, chloride of calcium or quicklime. Either of the forms of drying-tube represented in Fig. XXII. may be employed with these latter substances; in charging the horizontal tubes, bits of loose cotton-wool should first be placed against the exit-tube to prevent any particles of the chloride of calcium, or quicklime, from entering that tube; pieces of the perfectly dry solid are then introduced in such a way that the tube may be compactly filled with fragments which leave room for the gas to pass very deviously between them, but offer no direct channels through which the gas could find straight and quick passage. Quicklime must be charged much more loosely than chloride of calcium, because of its great expansion when moistened. Fused chloride of calcium is not so well adapted for drying gases as the unfused substance. It is not at all necessary that the fragments of chloride of calcium, or quicklime, should be of uniform size. When the tube is nearly full, a plug of loose cotton should be inserted before putting in the cork. A chloride-of-calcium tube, once filled, will often serve for many experiments; whenever out of use, its outlets should be covered with paper caps; or, better, caoutchouc connectors may be slipped upon the exit-tubes, and bits of glass rod thrust into these connectors. The moisture of the air is thus kept from the chloride of calcium. The dimensions of drying-tubes are of course very various; the bulb-tube shown in Fig. XXII. is seldom used with a greater length than 25 c.m.; when this form of tube is employed the gas should invariably enter by the end without a cork, where the small size of the tube permits direct connexion with a common gas-delivery-tube by means of a caoutchouc connector; the other horizontal tube, shown in the figure, may be of any length; but whenever a great extent of drying-surface is necessary, U-tubes have the advantage of compactness; for many can be hung upon one short frame. The upright cylinder may be from 25 c.m. to 40 c.m. in height. A good U-tube, with an addition which has the merit of economizing chloride of calcium, is shown in Fig. XXIII.; the addition consists of a short test-tube, into which the tube by which the gas comes in barely enters; a quantity of water is often

Fig. XXIII.

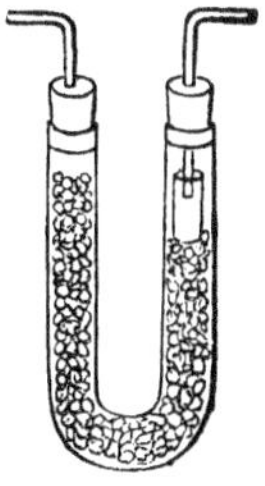

caught in this test-tube which otherwise would wet and spoil a considerable amount of chloride of calcium; the little test-tube may, of course, be taken out and emptied at will.

The choice between one or other of the three drying-substances is determined in each special case by the chemical relations of the gas to be dried; thus ammonia-gas, which is absorbed by sulphuric acid and by chloride of calcium, must be dried by passing it over quicklime, while sulphurous acid gas, which would combine with quicklime, must be dried by contact with sulphuric acid.

Fig. XXIV.

Fig. XXV.

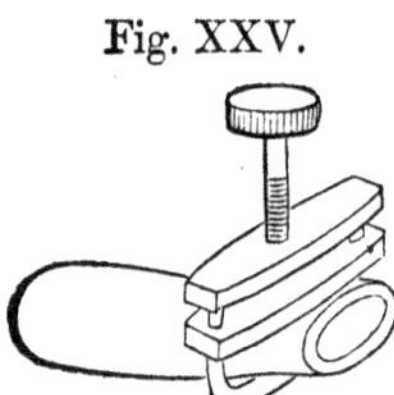

16. *Spring Clip and Screw Compressor.*—These are very convenient substitutes for the ordinary stopcock, and as such are in constant use in the laboratory. Their form, and the manner of their use, will be readily understood from the figures. As glass stopcocks are expensive and fragile, and metal stopcocks are usually out of the question, because so many gases and liquids attack the common metals, these excellent substitutes are used whenever a caoutchouc tube is not inadmissible; they cannot be used unless a bit of elastic tubing can be inserted into the apparatus which requires a cock.

Another effective mode of temporarily stopping or partially closing a caoutchouc tube, is to slip over the tube a common brass ring of about the same diameter as that of the tube, and then to thrust a slightly conical plug of hard wood or ivory between the ring and the flexible tube.

17. *Water-bath.*—It is often necessary to evaporate solutions at a moderate temperature which can permanently be kept below a certain known limit; thus, when an aqueous solution is to be quietly evaporated without spirting or jumping, the temperature of the solution must never be suffered to rise above the boiling-point of water, nor even quite to reach this point. This quiet evaporation is best effected by the use of a water-bath—a copper cup whose top is made of concentric rings of different diameters to adapt it to dishes of various size (Fig. XXVI.). This cup, two-thirds full of water,

Fig. XXVI.

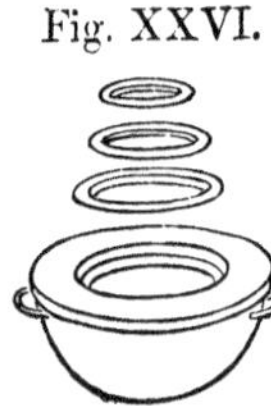

is supported on the iron stand over the lamp, and the dish containing the solution to be evaporated is placed on that one of the several rings which will permit the greater part of the dish to sink into the copper cup. The steam rising from the water impinges upon the bottom of the dish, and brings the liquid within it to a temperature which insures the evaporation of the water, but will not cause any actual ebullition. The water in the copper cup must never be allowed to boil away. Whenever a constant supply of steam is at hand, as in buildings warmed by steam, the copper cup above described may be converted into a steam-bath by attaching it to a steam-pipe by means of a small tube provided with a stopcock.

A cheap but serviceable water-bath may be made from a quart milk-can, oil-can, tea-cannister, or any similarly shaped tin vessel, by inserting the stem of a glass funnel into the neck of the can through a well-fitting cork. In this funnel the dish containing the liquor to be evaporated rests. The can contains the water, which is to be kept just boiling. On account of the shape of the funnel, dishes of various sizes can be used with the same apparatus.

The copper vessel first described may be conveniently employed when it is required to expose substances to a constant temperature higher than 100°. For this purpose the cup is filled with oil, wax, paraffine, or a solution of chloride of zinc or chloride of calcium; the flask or dish containing the substance to be heated should, in this case, be immersed in the fluid to about two-thirds of its depth; a thermometer must be used to indicate the temperature of such a bath. When oil, wax, or paraffine is used, the temperature must not be carried so high as to burn or decompose these organic matters, else a very disgusting vapor will be produced.

18. *Iron Retort.*—A retort, made of iron, of the form shown in Fig. XXVII., is a convenient utensil in making large quantities of oxygen, and in preparing illuminating-gas or marsh-gas. The iron top is fitted to the retort with a ground joint, fastened by a screw-clamp. When the top is removed, the whole inner surface of the retort is exposed—a decided advantage wherever the residue left in the retort after use is solid. A retort of about 300 c. c. capacity is amply large for most uses. A small iron kettle makes a serviceable retort; the lid must be luted on, and the nose becomes the exit-tube.

Fig. XXVII.

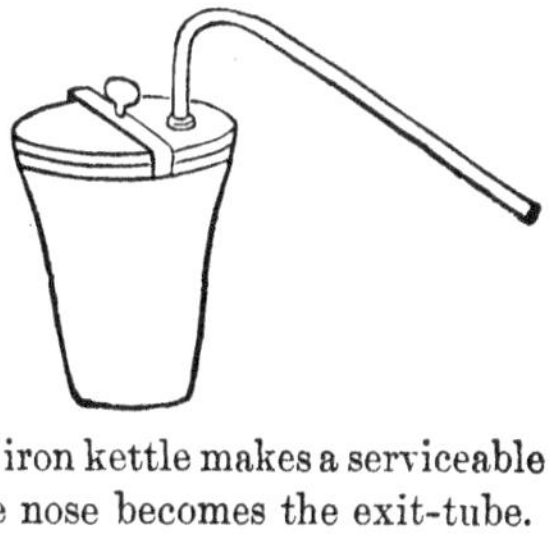

19 *Self-regulating Gas-generator.*—An apparatus which is always

ready to deliver a constant stream of hydrogen, and yet does not generate the gas except when it is immediately wanted for use, is a great convenience in an active laboratory or on a lecture-table. The same remark applies to the two gases sulphydric acid and carbonic acid, which are likewise used in considerable quantities, and which can be conveniently generated in precisely the same form of apparatus which is advantageous for hydrogen. Such a generator may be made of divers dimensions. The following directions, with the accompanying figure (Fig. XXVIII.), will enable the student to construct an apparatus of convenient size. Procure a glass cylinder 20 or 25 c.m. in diameter and 30 or 35 c.m. high; ribbed candy-jars are sometimes to be had of about this size; procure also a stout tubulated bell glass 10 or 12 c.m. wide and 5 or 7 c.m. shorter than the cylinder. Get a basket of sheet-lead 7·5 c.m. deep and 2·5 c.m. narrower than the bell-glass, and bore a number of small holes in the sides and bottom of this basket. Cast a circular plate of lead 7 m.m. thick and of a diameter 4 c.m. larger than that of the glass cylinder; on what is intended for its under side solder three equidistant leaden strips, or a continuous ring of lead, to keep the plate in proper position as a cover for the cylinder. Fit tightly to each end of a good brass gas-cock a piece of brass tube 8 c.m. long, 1·5 to 2 c.m. wide, and stout in metal. Perforate the centre of the leaden plate, so that one of these tubes will snugly pass through the orifice, and secure it by solder, leaving 5 c.m. of the tube projecting below the plate. Attach to the lower end of this tube a stout hook on which to hang the leaden basket. By means of a sound cork and common sealing-wax, or a cement made of oil mixed with red and white lead, fasten this tube into the tubulature of the bell glass airtight, and so firmly that the joint will bear a weight of 2 or 3 kilos. Hang the basket by means of copper wire within the bell 5 c.m. above the bottom of the latter. To the tube which extends above the stopcock attach by a good cork the neck of a tubulated receiver of 100 or 150 c. c. capacity, the interior of which has been loosely stuffed with cotton. Into the second tubulature of the receiver fit tightly the delivery-tube carrying a caoutchouc connector; into this connector can be fitted a tube adapted to convey the gas in any desired direction

Fig. XXVIII.

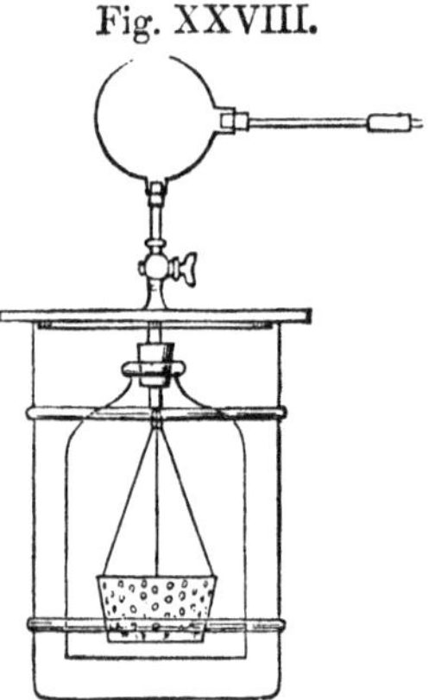

To charge the apparatus, fill the cylinder with dilute acid to within 10 or 12 c.m. of the top, place the zinc, sulphide of iron, or marble, as the case may be, in the basket, hang the basket in the bell, and put the bell-glass full of air into its place with the stopcock closed. On opening the cock, the weight of the acid expels the air from the bell, the acid comes in contact with the solid in the basket, and a steady supply of gas is generated until either the acid is saturated or the solid dissolved; if the cock be closed, the gas accumulates in the bell, and pushes the acid below the basket so that all action ceases. In cold weather the apparatus must be kept in a warm place. For generating hydrogen, sulphuric acid diluted with four or five parts of water is used; for sulphydric acid, sulphuric acid is diluted with fourteen parts of water; for carbonic acid, chlorhydric acid diluted with two or three parts of water is to be preferred.

20. *Glass Retorts, Flasks, Beakers, Test-tubes and Test-glasses.*—All glass vessels which are meant for use in heating liquids must have uniformly thin bottoms. Tubulated retorts are much more generally useful than those without a tubulature. As retorts are expensive in comparison with flasks, they are less used than formerly.

The neck of a flask should have such a form that it can be tightly closed by a cork, and the lip must be strengthened to resist the force used in pressing in the cork, either by a rim of glass added on the outside, or, better, by causing the rim itself to flare outward. The actual edge of the rim must never be sharp or rough, but always smooth and rounded by partial fusion. The thin-bottomed flasks in which olive-oil is sometimes imported from Italy are excellent for chemical uses; their edges always require, however, to be rounded in the lamp. These Florence flasks, which are very much to be recommended both on the ground of cheapness and of durability, may be cleaned by soaking them 24 hours in a weak caustic lye, and then washing them with boiling water.

Beakers are thin flat-bottomed tumblers with a slightly flaring rim. They are to be bought in sets or nests which sometimes include a large range of sizes. The small sizes are very useful vessels; the large are so fragile as to be almost worthless. Up to the capacity of about a litre, beakers are to be recommended for heating liquids whenever it is an object to have the whole interior of the vessel readily accessible.

Test-tubes are little cylinders of thin glass, with round, thin bottoms, and lips slightly flared. Their length may be from 12 c.m. to 18 c.m., and their diameter 1 c.m. to 2 c.m.; they should never have a diameter so large that the open end cannot be closed by the ball of the thumb. To hold the tubes upright a wooden rack is necessary. Besides the row of holes to receive a dozen test-tubes bottom down, the rack should have a row of pegs on which the test-tubes may be inverted when not in use; in this position the water in which they are rinsed drains off, and dust cannot be deposited within the tubes. Test-tubes are much used for heating small quantities of liquid over the gas- or spirit-lamp: they may generally be held by the upper end in the fingers without inconvenience, but if a liquid is to be boiled for some minutes in a test-tube, the tube must be held in wooden nippers, or in a strip of thick folded paper, nipped round the tube and grasped between the thumb and fore finger just outside the tube. The wooden nippers above mentioned are made of two bits of wood about 30 c.m. long hinged together at the back, and at once connected and kept apart by a sliding steel or brass spring, somewhat like those used on certain pruning-shears and some kinds of steel nippers. When a liquid is boiling actively in a test-tube, it sometimes happens that portions of the hot liquid are projected out of the tube with some force; the operator should therefore hold the tube in an inclined position, rolling it to a slight extent between the thumb and fore finger in order that all sides of the tube may be equally exposed to the flame; while thus using the tube, he should be careful not to direct it either towards himself or towards any other person in his neighbourhood. Test-tubes are cleaned by the aid of cylindrical brushes made of bristles caught between twisted wires, like those used for cleaning lamp-chimneys: they should have a round end of bristles.

Two precautions are invariably to be observed in heating glass and porcelain vessels of whatever form: first, the outside of the vessel to be heated must be made perfectly dry; secondly, the temperature must not be raised too rapidly. When a large flask or beaker containing a cold liquid is first warmed over a lamp, moisture almost invariably condenses upon the bottom of the vessel: this moisture should be wiped off with a cloth.

Stout conical glasses with strong stems and feet are convenient for many uses not involving the application of heat. They are called test-glasses, and may be had of various shapes and sizes. It is obvious that cheap wine- or beer-glasses and common jelly-tumblers would answer the purposes which these test-glasses serve.

21. *Measuring-glasses and Burettes.*—Measuring-glasses, divided into cubic centimetres, are made in the cylindrical form, and also in the flaring shape common in druggists' measuring-glasses; the cylindrical form is to be preferred. Such a glass of 250 c. c., or better of 500 c. c. capacity, is a very useful implement; a flask holding just one litre when filled to a mark upon its neck is also convenient. Smaller

quantities of liquid are measured with burettes. Mohr's burette is the most generally useful of all forms of this instrument (see Fig. XXIX., the right-hand instrument); it is a graduated tube drawn to a small bore at the bottom; a caoutchouc connector is slipped upon the bottom of the tube, a short bit of tube drawn to a fine point is thrust into the lower end of the connector, and a spring clip nips the connector between the two glass tubes. The spring clip closes the bottom of the burette, but it can of course be opened at will to permit the liquid in the burette to flow or drop out. The caoutchouc affects injuriously some of the liquids which are used in burettes; so that this common form of Mohr's burette is not always applicable. To avoid this difficulty the instrument may be made with a glass cock, but is then rather costly. Gay-Lussac's burette is available whenever the caoutchouc in Mohr's burette is objectionable. The construction of Gay-Lussac's burette is plainly to be seen in the figure (see Fig. XXIX., the left-hand instrument); a narrow tube runs up beside the large graduated tube to the top of the latter, and the liquid can be poured out in drops by gently inclining the instrument. Its fragility is a serious objection to this form of burette; the danger of breaking off the small tube is lessened, if a small piece of cork be inserted between the two tubes at the top, and a string tied round them both. A wooden foot in which it may stand upright upon a table is a convenient addition to Gay-Lussac's burette. Mohr's burette must be held upright in a suitable screw clamp, or be fastened on to a wooden frame in such a manner that the tube shall be vertical, firm, and at the same time easily detached. The fineness of measurement may be increased, without impairing the distinctness of the scale, by reducing the bore of the burette. For delicate work, burettes divided into tenths of a cubic centimetre are employed. The way in which the *reading-off* is effected is a matter of importance in using a burette; it is essential, 1st, that the burette should be vertical; 2nd, that the eye should be brought to a level with the surface of the fluid; 3rd, that a fixed standard should be adopted of what is to be considered the surface. If a burette, partly filled with a liquid, be held between the eye and a white wall, the surface of the liquid presents a light line which is nearly level, and just below this line a second line, which is dark and curved with the convexity downward. If a sheet of white paper be held immediately behind the tube, these two lines, though somewhat altered in appearance, are still distinctly

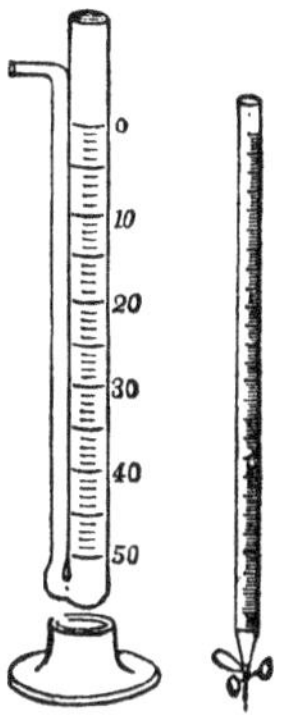

Fig. XXIX.

visible. They may be made still more distinct by using instead of white paper a card half white, half black, with a straight dividing line between the two colors. On holding this card with the white half uppermost, and the border line between white and black from 2 to 3 m.m. below the lowest visible dark line, two zones are brought out, a light zone and a dark zone, and the lower limit of the dark zone is made very distinct. Care must be taken to hold the card invariably in the same relative position, since, if it be held lower down, the lower border of the dark zone will move higher up. In practice the lower border of the dark zone is read off as the surface of the liquid, this being the most distinctly marked line. There is one exception to this rule; when an opaque solution of permanganate of potassium is to be measured in Gay-Lussac's burette, the upper border of the dark zone must be held to be the surface of the liquid; in this case it is best to place the burette against a white background.

The zero of the graduated scale on a burette is always near the top of the tube. In order to fill Mohr's burette, the point of the instrument is dipped into the liquid, the spring clip opened, and a little liquid, sufficient at least to reach into the burette tube, is sucked up by applying the mouth to the upper end; the spring clip is then closed, and the liquid poured into the burette through the upper end until it has risen a little above the zero-line. By opening the spring clip, the liquid is then allowed to drop out until the exact level of the zero-line is reached. The instrument is then ready for use. When a quantity of liquid has been allowed to flow out of a burette thus filled, and the operator desires to read off the amount used, he must wait a few moments to give the particles of fluid adhering to the sides of the emptied portion of the tube time to run down. This remark applies to all forms of burette. Erdmann's swimmer is an excellent addition to Mohr's burette. It is a cylindrical glass float of such a width as nearly to fill the burette, but yet so loosely as to float freely up and down with the liquor. To set the instrument at zero a ring cut round the swimmer is brought to coincide with the line 0 engraved on the burette. The absolute height of the liquor in the burette is to be disregarded. In order to read the height of the liquor in the burette at any time, it is only necessary to note that line on the scale with which the mark cut round the float coincides.

Fig. XXX.

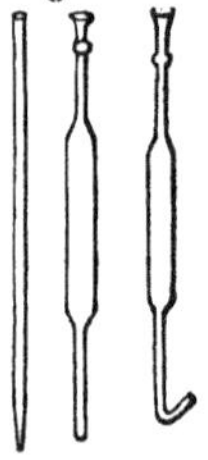

22. *Pipettes.*—Pipettes are tubes drawn to a point, and sometimes furnished with a bulb or a cylindrical enlargement. They are chiefly used to suck small quantities of fluid out of a vessel without disturbing the bulk of the liquid. Figure XXX. re-

presents three forms of pipette; the form with the lower end bent upwards is used to introduce liquids into a bell or bottle of gas standing over mercury. Pipettes graduated into cubic centimetres, or holding a certain number of cubic centimetres when filled to a mark on the stem, are often convenient.

23. *Wash-bottle.*—A wash-bottle is a flask with a uniformly thin bottom, closed with a sound cork or caoutchouc stopper, through which pass two glass tubes as shown in Fig. XXXI. The outer end of the longer tube is drawn to a moderately fine point. A short bend near the bottom of this longer tube in the same plane and direction as the upper bend is of some use, because it enables the operator to empty the flask more completely by inclining it. By blowing into the short tube, a stream of water will be driven out of the long tube with considerable force. This force with which the stream is projected adapts the apparatus to removing precipitates from the sides of vessels as well as to washing them on filters. For use in analytical operations, it is often convenient to attach a caoutchouc tube 12 or 15 c.m. long to the tube through which air is blown; this flexible tube should be provided with a glass mouthpiece about 3 c.m. long. As the wash-bottle is often filled with hot or even boiling water, it may be improved by binding about its neck a ring of cork, or winding closely round the neck a smooth cord. It may then be handled without inconvenience when hot.

Fig. XXXI.

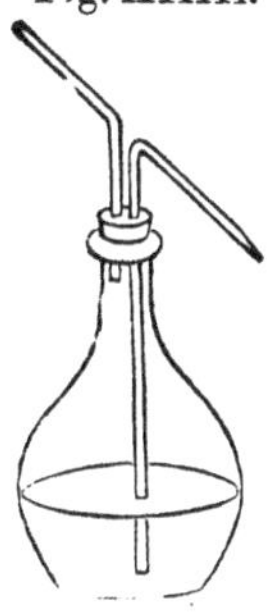

24. *Porcelain Dishes and Crucibles.*—Open dishes which will bear heat without cracking are necessary implements in the laboratory for conducting the evaporation of liquids. The best evaporating-dishes are those made of Berlin porcelain, glazed both inside and out, and provided with a little lip projecting beyond the rim. The dishes made of Meissen porcelain are not glazed on the outside, and are not so durable as those of Berlin manufacture; but they are much cheaper, and with proper care last a long time. The small Berlin dishes will generally bear an evaporation to dryness on the wire gauze over the open flame of the gas-lamp; the Meissen dishes do not so well endure this severe treatment. Evaporating-dishes are made of all diameters, from 3 c.m. to 45 c.m.; they should be ordered by specifying the diameter desired. The large sizes are expensive, and not very durable; they should never be used except on a sand-bath. Dishes of German earthenware are as good as porcelain for many uses, and are much to be recommended in place of the large sizes of porcelain dishes.

Very thin, highly glazed porcelain crucibles with glazed covers are made both at Berlin and at Meissen near Dresden; they are indispensable implements to the chemist. In general, the Meissen crucibles are thinner than the Berlin, but the Berlin crucibles are somewhat less liable to crack; both kinds are glazed inside and out, except on the outside of the bottom. Crucibles should be ordered by specifying the diameters of the sizes desired; they are to be had of nearly a dozen different sizes, with diameters varying from 2 c.m. to 9 c.m. The smallest and largest sizes are little used; for most purposes the best sizes are those between 3 c.m. and 5 c.m. in diameter. As the covers are much less liable to be broken than the crucibles, it is advantageous to buy more crucibles than covers, whenever it is possible so to do. Porcelain crucibles are supported over the lamp on an iron-wire triangle; they must always be gradually heated, and never brought suddenly into contact with any cold substance while they are hot.

25. *Rings to support round-bottomed vessels.*—It is often necessary to support globes, round-bottomed flasks, evaporating-dishes, and round receivers in a stable manner upon the table or other flat surface. For this purpose rings are used made of braided straw, or of straw wound about a core of straw, or of tin wound with listing or coarse woollen cloth. The material of which these rings are made, or with which they are covered, ought to be a substance which does not conduct heat well, because one of the chief uses of these rings is to receive hot vessels just removed from the lamp or sand-bath. A hot flask or dish would almost certainly be broken, if it were placed upon the cold surface of a good conductor of heat. The student must never touch a hot vessel with cold water, or bring it into sudden contact with a surface of marble, iron, copper, or other conductor of heat.

26. *Crucibles.*—For use in a coal fire there are three good kinds of crucibles, each of which has its own merits which recommend it for certain purposes. The Hessian crucibles are sold in nests containing from 3 to 10 crucibles; there are 10 sizes, which vary from 3 to 25 c.m. in height. They generally have a triangular form, and will withstand a very high temperature, if they are warmed before being put in the fire. They are not sold with covers; but covers may be bought separately, or a triangular piece of soapstone may be very conveniently used as a cover for these crucibles. Hessian crucibles are cheaper than any other kind, and are therefore the most used. The French crucibles, called Beaufay crucibles, are admirable, but too dear for common use. They have a tall, narrow form, a smooth surface, and a small lip; covers for the crucibles are sold separately. The crucibles are sold, not in nests, but singly; there are 22 sizes, which vary from

4 to 40 c.m. in height. They are highly refractory. Plumbago crucibles are used for the fusion of the most refractory metals—gold, silver, copper, brass, steel, iron, and so forth; they resist better than any other crucibles the combined action of a very high temperature and a strong flux; and as they are not liable to crack, they may often be used many times without risk. Their first cost is higher than that of any other crucibles. Crucibles are mainly used for the fusion and reduction of metals; but there are also many chemical compounds which can only be prepared at the very high temperatures which by the use of crucibles we are able to command. Although crucibles often withstand the most sudden changes of temperature, it is nevertheless expedient as a general rule to heat up a crucible gradually, and to previously warm a charge which is to be placed in a crucible already hot. If a cold crucible is to be introduced into a fire, it should first be placed in the least hot part of the fire and gradually brought into the hottest part.

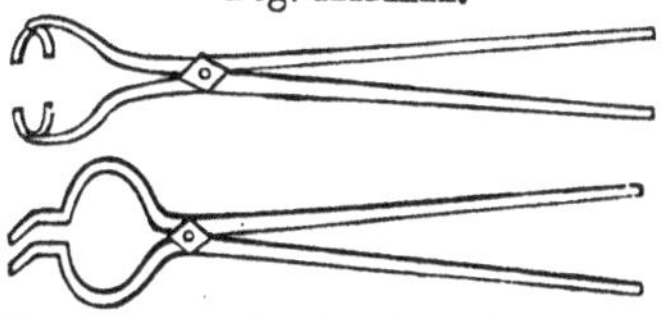

27. *Tongs and Pincers.*—Hot crucibles are handled by means of tongs of various shapes and sizes, according to the weight and nature of the vessels to be lifted. Fig. XXXII. represents two good forms of stout iron tongs for lifting large crucibles out of a coal fire. The manner of using them is readily understood from the figure.

Fig. XXXII.

Small porcelain crucibles are handled, when hot, by means of small steel or iron tongs, such as are represented in Fig. XXXIII.

Fig. XXXIII.

Small steel pincers (jewellers' tweezers) are applied in the laboratory to a great variety of uses.

28. *Mortars.*—Iron, porcelain, and agate mortars are used by chemists to reduce solids to powder. An iron mortar is useful for coarse work, and for effecting the first rough breaking up of substances which are subsequently powdered in the porcelain or agate mortar. If there be any risk of fragments being thrown out of the mortar, it should be covered with a cloth or piece of stiff paper, having a hole in the middle through which the pestle may be passed. Pieces of stone, minerals, and lumps of brittle metals may be safely broken into fragments suitable for the mortar by wrapping them in strong paper, laying them so enclosed upon an anvil, and striking them with a heavy hammer. The paper envelope retains the broken particles which might otherwise fly about in a dangerous manner and be lost.

The best porcelain mortars are those known by the name of Wedgewood-ware; but there are many cheaper substitutes. Porcelain mortars will not bear sharp and heavy blows; they are intended rather for grinding and trituration than for hammering; the pestle may either be formed of one piece of porcelain, or a piece of porcelain cemented to a wooden handle; the latter is the less desirable form of pestle. Unglazed porcelain mortars are to be preferred. In selecting mortars, the following points should be attended to:—1st, the mortar should not be porous; it ought not to absorb strong acids or any colored fluid, even if such liquids be allowed to stand for hours in the mortar; 2nd, it should be very hard, and its pestle should be of the same hardness; 3rd, it should be sound; 4th, it should have a lip for the convenience of pouring out liquids and fine powders. As a rule, porcelain mortars will not endure sudden changes of temperature. They may be cleaned by rubbing in them a little sand soaked in nitric or sulphuric acid, or, if acids are not appropriate, in caustic soda.

Agate mortars are only intended for trituration; a blow would break them. They are exceedingly hard, and impermeable. The material is so precious and so hard to work, that agate mortars are always small. The pestles are generally inconveniently short,—a difficulty which may be remedied by fitting the agate pestle into a wooden handle.

In all grinding-operations in mortars, whether of porcelain or agate, it is expedient to put only a small quantity of the substance to be powdered into the mortar at once. The operation of powdering will be facilitated by sifting the matter as fast as it is powdered, returning to the mortar the particles which are too large to pass through the sieve.

29. *Spatulæ.*—For transferring substances in powder, or in small grains or crystals, from one vessel to another, spatulæ and scoops made of horn or bone are convenient tools. A coarse bone paper-knife makes a good spatula for laboratory use. Cards free from glaze and enamel are excellent substitutes for spatulæ.

30. *Thermometers.*—Thermometers intended for chemical use must have no metal, and no wood or other organic material upon their outer surfaces; their external surfaces must be wholly of glass. The best thermometers are straight glass tubes, of uniform diameter, with cylindrical instead of spherical bulbs, and having the scale engraved upon the glass; such instruments can be passed tightly through a cork, and are free from many liabilities to error to which thermometers with paper or metal scales are always exposed. A cheaper kind of thermometer, having a paper or, better, enamel scale enclosed in a glass envelope, will answer for most experiments.

31. *Furnaces.*—For all common fusions, an anthracite or coke fire

in an ordinary cylinder stove will suffice. The chafing-dish, or open portable stove, such as is used by plumbers for example, is very convenient for operations which require less heat. The clay buckets used as open furnaces are better than the iron ones, because they hold the heat better.

Charcoal is the fuel used in these open fires. A very useful accompaniment to these portable furnaces is a piece of straight stove-pipe, about 60 c.m. long and 10 c.m. wide, and flaring out below like a funnel until it is wide enough to cover the top of the furnace. This contrivance powerfully increases the draught, and is used to urge the fire during kindling, or to intensify it while a fusion is in progress. With a furnace of this description there is no difficulty in keeping a small crucible white-hot for a short time.

THE METRICAL SYSTEM OF WEIGHTS AND MEASURES.

The metrical system, employed in the affairs of every-day life by most of the nations of continental Europe and by scientific writers throughout the world, is based upon a fundamental unit or measure of length, called a *metre*. This metre is defined as the 40-millionth part of the circumference of the earth, or, in other words, of a "great circle" or meridian; its length was originally determined by actual measurement of a considerable arc of a meridian; but the various measurements heretofore made of the length of the earth's meridian differ slightly from each other, and it is to be expected, and indeed hoped, that the steady improvements of methods and instruments will make each successive determination of the length of the meridian better than, and therefore different from, the preceding. It is therefore necessary to define the standard of length, by legislation, to be a certain rod of metal, deposited in a certain place under specified guaranties, and to secure the uniformity and permanence of the standard by the multiplication of exact copies in safe places of deposit.

From this single quantity, the metre, all other measures are decimally derived. Multiplied or divided by 10, 100, 1000, and so forth, the metre supplies all needed linear measures, and the square metre and cubic metre, with their decimal multiples, supply all needed measures of area or surface on the one hand, and of solidity or capacity on the other.

From the unit of measure to the unit of weight the transition is admirably simple and convenient. The cube of the 1-hundredth of the linear metre is, of course, the millionth of the cubic metre; its bulk is about that of a large die of the common back-gammon board. This little cube of pure water is the universal unit of weight, a *gramme*, which, decimally multiplied and divided, is made to express all weights. The numbers expressing all weights, from the least to the greatest, find direct expression in the decimal notation; the weights used in different trades only differ from each other in being different decimal multiples of the same fundamental unit; and in comparing together weights and volumes, none but easy decimal computations are ever necessary.

The nomenclature of the metrical system is extremely simple; one general principle applies to each of the following tables. The Greek prefixes for 10, 100, and 1000, viz. deca, hecto, and kilo, are used to signify multiplication, while the Latin prefixes for 10, 100, and 1000, viz. deci, centi, and milli, are employed to express subdivision. Of the names thus systematically derived from that of the unit in each table, many are not often used; the names in common use are those printed in small capitals. Thus, in the table for linear measure, only the metre, kilometre, centimetre, and millimetre are in common use,—the first for such purposes as the English yard subserves, the second instead of the English mile, the third and fourth in lieu of the fractions of the English foot and inch.

LINEAR MEASURE.

			Metre.
Divisions...	MILLIMETRE	=	0·001 or 1-1000th of a metre.
	CENTIMETRE	=	0·01 or 1-100th of a metre.
	Decimetre	=	0·1 or 1-10th of a metre.
Unit...........	METRE	=	1·
Multiples...	Decametre	=	10·
	Hectometre	=	100·
	KILOMETRE	=	1,000·

SURFACE MEASURE.

Divisions...	Millimetre square	=	0·000,001	of a metre square.
	Centimetre square	=	0·000,1	of a metre square.
	Decimetre square	=	0 01	of a metre square.
Unit...........	METRE SQUARE	=	1·	metre square.

CUBIC MEASURE.

			Cubic Metre.
Divisions...	Cubic Millimetre	=	0·000,000,001
	Cubic Centimetre	=	0·000,001
	Cubic Decimetre	=	0·001
Unit...........	CUBIC METRE	=	1·
Multiples...	Cubic Decametre	=	1,000·
	Cubic Hectometre	=	1,000,000·
	Cubic Kilometre	=	1,000,000,000·

The table for land-measure we omit, as having no connexion with our subject. For the measurement of wine, beer, oil, grain, and similar wet and dry substances, a smaller unit than the cubic metre is desirable. The cubic decimetre has been selected as a special standard of capacity for the measurement of substances such as are bought and sold by the English wet and dry measures. The cubic decimetre thus used is called a *litre.*

CAPACITY MEASURE.

			Litres.		Cubic Metre.		
Divisions...	Millilitre	=	0·001	=	0·000,001	=	1 cubic centimetre.
	Centilitre	=	0·01	=	0·000,01		
	Decilitre	=	0·1	=	0·000,1		
Unit...........	LITRE	=	1·	=	0·001	=	1 cubic decimetre.
Multiples..	Decalitre	=	10·	=	0·01		
	HECTOLITRE	=	100·	=	0·1		
	Kilolitre	=	1,000·	=	1·	=	1 cubic metre.

The table of weights bears an intimate relation to this table of capacity. As already mentioned, the weight of that die-sized cube, a cubic centimetre or millilitre of distilled water (taken at 4°, its point of greatest density) constitutes the metrical unit of weight. This weight is called a *gramme.* From the very definition of the gramme, and from the table of capacity-measure, it is clear that a litre of distilled water at 4° will weigh 1000 grammes.

WEIGHTS.

			Grammes.	
Divisions...	MILLIGRAMME	=	0·001	
	CENTIGRAMME	=	0·01	
	DECIGRAMME	=	0·1	
Unit............	GRAMME	=	1·	= 1 cubic centimetre of water at 4°.
Multiples..	Decagramme	=	10·	
	Hectogramme	=	100·	
	Kilogramme	=	1,000·	= 1 cubic decimetre of water at 4°.

The simplicity and directness of the relations between weights and volumes in the metrical system can now be more fully explained. The chemist ordinarily uses the gramme as his unit-weight, and for his unit of volume a cubic centimetre, which is the bulk of a gramme of water. For coarser work, the kilogramme becomes the unit of weight, and the corresponding unit of measure is the litre, which is the bulk of a kilogramme of water. In commercial dealings, in manufacturing-processes, and, above all, in scientific investigations, these simple relations between weights and measures have been found to be an inestimable advantage. The numerical expressions for metrical weights and measures may always be read as decimals. Thus 5·126 metres will be read five metres and one hundred and twenty-six thousandths, and not five metres, one decimetre, two centimetres, and six millimetres. The expression 10·5 grammes is read ten and five-tenths grammes; just as we say one hundred and five dollars, not ten eagles and five dollars; or sixty-five cents, not six dimes and five cents. All computations under the metrical system are made with decimals alone.

The abbreviations commonly met with in chemical literature are:—

m.m. for millimetre;	c.m. for centimetre;
m. for metre;	c. c. for cubic centimetre;
grm. for gramme;	kilo. for kilogramme.

The use of the metrical system of weights and measures in the arts and trades has been legalized both in the United States and in Great Britain. The United States coin, composed of copper and nickel, of the denomination 5 cents, and date 1866, is 2 c.m. in diameter, and weighs 5 grms.

TABLE *for the Conversion of Degrees on the Centigrade Thermometer into Degrees of Fahrenheit's Scale.*

Cent.	Fahr.	Cent.	Fahr.	Cent.	Fahr.	Cent.	Fahr.
−50°	−58°·0	−29°	−20°·2	− 8°	17°·6	13°	55°·4
−49	−56·2	−28	−18·4	− 7	19·4	14	57·2
−48	−54·4	−27	−16·6	− 6	21·2	15	59·0
−47	−52·6	−26	−14·8	− 5	23·0	16	60·8
−46	−50·8	−25	−13·0	− 4	24·8	17	62·6
−45	−49·0	−24	−11·2	− 3	26·6	18	64·4
−44	−47·2	−23	− 9·4	− 2	28·4	19	66·2
−43	−45·4	−22	− 7·6	− 1	30·2	20	68·0
−42	−43·6	−21	− 5·8	0	32·0	21	69·8
−41	−41·8	−20	− 4·0	+ 1	33·8	22	71·6
−40	−40·0	−19	− 2·2	2	35·6	23	73·4
−39	−38·2	−18	− 0·4	3	37·4	24	75·2
−38	−36·4	−17	+ 1·4	4	39·2	25	77·0
−37	−34·6	−16	3·2	5	41·0	26	78·8
−36	−32·8	−15	5·0	6	42·8	27	80·6
−35	−31·0	−14	6·8	7	44·6	28	82·4
−34	−29·2	−13	8·6	8	46·4	29	84·2
−33	−27·4	−12	10·4	9	48·2	30	86·0
−32	−25·6	−11	12·2	10	50·0	31	87·8
−31	−23·8	−10	14·0	11	51·8	32	89·6
−30	−22·0	− 9	15·8	12	53·6	33	91·4

TABLE (*continued*).

Cent.	Fahr.	Cent.	Fahr.	Cent.	Fahr.	Cent.	Fahr.
34°	93°·2	78°	172°·4	122°	251°·6	166°	330°·8
35	95·0	79	174·2	123	253·4	167	332·6
36	96·8	80	176·0	124	255·2	168	334·4
37	98·6	81	177·8	125	257·0	169	336·2
38	100·4	82	179·6	126	258·8	170	338·0
39	102·2	83	181·4	127	260·6	171	339·8
40	104·0	84	183·2	128	262·4	172	341·6
41	105·8	85	185·0	129	264·2	173	343·4
42	107·6	86	186·8	130	266·0	174	345·2
43	109·4	87	188·6	131	267·8	175	347·0
44	111·2	88	190·4	132	269·6	176	348·8
45	113·0	89	192·2	133	271·4	177	350·6
46	114·8	90	194·0	134	273·2	178	352·4
47	116·6	91	195·8	135	275·0	179	354·2
48	118·4	92	197·6	136	276·8	180	356·0
49	120·2	93	199·4	137	278·6	181	357·8
50	122·0	94	201·2	138	280·4	182	359·6
51	123·8	95	203·0	139	282·2	183	361·4
52	125·6	96	204·8	140	284·0	184	363·2
53	127·4	97	206·6	141	285·8	185	365·0
54	129·2	98	208·4	142	287·6	186	366·8
55	131·0	99	210·2	143	289·4	187	368·6
56	132·8	100	212·0	144	291·2	188	370·4
57	134·6	101	213·8	145	293·0	189	372·2
58	136·4	102	215·6	146	294·8	190	374·0
59	138·2	103	217·4	147	296·6	191	375·8
60	140·0	104	219·2	148	298·4	192	377·6
61	141·8	105	221·0	149	300·2	193	379·4
62	143·6	106	222·8	150	302·0	194	381·2
63	145·4	107	224·6	151	303·8	195	383·0
64	147·2	108	226·4	152	305·6	196	384·8
65	149·0	109	228·2	153	307·4	197	386·6
66	150·8	110	230·0	154	309·2	198	388·4
67	152·6	111	231·8	155	311·0	199	390·2
68	154·4	112	233·6	156	312·8	200	392·0
69	156·2	113	235·4	157	314·6	201	393·8
70	158·0	114	237·2	158	316·4	202	395·6
71	159·8	115	239·0	159	318·2	203	397·4
72	161·6	116	240·8	160	320·0	204	399·2
73	163·4	117	242·6	161	321·8	205	401·0
74	165·2	118	244·4	162	323·6	206	402·8
75	167·0	119	246·2	163	325·4	207	404·6
76	168·8	120	248·0	164	327·2	208	406·4
77	170·6	121	249·8	165	329·0	209	408·2

TABLE (*continued*).

Cent.	Fahr.	Cent.	Fahr.	Cent.	Fahr.	Cent.	Fahr.
210°	410·0°	238°	460·4°	266°	510·8°	294°	561·2°
211	411·8	239	462·2	267	512·6	295	563·0
212	413·6	240	464·0	268	514·4	296	564·8
213	415·4	241	465·8	269	516·2	297	566·6
214	417·2	242	467·6	270	518·0	298	568·4
215	419·0	243	469·4	271	519·8	299	570·2
216	420·8	244	471·2	272	521·6	300	572·0
217	422·6	245	473·0	273	523·4	301	573·8
218	424·4	246	474·8	274	525·2	302	575·6
219	426·2	247	476·6	275	527·0	303	577·4
220	428·0	248	478·4	276	528·8	304	579·2
221	429·8	249	480·2	277	530·6	305	581·0
222	431·6	250	482·0	278	532·4	306	582·8
223	433·4	251	483·8	279	534·2	307	584·6
224	435·2	252	485·6	280	536·0	308	586·4
225	437·0	253	487·4	281	537·8	309	588·2
226	438·8	254	489·2	282	539·6	310	590·0
227	440·6	255	491·0	283	541·4	311	591·8
228	442·4	256	492·8	284	543·2	312	593·6
229	444·2	257	494·6	285	545·0	313	595·4
230	446·0	258	496·4	286	546·8	314	597·2
231	447·8	259	498·2	287	548·6	315	599·0
232	449·6	260	500·0	288	550·4	316	600·8
233	451·4	261	501·8	289	552·2	317	602·6
234	453·2	262	503·6	290	554·0	318	604·4
235	455·0	263	505·4	291	555·8	319	606·2
236	456·8	264	507·2	292	557·6	320	608·0
237	458·6	265	509·0	293	559·4		

TABLE FOR CONVERSION OF GRAMMES INTO GRAINS AND CENTIMETRES INTO INCHES.

The equivalents in English weights and measures of those metrical weights and measures which are used in chemistry can be readily found by the aid of the following table, which is available not only for grammes and centimetres, but, by mere change of the position of the decimal point, for all decimal multiples or subdivisions of these quantities:—

Grammes	1	2		4	5	6	7	8	9
into Grains.	15·4346	30·8692	46·3038	61·7384	77·1730	92·6076	108·0422	123·4768	138·9114
Centimetres	1	2	3	4		6	7	8	9
into Inches.	·3937079	·7874158	1·1811237	1·5748316	1·9685395	2·3622474	2·7559553	3·1496632	3·5433711

One cubic metre	= 35·31660 cubic feet.	One pound avoirdupois	= 7000· grains.
„ „ decimetre (a litre)	= 61·02709 „ inches.	„ „ troy	= 5760· „
„ „ centimetre	= 0·06103 „ „	„ ounce avoirdupois	= 437·5 „
„ litre	= 0·22017 imp. gallon.	„ „ troy	= 480· „
„ „	= 0·88066 „ quart.	„ imperial gallon	= 277·274 cub. ins.
„ „	= 1·76133 „ pint.	„ old English wine-measure gall. (Still used in the United States.)	= 231· „ „

INDEX.

*** Compound names are to be found under the first word of the name—chloride of sodium, for example, under chloride, sulphuric acid under sulphuric. *Occ.* stands for occurrence, *prep.* for preparation, *prop.* for properties, *comp.* for composition. The numbers refer to pages. The Roman numerals refer to the Appendix.

ACETATE of aluminum, 518.
lead, 499.
Acetates of copper, 569.
Acetic acid, prep. from wood, 298.
Acid, meanings of the term, 53, 191, 451.
Acids, monobasic, bibasic, &c., 191.
Affinity, use of the term, 101.
Air, comp. of, 6–10.
not an element, 10.
displacing of, 5.
a mixture, 67.
physical prop. of, 6.
presence of, 4.
quantity necessary for complete combustion, 361.
reaction with nitric oxide, 59, 67.
weight of, 5.
Alabaster, 476.
Alcohol, from the fermentation of sugar, 328.
Alcohol-lamp flame, 347.
Alkalies, the caustic, 402, 416, 418.
definition of, 53.
Alkalimetry, 419.
Alkaline-earths, 489.
Alkaline reaction, 53.
Alkaloids, organic, absorbed by charcoal, 308.
Allotropism, 139.
Alum, 517.
in bread, 399.
ammonium, 518.
Alum-cake, 517.
Alumina, 514.
Aluminates of the alkalies, 515.
Aluminum, abundance of, 512.
alloys of, 513.
bronze, 513.
prop. of, 512.
Alums, atomic volumes of, 560.
class of, 529.
Amalgam of ammonium, 437.
gold, 588, 590.
sodium, 94.
tin, 577.
Amalgams, 576.
Ammonia, analysis of, 77.
comp. of, 81.
electrolysis of, 78.
occ. of, 83.
physical prop. of, 76, 77.
prep. of, 75.
solubility in water, 76.
sources of, 84.
synthesis of, 79.
Ammoniacal liquor of gas-works, 84.
Ammonia-water precipitates metallic oxides, 439.
prep. of, 84.
uses of, 84.
Ammonium, 82.
amalgam, 437.
the supposed metal, 437.
Ammonium-salts, prop. of, 438.
test for, 594.
Analysis, definition of, 3.
qualitative, 419.
a method of separating Ca, Sr, and Ba, 526.

Analysis of organic compounds, 566.
volumetric, 420.
Animal charcoal, decolorizing power of, 309.
Animals and plants, reciprocal action of, 330.
Annealing, 412.
Anthracite, 293, 294.
conducts heat, 294.
getting flame from, 360.
Antimoniate of antimony, 273.
Antimoniates, 275.
Antimonic acid, 274.
Antimoniuretted hydrogen, 269.
Antimony, alloys, 268.
alloys with zinc, 269.
amorphous, 268.
bloom, 272.
distinguished from arsenic, 270.
extraction of, 266.
glance, 277.
mirrors, 270.
occ. of, 266.
prop. of, 267.
the commercial metal, 267.
vermilion, 279.
Antozone, 139, 147.
distinguished from ozone, 153.
fogs and smokes, 149, 151.
makes air hard to dry, 150.
oxidizes water, 152.
prep. of, 148.
Antozonides, 153.
Aqua regia, 104.
Argand burner, 345, 346.
Arragonite and calc-spar compared, 469
Arseniates, isomorphous with phosphates, 252.
Arsenic, detection of the poison, 252.
determination of its atomic weight, 262.
distinguished from antimony, 270.
eating, 250.
extraction of, 241.
greens, 260.
mirrors, 260.
occ. of, 241.
prop. of, 242.
unit-volume weight, 242.
Arsenic acid, hydrates of, 251.
prep. of, 250.
reduction of, 252.
salts of, 252.
uses of, 252.
Arsenical pyrites, 241.
Arsenious acid, antidote for, 250.
a poison, 250.
reduction of, 248.
solubility of, 249.
sources of, 246.
two varieties of, 246.
uses of, 250.
Arseniuretted hydrogen, analysis of, 245.
decomposed by heat, 260.
prep. of, 243.
prop. of, 244.
Asbestos, platinized, 193.
Atom, definition of, 29.
Atomic heats of compounds, 601.
elements, 599.
Atomic volume, 201.
Atomic weight, analogy a guide in determining, 514.
Atomic weights, definition of, 31.
mutual comparison of, 462.
practical use of, 100.
table of, 597.
two sets in use, 603.
value of, 64.
Atomicity of the elements, 465.

BARIUM, 486.
compounds, 486–488.
flame, 489.
Barytes, 487, 488.
Beakers, xxviii.
Bell-metal, 564.
Bessemer steel, 541.
Biatomic metals, 73.
Bibasic acids, defined, 191.
Biborate of sodium, 408.
Bicarbonate of ammonium, 442.
potassium, 416.
sodium, 398.
Bichloride of platinum, 593.
tin, 584.
Bichromate of potassium, 523.
, use in photolithography, 526.
Binoxide of manganese, uses of, 530.

Binoxide of tin, 581.
Bismuth, extraction of, 281.
 makes fusible alloys, 282.
 prop. of, 281.
 salts, 283.
Bismuthic acid, 283.
Bisulphate of potassium, 428.
Bisulphide of carbon, 363.
 and nitric oxide, 364.
Bisulphide of iron, 549.
Bisulphite of calcium, checks fermentation, 486.
Bivalent metals, 73.
Black ball, 397.
Black-lead, 290.
Blast furnace, iron, working of, 535–538.
Bleaching by chloride of lime, 483.
 chlorine, 112.
 ozone, 143.
 sulphurous acid, 180.
Bleaching-powder, 482.
Blistered steel, 541.
Bloom, 535.
Bloomary, 535.
Blowers, viii.
Blowpipe with gas and air, 350.
Blowpipe, Bunsen's gas-, viii.
 mouth-, 351.
 oxidizing flame of, 352.
 oxyhydrogen, 46.
 reducing flame of, 353.
Boiler-scale, 468, 479.
Bone-ash, 214.
Bone-black, prep. of, 310.
Bone-phosphate of calcium, 481.
Bones, distillation of, 84, 310.
Boracic acid, comp. of, 367.
 extraction of, 365.
 prop. of, 368.
 strong at high temperatures, 369.
Borax, as a blowpipe-test, 409.
 two forms of, 408.
 uses of, 409.
Boron, allotropic states of, 366.
 occ. of, 365.
 resemblance to carbon and silicon, 372.
Brass, 564.
Bread, raising with chemicals, 399.
Breaking up stones, xxxiv.
Britannia metal, 268.
Bromhydric acid, 120.
Bromic acid, 122.
Bromide of ammonium, used in photography, 457.
 arsenic, 263.
 cadmium, used in photography, 457.
 nitrogen, 123.
 potassium, 425.
 silicon, 386.
 silver, 463.
 sodium, 394.
Bromides, formation of, 120.
 of iodine, 133.
 phosphorus, 240.
Bromine, mother-liquors of salt-works, a source of, 393.
 occ. of, 119.
 prop. of, 119.
 uses of, 120.
Bronze, 564.
Bronze-powder, 582.
Bulbs, blowing, v.
Bunsen's burner, 344, vi.
Burettes, xxix.
 how to fill, xxxi.
 read, xxx.

Cadmium, extraction of, 509.
 makes fusible alloys, 510.
 prop. of, 510.
 unit-volume weight half its atomic weight, 510.
Cæsium, 446.
Calcareous waters, 468, 479.
Calcined magnesia, 502.
Calcium-flame, 489.
 light, 47.
 the metal, 486.
Calc-spar and arragonite compared, 469.
Calomel, 573.
Candle-flames, 349.
Candles are gas-burners, 347.
Caoutchouc in sheets, xi.
 stoppers, xi.
 tubing, x.
Carbon, allotropic modifications of, 288.
 atomic weight of, 317.
 its atom hypothetical, 318.
 compounds with nitrogen, 363.
 dimorphous, 292.
 for galvanic batteries, 293.
 occ. of, 287.

Carbon, vapor-density hypothetical, 316.
Carbonate of ammonium, commercial, 441.
calcium, dissolves in water containing carbonic acid, 467.
dissolves in ammonium-salts, 468.
occ. of, 467.
lead, 499.
magnesium, 504.
potassium, prop. of, 415.
pure, source of, 424.
uses of, 416.
and hydrogen, 416.
sodium, sources and uses, 396.
and hydrogen, 398.
Carbonates, 334.
alkaline, impurities of, 419.
Carbonic acid, 321.
absorbents for, 324.
comp. of, 333.
decomposed by
carbon, 335.
potassium, 332.
dissociation of, 331.
formed in combustion, 321.
fermentation, 328.
respiration, 328.
generator, 323.
liquid, 330.
obtained from carbonates, 322.
occ. of, 327.
prop. of, 323.
solid, 330.
solubility of, 326.
sources of, in the air, 329.
synthesis of, 333.
test for, 321.
Carbonic oxide, comp. of, 341.
dissociation of, 340.
economy in burning it, 361.
heating value of, 339, 361.
a poison, 338.
Carbonic oxide, prep. of, 335.
prop. of, 338.
a reducing agent, 339.
Carburetted hydrogen, decomposed by heat, 344.
Carmine-lake, 516.
Cast iron, impurities of, 539.
varieties of, 538.
Caustic alkalies, strong bases, 418.
Caustic potash, 417.
soda, 403.
advantageous use of, 405.
Cement copper, 563.
Cement, Roman, 521.
Cerium, 561.
Chalk, 322, 467.
Chameleon-mineral, 532.
Charcoal absorbs gases, 303.
different gases in different proportions, 305.
causes combination of gases, 306.
a disinfectant, 307.
hot, decomposes steam, 301.
physical prop. of, 300.
prep. of, 296, 297.
a reducing agent, 301.
removes colors, 308.
satbility of, 302.
Chemical changes, 2.
combination, 32.
and solution compared, 38.
force, 32.
strength, 418.
Chemistry, subject-matter of, 1.
Chili-saltpetre, 51.
Chimneys create draughts, 355.
on fire, how to put out, 178.
use of, 354.
Chlorate of potassium, an ingredient of explosive mixtures, 435.
prep. of, 434, 435.
Chloraurates, 590.
Chlorhydric acid, 40.
analysis of, 92.
comp. of, 96.
electrolysis of, 94.

Chlorhydric acid gas, prep. of, 92.
Chlorhydric acid, prep. of, 1[illegible].
  prop. of, 92.
  synthesis of, 9[illegible].
  uses of, 103.
Chloric acid, 117.
Chloride, definition of term, 109.
  of aluminum, 516.
  ammonium, 86.
    dissociation of, 239.
    prep. of, 114.
    sources of, 439.
    uses of, 440.
  arsenic, 262.
  boron, 369.
  bromine, 123.
  calcium, 482.
    used for drying gases, 482.
    tube, xxiv.
  carbon, 362.
  gold, used in photography, 457, 459.
  lead, 498.
  lime, 482.
  magnesium, 503.
  nitrogen, 114.
  potassium, 424.
  silicon, decomposed by water, 384.
    prep. of, 383.
  silver, effect of light on, 456.
    occ. of, 460.
    reduction of, 461.
    solvents for, 456.
  sodium, 392.
    solubility of, 393.
  sulphur, 197.
  zinc, 508.
    a disinfectant, 509.
Chlorides of antimony, 275.
    terchloride, 275.
    quinquichloride, 276.
  chromium, 523.
Chlorides of gold, 589, 590.
  iodine, 132.
  iron, 550.
  mercury, 573, 574.
  phosphorus, 236.
    terchloride, 236.
    quinquichloride, 237.
  platinum, 593.
  selenium, 201.
  tin, 583.
Chlorimetry, 485.
Chlorine, action on ammonia, 113.
  atomic weight of, 97.
  bleaches, 112.
  burns in hydrogen, 110.
  combustion in, 110.
  decomposes water, 111.
  disinfects, 113.
  explosive mixture with hydrogen, 108.
  occ. of, 106.
  oxidizing action of, 112.
  physical prop. of, 107.
  prep. of, 105, 106.
  prep. from bleaching-powder, 483.
  prep. with bichromate of potassium, 525.
  unites with metals, 109.
  -water, 107.
Chloroform, formula of, 319.
Chloroplatinates, 596.
Chlorous acid, 117.
Chromates, 526.
Chrome-alum, 524.
  -green, 523.
  -iron ore, 522.
  -yellow, 527.
Chromic acid, 525.
Chromium, occ. of, 522.
Cinnabar, 570.
Citrate of magnesium, 504.
Classification, 598.
Clay, 519.
Cleavage, 157.
Coal, bituminous, 294.
  distillation of, 293.
  varieties of, 293, 294.
Coal-gas, comp. of, 320.
  purification of, 473, 546.
Coal-tar, 312.
Cobalt, 556.

Cochineal, tincture, preparation of, 421.
Coke, 293, 294.
conducts heat, 294.
Cold, greatest artificial, 57.
Collodion. 457.
Colloids, 255.
Columbium. 585.
Combination by volume, 203.
Combining weights of compounds, 71.
of organic compounds, determination of, 596.
and unit-volume weights compared, 205, 206.
Combustion, definition of, 15.
ordinary, 342–362.
spontaneous, 229.
Compound radicals, 363.
Condensation-ratios, 205.
Condenser, 34, 35.
Cooling flames by good conductors, 358.
Copper, alloys of, 564.
extraction of, 562.
occ. of, 562.
prop. of, 563.
pyrites, 562.
Copperas, 551.
Cork-cutters or borers, xii.
Corks, xi.
Correlation of force, 507.
Corrosive sublimate, 574.
antidote for, 575.
Cream of tartar, 424.
Crucibles, Beaufay, Hessian, plumbago, xxxiii.
porcelain, xxxii.
how to use, xxxiv.
material of, 520.
Cryolite, 391.
Crystalline form, a guaranty of purity, 435.
structure, 157.
Crystallization, the chief means of purification, 397, 434.
by fusion, 156.
six systems of, 158.
Crystalloids, 254.
Crystals, four methods of obtaining, 161.
Cupellation, 494.
Cyanhydric acid, 363.
Cyanide of potassium, formation of, 426.
in blast furnaces, 427.
uses of, 427.
iron, 555.
silver, 463.
Cyanogen, 363.

Daguerreotype, 456.
Decolorizing power of charcoal, 308–311.
Deflagrating-spoon, xxi.
Deflagration, 302.
Deodorizing by charcoal, 306.
chlorine, 113.
Destruction of organic matter in cases of poisoning, 257.
Detection of arsenic, 252–262.
Developers, a term of photography, 458.
Devitrification, 413.
Dialysis, 254.
applied to detection of poisons, 256.
Dialyzer, 255.
Diamond, 289.
Dichloride of copper, 567.
sulphur, 197.
Dichromate of lead, 527.
Didymium, 561.
Diffusion of carbonic acid, 325.
gases, 43, 326.
rapidity of, 325.
Dimorphous substance, defined, 159.
Dioxide of copper, 564.
mercury, 571.
Direct combination, 404.
Disinfectant, charcoal, 306.
chloride of zinc, 508.
chlorine, 112, 113.
hypochlorite of calcium, 482.
ozone, 144.
permanganate of potassium, 533.
sulphurous acid, 181.
Displacement, collection of gases by, 12.
Dissociation, 238.

Distillation, the process of, 34.
Dolomite. 503.
Drying-apparatus, xxiii.
gases, xxii.
substances used for, xxiii.
plastering, 475.
Dualistic formulæ, 87.

EARTHENWARE, 520.
Effervescing liquids, 327.
Electrolysis of ammonia, 78.
chlorhydric acid, 94.
marsh-gas, 317.
water, 24.
Electro-metallurgy, 568.
batteries, 507.
silver-plating, 463.
use of graphite, 290.
Electro-plating, 463, 568, 590.
Element, definition of, 3.
Elements are bodies incapable of decomposition, 33.
Empirical formulæ, 86.
Epsom salts, 503.
Equations, chemical, 90.
Erbium, 561.
Erdmann's swimmer, xxxi.
Etching glass, 138.
Evaporating-dishes, xxxii.
Explosion of oxygen and hydrogen, 48.
Explosions in coal-mines, 314.

FELDSPAR, 375, 519.
Fermentation, 328.
Ferric chloride, 550.
hydrate, used in purifying gas, 546.
nitrate, 554.
oxide, 543.
corrodes organic matter, 544.
reduction of, 544.
salts, 547.
sulphate, 554.
Ferricyanide of potassium, 555.
Ferrocyanide of potassium, 427.
Ferrous chloride, 550.
nitrate, 554.
oxide, 543.
salts, changed into ferric, 556.
Ferrous sulphate, 551.
dyeing black with, 553.
sulphide, prep. of, 547–549.
Filtering, xxii.
Filters, how to fold, xxii.
Filtrate, definition of, xxii.
Fire-brick, 520.
Firedamp, 314.
Fireworks, a green color, 488.
purple, 436.
red, 488.
white, 263.
Flame, dissection of, 348–350.
intensity, 346.
luminosity of, 343.
oxidizing, 352.
put out by good conductors, 357–359.
quantity, 346.
reducing, 353.
structure of, 48, 348–350.
Flaming fires, 360.
Flasks, xxviii.
Florence flasks, xxviii.
Fluoboric acid, 371.
Fluorhydric acid, 136.
action on silica, 137.
Fluoride of boron, 370.
calcium, 136.
silicon, its reaction with water, 388.
prep. of, 386.
volumetric comp. of, 387.
Fluorine, 135.
hard to get and keep, 136.
occ. of, 136.
Fluosilicates, 389.
Fluosilicic acid, prep. of, 388.
a reagent for potassium salts, 390.
a use of, 118.
Flux, used in smelting iron-ores, 536.
Fluxes—alkaline carbonates, 415.
borax, 409.
cyanide of potassium, 427.
limestone, 536.
sulphate of potassium and hydrogen, 428.
tartar, ignited, 424.
Formulæ, empirical and rational, 86.
molecular, 208.
Foundry-facings, 291.
Free gases exist as molecules, 208.

Freezing-apparatus, 77.
mixtures, see Refrigerating.
Friction-matches, 211.
Fulminating-silver, 455.
Fusible alloys, 510.
Furnaces, xxxv.

Galena, 490, 497.
Galvanic current, 506.
Gas, illuminating, 293, 297, 320.
Gas-burners, 346.
-carbon, 292.
-generator, self-regulating, xxvi.
-holders, xviii.
-lamps for heating, 344.
Gay-Lussac's burette, xxx.
German silver, 557, 564.
Gilding, 590.
Glass of antimony, 277.
Glass, colored, 413.
comp. of, 412.
etching of, 138.
not insoluble, 411.
varieties of, 412.
Glass beakers, xxviii.
cutting and cracking of, i.
flasks, xxviii.
retorts, xxviii.
tubing, bending, drawing and closing, iii.–v.
heating of, iii.
sizes and qualities of, i.
Glauber's salt, 395.
solubility of, 395.
supersaturated solution of, 396.
Glazes—lead, feldspar, salt, 520.
Glucinum, 522.
Gold, alloys of, 591.
coin, 591.
occ. of, 587.
prop. of, 588.
salts, 590.
separation from silver alloys, 464.
silver and copper, 589.
solution of, 589.
Golden sulphide of antimony, 279.
Graduation, 392.
Gramme, definition of, 19, xxxviii.
Graphite, 290.
artificial, 292.
in cast iron, 538.
Graphite, use in electro-metallurgy, 290.
Graphitic acid, 292.
Graphon, 292.
Gray iron, 538.
Green vitriol, 551.
Grey antimony, 277.
Group, the alkali, 464.
calcium, 489.
carbon, 390.
chlorine, 133.
magnesium, 511.
nitrogen, 285.
platinum, 596.
sesquioxide, 560.
sulphur, 202.
tin, 585.
Groups, principles concerning, 135.
table of, 598.
Gun-metal, 564.
Gunpowder, comp. of, 432.
manufacture of, 433.
products of combustion of, 432.
Gypsum, 476.

Hard water, 468, 479.
Hartshorn, 84.
Heat, relation to chemical force, 418.
unit of, 45.
Homologous series, 313.
Hydrate of aluminum, 514.
calcium, 472.
copper, 566.
iron, 545.
potassium, 416.
sodium, prep. of, 402.
type of bases, 403.
Hydraulic cement, 504, 521.
Hydrocarbons, 312.
Hydrogen, chemical effect of, 129.
derived from water, 22.
diffusive power of, 43, 44.
explosive mixture with oxygen, 48, 49.
heating power of, 45.
inflammable, 44.
lightness of, 42.
peroxide of, 50, 51.
physical prop. of, 41.
precautions in making, 40.
prep. of, 39.

Hydrogen, standard of specific gravity for gases, 41.
Hypobromous acid, 122.
Hypochloric acid, 117.
Hypochlorite of calcium, 482.
Hypochlorous acid, 116.
Hyponitric acid, comp. of, 61.
prep. of, 62.
prop. of, 62.
Hypophosphites, 228.
Hypophosphorous acid, 227.
Hyposulphite of sodium, prep. of, 414.
use in photography, 413, 456.
Hyposulphites, 196.
Hyposulphurous acid, 196.

Ice, crystals of, 21.
lighter than water, 21.
Ignition-tube, definition of, 8.
manner of using, 8, 11.
Illuminating gas, 293.
description of, 320.
manufacture of, 297.
purification of, 473, 546.
Incrustations in boilers, 468, 479.
Indelible ink, 453.
Indestructibility of matter, 356.
Indian fire, 263.
Ink, 553.
indelible, 453.
Intensity flames, 346.
Iodic acid, 130.
Iodide of arsenic, 263.
lead, 499.
nitrogen, 131.
potassium, 425.
silver, 463.
sodium, 394.
Iodides of phosphorus, 240.
Iodine, extraction of, 124.
manufacture of, 132.
occ. of, 123.
prop. of, 124.
reaction with starch, 126.
testing for, 126.
uses of, 125.
Iodohydric acid, prep. of, 128.
prop. of, 128, 129.
Iodo-starch paper, 127.
Iridium, 596.
Iron, cast, impurities of, 539.
varieties of, 538.
extraction of, 535.
galvanized, 506.
mordant, 555.
occ. of, 534.
ores — brown, clay, red hæmatite, magnetic, spathic, specular, 534.
puddling of, 539.
pyrites, 549.
replaces copper, 551.
wrought, 539.
Iron retort, xxvi.
Iron stand for supporting vessels, xiii.
Isomerism, 247.
Isomorphism, 200.

Kaolin, 519.
Kelp, 124.
Kermes mineral, 279.
Kindling-temperature, 357.

Lakes, 516.
Lampblack, 295.
manufacture of, 298.
Lamp-flames are gas-flames, 347.
Lamps for laboratory use, vi–viii.
Lanthanum, 561.
Laughing-gas, 57.
Lavoisier, his analysis of air, 9.
Lead, action of acids on, 491.
water on, 492.
classed with the calcium group, 500.
crystallization of, 491.
extraction of, 490.
forms basic salts, 495.
metallic, prop. of, 490.
testing for, 497.
tree, 507.
use of, for water-pipes and cisterns, 493.
Leblanc's process, 397.
wastefulness of, 400.
Light, action of, on silver salts, 458.
bichromate of potassium, 526.
Light carburetted hydrogen, 313.
Lime, air-slaked, 471.
the cheapest base, 476.
heat evolved in slaking, 472.
milk or cream of, 472.

Lime, slaked, absorbs carbonic and sulphydric acids, 473.
uses of, 476.
Limestone, 467.
Litharge, 494.
Lithium-flame, 444.
occ. of, 443.
resembles sodium and potassium, 443.
Litmus-paper. 53.
tincture, prep. of, 420.
Liver of sulphur, 428.
Luminosity of flames, 343.
Luminous flames, form of, 350.
Lunar caustic, 453.

MAGNESIA alba, 504.
crucibles, 503.
hydraulic, 502.
Magnesite, 503.
Magnesium light, 501.
the metal, prop. of, 501.
occ. of, 501.
salts, from mother-liquor of salt-works, 393.
Manganate of potassium, 531.
Manganates, 531.
Manganese, occ. of, 527.
Manganese-alum, 529.
Manganic acid, 531.
Marsh-gas, 313.
analysis of, 316.
prep. of. 314.
two vols. yield four vols. of hydrogen, 315.
Marsh's test for arsenic, 259.
Matches, 211, 219.
Matter, indestructibility of, 355.
Measuring-glasses, xxix.
Mercuric chloride, 574.
an antiseptic, 575.
forms double chlorine salts, 575.
iodide, changes of color of, 575.
oxide, 572.
as a source of oxygen, 572.
Mercurous chloride, 573.
oxide, 571.
Mercury, alloys of, 576.
detection of, 577.
extraction of, 570.
pneumatic trough, xvi.
Mercury, tendency to form double compounds, 573.
unit-volume weight half its atomic weight, 570.
uses of, 571.
Metal, meaning of the term, 451.
Metantimoniates, 275.
Metantimonic acid, 274.
Metaphosphate of silver, 407.
sodium, 407.
Metaphosphoric acid, 233.
Metastannic acid, 581.
Meteoric iron, 534.
Metre, definition of, xxxvi.
Metrical system of weights and measures, xxxvi.
Mica, 375.
Microcosmic salt, 407.
Minium, 496.
Mohr's burette, xxx.
Molecular condition of elementary gases, 207.
hypothesis. 30.
Molecule, definition of, 29.
Molybdate of ammonium, 586.
Molybdenum, 585.
Monoatomic metals, 73.
Mordant, use of, 519.
Mordants, 516.
Mortar, 474.
Mortars, xxxiv.
Mosaic gold, 582.
Multiple proportions, law of, 66.
Muriatic acid, 40, 91.
manufacture of, 99.

NASCENT state, 80.
Neutralization, 54.
Nickel, 556.
Nitrate of ammonium, 441.
prep. of, 53.
calcium, 485.
copper, 569.
lead, decomposition of, 62.
made, 61, 62.
potassium, occ. of, 429.
prop. of, 430.
silver, 453.
stains of, removed, 426, 427.
sodium, 392.
Nitrates of iron, 554.
mercury, 576.
Nitre, 429.

Nitre-paper. 354, 432.
Nitric acid, analysis of, 63.
 anhydrous, 69, 70.
 prep. of, 460.
 combining weight of, 73.
 monohydrated, 70.
 prep. of, 52.
 sources of, 51.
 uses of, 71.
Nitric oxide, analysis of, 60.
 prep. of, 58.
 a test for free oxygen, 59, 60.
Nitride of boron. 371.
Nitrogen, a constituent of air, 10.
 dilutes the oxygen in air, 18.
 obtained from air, 16.
 physical prop. of, 17.
 prep. by copper, 16.
 phosphorus, 17.
 widely diffused, 18.
Nitrous acid, 63.
 oxide, comp. of, 55, 56.
 physical prop. of, 57.
 prep. of, 54.
 supports combustion, 58.
Nomenclature, prefixes derived from Greek and Latin numerals, 174.
 the prefix *hypo-*, 61.
 the termination *ide*, 174.
 the terminations *ous* and *ic*, 60.
Nordhausen acid, 192.
Norium, 561.

Ochre, red, 543.
 yellow, 543.
Oil-bath, xxvi.
Oil of vitriol, manufacture of, 184–186.
Organic analysis, 566.
 chemistry, defined, 312.
Orpiment, 263.
Osmium, 596.
Oxidation by platinum-black, 595.
Oxide of aluminum, 514.
 calcium, 470.
 magnesium, 502.
 silicon, 375.
 silver, a strong base, 455.
Oxide of zinc, 508.
Oxides, definition of, 14.
 of antimony, 272.
 teroxide, 272.
 quadroxide, 273.
 quinquioxide, 274.
 arsenic, 245.
 bismuth, 282.
 teroxide, 282.
 quinquioxide, 283.
 chlorine, series of, 116.
 chromium, 522.
 cobalt, 557.
 iron, 542.
 lead, 492.
 manganese, 527.
 mercury, 571.
 molybdenum, 585.
 nickel, 557.
 nitrogen, series of, 65.
 platinum, 593.
 silver, 455.
 sulphur, series of, 175.
 tin, 580.
 tungsten, 586.
 uranium, 558.
Oxidizing agents, defined, 71.
Oxychloride of bismuth, 284.
Oxygen, abundance and importance of, 15.
 burning charcoal in, 13.
 iron in, 13.
 phosphorus in, 13.
 sulphur in, 13.
 burns in hydrogen, 50.
 a constituent of air, 9.
 explosive mixture with hydrogen, 48, 49.
 physical properties of, 12.
 precautions in making, 11.
 prepared from bichromate of potassium, 525.
 bleaching-powder, 484.
 chlorate of potassium, 11.
 peroxide of barium, 487.
 sulphuric acid, 196.
 supports combustion, 13.

Oxyhydrogen blowpipe, 46.
Ozone, 139.
atmospheric, 145.
a disinfecting agent, 144.
distinguished from antozone, 153.
influence on health, 146.
prep. by electricity, 141.
ether, 142.
permanganate of potassium, 142.
phosphorus, 140.
prop. of, 143.
resembles chlorine, 143.
tests for, 143, 144.
Ozonides, 147.

Palladium, 596.
Paraffine-bath, xxvi.
Parchment paper, 255.
Pearlash, 415.
Pearl-white, 284.
Perchloric acid, 118.
Periodic acid, 131.
Permanganate of potassium, 532.
a disinfectant, 533.
Permanganates, 533.
Permanganic acid, 532.
Peroxide of barium, prep. of, 487.
hydrogen, prep. of, 152.
lead absorbs sulphurous acid, 496.
an oxidizing agent, 495.
prep. of, 495.
Persulphide of hydrogen, 173.
Petrifactions, calcareous, 468.
siliceous, 378.
Pewter, 585.
Phosphate of magnesium and ammonium, 504.
silver, 406.
sodium, ammonium, and hydrogen, 407.
Phosphates of calcium, 481.
sodium, 405.
rhombic, 405.
pyro-, 406.
meta-, 407.
varieties of, 234.
Phosphide of calcium, 221.
Phosphorescence, 212.
Phosphoric acid, anhydrous, 231.
Phosphoric acid, anhydrous, affinity of, for water, 232.
glacial, 232.
hydrates of, 232.
strength of, 235.
terbasic, 234.
Phosphorous acid, 228.
hydrated, 229, 231.
Phosphorus, allotropism of, 210.
burnt under water, 436.
combinations of, 219.
common, 212.
comparison of red with common, 216.
compounds with hydrogen, 220.
inflammability of, 210.
manufacture of, 214.
occ. of, 209.
oxides of, 226.
poisonous, 214.
red, 216.
converted into common, 216.
crystallized, 217.
oxide of, 226.
prep. of, 218.
used on safety-matches, 218.
shines in the dark, 212.
solutions of, 213.
unit-volume weight of, 224.
Phosphuretted hydrogen, analysis of, 223.
comp. of, 224.
liquid, 225.
prep. of, 220.
from phosphide of calcium, 221.
solid, 226.
Photography, 456-460.
on glass, 457.
paper, 459.
Photolithography, 526.
Physical changes, 1.
Pincers, xxxiv.
Pink salt, 584.
Pipettes, xxxi.
Plants and animals, reciprocal action of, 330.
Plaster of Paris, 477.

Plaster-casts, 477.
Plastering, comp. of, 475.
dampness of, 475.
Platinized asbestos, 193.
Platinum, alloys of, 592.
black, 595.
cleaning, xxii.
foil and wire, xxi.
induces combination, 182, 193, 306, 593, 595.
melting of, 353.
metals, 596.
occ. of, 591.
prop. of, 592.
sponge, prep. of, 594.
uses of, 593.
vessels, precautions in using, 592.
working, one method of, 595.
Plumbate of lead, 497.
Pneumatic troughs, construction and use of, xiv–xviii.
Porcelain, 520.
dishes, xxxii.
Potash, obtained from ashes of plants, 415.
Potassium-flame, 423.
metallic, 422.
occ. of, 414.
salts, test for, 594.
Product-volume, defined, 207.
Protochloride of copper, 567.
tin, 583.
a reducing agent, 583.
Protoxide of chromium, salts of, 523.
copper, 565.
lead, 494.
manganese, 528.
tin, 580.
Prussian blue, 555, 556.
Prussic acid, 363.
Puddled steel, 541.
Puddling iron, process of, 539.
Pulverizing, xxxv.
Pyrites, 549, 562.
Pyritous shales, use of, 550.
Pyroligneous acid, 298.
Pyrophorus, of carbon and lead, 492.
Pyrophosphate of silver, 406.
sodium, 406.
Pyrophosphoric acid, 233.

QUANTITY flame, 345.
Quantivalence, 465.
Quartation, 590.
Quartz, 375.
Quicklime, 470.
manufacture of, 471.
Quicksilver, 570.
Quinquichloride of antimony, 276.
phosphorus, 237.
dissociation of, 238.
Quinquichlorides of antimony and phosphorus, chloridizing agents, 277.
Quinquioxide of antimony, 274.
bismuth, 283.
Quinquisulphide of antimony, 280.

RADICALS, compound, 363.
Rational formulæ, 87, 89.
Realgar, 263.
Red lead, manufacture and uses of, 496.
Reducing agent, defined, 71.
Reduction of metals by carbonic oxide, 339.
charcoal, 300, 321.
Refrigerating mixture—ice and salt, 180.
nitric oxide and bisulphide of carbon, 57.
saltpetre, 432.
sulphate of sodium and chlorhydric acid, 396.
Reinsch's test for arsenic, 253.
Replacement, 41, 404, 508.
Respiration, 329.
Retort, iron, xxvi.
use of, 34.
Retorts, xxviii.
Revivification of animal charcoal, 309.
Rhodium, 596.
Rochelle-powders, 399.
Rock-crystal, 376.
Rouge, jewellers', 543.
Rubidium, 446.
Ruby, 514.
Rust, of tin, iron, mercury, &c., contains something derived from the air, 7–9.
Rusts are oxides, 14.
Ruthenium, 596.
Rutile, 577.

SAFETY-lamps, 359.
-matches, 218.
Saline taste, and substance, 392.
Salt, decrepitation of, 393.
glaze, 394.
manufacture of, 392.
solubility of, 393.
sources of, 391.
uses of, 394.
Saltpetre, 429.
manufactured from nitrate of sodium, 430.
not explosive, 431.
plantations, 429.
purification of, 429.
Sand-bath, xiii.
Sapphire, 514.
Saturated solutions, 37.
Schweinfurt green, 570.
Screw compressor, xxv.
Selenhydric acid, 201.
Selenites and seleniates isomorphous with sulphites and sulphates, 200.
Selenium, occ. of, 198.
prop. of, 199.
sources of, 199.
Serpentine, 503.
Sesquicarbonate of ammonium, 442.
Sesquioxide, the term, 513.
of chromium, 523.
salts of, 524.
iron, 543.
manganese, 528.
Shot, arsenic added to, 243.
Silicates, 379.
alkaline, soluble, 380.
classes of, 380.
decomposition of, 381.
fused with alkaline carbonates, 415.
in glass, 412.
of aluminum, 519.
iron, 555.
sodium, making of, 410.
Siliceous petrifactions, 377.
Silicic acid, composition of, 382.
double salts of, 380.
in natural waters, 377.
isomeric modifications of, 376.
occ. of, 375.
prop. of, 378.
soluble, 376.
Siliciuretted hydrogen, 374.
Silicon, abundance of, 372.
allotropic conditions of, 372.
atomic weight of, 382, 385.
compound with hydrogen and chlorine, 386.
prep. of the three modifications of, 373.
prop. of, 373, 374.
Silver, atomic weight of, 461.
coin, 453.
extraction of, 449.
melted, absorbs oxygen, 452.
occ. of, 448.
precipitation of, 454.
quantitative determination of, 460.
separation from lead by crystallization, 491.
separation from lead by cupellation, 494.
solvents for, 452.
Silvering of mirrors, 577.
Slag, finery, 551.
of iron furnaces, 536.
Smalt, 557.
Soap, hard, 405.
soft, 417.
Soda-ash, manufacture of, 394, 397.
-crystals, 397.
grocers', 398.
-powders, 399.
-water, 327.
Sodium-amalgam, 94.
-flame, 402.
occ. of, 391.
prep. of, 400.
prop. of, 401.
Solder, 585.
Soldering, use of chloride of zinc in, 506, 509.
sal-ammoniac in, 579.
Soluble glass, 380.
Solution and chemical combination compared, 38.
defined, 36, 37.
heat either absorbed or set free during, 395.
how to apply the solvent, 37, 594.
of gases in water, 67.
Spatulæ, xxxv.
Specific gravity, definition of, 20.

Specific heat, definition of, 20.
heats, value of, 602.
Spectrum-analysis, 444.
Spongy platinum, 594.
Spontaneous combustion, 229.
of coal, 550.
Spring clip, xxv.
Stalactites, 468.
Stalagmites, 468.
Stannate of sodium, prep. of, 582.
Stannates, 581.
Stannic acid, 581.
Starch-paste, prep. of, 126.
Steam, physical prop. of, 21.
superheated, 21.
volumetric comp. of, 29.
Steel, Bessemer, 541.
blistered, 540.
cast, 541.
puddled, 541.
tempering, 542.
Stereotype-metal, 268.
Stone, artificial, 410.
Stove-polish, 290.
Straw rings, xxxiii.
Strontianite, 488.
Strontium, 486.
compounds, 486–488.
flame, 489.
Stucco, 478.
Suboxide of lead, 493.
Substitution, defined, 41.
Sugar, purification of, 308.
Sugar of lead, 499.
Sulphantimoniates, 280.
Sulphantimonic acid, 280.
Sulphantimonites, 278, 279.
Sulpharseniates, 265.
Sulpharsenites, 264.
Sulphate of aluminum, 517.
ammonium, 440.
barium, 487.
calcium, 476.
reduction of, 481.
stuffing for paper, 435.
chromium, 524.
copper, 568.
lead, 499.
magnesium, 503.
potassium, 428.
and hydrogen, 428.
Sulphate of silver, 464.
sodium, manufacture of, 394.
and hydrogen, 396.
zinc, 509.
reduction of, 509.
Sulphates, 191.
of iron, 551–554.
mercury, 576.
Sulphide of arsenic, precipitation of, 258.
reduction of, 259.
boron, 371.
lead, volatility of, 498.
silicon, 390.
silver, 463.
Sulphides of ammonium, 442.
antimony, 277.
tersulphide, 277.
quinquisulphide, 280.
arsenic, bisulphide, 263.
tersulphide, 263.
quinquisulphide, 265.
calcium, 173.
copper, 567.
iron, 547.
lead, 497.
mercury, 573.
phosphorus, 241.
potassium, 427.
sodium, 399.
tin, 582.
Sulphites, 183.
Sulphur, change of prismatic into octahedral, 160.
crystallization of, 156, 157.
dimorphous, 159.
extraction of, 155.
a kindling material, 165.
melting of, 156.
metals burn in, 164.
milk of, 163.
occ. of, 154.
salts, 264.
compared with oxygen salts, 265.
soft, 162.
solutions of, 163.

Sulphur, uses of, 165.
Sulphuretted hydrogen, see Sulphydric acid.
Sulphuric acid, absorbs water, 188.
  action on metals, 190.
    organic matter, 190.
  anhydrous, 192.
    decomposition of, 195.
    prep. of, 193.
    prop. of 194.
  bibasic, 191.
  freezing by, 189.
  fuming, 191.
  how to distil, 187.
  how to mix with water, 188.
  hydrates of, 189.
  importance of, 184.
  manufacture of, 184–186.
Sulphurous acid, antiseptic, 181.
  bleaches, 180.
  comp. of, 176.
  liquid, 180.
  oxidation of, 182.
  prep. of, 176, 177, 178.
  prop. of, 178, 179.
  stops combustion, 179.
Sulphydrate of calcium, 486.
Sulphydric acid, 165.
  analysis of, 168.
  comp. of, 169.
  decomposition of, 172.
  inflammable, 170.
  in mineral waters, 169.
  poisonous, 169.
  prep. of, 166.
  prop. of, 167.
  pure, 168.
  as a reagent, 171.
  solution in water, 170.
Supersaturated solution, 396.
Swimmer, Erdmann's, xxxi.
Symbolic formulæ, value of, 90.
Symbols, chemical, 41.
Synthesis, definition of, 3.
Systems of crystallization, 158.

TABLE of atomic heats of compounds, 601.
    gaseous elements, 600.
    solid elements, 599.
  atomic weights and symbols of the elements, 597.
  for conversion of centigrade into Fahrenheit degrees, xxxix–xli.
  for conversion of French into English measures and weights, xlii.
  of groups of the elements, 598.
Tables of metrical weights and measures, xxxvii, xxxviii.
Tantalum, 585.
Tar, product of the distillation of wood, 298.
Tartar, 423.
Tartar-emetic, 273.
Tartrate of lead, 492.
Tellurium, compounds of, 202.
  occ. of, 201.
  prop. of, 202.
Terbium, 561.
Terbromide of antimony, 277.
Terchloride of antimony, 275.
  bismuth, 283.
  phosphorus, 236.
Teriodide of antimony, 277.
Terminations *ous* and *ic*, 558.
Teroxide of antimony, 272.
    both an acid and a base, 273.
  bismuth, 282.
    a base, 283.
Tersulphide of antimony, 277–280.
  bismuth, 284.
Test-glasses, xxix.
Test-tubes, xxviii.
Thallium, 447.
Thermometers, xxxv.
Thermometer-scales, centigrade and Fahrenheit, compared, xxxix–xli.
Thionic acids, 175.
Thorium, 561.
Tin, alloys of, 585.

Tin, crystallization of, 578.
extraction of, 578.
occ. of, 578.
prop. and uses of, 578, 579.
Tin-salt, 583.
Tin-stone, 578.
Titanic acid, 577.
Titanium, occ. of, 577.
resembles tin, 578.
Tongs, xxxiv.
Tubing, sizes and qualities of, i.
Tungstate of sodium, 587.
Tungstates, 586.
Tungsten, 586.
steel, 586.
Tungstic acid, 586.
Type-metal, 268.
Types—chlorhydric acid, water, and ammonia, 205.
doctrine of, 88.
Typical hydrogen compounds, 319.

UNIT of heat, 45.
Unit-volume not absolute, 204.
the bulk of 1 grm. of hydrogen, 204.
a litre, 205.
weights, 205.
Univalent metals, 73.
Uranium, 558.
salts, peculiarity of, 559.

VANADIUM, 586.
Vegetable parchment, 255.
Ventilation of wells, 324.
Verdigris, 569.
Vermilion, 573.
Volume, combination by, 203.

WASH-BOTTLE, xxxii.
Water, analyzed by iron, 23.
sodium, 22.
blue color of, 19.
densest at 4°, 19.
dissolves air, 35.
what it dissolves from air 68.
Water, distillation of, 33.
electrolysis of, 23.
hardness of, 468, 480.
method of determining hardness of, 480.
occ. of, 33.
removal of gases from, 35.
the common solvent, 37.
standard of specific gravity, 20.
heat, 20.
symbol of, 32.
synthesis of, 25–29.
volumetric comp. of, 29.
Water-bath, xxv.
Waterglass, 380.
uses of, 410.
Wedgewood mortars, xxxiv.
Weights and measures, metrical, xxxvi.
Weights, table of, xxxviii.
White iron, 538.
lead, 500.
vitriol, 509.
White-wash, 474.
Wire gauze, use of, xiii.
Witherite, 488.
Wolfram, 586.
Wood, distillation of, 296.
Woulfe-bottles, 85.
Wrought iron, 539.
impurities of, 540.
making of, 539.

YEAST-powders, 399.
Yellow-metal, 564.
Yttrium, 561.

ZINC, alloys with antimony, 269.
extraction of, 504.
granulated, 505.
prop. of, 505.
replaces lead, 507.
white, 508.
Zirconium, 561.

ORDER-LIST OF CHEMICALS.

The quantities here given are the quantities which one person will use in performing the numbered experiments of the manual according to the directions. In ordering chemicals for a class of several students, a small reduction should be made upon the multiplied quantities. Less than twenty times the assigned quantities will suffice for twenty students. Teachers may get some idea of the cost of these chemicals by referring to the price-lists of the "Druggists' Circular," published monthly at 36 Beekman St., New York, price 13 cents.

Alcohol	1 oz.
Alum	¼ oz.
Ammonia-water (Aqua Ammonia)	6 oz.
Ammonium, chloride of	2 oz.
Antimony, chloride of, a strong solution of	¼ oz.
Antimony, metallic	¼ oz.
Antimony, tartrate of and potassium (Tartar emetic)	40 grains
Arsenious acid	¼ oz.
Asbestos	a few fragments
Barium, chloride of	8 grains
Bismuth	10 grains
Bleaching powder	¼ lb.
Bone-black	3 oz.
Borax	¼ oz.
Bromine	1/16 oz.
Calcium, chloride of	1½ oz.
Calcium, sulphate of (Plaster of Paris)	¼ lb.
Carbon, bisulphide of	1½ oz.
Caustic potash (solid)	¼ oz.
Caustic soda ("concentrated lye")	½ oz.
Chlorhydric (muriatic) acid	1½ lbs.
Cobalt, nitrate of	8 grains
Cochineal	15 grains
Copper filings	2 oz.
Copper, oxide of	30 grains
Copper, sulphate of	¼ oz.
Dutch metal	several leaves
Ether	a few drops
Fluor-spar	½ oz.
Gold-leaf	1 sq. inch
Graphite	¼ oz.
Gum arabic	¼ oz.
Gum guaiacum	15 grains
Iodine	10 grains
Iron filings	2 oz.
Iron, red oxide of	¼ oz.
Iron, sulphate of	¼ oz.
Iron, sulphide of	1 oz.
Iron wire (fine piano wire)	1 foot
Lead, acetate of	1 oz.
Lead, peroxide of	15 grains
Litharge	1 oz.
Litmus	15 grains
Logwood, extract of	15 grains
Magnesium wire	4 inches
Manganese, black oxide of	1½ oz.

Manganese, sulphate of 15 grains
Mercury, chloride of (corrosive sublimate) . . 8 grains
Mercury, metallic . . . $3\frac{1}{2}$ oz.
Mercury, red oxide of . $\frac{3}{4}$ oz.
Nitric acid . . . . . 1 oz.
Nutgalls . . . . . . $\frac{1}{2}$ oz.
Oxalic acid . . . . . $\frac{1}{2}$ oz.
Phosphorus, 1 stick, $3\frac{1}{2}$ inches long
Platinum, scrap . . 10 grains
Potassium, 1 piece as large as a pea.
Potassium, bichromate of $1\frac{1}{2}$ oz.
Potassium, chlorate of . . 2 oz.
Potassium, iodide of . . $\frac{1}{16}$ oz.
Potassium, yellow prussiate of $\frac{1}{4}$ oz.
Salt . . . . 2 oz.
Saltpetre . . . . . . 5 oz.
Silver, nitrate of . . . 30 grains
Sodium, 4 pieces of the size of a pea.
Sodium, acetate of . 30 grains
Sodium, carbonate of (anhydrous) . . $\frac{1}{4}$ oz.
Sodium, carbonate of (Sal Soda) . . . $\frac{1}{4}$ lb.
Sodium, hyposulphite of . $\frac{1}{2}$ oz.
Sodium, phosphate of . 1 oz.
Sodium, sulphate of . . $\frac{1}{2}$ lb.
Starch . . . . $\frac{1}{4}$ oz.
Strontium, nitrate of . . 8 grains
Sulphur, flowers of . . 6 oz.
Sulphur, roll brimstone . $\frac{1}{4}$ lb.
Sulphuric acid . . . . 2 lbs.
Tartaric acid . . . 15 grains
Tin foil . . . . 1 oz.
Tin, granulated . . . $\frac{1}{2}$ oz.
Turmeric paper, 2 or 3 square in.
Turpentine, oil of . . . $\frac{1}{2}$ oz.
Zinc, granulated . . . 3 oz.
Zinc, solid metal or scraps $\frac{1}{2}$ lb.
Zinc, sulphate of . . 15 grains.

ORDER-LIST OF UTENSILS.

The following list includes the utensils which one person will need in performing all the numbered experiments in this manual. The principal articles of steady consumption are glass tubing, retorts, flasks, corks, and filter-paper. Many of the other articles, once obtained, last a long time. It is evidently not necessary to provide all this apparatus for every member of a large class. Six retorts, as many gallon bottles, six Woulfe-bottles, four soda-water bottles, two measuring-glasses, two mortars, two pipettes, four wire-gauze lamps, one blast-lamp and bellows, one blowpipe jet, three or four pieces of platinum foil, two thermometers, one burette, one pair of scales, and one set of weights, will suffice, if used with method, for a class of twenty or thirty. The wholesale druggists keep suitable cheap bottles and porcelain ware.

Many of the articles may be made by a tinman or gas-fitter. For the rest, teachers may consult the priced catalogues of the dealers in philosophical apparatus and chemical ware.

Glass tubing (App. Fig. 1) —

2 sticks about 3 feet long of No. 1
3 " " " " " 2
2 " " " " " 3
1 " " " " " 4
1 " " " " " 5
3 " " " " " 6
2 " " " " " 7
1 " " " " " 8

1 tube about 1 foot long and 1½ inches in inside diameter.

1 tube about 1 foot long and 1 inch in inside diameter.

1 tube about 3 feet long and 1 inch in inside diameter.

[If ignition-tubes can be bought ready-made, a dozen may be bought instead of 1 stick of No. 1 and 1 stick of No. 2.]

Retorts —

1 of 12 oz. capacity without tubulature.

1 of 6 oz. capacity, with glass stopper.

1 receiver of 12 oz. capacity, with tubulature.

Phials and bottles —

1 glass-stoppered, 4 oz.
2 narrow-mouthed, gallon.
1 wide-mouthed, ½ gallon.
1 " " quart.
1 " " pint.
1 narrow-mouthed, pint.
4 wide-mouthed, 8 oz.
3 " " 4 oz.
2 " " 2 oz.

[All these bottles should be of a very common quality. The glass should be tolerably white.]

Funnels —

1 three inches in diameter.
1 two inches in diameter.

3 Woulfe-bottles of about 8 oz. capacity.

Glass flasks —

1 flat-bottomed of 1½ pint capacity.

1 flat-bottomed of 4 oz. capacity.

1 flat-bottomed of 2 oz. capacity.

3 cheap round-bottomed flasks such as olive oil used to be imported in.

1 thistle-tube.

1 Mohr's burette.

2 soda-water bottles, stout.

2 wine-glasses.

3 test-tubes.

1 chloride of calcium tube.

1 graduated measuring-glass of 250 c. c. capacity.

1 pipette.

1 nest of beakers, of which the largest is of about 500 c. c. capacity.

3 bits of window-glass, 3 inches square.

Porcelain dishes —

1 about 2½ inches in diameter.
1 about 3½ inches in diameter.

1 small iron mortar.

1 Wedgewood mortar about 4 inches in diameter.
1 Bunsen gas-lamp (or spirit-lamp where gas is not to be had).
1 iron ring-stand.
1 piece of iron wire-gauze about 4 inches square.
1 piece of fine copper gauze about 2½ inches square.
2 wire-gauze lamps (App., Fig. VIII.).
1 piece of thin iron gas-pipe about 15 inches long and ½ inch in diameter.
1 iron sand-bath 4½ inches in diameter.
1 sheet-iron trough 8 inches long and 1¼ wide.
1 glass-blower's lamp or Bunsen's gas blast-lamp.
1 small double-acting bellows.
1 mouth-blowpipe.
1 triangular file.
1 round file.
1 pair of jewellers' tweezers.
1 jet for oxyhydrogen blowpipe, p. 46.
1 piece of platinum-foil 1½ inches square.
1 piece of finest platinum wire 4 inches long.
1 piece of platinum wire 8 inches long and not thicker than the wire of a very small pin.
1 stone-ware milk-pan.
1 flower-pot saucer.
1 Hessian crucible of about 8 oz. capacity.
2 common plates, of which one a soup-plate.
2 spring-clips or spring clothes-pins.
1 thermometer (glass).
1 pair of small scales ("Tea"-scales for grocers' use).
1 set of gramme weights, 1 grm to 250 grms.
Corks — an assortment of three or four sizes, small for ignition tubes, medium for flasks, and large for the 8 oz. and 4 oz. wide-mouthed bottles.
Caoutchouc tubing —
2 feet of ¼ inch.
3 feet of $\frac{3}{16}$ inch.
2 feet to fit glass tubing No. 6.
1 foot to fit No. 7.
1 pair of wooden nippers.
½ quire of common filter-paper.
½ sheet of parchment paper.

THE END.

www.ingramcontent.com/pod-product-compliance
Lightning Source LLC
LaVergne TN
LVHW021048110826
845150LV00001B/18

* 9 7 8 1 4 2 5 5 6 8 4 4 3 *